Liquid-Phase Reaction Rate Constants

Liquid-Phase Reaction Rate Constants

E.T. Denisov
Institute of Chemical Physics
Moscow, USSR

Translated from Russian by
R. K. Johnston

IFI/PLENUM NEW YORK-WASHINGTON - LONDON

Library of Congress Cataloging in Publication Data

Denisov, Evgeniĭ Timofeevich.
Liquid-phase reaction rate constants.

Translation of Konstanty skorosti gomoliticheskikh zhidkofaznykh reakt͡siĭ.
Includes bibliographies.
1. Chemical reaction, Rate of–Tables, etc. I. Title.
QD502.D4613 541'.39 73-79419
ISBN 978-1-4684-8302-4 ISBN 978-1-4684-8300-0 (eBook)
DOI 10.1007/978-1-4684-8300-0

The original Russian text, published by Nauka Press in Moscow in 1971, has been corrected by the author for the present edition. This translation is published under an agreement with Mezhdunarodnaya Kniga, the Soviet book export agency.

Константы скорости гомолитических жидкофазных реакций.
Е.Т. Денисов

KONSTANTY SKOROSTI GOMOLITICHESKIKH ZHIDKOFAZNYKH REAKTSII
E. T. Denisov

Softcover reprint of the hardcover 1st edition 1974
A Division of Plenum Publishing Corporation
227 West 17th Street, New York, N.Y. 10011

United Kingdom edition published by Plenum Press, London
A Division of Plenum Publishing Company, Ltd.
4a Lower John Street, London W1R 3PD, England

PREFACE

The past 25 years in chemical kinetics have seen major advances in studying the mechanisms of complex chemical reactions, in particular free radical reactions. Many different methods have been developed for quantitative studies of elementary chemical reactions. Thousands of rate constants have been measured, for hundreds of diverse chemical reactions. It is becoming more and more difficult for the chemist to orient himself in the voluminous and rapidly growing literature of chemical reaction kinetics. This leads to major expenditures of time in searching out, collecting, and evaluating quantitative kinetic data; to unnecessary repetition (duplication) of research; and to a situation in which the rich material already accumulated in the field of chemical kinetics is very often not fully utilized in comparing, interpreting, and analyzing new experimental data. There is a pressing need for the creation of a series of handbooks on reaction rate constants. Such work was begun several years ago at the initiative of V. N. Kondrat'ev, and is now going forward under his direction at the Institute of Chemical Physics of the USSR Academy of Sciences.

This book is devoted to liquid-phase, homolytic reactions. Part One contains data on monomolecular reactions in which molecules decompose to form radicals, as well as data on bimolecular and trimolecular reactions that form free radicals. Also collected in Part One are data on the probability of radical escape into the bulk volume from a "cage" of solvent with various initiators, as well as values of the stoichiometric inhibition coefficients for various phenols and aromatic amines that are used as free radical acceptors. In Part Two, rate constants are collected for reactions of decomposition, substitution, and addition of radicals to molecules, as well as radical recombination and disproportionation reactions. Part Three contains data on ion--molecule reactions that are accompanied by radical formation, as well as reactions of free radicals with metal ions and acid anions, reactions of ion-radicals and solvated electrons, and ionic oxidation-reduction reactions with electron transfer. In this book, rate constant values are presented along with a statement of the conditions and a brief characterization of the method of measurement.

I wish to express deep gratitude to Academician V. N. Kondrat'ev for having inspired me to this not easy but useful task, and for having aided me with counsel. The last chapter of this book was written by V. N. Berdnikov. For assistance in collecting the reaction rate constants, I thank A. L. Buchachenko, R. L. Vardanyan, S. S. Ivanchev, N. F. Kazanskaya, G. M. Nazin, A. A. Shteinman, and V. Ya. Shlyapintokh. I also wish to express gratitute to Z. A. Denisova for painstaking work in preparing the manuscript for publication.

E. T. Denisov

CONTENTS

PART TWO - REACTIONS OF FREE ATOMS AND RADICALS

LIST OF TABLES

CHAPTER I. MONOMOLECULAR REACTIONS

CHAPTER VI. RADICAL SUBSTITUTION REACTIONS

SYMBOLS

k	Reaction rate constant, expressed in sec^{-1} for first-order reaction, liter/mole·sec for second-order reaction, and $liter^2/mole^2 \cdot sec$ for third-order reaction
K	Equilibrium constant
E	Activation energy, kcal/mole
$t°$	Temperature in degrees Celsius
$T°$	Temperature in degrees Kelvin $\theta = 4.575\ T/1000$
A	Preexponential factor, with dimensionality corresponding to that of k for the reaction
W_ν	Rate of chain reaction
k_p	Chain propagation rate constant in chain reaction
k_t	Rate constant for reaction between two radicals
W_i and k_i	Respectively, rate and rate constant for reaction of radical formation from molecules. If radicals R· are formed at a rate W_i and are destroyed only by reaction with each other (rate constant k_t), then in the stationary regime $W_i = 2k_t[R\cdot]^2$
e	Probability of escape of radicals into the bulk volume from a "cage" of solvent; if k is the rate constant for decomposition of initiator into two radicals, then $k_i = 2\ ek$
C	Concentration, mole/liter

InH	Inhibitor or acceptor of free radicals
f	Inhibition coefficient, equal to the number of radicals successively reacted with one molecule of inhibitor and conversion products from the inhibitor molecule

ABBREVIATIONS

The following abbreviations, which denote methods for measuring rate constants, are not listed in the original Russian text but have been compiled here for convenience. Abbreviations used here (in the English translation) are listed in the left-hand column; abbreviations used in the original Russian text are transliterated and listed in the right-hand column for reference.

Catal.	Ionic catalytic reactions	Kataliz
CINH	Inhibited chain reaction	TsING
CL	Chemiluminescence	KhL
CSM	Consumption of starting material	RIV
DP	Degree of polymerization	SP
EP	Emulsion polymerization	ÉP
EPR	Electron paramagnetic resonance	ÉPR
FM	Flash method	FM
ICA	Initial consumption of (radical) acceptor	NRA
IIM	Intermittent illumination method	PO
IIP	Introduction of initiator fragments into polymer	VIP
IM	Isotope method	IM
JM	Jet method	JM
KEPF	Kinetics of end product formation	KOKP
KICR	Kinetics of initiated chain reaction	KTsIR
Kin. Racem.	Kinetics of racemization	Kin. Ras.
KNP	[not identified]	KNP
KRAC	Kinetics of radical acceptor consumption	KRAR
KRR	[not identified]	KRR
MCR	Method of competing reactions	MKR
NK	[not identified]	NK
NKR	Nonstationary kinetics of radical chain reaction	NKR

NMR	Nuclear magnetic resonance	YaMR
NR	[not identified]	NR
Pol.	Polarographic method	Pol.
POL	[not identified]	POL
PR	Pulse radiolysis	IP
PSD	Photochemical space discontinuity	FPP
RICR	Rate of initiated chain reaction	STsIR
RRC	Products of radical recombination in cage	PR
RUCR	Rate of unbranched chain reaction	STsNR
SUNR	[not identified]	SUNR
TIM	Tritium isotope method	ITM
TJ	Temperature jump	TS
TsINR	[not identified]	TsINR

Note: For a listing of abbreviations for names of ligands in metal complexes, see Chapter X, p. 482.

PART ONE

REACTIONS OF MOLECULES

CHAPTER I

MONOMOLECULAR REACTIONS

§1. Methods for Measuring Rate Constants of Monomolecular Reactions

a. Kinetics of Consumption of Starting Material (CSM)

A simple and widely used method for measuring rate constants is the calculation of k from the kinetic curve of starting material consumption. Monomolecular decomposition is described by an exponential law, $C = C_0 e^{-kt}$ or $\ln(C_0/C) = kt$. The constant is found from the slope of the straight line plotted on coordinates of $\log C_0$ vs. t. It is evident that k should not depend on C_0. Nevertheless, in some cases, along with the monomolecular decomposition, an induced decomposition occurs. For example, benzoyl peroxide in certain solvents decomposes monomolecularly with a velocity kC, and under the action of free radicals with a velocity $k'C^{3/2}$ (k' is the effective rate constant of the induced decomposition). The constant found experimentally from the initial section of the kinetic curve is $k_e = k + k'C_0^{1/2}$. The rate constant for monomolecular decomposition, k, is found by extrapolation: $k = \lim k_e$ as $C_0^{1/2} \to 0$, where $k_e = k + k'C_0^{1/2}$.

b. Kinetics of End Product Formation (KEPF)

The kinetics of the consumption of starting material can be followed from the formation of a stable product of the conversion. The rate of end-product accumulation is

$$\frac{dC}{dt} = k[A] = k(C_\infty - C),$$

where A is the starting material and C_∞ is the concentration of the end product when $t \to \infty$.

The kinetics of end-product accumulation are described by the formula

$$\ln \frac{C_\infty}{C_\infty - C} = kt.$$

The rate constant is found from the slope of the straight line on coordinates of

$$\log \frac{C_\infty}{C_\infty - C} \text{ vs. } t.$$

Such a method is widely used, particularly in studying azo compounds, which decompose with N_2 evolution. The reaction kinetics are followed by measuring the volume of evolved gas and the rate constant is found from the relation

$$\log \frac{V_\infty}{V_\infty - V} = kt.$$

c. <u>Initial Rate of Consumption of Free Radical Acceptor (ICA)</u>

In the decomposition of a molecule, two free radicals are formed; and for a certain time (10^{-10} sec) they exist side by side, surrounded by molecules of solvent (cage effect). Part of the free radicals will react with each other to form molecular products, and the remainder (which we will designate as "e") escape into the bulk volume as a result of diffusion. If a free radical acceptor is present in the solution, the rate of free radical formation can be measured by the rate of acceptor consumption. Materials used as acceptors include iodine, which reacts rapidly with radicals having a free valence on a C atom; oxidation inhibitors such as phenols, naphthols, and aromatic amines, which react rapidly with radicals of the $RO^\bullet$ and $RO_2^\bullet$ type [1], and stable free radicals. If the acceptor is itself a free radical, then it reacts with one radical; the rate of change of acceptor concentration $W = W_i$, where W_i is the rate of radical formation. If the acceptor is, for example, molecular iodine, the disappearance of one molecule of acceptor is accompanied by the destruction of two free radicals:

$$R^\bullet + I_2 \longrightarrow RI + I^\bullet; \quad I^\bullet + R^\bullet \longrightarrow RI; \quad W = 1/2 W_i.$$

In the general case, W_i - fW.

Radical acceptors react only with those radicals that have escaped into the bulk volume. Since the decomposition of a molecule with the rupture of one bond will form two radicals, then, when the cage effect is taken into account, the rate constant for the formation of radicals by initiator decomposition $k_i = 2ek$. Since $W_i = k_iC_o$ and $W_i = fW$, then k_i - fW/C_o, where W is the rate of consumption of the free radical acceptor. The experiment is designed so that the initiator concentration remains practically constant during the course of the experiment. Having measured W and knowing f and C_o, one can find k. The radical acceptor is introduced in a concentration such that practically all the radicals will react with the acceptor and not with each other. The acceptor and the products of its conversion must not react with the initiator.

Sometimes it is not the radical acceptor consumption that is measured, but rather the induction period of a chain reaction (e.g., oxidation or polymerization) when known concentrations of initiator and inhibitor (free radical acceptor) are introduced into the system. If, during the induction period, the initiator remains the sole source of free radicals, if the inhibitor is consumed only in reaction with free radicals, and if all chains are broken in reactions of radicals with inhibitor, then $W_i = f[InH]_o/\tau$, where τ is the induction period and $[InH]_o$ is the initial inhibitor concentration; also, $k_i = f\,[InH]_o/C_o\tau$. The constants k_i and k are little different, since in the majority of cases e = 0.4 to 0.8, and k_i is 0.8 to 1.6 times k. The difference in activation energies $E_i - E \cong E_e$, about 2 to 4 kcal/mole.

d. Kinetics of Radical Acceptor Consumption (KRAC)

When the initial rate acceptor consumption is measured, only k_i can be determined. If the experiment is conducted in such a manner that all of the initiator has decomposed during this time, then the kinetic curve for consumption of the free radical acceptor can be used to determine both k and e [2]. Here the acceptor is added in large excess, so that the radicals formed from the initiator will react with the acceptor. The rate of acceptor consumption is

$$-\frac{d\,[InH]}{dt} = \frac{2\,ek}{f}\;C_o e^{-kt},$$

and the kinetics of acceptor consumption are described by the formula

$$[\mathrm{InH}]_o - [\mathrm{InH}] = \frac{2eC_o}{f}(1 - e^{-kt})$$

or

$$\ln \frac{[\mathrm{InH}]_o - [\mathrm{InH}]_\infty}{[\mathrm{InH}] - [\mathrm{InH}]_\infty} = kt.$$

From the straight-line slope on coordinates of

$$\log \frac{[\mathrm{InH}]_o - [\mathrm{InH}]_\infty}{[\mathrm{InH}] - [\mathrm{InH}]_\infty} \text{ vs. } t$$

k is found; from the difference $[\mathrm{InH}]_o - [\mathrm{InH}]_\infty$, e is determined:

$$e = \frac{f}{2C_o}([\mathrm{InH}]_o - [\mathrm{InH}]_\infty).$$

e. Rate of Chain Reaction in Presence of Initiator (RICR)

Substances that decompose into free radicals are often used as initiators of chain reactions. In the liquid phase in the absence of inhibitors, chains are broken as a result of encounters between two free radicals, at a constant rate k_t. Therefore, the concentration of free radicals is $(W_i/2k_t)^{1/2}$, and the rate of the chain reaction $W_\nu = k_p(2k_t)^{-1/2}[RH]\sqrt{W_i}$, where k_p is the rate constant for the chain propagation reaction, and [RH] is the concentration of the reacting substance. If the ratio $k_p/(2k_t)^{1/2}$ is known, then, having measured W_ν, one can determine W_i according to the formula

$$W_i = \left(\frac{W_\nu\sqrt{2k_t}}{k_p[RH]}\right)^2,$$

whence one finds $k_i = W_i/C_o$. The experiment is generally conducted in such a manner that the inhibitor concentration remains practically constant during the time of measurement. If the chains are not very long ($\nu < 20$), one should take into account that the experimentally measured chain reaction rate $W = W_\nu + W_i$, and $W_\nu = W - W_i$. The ratio of the constants $k_p/\sqrt{2k_t}$ can be determined if the reaction is conducted with an initiator for which k_i is known.

f. Kinetics of Chain Reaction in Presence of Initiator (KICR)

If the initiator is the sole source of radical formation in the system, and if the ratio $k_p/\sqrt{2k_t}$ does not change during the course of the experiment, then the chain reaction kinetics can be used to determine k and e for the initiator, carrying the experiment up to complete consumption of initiator [3]. The rate of the chain reaction, measured as the rate of accumulation of product P, is

$$\frac{dP}{dt} = k_p(2k_t)^{-1/2}[RH](keC)^{1/2},$$

and $C = C_o e^{-kt}$. Consumption of the starting material is subject to the kinetic law

$$[P] = [P]_{max}(1 - e^{-0.5kt})$$

or

$$\ln\frac{[P]_{max}}{[P]_{max} - [P]} = \frac{1}{2}kt, \qquad [P]_{max} = \frac{2k_p[RH]}{\sqrt{k_t k}}\sqrt{eC_o}.$$

From the slope of the straight line on coordinates of $\log[1/(1 - x)]$ vs. t, where $x = [P]/[P]_{max}$, we find k; from $[P]_{max}$ we find e.

Another method for determining k is to measure the rate of the chain reaction as it changes along the course of the experiment (owing to initiator consumption):

$$W_v = k_p(2k_t)^{-1/2}[RH](keC_o)^{1/2}e^{-0.5kt},$$

whence

$$\ln W_v = \ln W_o - 1/2\,kt.$$

g. Chemiluminescence Method (CL)

The oxidation of hydrocarbons is accompanied by a weak chemiluminescence generated by the reaction between two peroxy radicals [4]. The intensity of chemiluminescence $I = 2\eta k_t[RO_2^\cdot]^2$, where η is the quantum yield of light in this reaction. If an initiator is added to the hydrocarbon, and if radicals are formed only from this initiator during the entire period of the experiment, then, by following the change in chemiluminescence intensity, one can measure the rate constant for initiator decomposition k [4]. In the stationary state

$$I = 2\eta k_t [RO_2\cdot]^2 = \eta w_i = 2e\eta k C = 2e\eta k C_0 e^{-kt}$$

$$\text{and } I = I_0 e^{-kt} \quad \text{or} \quad \log\frac{I_0}{I} = kt.$$

In the experiment, the light intensity $\dot{I}$ is recorded. From a plot of $\log(I_0/I)$ vs. t, we find the decomposition rate constant k. This method can be used to determine k in substances that are subject to oxidation accompanied by chemiluminescence. The conditions are selected so that the oxidation products (peroxides) give practically no chain initiation in comparison with that of the added initiator.

§2. Rate Constants for Decomposition of Peroxide Compounds

Benzoyl peroxide decomposes as a result of O-O bond rupture:

$$C_6H_5CO\text{-}OO\text{-}COC_6H_5 \rightarrow 2\ C_6H_5COO\cdot$$

Along with molecular decomposition, inductive decomposition of peroxide under the action of free radicals is often observed. Therefore, to measure k, either the peroxide decomposition is carried out in the presence of a free radical acceptor, or the decomposition rate constant measured in the experiment is extrapolated to zero concentration of peroxide (see §1). The introduction of substituents into the benzene ring changes k: Electropositive substituents accelerate, and electronegative substituents retard the peroxide decomposition (see Table 3). This is explained on the basis that the two benzoyloxy groups in the peroxide molecule are two mutually repelling dipoles. Substituents affect the dipole moment of each of the groups and consequently affect the peroxide decomposition.

For aromatic solvents (benzene, styrene, alkylaromatic hydrocarbons), the averaged rate constant for benzoyl peroxide decomposition is

$$\log k = 14.48 - 30.8/\theta.$$

The averaged value of the initiation rate constant for tertiary butyl peroxide in aromatic solvents (benzene, toluene, styrene, cumene) is

$$\log k = 14.86 - 35.0/\theta.$$

The averaged value of k for cumyl peroxide in aromatic and paraffinic hydrocarbons is

$$\log k = 14.60 - 34.4/\theta.$$

The rate constants and activation energies for decomposition of ROOR peroxides (where R is ethyl, propyl, or primary, secondary, or tertiary butyl) are very similar.

Peroxides of the type

$$o\text{-}IC_6H_4\overset{O}{\overset{\|}{C}}OOC(CH_3)_3,\quad o\text{-}RSC_6H_4C\begin{matrix}\nearrow O\\ \searrow OOC(CH_3)_3\end{matrix}\quad \text{and}$$

$$o\text{-}CR_2{=}CC_6H_4COOOC(CH_3)_3,$$

as is evident from Table 9, decompose with rate constants that are several orders of magnitude larger than k for the decomposition of analogous peroxides that do not have any substituents in the ortho position or have only alkyl substituents in the ortho position.

This is explained on the basis that the I and S atoms and the double bond do participate in the act of decomposition, forming a bond with oxygen:

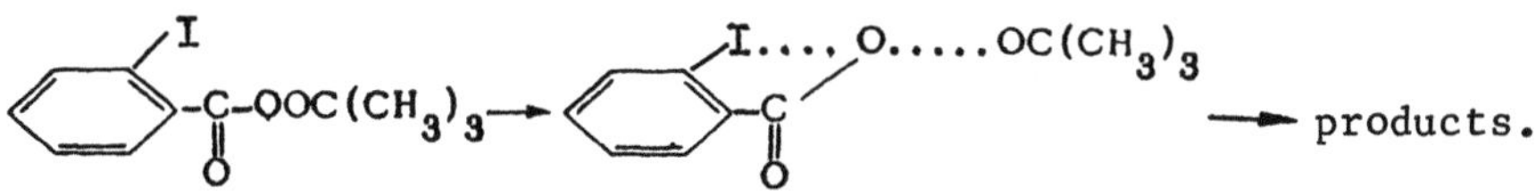

An analogous picture is observed in the decomposition of substituted benzoyl peroxides (see Table 3).

Hydroperoxides react actively with solvents having weak C-H bonds, double bonds, or alcohol, or carboxyl groups. Evidently it is only in inert solvents such as benzene, CCl_4, or fluorocarbons that one may observe true monomolecular decomposition of hydroperoxides at the O-O bond.

§3. Decomposition of Azo Compounds

The decomposition of an azo compound, $R\text{-}N_2\text{-}R$, results in the formation of one molecule of nitrogen and two $R^{\bullet}$ radicals. Azo compounds are widely used in the laboratory as initiators and photoinitiators of radical chain reactions. They are also convenient objects for studying the cage effect (see Chapter III). Among the initiators of this class, the one that has been studied in the most detail is azobisisobutyronitrile, the decomposition rate constant of which changes very little from solvent to solvent (Table 12). The average value of decomposition rate constant for the azobisisobutyronitrile in aromatic solvents (according to the data of [108] is

$$\log k = 15.00 - \frac{30.45}{\theta}.$$

§4. Decomposition at C-C, N-N, N-C, N-O, and C-Metal Bonds

The C-C bond strength in paraffinic hydrocarbons is approximately 80 kcal/mole, and these hydrocarbons do not decompose at C-C bonds at temperatures below 400°. The introduction of phenyl groups into the hydrocarbon molecule weakens the C-C bond, owing to stabilization of the free radicals that are formed upon decomposition. For example, the strength of the C-C bond in butane, $CH_3CH_2\text{-}CH_2CH_3$, is Q = 80 kcal/mole; in the hydrocarbon $(CH_3)_2(C_6H_5)C\text{-}C(C_6H_5)(CH_3)_2$, Q = 50 kcal/mole; in the hydrocarbon $(CH_3)(C_6H_5)_2C\text{-}C(C_6H_5)_2(CH_3)$, Q = 30 kcal/mole; and in hexaphenylethane, $(C_6H_5)_3C\text{-}C(C_6H_5)_3$, Q = 19 kcal/mole.

In an analogous manner, the introduction of phenyl groups into compounds containing nitrogen will lead to a weakening of N-N and N-C bonds (Tables 16 and 17). The decomposition of diazo compounds (Table 17) and triazene (Table 16) proceeds with the formation of a biradical and is facilitated to a considerable degree by the formation of the very stable nitrogen molecule.

The decomposition of N-nitrosoacylarylamines (Table 18) is apparently preceded by an intramolecular rearrangement with the formation of an azo compound:

$$C_6H_5N(N{=}O)COR \longrightarrow C_6H_5\text{-}N{=}N\text{-}OCOR \longrightarrow C_6H_5\cdot + N_2 + RCOO\cdot$$

Nitrogen pentoxide decomposes with the rupture of an N-O bond:

$$O_2N{-}O{-}NO_2 \longrightarrow NO_3 + NO_2.$$

The end products of the decomposition are O_2 and N_2O_4. Decomposition in solution proceeds with the same rate constant as in the gas phase.

Nitro compounds decompose in accordance with first-order kinetics. It has been postulated (but not strictly proved) that the decomposition is limited by the rate of O-N bond rupture in the case of nitrate esters and by the rate of C-N bond rupture in the case of nitro compounds (the mechanism of nitro compound decomposition is discussed in [179]).

§5. Decomposition of Iodobenzene Dichloride

Iodobenzene dichloride decomposes with the formation of C_6H_5I and one molecule of chlorine; it is used in the laboratory as a chlorinating agent. The reaction proceeds with a change in the valence state of the I atom. If the Cl_2 is not absorbed by the solvent or lost into the gas phase, an equilibrium is established:

$$C_6H_5ICl_2 \rightleftharpoons C_6H_5I + Cl_2,$$

for which ΔH = 9.9 kcal/mole and ΔS = 20.4 cal/degree [194]. Polar ortho-substituents on the benzene ring accelerate the decomposition reaction by approximately two orders of magnitude, apparently taking a direct part in the formation of an active complex.

§6. Correlation Equations

Hammett Equation. In the decomposition of compounds having structures that include a substituted benzene ring, a linear relation is often observed between log k and the Hammett function, σ ; i.e., $\log(k/k_0) = \rho\sigma$. The Hammett function is $\sigma = \log(K/K_0)$ where K is the dissociation constant of benzoic acid with a substituent in the para or meta position, in aqueous solution at 25°, and K_0 is the corresponding value without substituent. For electropositive substituents such as CH_3 or CH_3O, $\sigma < 0$; and for electronegative substituents such as Cl, NO_2, or CN, $\sigma > 0$. In the case of decomposition of substituted benzoyl peroxides and substituted peroxides of the general formula $XC_6H_4COOOCOCH_3$, the coefficient ρ is negative (see Table 22); i.e., electropositive substituents accelerate the peroxide decomposition. For decomposition of peroxides $XC_6H_4CH_2COOOC(CH_3)_3$, a linear correlation exists between log k and σ^+ (Table 22). The function

$$\sigma^+ = \frac{\log(k/k_o)}{\rho_o},$$

where k is the rate constant for solvolysis of $XC_6H_4C(CH_3)_2Cl$ in 90% acetone at 25°, k_o is the solvolysis constant for $C_6H_5C(CH_3)_2Cl$, and

$$\rho_o = \frac{\log(k/k_o)}{\sigma - \sigma_o}$$

for substituents in the meta position. In this case also, ρ is negative. However, the correlation between k and σ is not always observed. In the decomposition of N-nitrosoacylarylamines, the decomposition rate constant is independent of the substituent X (independent of σ).

Compensation Effect. For a group of monotypical reactions, an increase in the preexponential factor is observed as the activation energy increases:

$$\Delta \log A = \alpha \cdot \Delta E \text{ or } \Delta H^{\neq} = \beta \Delta S^{\neq}.$$

At the isokinetic temperature $\beta = 1/4.575\alpha$, the rate constants of all reactions of a given series are identical. Since the relation between log A and E is approximately linear, then at the isokinetic temperature the differences among the rate constants of the series are minimal.

The linear relation between log A and E is followed rather well for the decomposition of diacyl peroxides, dialkyl peroxides, dinitriles of azo acids, etc. (see Table 23). However, in certain cases, for example, in the decomposition of polyphenylethanes, this relation is not followed; log A ≈ 16 for a change in E from 30 to 50 kcal/mole. Apparently the linear relation between $\Delta S^{\neq}$ and $\Delta H^{\neq}$ is fulfilled only for a variation of $\Delta H^{\neq}$ within a certain interval, and at very high or low values of $\Delta H^{\neq}$ this relation is not valid.

§7. Effect of Pressure and Solvent on Monomolecular Reactions

Effect of Pressure. The theory of absolute reaction rate gives the following relation between the pressure P and the change in molar volume $\Delta V^{\neq}$ in the formation of an activated complex:

$$\left(\frac{\partial \log k}{\partial P}\right)_T = -\frac{\Delta V^{\neq}}{RT}.$$

For a monomolecular reaction, in which one bond is broken, the volume of the activated complex must be greater than that of the original molecule, and k must decrease with increasing pressure. This is indeed observed in the majority of cases (see Table 24). From the magnitude of $\Delta V^{\neq}$, one can calculate approximately the lengthening of the broken bond in the activated complex. For example, in the decomposition of benzoyl peroxide, $\Delta V^{\neq} = 10$ cm^3/mole, which corresponds to an extension of the O-O bond in the activated complex by 0.4 Å [200]. In the decomposition of the peroxide of isobutyric acid, $\Delta V^{\neq}$ has a negative value, i.e., the decomposition is accompanied by a compression of the molecule. This is explained by the rupture of several bonds and by the following structure of the activated complex:

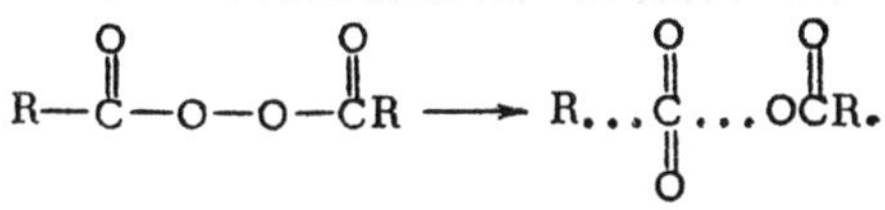

It is possible that the activated complex in the decomposition of

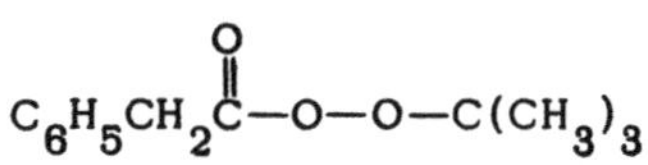

(see Table 24) has a similar structure.

Effect of Solvent. Three factors affect the decomposition of a molecule in solution: the internal pressure of the solvent, the cage effect, and the polarity of the solvent. The internal pressure in organic solvents can vary over the range $1\text{-}7 \cdot 10^3$ atm. If $\Delta V^{\neq} = 10$ cm^3/mole and P changes by $5 \cdot 10^3$ atm, then at 400°K

$$\Delta \log k \cong -\frac{\Delta P\, \Delta V^{\neq}}{RT} \text{ or } \Delta \log k = -0.66,$$

i.e., $k/k_0 = 0.22$. Thus, the internal pressure of the solvent can change k as much as twofold or fivefold.

If only one bond is broken in the decomposition of a compound, and if the radicals that are formed do not undergo instantaneous rearrangement or decomposition, then the radicals may recombine in the solvent cage, reverting back to the original compound. In this case, one determines experimentally the apparent rate constant for consumption of the starting material, $k_{app} = k(1 - e)$, where k is the true constant of decomposition. For acetyl peroxide, the existence of such a form of recombination of acetyl radicals was demonstrated in [206] by means of the isotope O^{18} (for more detail on the cage effect, see Chapter III). Such a back-recombination apparently can take place for peroxides that decompose with the rupture of only one O-O bond, for polyphenylethanes, and for polyphenylhydrazines.

Apparently these two factors are the extent of the solvent effects on the decomposition of such compounds as hexaphenylethane and a number of peroxides (see Table 25). Azo compounds R-N=N-R decompose simultaneously at two C-N bonds. Therefore, the cage effect should not affect k as measured by the kinetics of starting-material consumption or nitrogen evolution. As is evident from Table 25, k for the decomposition of the azobisisobutyronitrile or azoethylbenzene is very little changed from solvent to solvent.

The decomposition of benzoyl peroxide or the peroxide of isobutyric acid is strongly dependent on the solvent, which is explained by the influence of solvent polarity. The benzoyl peroxide molecule may be regarded as two mutually repelling dipoles, each of which is readily polarized by the surrounding molecules (easy polarizability is caused by the π-electrons of the benzene ring). This leads to a certain change in the bond energy in the benzoyl peroxide molecule when the polarity of the medium is changed. The effect of solvent polarity on k for decomposition of the peroxide of isobutyric acid is explained by the polarity of a transitional complex having the structure

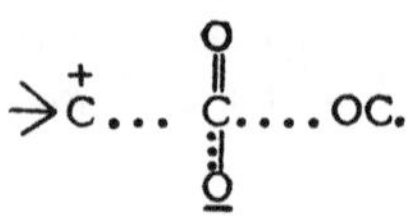

The formation of such a complex is facilitated in a polar solvent, since the solvation energy of the transitional complex is greater than that of the original peroxide. A linear relation is observed between log k and $(\varepsilon - 1)/(2\varepsilon + 1)$ [29], where ε is the dielectric constant of the medium.

Comparison of Gas- and Liquid-Phase Decomposition. In the decomposition of a molecule in the gas phase, the free radicals that are formed are immediately scattered. The decomposition of a molecule in a liquid requires an additional expenditure of energy in overcoming the internal pressure of the liquid in the formation of an activated complex. As an approximation, the change in the decomposition rate constant due to an increase in pressure will be

$$\Delta \ln k \cong - \frac{\Delta V^{\neq} \Delta P}{RT}.$$

Therefore, one should expect a decrease in k when going from a gaseous medium to a nonpolar solvent. A second factor that will decrease k in the liquid phase (as compared with the gas phase) is the cage effect, when a pair of radicals recombines to form the original molecule. The cage effect does not play any role if the decomposition proceeds through the simultaneous rupture of several bonds with the formation of three or more particles, as evidently occurs in the decomposition of azo compounds:

$$R{-}N{=}N{=}R \longrightarrow R\cdot + N_2 + R\cdot$$

Apparently these two factors explain why acetyl peroxide in the gas phase decomposes more rapidly than in liquid: In the gas phase, at 85.2°, $k = 2.24 \cdot 10^{-4}$ sec^{-1}, and $E = 29.5$ kcal/mole [23]; in CCl_4 at 85.2°, $k = 1.17 \cdot 10^{-4}$ sec^{-1}, and $E = 33.4$ kcal/mole [27].

The reverse picture is observed for tertiary butyl peroxide: In benzene, cumene, or tri-n-butylamine solution, it decomposes somewhat more rapidly ($\log k = 15.63 - 37.2/\theta$, with $k = 1.5 \cdot 10^{-5}$ at 125°) [43] than in the gas phase, where $\log k = 16.505 - 39.1/\theta$, with $k = 1.0 \cdot 10^{-5}$ at 125° [50], or $\log k = 15.845 - 38.0/\theta$, with $k = 0.9 \cdot 10^{5}$ at 125° [207], or $\log k = 14.78 - 36.0/\theta$, with $k = 1.0 \cdot 10^{-5}$ at 125° [208]. The more rapid decomposition of a polar molecule in solution than in the gas phase can be explained by strong solvation of the activated complex (in comparison with the original state) by polar or readily polarized solvent molecules.

On the whole, based on the two examples cited, it can be stated that transition from the gas phase to the liquid phase has only a weak effect on the rate constant for decomposition of a molecule -- at least, in certain cases.

TABLE 1. Rate Constants for Benzoyl Peroxide Decomposition

Solvent	t°	log k 80°	log A	E	Method	Ref.
C_6H_6	60–80	$\bar{5}$,68	16,28	33,3	CSM	[5]
	80–90	$\bar{5}$,02	14,04	30,7	CSM	[6]
	50–76	$\bar{5}$,18	13,48	29,6	ICA	[7]
	60–80	$\bar{5}$,34	13,64	29,6	CSM	[8]
$C_6H_5C(CH_3)_3$	80–90	$\bar{5}$,68	14,48	30,4	CSM	[6]
$C_6H_5CH(CH_3)_2$	45–80	$\bar{5}$,25	13,08	28,8	CSM	[9]
$C_6H_5CH{=}CH_2$	40–70	$\bar{5}$,50	16,30	33,6	RICR	[10]
	60–80	$\bar{5}$,32	14,27	30,6	RICR	[11]
	73–85	$\bar{5}$,56	14,76	31,0	RICR	[12]
$C_6H_5COCH_3$	70–95	$\bar{5}$,66	14,34	30,2	CSM	[13]
n-C_8H_{18}	80–90	$\bar{4}$,59	14,54	29,0	CSM	[6]
CH_3COCH_3	100–120	$\bar{5}$,96	15,16	31,0	KICR	[3]
$(CH_3CO)_2O$	60–80	$\bar{6}$,47	13,49	30,7	CSM	[5]
$CH_3CO{-}OCH{=}CH_2$	45–80	$\bar{5}$,90	13,42	28,3	CSM	[14]
cyclo-C_6H_{12}	80–90	$\bar{4}$,03	13,49	28,2	CSM	[6]
cyclo-$C_6H_{11}CH_3$	80–90	$\bar{5}$,81	14,83	30,7	CSM	[6]
Dioxane	60–80	$\bar{5}$,60	14,18	30,0	CSM	[15]
CCl_4	59–77	$\bar{5}$,45	15,68	32,7	ICA	[16]

TABLE 2. Decomposition of Benzoyl Peroxide in Various Solvents (Temperature 80°) [5, 15, 17]

Solvent	log k	Solvent	log k
$C_2H_5OCOCH_3$	$\bar{5},955$	$C_6H_5NO_2$	$\bar{5},52$
CH_3COOH	$\bar{5},91$	$C_6H_5C(CH_3)_3$	$\bar{5},52$
$(CH_3CO)_2O$	$\bar{5},875$	$C_6H_5CONH_2$	$\bar{5},505$
cyclo-C_6H_{12}	$\bar{5},805$	$CH_3COOCH{=}CH_2$	$\bar{5},45$
C_2H_5I	$\bar{5},60$	$(C_6H_5CHCH_2-)_n$	$\bar{5},44$
Dioxane	$\bar{5},625$	$C_6H_5CH(CH_3)_2$	$\bar{5},416$
		$C_6H_5CH(CH_3)_2$	$\bar{5},29$
		$C_6H_5C_2H_5$	$\bar{5},27$
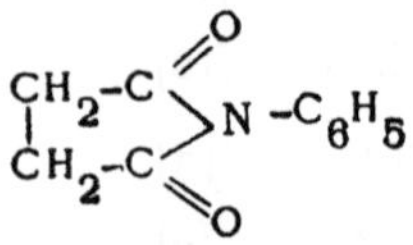	$\bar{5},625$	p-$CH_3C_6H_4CH_3$	$\bar{5},24$
		C_6H_5COOH	$\bar{5},38$
CH_3CONH_2	$\bar{5},625$	$CH_3CONHCN$	$\bar{5},36$
$CH_3COCH_2COOC_2H_5$	$\bar{5},625$	cyclo-$C_6H_{10}O$	$\bar{5},285$
$H_2NCOOC_2H_5$	$\bar{5},625$	cyclo-C_6H_{10}	$\bar{5},285$
		CCl_4	$\bar{5},32$
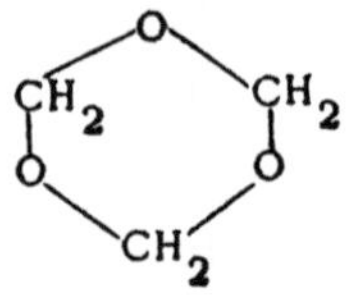	$\bar{5},625$	$C_6H_5COOCH_2C_6H_5$	$\bar{5},285$
		$C_6H_5NHCOOH$	$\bar{5},20$
C_6H_6	$\bar{5},57$	$CH_3CONHC_6H_5$	$\bar{5},14$
C_6H_6	$\bar{5},52$	$(C_6H_5)_3CH$	$\bar{5},11$
$C_6H_5CH_3$	$\bar{5},52$	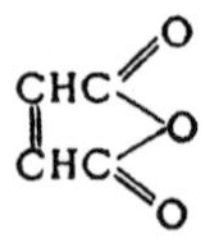	$\bar{5},03$
$C_6H_5CH_3$	$\bar{5},53$		
$C_6H_5OCH_3$	$\bar{5},555$	$C_6H_5CH{=}CHC_6H_5$	$\bar{6},96$
$CH_3COOCH_2CH{=}CH_2$	$\bar{5},53$	$CH_3COCOCH_3$	$\bar{6},945$

Table 2 (continued)

Solvent	log k	Solvent	log k
CH_3NO_2	$\bar{6}$,925	CH_2 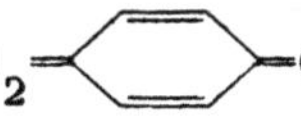 CH_2	$\bar{6}$,56
m-$CH_3C_6H_4OH$	$\bar{6}$,87	CH_2=CH-CN	$\bar{6}$,54
p-$CH_3C_6H_4NO_2$	$\bar{6}$,87	3,4-$Cl_2C_6H_3CH$=CH_2	$\bar{6}$,54
$C_6H_5CH_2CN$	$\bar{6}$,74	C_6H_5CH=CH_2	$\bar{6}$,54
$C_6H_5SSC_6H_5$	$\bar{6}$,63	C_6H_5CH=CH-CH=CHC_6H_5	$\bar{6}$,54
C_2H_5OCOCH=$CHCOOC_2H_5$	$\bar{6}$,62		

TABLE 3. Decomposition of Symmetrically Substituted Benzoyl Peroxides (CSM Method)

Substituent X	In acetophenone at 80° [13]		In dioxane at 80° [15]	In styrene at 70° [10]
	log k	E	log k	log k
p-OCH_3	$\bar{4}$,19	28,7	$\bar{4}$,07	$\bar{5}$,27
m-OCH_3	$\bar{5}$,81	28,9	$\bar{5}$,76	$\bar{6}$,95
o-OCH_3	$\bar{3}$,33	27,2	-	$\bar{4}$,14
p-CH_3	$\bar{5}$,77	29,9	$\bar{5}$,79	$\bar{4}$,04
m-CH_3	$\bar{5}$,67	30,2	$\bar{5}$,64	$\bar{6}$,78
o-CH_3	$\bar{4}$,50	30,2	-	$\bar{5}$,44
p-$C(CH_3)_3$	-	-	$\bar{5}$,70	$\bar{5}$,10
p-iso-C_3H_7	-	-	-	$\bar{5}$,03
p-C_2H_5	-	-	-	$\bar{5}$,00
p-C_6H_5	-	-	-	$\bar{6}$,93
p-n-$C_8H_{17}O$	-	-	-	$\bar{5}$,26

Table 3 (continued)

Substituent X	In acetophenone at 80° [13]		In dioxane at 80° [15]	In styrene at 70° [10]
	log k	E	log k	log k
o-OC_2H_5	-	-	-	$\bar{4}$,08
o-$OCH_2C_6H_5$	-	-	-	$\bar{4}$,08
p-CO_2CH_3	-	-	-	$\bar{6}$,78
o-CO_2CH_3	-	-	-	$\bar{5}$,00
o-OC_6H_5	$\bar{4}$,70	29,0	-	-
H	$\bar{5}$,64	30,2	$\bar{5}$,62	$\bar{6}$,84
p-F	-	-	-	$\bar{6}$,72
m-F	-	-	-	$\bar{6}$,56
o-F	-	-	-	$\bar{5}$,00
p-Cl	$\bar{5}$,58	30,4	$\bar{5}$,56	$\bar{6}$,73
m-Cl	$\bar{5}$,45	30,7	$\bar{5}$,42	$\bar{6}$,56
o-Cl	$\bar{4}$,59	29,4	-	$\bar{5}$,45
p-Br	-	-	$\bar{5}$,51	$\bar{6}$,72
m-Br	-	-	$\bar{5}$,41	$\bar{6}$,58
o-Br	-	-	-	$\bar{5}$,93
p-I	-	-	-	$\bar{6}$,73
m-I	-	-	-	$\bar{6}$,56
o-I	-	-	-	$\bar{7}$,04
p-NO_2	$\bar{5}$,63	30,3	-	$\bar{7}$,30
m-NO_2	$\bar{5}$,58	30,2	-	$\bar{7}$,85
o-NO_2	$\bar{3}$,13	28,6	-	$\bar{7}$,90
p-CN	$\bar{5}$,38	31,2	$\bar{5}$,30	$\bar{6}$,56
m-CN	-	-	$\bar{5}$,23	-
p-CH_2=CH	-	-	-	$\bar{6}$,81
3,5-$(NO_2)_2$	$\bar{5}$,27	31,2	-	-
3,5-$(CH_3)_2$	-	-	-	$\bar{6}$,89
3,4-$(OCH_3)_2$	-	-	-	$\bar{6}$,95
2,3-$(OCH_3)_2$	-	-	-	$\bar{5}$,68
2,3,4,5-Cl_4	-	-	-	$\bar{6}$,93

TABLE 3A. Decomposition of Unsymmetrically Substituted Benzoyl Peroxides and Phthaloyl Peroxide

Substituent	Solvent	t°	log k	log A	E	Method	Ref.
Monosubstituted benzoyl peroxides, $XC_6H_4COOOCOC_6H_5$							
p-OCH_3	Dioxane	80	$\bar{5}$,88			CSM	
m-OCH_3	- " -	80	$\bar{5}$,68			CSM	[15]
m-CN	- " -	80	$\bar{5}$,43			CSM	
m-NO_2	$C_6H_5CH{=}CH_2$	70	$\bar{6}$,08			CSM	[10]
m-NO	$C_6H_5CH{=}CH_2$	70	$\bar{6}$,36			CSM	[10]
o-CH_3	CCl_4	80	$\bar{4}$,04			CSM	
o-C_6H_5	- " -	80	$\bar{4}$,34			CSM	[18]
o-$C_6H_5CH_2$	- " -	80	$\bar{4}$,12			CSM	
p-$C_6H_5CH_2$	- " -	80	$\bar{4}$,41			CSM	
p-Cl	$C_6H_5CH{=}CH_2$	35–100	$\bar{4}$,67 (80°)	13,24	28,3	RICR	[19]
o-Cl, p-Cl	- " -	35–100	$\bar{4}$,49 (80°)	13,92	28,1	RICR	[19]
o-I*	CH_3I	25–45	$\bar{5}$,25 (25°)	9,95	20,0	CSM	[20]

Table 3A (continued)

Substituent	Solvent	t°	log k	log A	E	Method	Ref.
	Disubstituted benzoyl peroxides						
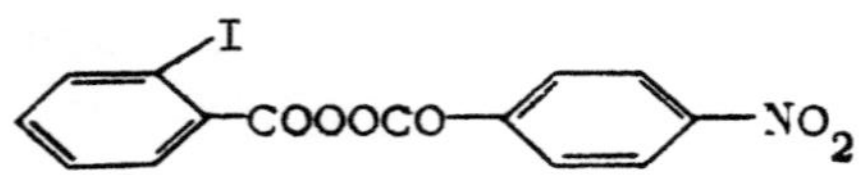	C_6H_6	25	$\bar{5},755$			CSM	[21]
	$C_6H_5NO_2$	25	$\bar{4},79$			CSM	
	CCl_4	25	$\bar{5},53$			CSM	
	$CHCl_3 + C_2H_5OH$	25	$\bar{4},45$			CSM	
	CH_3COCH_3	25	$\bar{4},48$			CSM	
	CH_3CN	25	$\bar{4},32$			CSM	
	Phthaloyl peroxide						
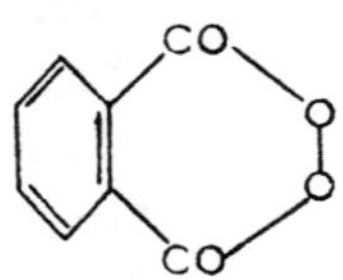	$CH_2{=}C(CH_3)COOCH_3$	75	$\bar{5},15$			CSM	[22]

*Conversion of peroxide goes through isomerization stage:

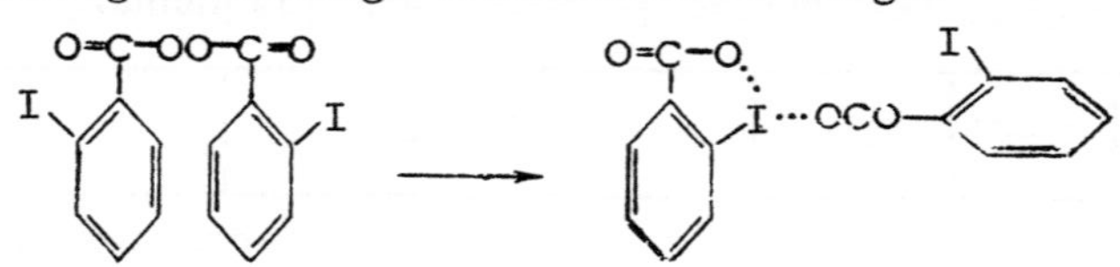

TABLE 4. Decomposition of Acetyl Peroxide in Various Solvents (CSM Method)

Solvent	t^{o}	log k* (80°)	log A	E	Ref.
C_6H_6	55–85	$\bar{5}$,91	15,91	32,3	[23]
	80	$\bar{5}$,94			[24]
$C_6H_5CH_3$	55–85	$\bar{5}$,90	15,70	32,0	[23]
	50–80	$\bar{5}$,91	16,31	33,0	[25]
	50–80	$\bar{5}$,85	15,05	31,0	[26]
$C_6H_5CH{=}CH_2$	70	$\bar{6}$,30			[10]
$CH_3C(CH_3)_2CH_2CH(CH_3)_2$	55–85	$\bar{5}$,76	15,81	32,2	[23]
$CH_2{=}CHCH_2CH_2CH_3$	70–90	$\bar{5}$,97	15,77	32,0	
$CH_2{=}C(CH_3)CH_2CH_2CH_3$	70–90	$\bar{4}$,01	14,76	30,3	[24]
$CH_2{=}CH(CH_2)_3CH_3$	70–90	$\bar{4}$,04	15,64	31,7	
cyclo-C_6H_{12}	55–85	$\bar{5}$,80	15,25	31,4	[23]
cyclo-C_5H_3	70–90	$\bar{5}$,89	16,29	32,8	[24]
cyclo-C_6H_{10}	60–100	$\bar{5}$,85	14,65	31,9	
CCl_4	60–100	$\bar{5}$,74	16,44	33,4	[27]
	85,2	$\bar{4}$,00			[25]
	80	$\bar{5}$,75			[24]
CH_3COOH	55–85	$\bar{5}$,81	14,51	30,2	[23]
	85,2	$\bar{4}$,07			[25]
CH_3CH_2COOH	85,2	$\bar{4}$,22			[23]

*Values of constant are cited at 80° for the cases in which E and A are known.

TABLE 5. Decomposition of Symmetrical Diacyl Peroxides, RCOOOCOR

R	Solvent	t°	log k	log A	E	Method	Ref.
C_2H_5	$C_6H_5CH{=}CH_2$	73–85	$\bar{5}$,26 (70°)	14,81	30,7	RICR	[12]
	$C_6H_5C_2H_5$	73–90	$\bar{5}$,55 (70°)	14,80	30,2	CSM	[28]
n–C_3H_7	$C_6H_5CH{=}CH_2$	73–85	$\bar{5}$,39 (70°)	13,36	28,2	RICR	[12]
	- " -	75–95	$\bar{4}$,80 (95°)	14,33	29,5	CL	[28]
	- " -	70	$\bar{6}$,28			RICR	[10]
	$C_6H_5C_2H_5$	73–90	$\bar{5}$,84 (73,5°)	14,68	29,8	CSM	[28]
$(CH_3)_2CH$	$C_6H_5CH{=}CH_2$	70	$\bar{4}$,56			RICR	[10]
	$C_6H_5C_2H_5$	40–55	$\bar{4}$,16 (40°)	11,84	24,8	CSM	[28]
	C_6H_6	40	$\bar{4}$,38			CSM	
	$C_6H_5NO_2$	40	$\bar{4}$,76			CSM	
	C_6H_5CN	40	$\bar{4}$,62			CSM	[29]
	$(CH_3)_2CHOH$	40	$\bar{4}$,48			CSM	
	$(CH_3)_3COH$	40	$\bar{4}$,40			CSM	

		40	$\bar{4}$,24				CSM	[29]
	C_6H_5Cl	40	$\bar{4}$,24				CSM	
	$C_6H_5CH_3$	40	$\bar{4}$,15				CSM	
	p-$CH_3C_6H_4CH_3$	40	$\bar{4}$,15				CSM	
	C_6H_5	40	$\bar{4}$,09				CSM	
	CCl_4	40–70	$\bar{5}$,89	(40°)	14,58	26,7	CSM	
	$CHCl_3$	40	$\bar{5}$,88				CSM	
	cyclo–C_6H_{12}	40	$\bar{5}$,65				CSM	
	$CH_3C(CH_3)_2CH_2CH(CH_3)_2$	40	$\bar{5}$,49				CSM	
n-C_4H_9	$C_6H_5CH{=}CH_2$	73–85	$\bar{5}$,41	(70°)	13,56	28,5	RICR	[12]
	$C_6H_5C_2H_5$	73–85	$\bar{5}$,83	(73,5°)	14,55	29,6	CSM	[28]
n-C_5H_{11}	$C_6H_5C_2H_5$	73–85	$\bar{5}$,83	(73,5°)	15,34	30,8	CSM	[28]
	- " -	73–85	$\bar{5}$,50	(70°)	14,40	29,7	RICR	[12]
	- " -	70	$\bar{5}$,27				RICR	[10]
	- " -	75–85	$\bar{4}$,82	(85°)	14,33	30,3	CL	[28]
	- " -	75–85	$\bar{4}$,74	(85°)	15,00	30,7	CSM	[28]

Table 5 (continued)

R	Solvent	t°	log k	log A	E	Method	Ref.
n-C_6H_{13}	$C_6H_5C_2H_5$	73–85	$\bar{5}$,51 (70°)	13,36	28,0	RICR	[12]
	- " -	73–85	$\bar{5}$,85 (73,5°)	14,89	30,1	CSM	
	$C_6H_5CH{=}CH_2$	75–95	$\bar{4}$,80 (95°)	15,06	30,7	CSM	[28]
	$C_6H_5CH{=}CH_2$	75–95	$\bar{4}$,81 (95°)	14,59	29,9	CL	
n-C_7H_{15}	$C_6H_5CH{=}CH_2$	73–85	$\bar{5}$,48 (70°)	13,68	28,6	RICR	[12]
	$C_6H_5C_2H_5$	73–85	$\bar{5}$,38 (73,5°)	14,78	29,9	CSM	
	$C_6H_5CH{=}CH_2$	75–95	$\bar{4}$,77 (95°)	14,55	29,9	CSM	[28]
	- " -	75–95	$\bar{4}$,79 (95°)	14,94	30,5	CL	
	- " -	70	$\bar{5}$,32			RICR	[10]
n-C_8H_{17}	$C_6H_5CH{=}CH_2$	75–95	$\bar{4}$,79 (95°)	14,46	29,7	CSM	
	- " -	75–95	$\bar{4}$,82 (95°)	14,72	30,1	CL	[28]
	$C_6H_5C_2H_5$	73–85	$\bar{5}$,89 (73,5°)	14,87	30,0	CSM	
n-C_9H_{19}	$C_6H_5CH{=}CH_2$	73–85	$\bar{5}$,42 (70°)	14,02	29,2	RICR	[12]

	- " -	75–95	$\bar{4}$,78 (95°)	14,68	30,1	CSM	[28]
	- " -	75–95	$\bar{4}$,79 (95°)	14,69	30,1	CL	[28]
	- " -	70	$\bar{5}$,27			RICR	[10]
	$C_6H_5C_2H_5$	73–85	$\bar{5}$,83 (73,5°)	14,98	30,2	CSM	[28]
n-$C_{11}H_{23}$	$C_6H_5CH{=}CH_2$	73–85	$\bar{5}$,40 (70°)	13,40	28,2	RICR	[12]
	- " -	75–95	$\bar{4}$,78 (95°)	14,68	30,1	CSM	[28]
	- " -	75–95	$\bar{4}$,82 (95°)	15,07	30,7	CL	[28]
	- " -	35–75	$\bar{5}$,76 (70°)	14,16	28,9	RICR	[30]
	$C_6H_5C_2H_5$	73–85	$\bar{5}$,89 (73,5°)	14,99	30,2	CSM	[28]
n-$C_{13}H_{27}$	$C_6H_5CH{=}CH_2$	70	$\bar{5}$,19			RICR	[10]
n-$C_{15}H_{31}$	$C_6H_5C_2H_5$	73–85	$\bar{5}$,89 (73,5°)	15,20	30,2	CSM	[28]
	$C_6H_5CH{=}CH_2$	75–95	$\bar{4}$,78 (95°)	14,63	30,0	CSM	[28]
	- " -	75–95	$\bar{4}$,75 (95°)	14,78	30,3	CL	[28]
n-$C_{17}H_{35}$	$C_6H_5CH{=}CH_2$	75–95	$\bar{4}$,81 (95°)	14,61	29,9	CSM	[28]
	- " -	75–95	$\bar{4}$,75 (95°)	14,83	30,4	CL	[28]
	- " -	73–85	$\bar{5}$,34 (70°)	15,09	31,0	RICR	[12]
	- " -	70	$\bar{5}$,17			RICR	[10]
	$C_6H_5C_2H_5$	73–85	$\bar{5}$,89 (73,5°)	14,86	30,0	CSM	[28]

Table 5 (continued)

R	Solvent	t°	log k	log A	E	Method	Ref.
$CH_3CH_2CH{=}CH$	$C_6H_5CH{=}CH_2$	70	$\bar{5}$,45			RICR	
$CH_3CH{=}CHCH_2$	- " -	70	$\bar{6}$,80			RICR	
$CH_3CH{=}CHCH{=}CH$	$C_6H_5CH{=}CH_2$	70	$\bar{5}$,38			RICR	
$CH_3CH{=}CH(CH_2)_7$	- " -	70	$\bar{5}$,28			RICR	[10]
$CH_3(CH_2)_7CH{=}CH(CH_2)_7$	- " -	70	$\bar{5}$,70			RICR	
$C_6H_5CH_2$	$C_6H_5CH{=}CH_2$	70	$\bar{7}$,98			RICR	
	$C_6H_5C_2H_5$	73–85	$\bar{5}$,43 (73,5°)	9,33	22,0	CSM	[28]
	C_6H_6	45–60		17,12	31,8	CSM	[31]
	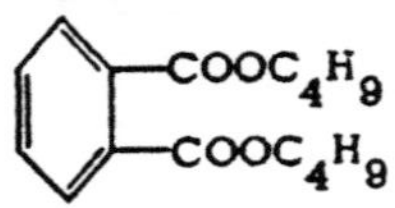	45–60		17,275	32,1	CSM	[31]
o-$IC_6H_4CH_2$	$CHCl_3$	0	$\bar{5}$,60			CSM	
	$C_6H_5CH_3$	0	$\bar{5}$,115			CSM	[32]
	CH_3COCH_3	0	$\bar{5}$,415			CSM	

$C_6H_5(CH_2)_2$	$C_6H_5CH{=}CH_2$	73–85	$\bar{5},50$ (70°)	11,50	25,1	RICR	[12]
	$C_6H_5C_2H_5$	73–85	$\bar{5},92$ (73,5°)	14,76	29,8	CSM	[28]
$C_6H_5(CH_2)_3$	$C_6H_5CH{=}CH_2$	73–85	$\bar{5},11$ (70°)	13,18	28,3	RICR	[12]
	- " -	70	$\bar{5},31$			RICR	[10]
	$C_6H_5C_2H_5$	73–85	$\bar{5},70$ (73,5°)	14,18	30,2	CSM	[28]
$C_6H_5(CH_2)_4$	CCl_4	70–85	$\bar{5},45$ (70°)	15,25	31,1	CSM	[33]
$C_6H_5CH{=}CH$	$C_6H_5CH{=}CH_2$	73–85	$\bar{5},49$ (70°)	13,19	27,8	RICR	[12]
	- " -	70	$\bar{5},21$			RICR	[10]
$C_6H_5CH{=}CH{-}CH{=}CH$	$C_6H_5CH{=}CH_2$	70	$\bar{7},93$			RICR	[10]
p-$CH_3OC_6H_4CH{=}CH$	- " -	70	$\bar{5},28$			RICR	[10]
p-$NO_2C_6H_4CH{=}CH$	- " -	70	$\bar{7},84$			RICR	[10]
$C_6H_5C{\equiv}C$	$C_6H_5CH_3$	70	$\bar{5},77$			CSM	[34]
$CH_2Br\,CHBr\,CH_2$	$C_6H_5CH{=}CH_2$	70	$\bar{4},26$			RICR	[10]
$CH_3(CHBr)_2(CH_2)_7$	- " -	70	$\bar{5},24$			RICR	[10]
$CH_3(CH_2)_7(CHBr)_2(CH_2)_7$	- " -	70	$\bar{5},18$			RICR	[10]
$(CH_3)_2CH$	- " -	40–58	$\bar{4},70$ (50°)	11,80	24,8	CSM	[35]
cyclo-C_3H_5	p-$CH_3C_6H_4CH(CH_3)_2$	70–113	$\bar{5},04$ (70°)	14,04	29,8	CSM	[36]
	$C_6H_5COOC_2H_5$	70	$\bar{6},90$			CSM	[36]

Table 5 (continued)

R	Solvent	t°	log k	log A	E	Method	Ref.
cyclo-C_3H_5	CCl_4	63–76	$\bar{6}$,97 (75°)	12,74	28,3	CSM	[37]
cyclo-C_4H_7	- " -	65–75	$\bar{4}$,15 (75°)	11,21	23,6	CSM	[37]
cyclo-C_5H_9	- " -	40–55	$\bar{4}$,59 (75°)	11,14	22,8	CSM	[37]
cyclo-C_6H_{11}	CCl_4	35–45	$\bar{3}$,31 (75°)	14,33	26,7	CSM	[37]
	C_6H_6	30–60	$\bar{4}$,00 (30°)	11,03	20,8	ICA	[38]
	CCl_4	35–60	$\bar{4}$,65 (50°)	13,25	24,5	CSM	[29]
	$C_6H_5CH{=}CH_2$	40–70	$\bar{4}$,87 (70°)	16,90	31,5	RICR	[19]
cyclo-C_7H_{13}	CCl_4	35–45	$\bar{4}$,17 (40°)	9,02	18,4	CSM	[37]
cis-$(CH_3)_3C$– 	- " -	40–51	$\bar{5}$,89 (40°)	9,04	18,8	CSM	[39]
trans-$(CH_3)_3C$– 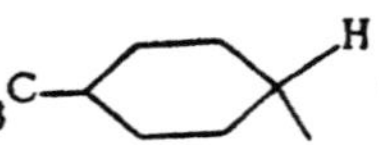	- " -	40–50	$\bar{5}$,63 (40°)	9,66	20,1	CSM	[39]

cyclo-$C_4H_7CH_2$	- " -	65–75	$\bar{5}$,58 (75°)	10,89	24,0	CSM	
cyclo-$C_5H_9CH_2$	- " -	60–75	$\bar{5}$,70 (75°)	7,50	18,5	CSM	[37]
cyclo-$C_6H_{11}CH_2$	- " -	60–75	$\bar{5}$,56 (75°)	5,77	16,0	CSM	
C_2H_5O	Carbonate	40–60	$\bar{5}$,46 (50°)	16,06	30,4	CSM	
$(CH_3)_2CHO$	- " -	40–60	$\bar{5}$,36 (50°)	14,36	28,1	CSM	
	$C_6H_5CH_3$	50	$\bar{5}$,485			CSM	[40]
$C_6H_5CH_2O$	- " -	50	$\bar{5}$,465			CSM	
$(CH_3)_2(NO_2)CCH_2O$	- " -	50	$\bar{5}$,35			CSM	
cyclo-$C_6H_{11}O$	$C_6H_5C_2H_5$	60–80	$\bar{4}$,83 (75°)	15,43	29,6	KICR	[41]
	C_6H_6	55–75	$\bar{3}$,18 (75°)	14,78	28,0	KICR	[2]
	- " -	45–60		16,01	29,7	CSM	[31]
	C_7H_{16}	55–75	$\bar{4}$,94 (75°)	12,04	24,0	KICR	
	$CH_3COC_2H_5$	55–75	$\bar{3}$,20 (75°)	10,40	21,0	KICR	[2]
	C_6H_5Cl	75	$\bar{3}$,26			KICR	
	cyclo-$C_6H_{10}O$	75	$\bar{3}$,15			KICR	
	$(CH_3)_2C(=O)NH_2$	25–60	$\bar{4}$,10 (40°)	11,45	22,0	CSM	[42]

TABLE 6. Decomposition of Unsymmetrical Diacyl Peroxides, $R_1COOOCOR_2$

R_1	R_2	Solvent	t^o	log k (70°)	log A	E	Method	Ref.
CH_3	C_2H_5	$C_6H_5CH{=}CH_2$	60–80	$\bar{5}$,25	15,75	32,2	RICR	[27]
	n-C_3H_7	- " -	60–80	$\bar{5}$,18	12,18	26,7	RICR	
	n-C_4H_9	- " -	60–80	$\bar{5}$,20	15,90	32,5	RICR	
	$CH_2CH(CH_3)_2$	- " -	60–80	$\bar{5}$,12	14,72	30,7	RICR	
	n-C_6H_{13}	- " -	60–80	$\bar{5}$,26	15,41	31,6	RICR	
	cyclo-C_6H_{11}	- " -	60–70	$\bar{5}$,29	7,56	18,4	RICR	
	cyclo-C_6H_{11}	- " -	60–80	$\bar{4}$,96	7,23	18,4	RICR	[11]
	C_6H_5	- " -	60–80	$\bar{5}$,05	13,15	28,4	RICR	
	- " -	- " -	70	$\bar{5}$,22			RICR	
	- " -	C_6H_5Cl	70	$\bar{5}$,30			CSM	[35]
	o-ClC_6H_4	$C_6H_5CH{=}CH_2$	60–80	$\bar{5}$,39	13,42	28,3	RICR	[11]
	m-ClC_6H_4	- " -	60–80	$\bar{5}$,81	13,91	28,4	RICR	
	p-ClC_6H_4	- " -	60–80	$\bar{6}$,57	12,87	28,7	RICR	
	o-Br C_6H_4	- " -	60–80	$\bar{5}$,55	12,00	25,8	RICR	
	m-Br C_6H_4	- " -	60–80	$\bar{6}$,73	14,49	31,0	RICR	

	p-Br C_6H_4	- " -	60–80	$\bar{5}$,97	14,34	28,8	RICR	
	m-$CH_3C_6H_4$	- " -	60–80	$\bar{5}$,12	14,62	30,6	RICR	
	p-$CH_3C_6H_4$	- " -	60–80	$\bar{5}$,10	14,50	30,4	RICR	
	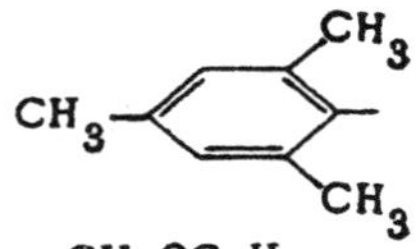	- " -	60–80	$\bar{4}$,29	12,99	26,2	RICR	
	o-$CH_3OC_6H_4$	- " -	60–80	$\bar{5}$,76	13,86	28,4	RICR	[11]
	p-$CH_3OC_6H_4$	- " -	60–80	$\bar{5}$,27	14,67	30,4	RICR	
	o-$CH_3OC_6H_4$	- " -	60–80	$\bar{5}$,12	14,02	29,7	RICR	
	p-$C_6H_5C_6H_4$	- " -	60–80	$\bar{5}$,11	14,10	29,7	RICR	
	m-$NO_2C_6H_4$	- " -	60–80	$\bar{6}$,37	17,02	36,5	RICR	
$ClCH_2$	C_6H_5	$C_6H_5CH{=}CH_2$	60–80	$\bar{5}$,14	14,04	29,7	RICR	[11]
C_2H_5	n-C_3H_7	$C_6H_5CH{=}CH_2$	60–80	$\bar{4}$,40	14,50	28,4	RICR	[43]
	n-C_4H_9	- " -	60–80	$\bar{4}$,40	15,65	30,2	RICR	[43]
	C_6H_5	- " -	60–80	$\bar{6}$,89	13,19	28,7	RICR	[11]
n-C_3H_7	C_6H_5	$C_6H_5CH{=}CH_2$	60–80	$\bar{6}$,89	13,94	29,9	RICR	[11]

Table 6 (continued)

R_1	R_2	Solvent	t°	log k (70°)	log A	E	Method	Ref.
n-C_3H_7	$C_6H_5CH{=}CH(CH_2)_2$	CCl_4	60–70	$\bar{4}$,01	13,15	26,9	CSM	[33]
	- " -	$C_6H_5COCH_3$	50–65	$\bar{4}$,86	12,03	23,8	CSM	[33]
	- " -	Polycarbonate	40–50	$\bar{3}$,12	10,77	21,4	CSM	[33]
$(CH_3)_2CHCH_2$	C_6H_5	$C_6H_5CH{=}CH_2$	60–80	$\bar{6}$,84	14,34	30,6	RICR	[11]
n-$C_{17}H_{35}$	- " -	- " -	70	$\bar{5}$,32			RICR	[11]
		- " -	70	$\bar{5}$,13			RICR	[11]
C_6H_5	C_6H_5O	C_6H_6	45–65	$\bar{4}$,40	16,55	31,6	CSM	[44]
	o-$(C_6H_5CH{=}CH)C_6H_4$	C_6H_5Cl	35–70	$\bar{3}$,48	9,09	18,2	CSM	[45]
	- " -	CH_3OH	35	$\bar{4}$,99			CSM	[45]
p-$NO_2C_6H_4$	o-$(C_6H_5CH{=}CH)C_6H_4$	C_6H_5Cl	35	$\bar{4}$,91			CSM	[45]
C_6H_5	o-(m-$NO_2C_6H_4CH{=}CH)C_6H_4$	C_6H_5Cl	70	$\bar{4}$,26			CSM	[45]
	o-(p-$NO_2C_6H_4CH{=}CH)C_6H_4$	C_6H_5Cl	70	$\bar{4}$,08			CSM	[45]
	- " -	cyclo-C_5H_8O	70	$\bar{5}$,79			CSM	[45]
	o-(p-$NO_2C_6H_4CH{=}C(p$-$NO_2C_6H_4))C_6H_4$	C_6H_5Cl	70	$\bar{3}$,19			CSM	[45]
p-$NO_2C_6H_4$	o-(p-$NO_2C_6H_4CH{=}CH)C_6H_4$	C_6H_5Cl	70	$\bar{4}$,58			CSM	[45]
p-$CH_3OC_6H_4$	- " -	C_6H_5Cl	70	$\bar{5}$,78			CSM	[45]

TABLE 7. Decomposition of Peroxides, ROOR

R	Solvent	t°	log k (120°)	log A	E	Method	Ref.
C_2H_5	$C_6H_6 + C_6H_5CH{=}CH_2$ (1:1)	60–80	$\bar{6}$,58	14,18	35,2	RICR	[46]
n-C_3H_7	$C_6H_5CH{=}CH_2$	60	$\bar{9}$,73			RICR	[47]
$(CH_3)_2CH$	$C_6H_6 + C_6H_5CH{=}CH_2$ (1:1)	60–80	$\bar{6}$,52	15,265	37,3	RICR	[46]
n-C_4H_9	- " -	60–80	$\bar{6}$,88	13,98	34,3	RICR	[48]
sec-C_4H_9	- " -	60–80	$\bar{6}$,66	13,56	34,0	RICR	[46]
$(CH_3)_3C$	$C_6H_6 + C_6H_5CH{=}CH_2$ (1:1)	60–80	$\bar{6}$,64	14,11	35,0	RICR	[49]
	$C_6H_6 + C_6H_5CH{=}CH_2$	125–145	$\bar{6}$,93	15,63	37,2	CSM	[50]
	$C_6H_5CH{=}CH_2$	90–115	$\bar{5}$,08	17,18	39,7	RICR	[28]
	- " -	80–110	$\bar{5}$,00	15,90	37,6	RICR	[51]
	- " -	60–98	$\bar{6}$,97	14,45	35,0	RICR	[52]
	- " -	80–100	$\bar{5}$,39	16,84	38,6	RICR	[53]
	- " -	60	$\bar{9}$,30			RICR	[54]
	C_6H_6	80–130	$\bar{5}$,58	14,08	34,0	RICR	[55]
	- " -	120	$\bar{5}$,14			CSM	[56]
	- " -	100–125	$\bar{5}$,05	15,10	36,0	ICA	[57]
	- " -	120–135	$\bar{5}$,51	15,11	36,1	CSM	[58]

Table 7 (continued)

R	Solvent	t°	log k (120°)	log A	E	Method	Ref.
$(CH_3)_3C$	cyclo-C_6H_{12}	120–135	$\bar{5}$,41	18,01	41,6	CSM	
	$(C_2H_5)_3N$	120–135	$\bar{5}$,50	17,96	41,4	CSM	
	$C_6H_5N(CH_3)_2$	120–135	$\bar{5}$,53	16,38	38,4	CSM	
	cyclo-C_6H_{12}	120–135	$\bar{5}$,45	16,15	38,1	CSM	
	$CH_2(CH_2)_3O$ (ring)	120–135	$\bar{5}$,53	16,13	37,9	CSM	
	CH_3CN	115–130	$\bar{5}$,75	13,03	31,8	CSM	[58]
	$C_2H_5(CH_3)_2COH$	120–135	$\bar{5}$,65	15,48	36,5	CSM	
	$C_6H_5NO_2$	120–135	$\bar{5}$,63	15,41	36,4	CSM	
	$C_6H_5CO_2C_2H_5$	120–135	$\bar{5}$,53	15,26	36,3	CSM	
	$(CH_3)_3COH$	120–135	$\bar{5}$,63	14,71	35,1	CSM	
	CH_3COOH	115–130	$\bar{5}$,80	14,38	34,2	CSM	
	C_6H_5Cl	95–125	$\bar{6}$,82	13,76	34,0	CSM	[59]
	$C_6H_5CH_3$	120	$\bar{5}$,13			CSM	
	CCl_4	120	$\bar{6}$,95			CSM	[56]
	cyclo-C_6H_{10}	120	$\bar{6}$,92			CSM	
	n-$C_{10}H_{22}$	110	$\bar{6}$,42			CSM	[60]
	$(CH_3)_2CHOH$	86–138	$\bar{5}$,16	14,64	35,0	ICA	[61]

	cyclo-$C_6H_{11}OH$	120–140	$\bar{5}$,19	12,89	32,8	ICA	[62]
$(CH_3)_2(C_2H_5)C$	$C_6H_5CH{=}CH_2$	80–100	$\bar{4}$,13	15,48	34,8	RICR	[53]
$(CH_3)_2(C_2H_5)C$	- " -	80–110	$\bar{5}$,19	16,19	37,7	RICR	[51]
$CH_3(C_2H_5)_2C$	- " -	80–110	$\bar{5}$,40	16,55	38,0	RICR	[51]
$C_6H_5(CH_3)_2C$	C_6H_6	100–125	$\bar{5}$,27	14,70	34,9	ICA	[57]
	C_6H_5Cl	95–125	$\bar{5}$,38	13,18	32,0	ICA	[59]
	$C_6H_5CH{=}CH_2$	95–105	$\bar{5}$,56	17,76	39,9	RICR	[9]
	$C_6H_5CH(CH_3)_2$	110–150	$\bar{5}$,43	14,63	34,5	CSM	[9]
	n-$C_{10}H_{22}$	110–140	$\bar{5}$,47	14,57	34,3	CSM	[60]
	n-$C_{12}H_{26}$	120–150	$\bar{5}$,52	14,17	33,5	RICR	[63]
	Paraffin oil	120–150	$\bar{5}$,51	13,51	32,4	RICR	[64]
	Mineral oil	120–150	$\bar{5}$,56	14,43	33,9	RICR	[65]
	$CH_2{=}C(CH_3)(CH_2)_3CH(CH_3)CH{=}CH_2$	120–150	$\bar{5}$,46	13,26	32,0	RICR	[66]
	cis-1,4-Polyisoprene	120–150	$\bar{5}$,05	13,15	32,5	RICR	[67]
	- " -	120–150	$\bar{5}$,71	15,76	36,0	RICR	[68]
	$C_6H_5CH_3$	125	$\bar{5}$,13			CSM	[69]
$C_6H_5(CH_3)_2C$	cyclo-C_6H_{12}	125	$\bar{6}$,92			CSM	
$C_6H_5(CH_3)_2C$	C_6H_6	125	$\bar{5}$,14			CSM	
	CCl_4	125	$\bar{6}$,95			CSM	
$(CH_3)_3Si$	n-C_7H_{16}	183–203	$\bar{8}$,14	15,04	41,2	CSM	[70]
$(CH_3)_3Si$	$CH_2{=}CH(CH_2)_5CH_3$	183–203	$\bar{8}$,09	15,59	42,3	CSM	[70]

TABLE 8. Decomposition of Unsymmetrical Peroxides, R_1OOR_2

R_1	R_2	Solvent	t°	log k (130°)	log A	E	Method	Ref.
$(CH_3)_3C$	$C(CH_3)_2C_2H_5$	n-$C_{10}H_{22}$	110–130	$\bar{5}$,68	12,08	30,3	CSM	
	$C_6H_5(CH_3)_2C$	n-$C_{10}H_{22}$	110–130	$\bar{5}$,56	15,40	36,3	CSM	[60]
	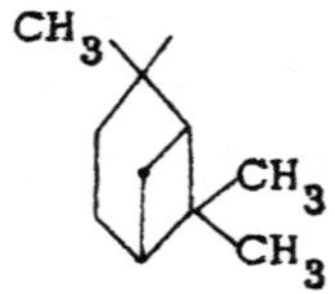	n-$C_{10}H_{22}$	110–130	$\bar{5}$,65	11,48	29,0	CSM	
	$(CH_3)_3Si$	n-C_7H_{16}	183–203	$\bar{8}$,69	15,04	41,2	CSM	[70]
	$(CH_3)_3Si$	$CH_2=CH(CH_2)_5CH_3$	183–203	$\bar{8}$,64	15,59	42,3	CSM	[70]
	$CH_3OCH_2OCH_2$	$C_6H_5CH=CH_2$	80–110	$\bar{5}$,87	12,77	31,1	RICR	
	$C_2H_5OCH_2OCH_2$	- " -	80–110	$\bar{5}$,90	13,20	31,8	RICR	
	$C_3H_7OCH_2OCH_2$	- " -	80–110	$\bar{5}$,88	13,65	32,7	RICR	[51]
	$(CH_3)_2CHOCH_2OCH_2$	- " -	80–110	$\bar{4}$,02	14,77	33,4	RICR	
	$C_4H_9OCH_2OCH_2$	- " -	80–110	$\bar{5}$,85	13,20	31,9	RICR	
	CH_2OH	$C_6H_5NO_2$	90–109	$\bar{6}$,06	7,66	25,0	ICA	[71]
	CH_2OH	n-$C_{14}H_{30}$	115–129	$\bar{7}$,31	11,09	32,7	ICA	[71]

$C_2H_5(CH_3)_2C$	$C_6H_{13}COOCH_2$	$C_6H_5NO_2$	90–100	$\bar{3}$,92	3,42	17,5	ICA	
	$CH_2OCOC_6H_{13}$	n–$C_{14}H_{30}$	110–129	$\bar{7}$,24	5,96	23,4	ICA	[71]
	$C_4H_9OCH_2OCH_2$	– " –	110–129	$\bar{4}$,10	14,56	33,9	ICA	
	$(CH_3)_2CHCH_2(OCH_2)_2$	– " –	110–129	$\bar{4}$,20	14,25	33,2	ICA	[51]
$C_6H_5(CH_3)_2C$	$(C_6H_5)_2CCH_3$	n–$C_{10}H_{22}$	110–130	$\bar{4}$,08	12,09	29,4	CSM	[51]
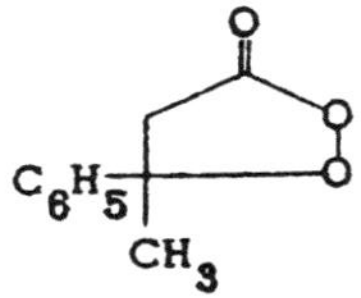		CCl_4	106–126	$\bar{5}$,88	13,34	32,2	CSM	[72]
$(CH_3)_3COOB(OCH_3)_2$		n–C_9H_{20}	160–190	$\bar{6}$,31	10,54	29,9	CSM	[73]
$(CH_3)_2NCH_2$	$C(CH_3)_2C_3H_7$	$C_6H_5NO_2$	100	$\bar{8}$,74			CSM	
	$C(CH_3)_2C_3H_7$	n-$C_{14}H_{30}$	110	$\bar{8}$,85			CSM	
n-$C_4H_9OCH_2$	$C(CH_3)_2C_2H_5$	$C_6H_5NO_2$	110–130	$\bar{7}$,33	7,08	25,3	CSM	
	$C(CH_3)_2C_2H_5$	n–$C_{14}H_{30}$	85–100	$\bar{6}$,15	12,35	32,2	CSM	[71]
n–$C_4H_9POOCH_2$	$C(CH_3)(C_2H_5)_2$	$C_6H_5NO_2$	95–109	$\bar{7}$,54	1,43	14,5	CSM	
$C_6H_{10}NCH_2$	$C(CH_3)_2C_2H_5$	$C_6H_5NO_2$	95–109	$\bar{7}$,54	4,36	18,9	CSM	
	$C(CH_3)_2C_2H_5$	n-$C_{14}H_{30}$	100–110	$\bar{7}$,90	7,10	24,3	CSM	

TABLE 9. Decomposition of Peroxides, $(CH_3)_3COOCOR$

R	Solvent	t°	log k (110°)	log A	E	Method	Ref.
CH_3	C_6H_5Cl	80–100	$\bar{5}$,71	17,81	38,7	CSM	[74]
	$C_6H_5C_2H_5$	85–110	$\bar{5}$,72	17,79	38,6	CSM	[28]
	$C_6H_5CH{=}CH_2$	73–95	$\bar{5}$,68	17,63	38,4	RICR	[12]
	$n\text{-}C_{11}H_{24}$	100–130	$\bar{5}$,85	14,10	31,9	CSM	[75]
C_2H_5	C_6H_6	70–90	$\bar{5}$,75	15,68	34,9	CSM	[76]
$(CH_3)_2CH\text{–}$	C_6H_6	70–90	$\bar{4}$,85	14,67	31,2	CSM	[76]
	$C_6H_5CH(CH_3)_2$	90–110	$\bar{5}$,80	14,37	32,5	CSM	[77]
	C_6H_5Cl	90–110	$\bar{5}$,90	15,50	34,3	CSM	[77]
	$C_6H_5C_2H_5$	85–110	$\bar{4}$,52	13,42	29,6	CSM	[28]
	$C_6H_5CH{=}CH_2$	85–110	$\bar{4}$,59	13,74	30,0	RICR	[28]
$n\text{-}C_3H_7$	$CCl_4 + C_6H_5CH{=}CH_2$	85–110	$\bar{5}$,75	12,90	30,0	CSM	[28]
$n\text{-}C_4H_9$	$C_6H_5C_2H_5$	85–110	$\bar{5}$,94	18,54	39,6	CSM	[28]
	$C_6H_5CH{=}CH_2$	85–110	$\bar{4}$,17	17,32	37,0	RICR	[28]

$(CH_3)_3C$	C_6H_5Cl	58–88	$\bar{2}$,15	15,30	30,0	CSM	
	$C_6H_5C_2H_5$	85–110	$\bar{3}$,94	14,62	29,2	CSM	[78]
	$C_6H_5CH{=}CH_2$	85–110	$\bar{3}$,43	11,33	24,0	RICR	
	$C_6H_5CH(CH_3)_2$	65–85	$\bar{3}$,915	14,015	28,2	CSM	[79]
	$C_6H_5CH(CH_3)_2$	45–65	$\bar{3}$,66	12,56	26,1	CSM	[80]
$(CH_3)_2CHCH_2$	$C_6H_5C_2H_5$	45–65	$\bar{4}$,21	18,61	39,2	CSM	[78]
n–C_5H_{11}	n–$C_{11}H_{24}$	100–130	$\bar{5}$,88	16,25	35,5	CSM	[75]
	$C_6H_5NO_2$	100–130	$\bar{5}$,82	16,37	36,0	CSM	
	C_6H_5Cl	110	$\bar{5}$,92			CSM	
	$C_6H_5OC_6H_5$	110	$\bar{5}$,805			CSM	
	CH_3COOH	110	$\bar{4}$,00			CSM	[81]
	C_2H_5COOH	110	$\bar{5}$,73			CSM	
	C_3H_7COOH	110	$\bar{5}$,73			CSM	
	$CH_3CHClCOOH$	110	$\bar{5}$,91			CSM	
	$(CH_3)_2CClCOOH$	110	$\bar{4}$,66			CSM	
$CH_3(CH_2)_2C(CH_3)_2$	$C_6H_5CH(CH_3)_2$	45–65	$\bar{4}$,85	11,59	25,8	CSM	[80]
$(C_2H_5)_3C$	– " –	45–65	$\bar{2}$,06	12,51	25,3	CSM	[80]
n–C_7H_{15}	$C_6H_5C_2H_5$	85–110	$\bar{5}$,91	18,53	39,6	CSM	[28]

Table 9 (continued)

R	Solvent	t^o	log k (110°)	log A	E	Method	Ref.
n-C_7H_{15}	$C_6H_5CH{=}CH_2$	85–110	$\bar{4}$,30	17,20	36,6	RICR	[28]
n-C_9H_{19}	$C_6H_5C_2H_5$	95–110	$\bar{5}$,95	18,55	39,6	CSM	[28]
$CH_3(CH_2)_{10}$	C_6H_5Cl	110	$\bar{5}$,955			CSM	[81]
n-$C_{11}H_{23}$	$C_6H_5C_2H_5$	95–110	$\bar{5}$,91	18,49	39,5	CSM	[28]
$CH_3(CH_2)_{12}$	C_3H_7COOH	110–130	$\bar{5}$,72	15,77	35,1	CSM	
	$CH_3CH_2ClCOOH$	110–130	$\bar{5}$,895	14,30	32,2	CSM	
	$(CH_3)_2CClCOOH$	110	$\bar{4}$,65			CSM	
	CH_3COOH	110	$\bar{4}$,00			CSM	[81]
	C_6H_5Cl	110	$\bar{5}$,94			CSM	
	$C_6H_5NO_2$	110	$\bar{5}$,805			CSM	
	$C_6H_5OC_6H_5$	110	$\bar{5}$,81			CSM	
	C_2H_5COOH	110	$\bar{5}$,74			CSM	
$CH_3(CH_2)_{14}$	C_6H_5Cl	110	$\bar{5}$,93			CSM	[81]
n-$C_{15}H_{31}$	$C_6H_5C_2H_5$	95–110	$\bar{5}$,94	18,69	39,8	CSM	[28]
$CH_2{=}CH$	C_6H_6	80	$\bar{6}$,125			CSM	[82]

$CH_2{=}CH$	- " -	90–110	$\bar{5}$,69	7,47	20,61	CSM	[83]
$CH_2{=}C(CH_3)$	- " -	90–110	$\bar{4}$,285	14,88	32,75	CSM	[83]
$CH_2{=}CH$	- " -	80	$\bar{6}$,125			CSM	[83]
n-C_3H_7 n-C_3H_7 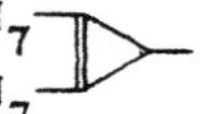	C_6H_5Cl	40–60	$\bar{1}$,92	14,37	25,3	CSM	[84]
C_6H_5 C_6H_5 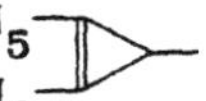	C_6H_5Cl	40–60	$\bar{1}$,88	13,76	24,3	CSM	[84]
	$C_6H_5CH(CH_3)_2$	45–65	$\bar{6}$,66	16,16	37,7	CSM	[80]
	- " -	45–65	$\bar{3}$,10	13,70	29,1	CSM	[80]
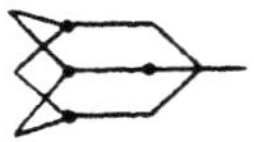	$C_6H_5CH(CH_3)_2$	45–65	$\bar{2}$,02	14,12	28,2	CSM	[80]
	- " -	65–85	$\bar{2}$,57	14,29	28,5	CSM	[79]
CH_3OCH_2	C_6H_6	40–60	$\bar{1}$,34	13,39	24,6	CSM	[85]
	- " -	40	$\bar{4}$,34			CSM	[86]
$C_2H_5OCH_2$	C_6H_6	40	$\bar{4}$,47			CSM	[86]

Table 9 (continued)

R	Solvent	t°	log k (110°)	log A	E	Method	Ref.
$(CH_3)_2CHOCH_2$	C_6H_6	40	$\bar{4},68$			CSM	[86]
$n\text{-}C_4H_9OCH_2$	- " -	40–60	$\bar{1},40$	12,30	22,6	CSM	[85]
$CH_3OCH(CH_3)CH_2$	- " -	40–60	$\bar{3},95$	15,20	33,7	CSM	[85]
Cl	$C_6H_5CH(CH_3)_2$	60–80	$\bar{2},57$	15,54	29,7	CSM	[87]
$ClCH_2$	$C_6H_5C_2H_5$	85–110	$\bar{5},48$	15,88	35,7	CSM	[28]
Cl_3C	C_6H_5Cl	85–110	$\bar{3},45$	14,75	30,3	CSM	[28]
$CH_3(CH_2)_3CHBr$	$C_6H_5C_2H_5$	85–110	$\bar{4},21$	16,96	36,3	CSM	[28]
C_6H_5	$p\text{-}ClC_6H_4CH_3$	60–90	$\bar{4},51$	16,11	34,3	CSM	[88]
	$C_6H_5C_2H_5$	95–110	$\bar{5},60$	15,36	34,6	CSM	[28]
	$CCl_4+C_6H_5CH{=}CH_2$	95–110	$\bar{6},62$	16,04	36,0	RICR	[28]
	$n\text{-}C_{11}H_{24}$	100–130	$\bar{5},63$	14,58	33,2	CSM	[89]
	$C_6H_5OC_6H_5$	100–130	$\bar{5},45$	16,85	37,5	CSM	[90]
$p\text{-}CH_3OC_6H_4$	$C_6H_5OC_6H_5$	100–130	$\bar{5},55$	16,00	35,8	CSM	[90]
$p\text{-}CH_3C_6H_4$	$C_6H_5OC_6H_5$	100–130	$\bar{5},51$	16,11	36,1	CSM	[90]
	$n\text{-}C_{11}H_{24}$	110	$\bar{5},68$			CSM	[75]

m-$CH_3C_6H_4$	n-$C_{11}H_{24}$	110	$\bar{5}$,70			CSM	[75]
o-$CH_3C_6H_4$	- " -	110–120	$\bar{5}$,93	15,17	32,6	CSM	[75]
p-ClC_6H_4	$C_6H_5OC_6H_5$	100–130	$\bar{5}$,21	17,66	39,3	CSM	[90]
	n-$C_{11}H_{24}$	90–130	$\bar{5}$,52	15,54	35,0	CSM	[75]
m-ClC_6H_4	n-$C_{11}H_{24}$	90–130	$\bar{5}$,56	14,26	32,7	CSM	[75]
o-ClC_6H_4	- " -	90–130	$\bar{5}$,80	14,40	32,5	CSM	[75]
2,4-$Cl_2C_6H_3$	- " -	90–130	$\bar{5}$,73	14,23	32,3	CSM	[75]
p-$NO_2C_6H_4$	$C_6H_5OC_6H_5$	100–130	$\bar{5}$,22	18,82	41,3	CSM	[90]
	n-$C_{11}H_{24}$	100–130	$\bar{5}$,38	15,36	35,0	CSM	[75]
o-CH=C[o,o′-$(C_6H_4)_2$]C_6H_4	n-$C_{11}H_{24}$	90	$\bar{3}$,39			CSM	[45]
[o-CH=C$(C_6H_5)_2$]C_6H_4	- " -	90–119	$\bar{4}$,82	12,24	27,0	CSM	[45]
o-$C_6H_5C_6H_4$	$C_6H_5CH(CH_3)_2$	110–130	$\bar{5}$,97	16,77	36,5	CSM	[91]
o-$C_6H_5SC_6H_4$	C_6H_5Cl	99–120	$\bar{4}$,53	10,88	23,7	CSM	[92]
(o-$C_6H_5S)_2C_6H_4$	C_6H_5Cl	40,1	$\bar{5}$,85			CSM	[92]
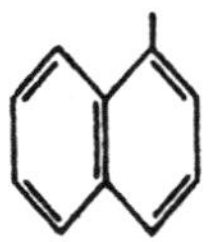	C_6H_5Cl	110	$\bar{5}$,94			CSM	[92]

Table 9 (continued)

R	Solvent	t°	log k (110°)	log A	E	Method	Ref.
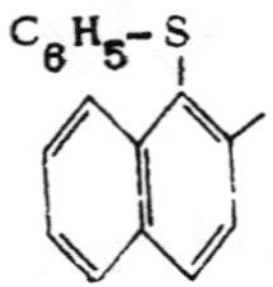	C_6H_5Cl	50–80	1,60	15,54	24,4	CSM	[92]
o–$C_6H_5SC_6H_4$	C_6H_5Cl	40–70	$\bar{5}$,95 (40°)	12,52	23,7	CSM	[93]
o–$CH_3SC_6H_4$	- " -	35–70	$\bar{5}$,87 (40°)	12,15	23,3	CSM	[93]
o–$(CH_3)_3CC_6H_4$	- " -	100–136	$\overline{12}$,73 (40°)	16,60	34,9	CSM	[93]
o–IC_6H_4	- " -	85–119	$\bar{7}$,08 (40°)	13,13	28,7	CSM	[93]
o–$C_6H_5SC_6H_4$	cyclo–C_6H_{12}	25–40	$\bar{6}$,84 (40°)	11,59	24,0	CSM	[94]
	$\overline{CH_2(CH_2)_3O}$	40	$\bar{5}$,48			CSM	[94]
	$C_6H_5NO_2$	40	$\bar{5}$,945			ICA	[94]
	CH_3COCH_3	25–40	$\bar{4}$,09 (40°)	12,16	23,0	CSM	[94]
	$(CH_3)_3COH$	25–40	$\bar{4}$,51 (40°)	12,16	22,4	CSM	[94]
	CH_3CN	25–40	$\bar{4}$,735 (40°)	11,19	20,7	CSM	[94]
	$(CH_3)_2SO$	25–40	$\bar{4}$,73 (40°)	11,30	20,8	CSM	[94]

	$(CH_3)_2CHOH$	25–40	$\bar{4},86$ (40°)	11,52	21,0	CSM	[94]
	C_2H_5OH	25–40	$\bar{3},22$ (40°)	14,19	24,3	CSM	
	CH_3OH	25–40	$\bar{3},67$ (40°)	12,775	21,6	CSM	
o–(p–$CH_3OC_6H_4S)C_6H_4$	C_6H_5Cl	34–65	$\bar{4},38$ (40°)	10,98	20,9	CSM	[94]
o–(p–$ClC_6H_4S)C_6H_4$	– " –	40–70	$\bar{5},61$ (40°)	11,68	23,0	CSM	
o–(p–$NO_2C_6H_4S)C_6H_4$	– " –	[illegible]0–84	$\bar{5},04$ (40°)	10,84	22,6	CSM	
o–$CH_3SO_2C_6H_4$	– " –	105–150	$\bar{6},54$	17,19	38,7	CSM	
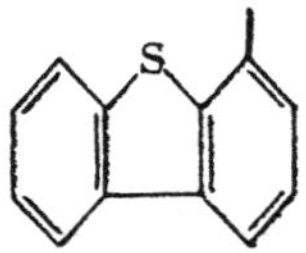	C_6H_5Cl	105–135	$\bar{5},26$	13,56	31,3	CSM	[94]
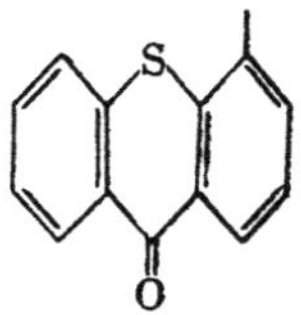	C_6H_5Cl	120–135	$\bar{5},16$	17,46	38,1	CSM	[94]
$C_6H_5CH_2$	C_6H_5Cl	58–88	$\bar{3},30$	13,34	28,1	CSM	[78]
	$C_6H_5CH_3$	80–100	$\bar{3},37$	13,30	27,9	CSM	[78]
	C_6H_5Cl	80–100	$\bar{3},35$	14,15	29,4	CSM	[74]
	$C_6H_5C_2H_5$	735–85	$\bar{3},27$	13,37	28,2	CSM	[28]

Table 9 (continued)

R	Solvent	t^{o}	log k (110°)	log A	E	Method	Ref.
p-$CH_3OC_6H_4CH_2$	$C_6H_5CH_3$	56–80	$\bar{2}$,46	12,86	25,2	CSM	[78]
	- " -	56	$\bar{5}$,11			ICA	[95]
p-$CH_3SC_6H_4CH_2$	$C_6H_5C_2H_5$	90,6	$\bar{3}$,03			CSM	[95]
	$C_6H_5CH_3$	56	$\bar{6}$,86			ICA	[95]
p-$CH_3C_6H_4CH_2$	- " -	70–90	$\bar{3}$,05	12,20	26,5	CSM	
m-$CH_3C_6H_4CH_2$	- " -	80–100	$\bar{3}$,34	14,02	29,2	CSM	
m-$ClC_6H_4CH_2$	- " -	80–100	$\bar{3}$,01	13,81	29,4	CSM	[78]
p-$ClC_6H_4CH_2$	- " -	80–100	$\bar{3}$,30	13,30	28,0	CSM	
p-$NO_2C_6H_4CH_2$	- " -	80–100	$\bar{4}$,83	13,83	29,8	CSM	
$C_6H_5CH_2CH_2$	$C_6H_5C_2H_5$	95–110	$\bar{5}$,79	15,89	35,2	CSM	[28]
	C_6H_5Cl	109–122	$\bar{5}$,80	16,23	35,76	CSM	[95a]
$C_6H_5(CH_2)_3$	$C_6H_5C_2H_5$	95–110	$\bar{5}$,85	15,70	34,7	CSM	[28]
	C_6H_5Cl	109–122	$\bar{5}$,815	16,84	36,76	CSM	[95a]
$(C_6H_5)_2CH$	C_6H_5Cl	80–100	$\bar{2}$,90	13,08	25,0	CSM	[74]
	$C_6H_5CH(CH_3)_2$	40–70	$\bar{2}$,68	13,31	25,6	CSM	[77]
	C_6H_5Cl	40–60	$\bar{2}$,77	13,05	25,0	CSM	[74]

$C_6H_5(CH_3)_2C$	C_6H_5Cl	40–60	$\bar{1}$,27	14,57	26,8	CSM	[74]
$(C_6H_5)_2(CH_3)C$	- " -	40–60	$\bar{1}$,52	14,02	25,4	CSM	[74]
$(C_6H_5)_3C$	$C_6H_5CH(CH_3)_2$	28–45	0,26	14,36	24,7	CSM	[96]
$C_6H_5OCH_2$	C_6H_6	40–60	$\bar{2}$,46	13,66	26,6	CSM	
$C_6H_5O(CH_2)_2$	- " -	40–60	$\bar{5}$,71	15,56	37,4	CSM	
$C_6H_5O(CH_2)_3$	- " -	40–60	$\bar{4}$,08	15,33	33,7	CSM	[85]
$C_6H_5OCH(CH_3)$	- " -	40–60	$\bar{1}$,64	12,54	22,6	CSM	
$C_6H_5OC(CH_3)_2$	- " -	40–60	$\bar{1}$,87	11,75	20,6	CSM	
p-$NO_2C_6H_4OCH_2$	- " -	70,5	$\bar{5}$,83			CSM	
2,4-$Cl_2C_6H_4OCH_2$	- " -	70,5	$\bar{5}$,15			CSM	
p-Br $C_6H_4OCH_2$	- " -	70,5	$\bar{4}$,41			CSM	
p-$ClC_6H_4OCH_2$	- " -	70,5	$\bar{4}$,46			CSM	
$C_6H_5OCH_2$	- " -	70,5	$\bar{4}$,64			CSM	[86]
p-$CH_3C_6H_4OCH_2$	- " -	70,5	$\bar{4}$,86			CSM	
p-$CH_3OC_6H_4OCH_2$	- " -	70,5	$\bar{3}$,16			CSM	
$C_6H_5CH_2SCH_2$	- " -	40	$\bar{5}$,42			CSM	
$C_6H_5CH_2OCH_2$	- " -	40	$\bar{4}$,12			CSM	
C_6H_5OCO	- " -	45–65	$\bar{5}$,665 (45°)	14,565	27,5	CSM	
$C_6H_5CH_2OCO$	- " -	45–65	$\bar{5}$,56 (45°)	14,26	27,2	CSM	
p-$CH_3OC_6H_4CH_2OCO$	- " -	45–65	$\bar{5}$,825 (45°)	14,275	26,8	CSM	[97]
p-$NO_2C_6H_4CH_2OCO$	- " -	45–65	$\bar{5}$,11 (45°)	14,71	28,5	CSM	

TABLE 10. Decomposition of Dibasic Peroxides

Peroxide	Solvent	t°	log k (90°)	log A	E	Method	Ref.
$(CH_3)_3COOCOCOOOC(CH_3)_3$	C_6H_6	35–55	$\bar{2}$,68	14,46	26,2	CSM	[98]
$(CH_3)_3COOCO(CH_2)_2COOOC(CH_3)_3$	$C_6H_5CH{=}CH_2$	95–105	$\bar{6}$,13	9,53	25,6	RICR	
$(CH_3)_3COOCO(CH_2)_3COOOC(CH_3)_3$	- " -	95–105	$\bar{6}$,31	12,11	29,6	RICR	
$(CH_3)_3COOCO(CH_2)_4COOOC(CH_3)_3$	- " -	95–105	$\bar{6}$,75	12,11	29,0	RICR	[28]
$(CH_3)_3COOCO(CH_2)_5COOOC(CH_3)_3$	- " -	95–105	$\bar{6}$,53	14,83	33,7	RICR	
$(CH_3)_3COOCO(CH_2)_7COOOC(CH_3)_3$	- " -	95–105	$\bar{6}$,66	16,81	36,8	RICR	
$(CH_3)_3COOCO(CH_2)_8COOOC(CH_3)_3$	- " -	95–105	$\bar{6}$,40	18,00	39,2	RICR	
1,2-$(C_6H_5COOOCO)_2$cyclo-C_6H_{10}	- " -	70	$\bar{6}$,90			RICR	[10]
1,1-$[(CH_3)_3COO]_2$cyclo-C_6H_{10}	n-$C_{10}H_{22}$	90–130	$\bar{5}$,04	15,39	33,8	RICR	[60]
$(CH_3)_3COOCOC(CH_3)_2COOOC(CH_3)_3$	$C_6H_5(CH_3)_2CH$	79,5–100	$\bar{4}$,37	14,42	30,0	CSM	[99]
$(CH_3)_3COOCOC(CH_3)_2COOOC_2H_5$	$C_6H_5(CH_3)_2CH$	79,5–100	$\bar{4}$,26	14,61	30,5	CSM	[99]
$(C_6H_5)_2C(COOC(CH_3)_3)_2$	$C_6H_5CH(CH_3)_2$	50–70	$\bar{2}$,40	14,30	26,4	CSM	[77]
$[C_6H_5CHCOOOC(CH_3)_3]_2$	$C_6H_5CH(CH_3)_2$	70–90	$\bar{3}$,08	15,58	30,7	CSM	[100]

$(CH_3)_3COOCH(C_2H_5)OOC(CH_3)_3$	$C_6H_5CH{=}CH_2$	80–90	$\bar{6}$,96	15,61	34,3	RICR	[101]
$[(CH_3)_3COO]_2CHCH_3$	- " -	80–100	$\bar{7}$,675	17,98	39,1	RICR	
$[(CH_3)_3COO]_2C(CH_3)_2$	- " -	80–100	$\bar{6}$,06	17,46	37,7	RICR	
$[(CH_3)_3COO]_2CHCH_2CH_2CH_3$	- " -	80–100	$\bar{6}$,32	16,12	35,1	RICR	
$[(CH_3)_3COO]_2C(CH_3)C_2H_5$	- " -	80–100	$\bar{6}$,425	16,00	34,7	RICR	
$[(CH_3)_3COO]_2CHCH(CH_3)_2$	- " -	80–100	$\bar{6}$,585	15,88	34,3	RICR	
$[(CH_3)_3COO]_2C(C_2H_5)_2$	- " -	80–100	$\bar{6}$,60	14,55	32,1	RICR	[53]
$[(CH_3)_2(C_2H_5)COO]_2CH_2$	- " -	80–100	$\bar{6}$,02	15,32	34,3	RICR	
$[(CH_3)_2(C_2H_5)COO]_2CHCH_3$	- " -	80–100	$\bar{6}$,09	15,24	34,0	RICR	
$[(CH_3)_2(C_2H_5)COO]_2C(CH_3)_2$	- " -	80–100	$\bar{6}$,29	13,56	31,0	RICR	
$[(CH_3)_2(C_2H_5)COO]_2C(CH_3)C_2H_5$	- " -	80–100	$\bar{6}$,57	13,88	31,1	RICR	
$[(CH_3)_2(C_2H_5)COO]_2C(C_2H_5)_2$	- " -	80–100	$\bar{6}$,72	13,99	31,0	RICR	
$(CH_3)_2C[OOC(C_2H_5)_2CH_3]_2$	C_6H_5Cl	106,6	$\bar{4}$,025	12,175	28,0	CL	
	n-$C_5H_{11}OH$	91,5	$\bar{4}$,245	13,345	28,5	CL	[71]
	CH_3COOH	86,8	$\bar{4}$,95	12,13	24,9	CL	
$(CH_3)[(CH_3)_2CH]C[OOC(CH_3)_3]_2$	n-$C_{14}H_{30}$	136	$\bar{4}$,165	12,60	30,7	CL	
	C_6H_5Cl	106,6	$\bar{4}$,22	10,34	24,5	CL	[71]
	n-$C_5H_{11}OH$	94,5	$\bar{4}$,24	9,95	23,0	CL	
	CH_3COOH	86,8	$\bar{4}$,98			CL	

Table 10 (continued)

Peroxide	Solvent	t^o	log k (90°)	log A	E	Method	Ref.
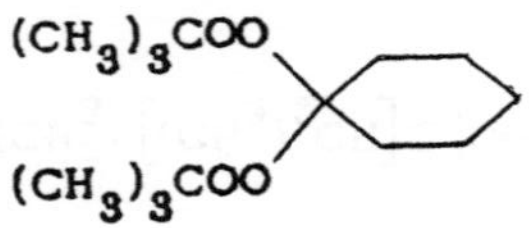	C_6H_5Cl	106,6	$\bar{4}$,045	13,515	31,3	CL	[71]
	CH_3COOH	86,8	$\bar{4}$,86			CL	[71]
$C_6H_5CH[OOC(C_2H_5)_2CH_3]_2$	n-$C_{14}H_{30}$	83,5	$\bar{4}$,02	8,41	20,2	CL	[71]
	C_6H_5Cl	106,6	$\bar{5}$,955	9,235	23	CL	
	n-$C_5H_{11}OH$	94,5	$\bar{4}$,355	6,35	16,7	CL	
	CH_3COOH	86,8	$\bar{4}$,56	6,07	15,6	CL	
$C_5H_4NCH_2N[OCH_2OOC(CH_3)_3]_2$	n-$C_{14}H_{30}$	132	$\bar{4}$,33		16,2	CL	[71]
	C_6H_5Cl	106,6	$\bar{4}$,155	6,545	18,0	CL	
	n-$C_5H_{11}OH$	94,5	$\bar{4}$,32	6,16	16,5	CL	
	CH_3COOH	86,8	$\bar{4}$,71	8,42	11,0	CL	
$(CH_3)_2Si[OCH_2OOC(CH_3)(C_2H_5)_2]_2$	n-$C_{14}H_{30}$	83,5	$\bar{4}$,43	7,16	17,5	CL	[71]
	C_6H_5Cl	106,6	$\bar{4}$,12	8,23	21	CL	
	CH_3COOH	86,8	$\bar{4}$,73			CL	

$[(CH_3)_3COO]_3B$	$n\text{-}C_9H_{20}$	170–190	$\bar{4},60$ (180°)	12,4	32,6	CSM	[102]
$C_6H_5C(O)OC(O)CH_2C(=CH_2)-C(O)OC(O)C_6H_5$ *	C_6H_6, C_6H_5Cl, CCl_4	60–85	$\bar{3},44$ $\bar{4},27$	9,495 12,84	20,0 27,5	CSM	[102a]
$C_6H_5C(O)OC(O)CH_2CH(Cl)-C(O)OC(O)C_6H_5$ *	- " -	60–85	$\bar{3},01$ $\bar{4},03$	8,93 12,60	19,8 27,5	CSM	[102a]
$C_6H_5C(O)OC(O)CH_2CH(Br)-C(O)OC(O)C_6H_5$ *	- " -	70–85	$\bar{4},70$ $\bar{5},97$	9,40 12,41	21,1 27,3	CSM	[102a]
$C_6H_5C(O)OC(O)CH_2CH(CH_3)-C(O)OC(O)C_6H_5$ *	- " -	70–85	$\bar{3},14$ $\bar{4},08$	9,06 12,83	19,8 27,8	CSM	[102a]
$C_6H_5C(O)OC(O)(CH_2)_2CH(Br)-C(O)OC(O)C_6H_5$ *	- " -	60–85	$\bar{3},08$ $\bar{4},20$	10,46 12,35	22,2 26,8	CSM	[102a]
$C_6H_5C(O)OC(O)(CH_2)_3CH(Br)-C(O)OC(O)C_6H_5$ *	- " -	70–85	$\bar{3},30$ $\bar{4},29$	10,50 12,44	21,9 26,8	CSM	[102a]
$C_6H_5C(O)OC(O)(CH_2)_4CH(Br)-C(O)OC(O)C_6H_5$ *	- " -	70–85	$\bar{3},505$ $\bar{4},45$	10,705 12,735	21,9 27,0	CSM	[102a]
$C_6H_5C(O)OC(O)(CH_2)_6CH(Br)-C(O)OC(O)C_6H_5$ *	- " -	70–85	$\bar{3},41$ $\bar{4},53$	10,06 12,49	21,0 26,5	CSM	[102a]

Table 10 (continued)

Peroxide	Solvent	t°	log k (90°)	log A	E	Method	Ref.
$C_6H_5C(O)OOC(O)(CH_2)_7CH(Br){-}C(O)OOC(O)C_6H_5$ *	C_6H_6, C_6H_5Cl, CCl_4	70–85	$\bar{3}$,405; $\bar{4}$,54	10,075; 12,505	21,0; 26,5	CSM	[102a]
$CH_3(CH_2)_6C(O)OOC(O)CH_2C(=CH_2){-}C(O)OOC(O)(CH_2)_6CH_3$ *	- " -	70–85	$\bar{3}$,30; $\bar{4}$,52	10,68; 14,90	22,2; 30,5	CSM	[102a]
$CH_3(CH_2)_6C(O)OOC(O)CH_2CH(CH_3){-}C(O)OOC(O)(CH_2)_6CH_3$ *	- " -	70–85	$\bar{3}$,56; $\bar{4}$,69	11,43; 15,675	23,0; 31,3	CSM	[102a]
$CH_3(CH_2)_6C(O)OOC(O)CH_2CH(Cl){-}C(O)OOC(O)(CH_2)_6CH_3$ *	- " -	70–85	$\bar{3}$,09; $\bar{4}$,45	10,82; 14,655	22,8; 30,2	CSM	[102a]
$CH_3(CH_2)_6C(O)OOC(O)(CH_2)_2CH(Br){-}C(O)OOC(O)(CH_2)_6CH_3$ *	- " -	70–85	$\bar{3}$,14; $\bar{4}$,61	11,24; 14,81	23,4; 30,2	CSM	[102a]
$CH_3(CH_2)_6C(O)OOC(O)(CH_2)_3CH(Br)C(O)OOC(O)(CH_2)_6CH_3$ *	- " -	60–85	$\bar{3}$,53; $\bar{4}$,57	11,45; 14,24	23,1; 29,3	CSM	[102a]
$CH_3(CH_2)_6C(O)OOC(O)(CH_2)_4CH(Br){-}C(O)OOC(O)(CH_2)_6CH_3$ *	- " -	70–85	$\bar{3}$,67; $\bar{4}$,83	11,95; 15,93	23,7; 31,7	CSM	[102a]

$CH_3(CH_2)_6C(=O)OOC(=O)(CH_2)_6CH(Br)-C(=O)OOC(=O)(CH_2)_6CH_3$ *	- " ..	60–85	$\bar{3}$,69 $\bar{4}$,82	12,02 15,50	23,8 31,0	CSM	[102a]
$C_6H_5C(=O)OOC(=O)(CH_2)_2C(=O)OOC(=O)C_6H_5$	- " -	70–85	$\bar{5}$,90	13,97	30,0	CSM	[102a]
$CH_3(CH_2)_6C(=O)OOC(=O)CH_2CH(Br)-C(=O)OOC(=O)(CH_2)_6CH_3$ *	- " -	70–85	$\bar{3}$,03 $\bar{4}$,35	10,94 14,55	23,1 30,2	CSM	[102a]
$(CH_3)_3COOC(=O)CH_2C(=CH_2)-C(=O)OOC(CH_3)_3$ *	- " -	95–115	$\bar{4}$,19 $\bar{6}$,55	10,84 16,405	24,3 36,3	CSM	[102a]
$(CH_3)_3COOC(=O)CH_2CH(Cl)-C(=O)OOC(CH_3)_3$ *	- " -	105–125	$\bar{5}$,93 $\bar{6}$,41	10,75 15,06	24,6 34,3	CSM	[102a]
$(CH_3)_3COOC(=O)(CH_2)_2CH(Br)-C(=O)(CH_3)_3$ *	- " -	110–125	$\bar{5}$,67 $\bar{5}$,67	10,725 16,87	25,0 35,2	CSM	[102a]
$(CH_3)_3COOC(=O)(CH_2)_3CH(Br)-C(=O)OOC(CH_3)_3$ *	- " -	95–125	$\bar{5}$,925 $\bar{6}$,92	10,505 15,52	24,2 34,2	CSM	[102a]

*In these peroxides, the peroxide groups decompose with different values of k; therefore, each peroxide is characterized by a pair of decomposition rate constants.

TABLE 11. Decomposition of Hydroperoxides in Aromatic Solvents

ROOH	Solvent	t°	log k (120°)	log A	E	Method	Ref.
$(CH_3)_3COOH$	C_6H_6	154–174	$\bar{7}$,33	16,03	40,8	CSM	[103]
$(CH_3)_3COOH$	C_6H_6Cl	100–126	$\bar{6}$,23	12,56	33,0	ICA	[104]
$C_6H_5CH_2C(CH_3)_2OOH$	C_6H_6	144–176	$\bar{6}$,18	10,85	30,0	CSM	[103]
$C_6H_5CH[CH(CH_3)_2]OOH$	C_6H_6	134–174	$\bar{6}$,42	10,62	29,2	CSM	[103]
$CH_3COCH(OOH)CH_3$	C_6H_6	30–70	$\bar{2}$,31	13,53	27,4	CL	[105]
$CH_3CH(OOH)(CH_2)_7CH_3$	C_6H_5Cl	130	$\bar{6}$,45			ICA	[106]
Polyethylene hydroperoxide	C_6H_5Cl	110–130	$\bar{6}$,70	14,20	35,0	ICA	[107]

TABLE 12. Decomposition of Azobisisobutyronitrile

Solvent	t°	log k (80°)	log A	E	Method	Reference
C_6H_6	50–70	$\bar{4}$,18	15,11	30,5	KEPF	[108]
$C_6H_6 + C_6H_5CH_3$	37–100	$\bar{4}$,10	15,20	30,8	KEPF	[110]
C_6H_6	40–78	$\bar{5}$,95	14,25	29,5	ICA	[111]
	40–70	$\bar{5}$,95	15,00	30,7	ICA	[7]
	60	$\bar{5}$,05			KEPF	[112]
	62,5	$\bar{5}$,15			KEPF	[113]

Table 12 (continued)

Solvent	t°	log k (80°)	log A	E	Method	Reference
$C_6H_5CH_3$	70–105	$\bar{4}$,13	15,43	31,1	CSM	[114]
	80	$\bar{4}$,19			KEPF	[112]
	80	$\bar{4}$,22			KEPF	[115]
$CH_3C_6H_4CH_3$	77–82	$\bar{4}$,00	16,03	32,3	KEPF	[116]
	70–90	$\bar{4}$,19	15,59	31,3	KEPF	[117]
$C_6H_5C_2H_5$	60–80	$\bar{4}$,05	15,42	31,2	KICR	[41]
$C_6H_5CH{=}CH_2$	50–77	$\bar{4}$,22	15,27	30,7	RICR	[19]
$(C_6H_5)_2CH_2$	66,8	$\bar{5}$,46			KEPF	[118]
C_6H_5Cl	62,5	$\bar{5}$,19			KEPF	[113]
$C_6H_5NO_2$	80–100	$\bar{4}$,32	15,62	31,1	KEPF	[119]
	80	$\bar{4}$,30			KEPF	[112]
$C_6H_5NH_2$	80	$\bar{4}$,225			KEPF	[112]
$C_6H_5N(CH_3)_2$	80	$\bar{4}$,26			KEPF	[112]
	66,8	$\bar{5}$,53			KEPF	[118]
$CH_3COOC_2H_5$	40–60	$\bar{5}$,82	15,19	31,2	ICA	[7]
	60	$\bar{5}$,055			KEPF	[112]

$CH_2=C(CH_3)COOCH_3$	35–100	$\bar{4}$,46	15,265	30,75	RICR	[120]
$C_6H_5CH(CH_3)_2$	52–73	$\bar{4}$,33	16,53	32,78	RICR	[121]
p-$CH_3C_6H_4CH(CH_3)_2$	52–73	$\bar{4}$,21	17,99	35,14	RICR	[121]
$CH_2=C(CH_3)COOC_2H_5$	35–100	$\bar{4}$,21	16,51	32,75	RICR	
$CH_2=C(CH_3)COOCH_2CH(CH_3)_2$	35–100	$\bar{4}$,23	16,93	33,40	RICR	[120]
$CH_2=C(CH_3)COOC_9H_{19}$	35–100	5,97	17,07	34,02	RICR	
cyclo -$C_6H_{11}OH$	50–90	$\bar{4}$,00	17,70	35,0	ICA	[122]
$(CH_3)_2CCH_2OH$	80	$\bar{4}$,235			KEPF	[115]
	82	$\bar{4}$,22			KEPF	[116]
$CH_3(CH_2)_2CH_2OH$	66,8	$\bar{5}$,43			KEPF	[118]
	82	$\bar{4}$,19			KEPF	[116]
$(CH_3)_2(C_2H_5)COH$	80	$\bar{4}$,15			KEPF	[115]
cyclo -$C_6H_{10}O$	82	$\bar{4}$,155			KEPF	[115]
	55–80	$\bar{4}$,24	17,30	34,0	ICA	[123]
CH_3COOH	80	$\bar{4}$,18			KEPF	[117]
	82	$\bar{4}$,18			KEPF	[116]
n-$C_3H_7NHCH_3$	66,8	$\bar{5}$,47			KEPF	[109]
n-$C_{12}H_{25}SH$	80	$\bar{4}$,16			KEPF	[117]
CCl_4	40–60	$\bar{5}$,70	14,60	30,4	ICA	[7]

TABLE 13. Decomposition of R-N=N-R (KEPF Method)

R	Solvent	t°	log k (80°)	log A	E	Ref.
$(CH_3)_3C$	$n\text{-}C_{100}H_{202}$	170–200	$\overline{10},28$	17,48	43,9	[124]
$n\text{-}C_3H_7(CH_3)_2C$	- " -	160–200	$\overline{11},88$	15,58	41,4	[124]
	C_6H_6	273–290	$\overline{20},96$	20,87	63,0	[124]
$C_6H_5(CH_3)CH$	$C_6H_5C_2H_5$	100–110	$\overline{6},63$	14,83	32,6	[125]
	$n\text{-}C_{12}H_{26}$	97,3	$\overline{5},50$			[118]
	Polycarbonate	97,3	$\overline{5},52$			
	$(n\text{-}C_4H_9)_2CHOH$	97,3	$\overline{5},525$			
	$(C_6H_5CH_2OCH_2)_2$	97,3	$\overline{5},57$			
	$C_2H_5CONHCH_3$	97,3	$\overline{5},57$			
	$(C_6H_5)_2CH_2$	97,3	$\overline{5},60$			
	$C_6H_5N(CH_3)CH_2C_6H_5$	97,3	$\overline{5},62$			
$C_6H_5CH(CH_3)$	$C_6H_5C_2H_5$	105	$\overline{5},95$			[126]
$C_6H_5CD(CH_3)$	- " -	105	$\overline{5},88$			[126]
$C_6H_5CH(C_2H_5)$	- " -	100–110	$\overline{6},31$	14,31	32,3	[125]

p-$CH_3C_6H_4CH(CH_3)$	- " -	100–110	$\bar{6}$,67	15,07	32,9	[125]
p-$XC_6H_4CH(CH_3)$						
X=CH_3	$C_6H_5CH_3$	44–54	$\bar{3}$,53	12,40	24,0	
H	- " -	44–54	$\bar{3}$,67	14,40	27,0	
CH_3O	- " -	44–54	$\bar{3}$,69	15,04	28,0	
HO	- " -	44–54	$\bar{3}$,66	15,63	29,0	
CH_3CONH	- " -	44–54	$\bar{3}$,63	16,23	30,0	
Br	- " -	44–54	$\bar{3}$,39	14,74	28,0	[127]
NO_2	- " -	44–54	$\bar{3}$,06	13,81	27,0	
H	CH_3COOH	44–54	$\bar{3}$,71	15,06	28,0	
HO	- " -	44–54	$\bar{3}$,74	17,56	32,0	
H	C_5H_5N	44–54	$\bar{3}$,77	16,97	31,0	
HO	- " -	44–54	$\bar{3}$,40	17,58	32,0	
p-$CH_3OC_6H_4CH(CH_3)$	$C_6H_5C_2H_5$	100–110	$\bar{6}$,645	16,845	35,8	[125]
$C_6H_5CH[CH_2CH(CH_3)_2]_2$	$C_6H_5C_2H_5$	100–110	$\bar{6}$,77	15,38	33,3	[125]
$(C_6H_5)_2CH$	$C_6H_5CH_3$	54–64	$\bar{3}$,27	13,755	26,6	[128]
$C_6H_5C(CH_3)_2$	- " -	40–69	$\bar{5}$,67	14,02	29,6	[129]
p-$XC_6H_4C(CH_3)_2$						
X=H	$C_6H_5CH_3$	41–58	$\bar{3}$,20	15,35	29,3	
Br	- " -	43–58	$\bar{3}$,53	14,93	28,1	[130]
CH_3	- " -	41–58	$\bar{3}$,07	15,22	28,8	

Table 13 (continued)

R	Solvent	t°	log k (80°)	log A	E	Ref.
p-$XC_6H_4C(CH_3)_2$						
X=C_2H_5	$C_6H_5CH_3$	43–58	$\bar{3}$,03	13,13	26,0	[130]
$(CH_3)_2CH$	- " -	43–58	3,26	15,66	29,7	[130]
$(CH_3)_3C$	- " -	43–58	$\bar{3}$,20	15,50	29,5	[130]
$C_6H_5CH_2(CH_3)(CN)C$	$C_6H_5CH_3$	80	$\bar{4}$,065			[131]
p-$ClC_6H_4CH_2(CH_3)(CN)C$	- " -	80	$\bar{5}$,945			[131]
p-$NO_2C_6H_4CH_2(CH_3)(CN)C$	- " -	80	$\bar{4}$,00			[131]
$(CH_3)_2C(COOC_2H_5)$	- " -	70–96,2	$\bar{5}$,60	14,875	30,2	[132]
$CH_3(C_2H_5)(CN)C$	$C_6H_5CH_3$	70–80	$\bar{5}$,95	14,15	29,4	[115]
	$CH_3C_6H_4CH_3$	70–80	$\bar{4}$,00	14,20	29,4	[117]
	$C_2H_5CONHCH_3$	67–72	$\bar{5}$,70	15,955	31,86	[118]
	Polycarbonate	67–72	$\bar{4}$,15	15,855	31,75	[118]
	$(C_6H_5)_2CH_2$	67–72	$\bar{5}$,18	14,535	31,21	[118]
	$C_6H_5N(CH_3)_2$	67–72	$\bar{5}$,41	14,06	30,08	[118]
$(C_2H_5)_2(CN)C$	$C_6H_5NO_2$	80–100	$\bar{4}$,10	17,10	34,2	[119]
n-$C_3H_7(CH_3)(CN)C$	$C_6H_5CH_3$	70–80	$\bar{4}$,18	16,635	33,0	[115]

$(n\text{-}C_3H_7)_2(CN)C$	$C_6H_5NO_2$	100	$\bar{4}$,06			[119]
sec -$C_3H_7(CH_3)(CN)C$	$C_6H_5CH_3$	70–80	$\bar{5}$,61	15,81	32,0	[115]
sec -$C_3H_7(C_2H_5)(CN)C$	- " -	80	$\bar{5}$,98			[133]
(sec-$C_3H_7)_2(CN)C$	- " -	80	$\bar{4}$,10			[133]
n-$C_4H_9(CH_3)(CN)C$	- " -	80	$\bar{4}$,20			[115]
sec-$C_4H_9(CH_3)(CN)C$	- " -	70–80	$\bar{4}$,83	14,80	29,0	[115]
(sec-$C_4H_9)_2(CN)C$	- " -	60–80	$\bar{3}$,71	16,48	30,3	[133]
$(CH_3)_2CHCH_2(CH_3)(CN)C$	$CH_3C_6H_4CH_3$	77	$\bar{4}$,76			[116]
$(CH_3)_3CCH_2(CH_3)CCN$	$C_6H_5CH_3$	40–50	$\bar{2}$,17	14,90	27,0	[133]
$(CH_3)_3CCH_2(^{\bullet}CH_3)CCN$(stereo-isomer)	$C_6H_5CH_3$	40–50	$\bar{2}$,11	16,08	29,0	[133]
n-$C_5H_{11}(CH_3)CCN$	$CH_3C_6H_4CH_3$	70–90	$\bar{2}$,25	14,95	30,2	[117]
cyclo-C_4H_6CN	$C_6H_5CH_3$	120–140	$\bar{7}$,22	13,115	32,1	[134]
cyclo-C_5H_8CN	- " -	80–95	$\bar{5}$,81	16,76	33,8	[134]
cyclo-$C_6H_{10}CN$	$C_6H_5CH_3$	80–102	$\bar{6}$,80	16,70	35,4	[134]
	- " -	80	$\bar{6}$,92			[115]
	C_6H_6	80	$\bar{6}$,92			[135]
	$CH_3C_6H_4CH_3$	80–100	$\bar{6}$,66	19,36	39,9	[117]

Table 13 (continued)

R	Solvent	t°	log k (80°)	log A	E	Ref.
cyclo-$C_6H_{10}CN$	$CH_3C_6H_4CH_3$	77	$\bar{6}$,725			[116]
	C_6H_5Cl	80–100	$\bar{6}$,90	15,49	33,2	[136]
	$C_6H_5CH(CH_3)_2$	74–97	$\bar{5}$,07	15,57	33,07	[121]*
	p-$CH_3C_6H_4CH(CH_3)_2$	74–97	$\bar{6}$,94	17,24	35,99	[121]*
cyclo-$C_7H_{12}CN$	$C_6H_5CH_3$	49–67	$\bar{3}$,04	14,09	27,5	[134]
cyclo-$C_8H_{14}CN$	- " -	37–50	$\bar{3}$,87	13,92	25,9	[134]
cyclo-$C_{10}H_{18}CN$	- " -	51–70	$\bar{3}$,25	14,60	28,0	[134]
cyclo-$C_3H_5C(CH_3)CN$	$C_6H_5CH_3$	80	$\bar{3}$,52			[137]
cyclo-$C_4H_7C(CH_3)CN$	- " -	80	$\bar{4}$,18			[137]
cyclo-$C_5H_9C(CH_3)CN$	- " -	80	$\bar{4}$,11			[137]
cyclo-$C_6H_{11}C(CH_3)CN$	- " -	80	$\bar{4}$,355			[137]

$CH_3(CN)C(CH_2CH_2COOH)$	$CH_3C_6H_4CH_3$	60–80	$\bar{5}$,95	16,955	34,0	[117]
	H_2O	15–50	$\bar{4}$,06	12,165	26,0	[138]
$NH{=}C(NH_2)C(CH_3)_2$	$(CH_3)_2SO$	70	$\bar{6}$,30			[139]
$^{+}H_2N{=}C(NH_2)C(CH_3)_2$	- " -	70	$\bar{5}$,78			[139]
$^{+}H_2N{=}C(\overset{+}{N}H_3)C(CH_3)_2$	- " -	70	$\bar{4}$,23			[139]
$CH_2-N{=}CC(CH_3)_2$ CH_2-NH	- " -	60	$\bar{5}$,00			[139]
$CH_2-N{=}CC(CH_3)_2$ $CH_2-\overset{+}{N}H_2$	- " -	60	$\bar{4}$,40			[139]
$CH_2-\overset{+}{N}H{=}CC(CH_3)_2$ $CH_2-\overset{+}{N}H_2$	- " -	60	$\bar{4}$,32			[139]
$(CH_3)_2C(COOH)$	iso-C_8H_{18}	53–95	$\bar{5}$,37	19,07	37,3	[140]

*k measured on the basis of kinetics of luminescence of hydrocarbon being oxidized.

TABLE 14. Decomposition of R_1-N=N-R_2 (CSM and KEPF Methods)

R_1	R_2	Solvent	t°	log k (80°)	log A	E	Ref.
CH_3	$C_6H_5(CH_3)CH$	$C_6H_5OC_6H_5$ + quinone	151–171	$\bar{5}$,345	15,545	32,6	[141]
$(CH_3)_2CH$	- " -	- " -	143–165	$\bar{8}$,88	15,48	36,5	
$(CH_3)_2CH$	$C_6H_5(CH_3)_2C$	- " -	120–143	$\bar{7}$,63	16,43	36,7	[142]
$(C_2H_5)_2CH$	$C_6H_5(C_2H_5)_2C$	- " -	120–147	$\bar{7}$,39	17,59	39,0	
$C_6H_5(CH_3)C(CH_2CH(CH_3)_2)$	$(CH_3)CH(CH_2CH(CH_3)_2)$	- " -	103–130	$\bar{6}$,71	16,53	35,2	
C_6H_5	$C(CH_3)_2CN$	o-ClC_6H_4Cl	140–180	$\bar{9}$,10	15,20	38,9	[143]
	$CH(C_6H_5)_2$	Decalin	124–144	$\bar{7}$,12	14,22	34,0	[144]
	$C(C_6H_5)_3$	C_6H_5CN	25–75	$\bar{3}$,73	16,08	29,6	
	- " -	$C_6H_5OCH_3$	25–75	$\bar{3}$,77	16,02	29,4	
	- " -	C_6H_5Cl	25–75	$\bar{3}$,80	15,70	28,9	[145]
	- " -	$C_6H_5NO_2$	25–75	$\bar{3}$,74	15,71	29,0	
	- " -	$CH_3OCOCH_2COOCH_3$	25–75	$\bar{3}$,71	15,38	28,5	
	- " -	C_6H_6	25–75	$\bar{3}$,65	14,65	27,4	
	- " -	cyclo-C_6H_{12}	25–75	$\bar{3}$,40	12,97	25,1	

XC_6H_5:							
X = H	$C(C_6H_5)_3$	CH_3NO_2	50–65	$\bar{3}$,61	15,04	28,0	
H	- " -	$C_6H_5NO_2$	50–65	$\bar{3}$,09	14,34	27,0	
H	- " -	CH_3COOH	50–65	$\bar{3}$,68	14,43	27,0	
H	- " -	n-C_3H_7OH	50–65	$\bar{3}$,81	17,04	31,0	
H	- " -	$CH_3C_6H_5$	50–65	$\bar{2}$,38	14,725	27,4	
p-CH_3	- " -	- " -	50–65	$\bar{3}$,87	13,29	25,2	[146]
p-Cl	- " -	- " -	50–65	$\bar{3}$,42	14,34	27,3	
p-Br	- " -	- " -	50–65	$\bar{3}$,70	14,62	27,8	
p-NO_2	- " -	- " -	50–65	$\bar{3}$,14	15,32	29,3	
m-CH_3	- " -	- " -	50–65	$\bar{3}$,84	15,94	29,2	
m-Cl	- " -	- " -	50–65	$\bar{3}$,45	15,945	29,8	
m-NO_2	- " -	- " -	50–65	$\bar{3}$,17	16,09	30,5	
m-CH_3	- " -	- " -	44–54	$\bar{3}$,76	14,645	27,2	
m-Br	- " -	- " -	44–54	$\bar{3}$,50	16,06	29,9	[147]
m-NO_2	- " -	- " -	44–54	$\bar{3}$,06	13,49	26,5	
$(p\text{-}CH_3C_6H_4)_3C$	C_6H_5	$C_6H_5CH_3$	50–65	$\bar{3}$,75	14,80	27,5	[146]
$(p\text{-}ClC_6H_4)_3C$	C_6H_5	$C_6H_5CH_3$	50–65	$\bar{2}$,11	14,23	26,0	[146]
$(p\text{-}CH_3OC_6H_4)_3C$	p-$NO_2C_6H_4$	C_6H_5CN	54–84	$\bar{3}$,30	15,00	28,55	[148]

Table 14 (continued)

R_1	R_2	Solvent	t^o	log k (80°)	log A	E	Ref.
$(p\text{-}CH_3OC_6H_4)_3C$	$p\text{-}NO_2C_6H_4$	Decalin	54–84	$\bar{3}$,16	15,32	30,08	[148]
$(p\text{-}CH_3OC_6H_4)_3C$	$p\text{-}NO_2C_6H_4$	$CH_3OC_6H_4OCH_3$	54–84	$\bar{3}$,26	16,30	30,7	[148]
$o\text{-}CH_3C_6H_4$	$C(CH_3)_2CN$	$o\text{-}ClC_6H_4Cl$	140–180	$\bar{9}$,68	13,80	35,7	[143]
$C_6H_5(CH_3)_2C$	$CH(CN)CH_2CH(CH_3)_2$	$o\text{-}ClC_6H_4Cl$	140–180	$\bar{8}$,50	18,90	42,6	[143]
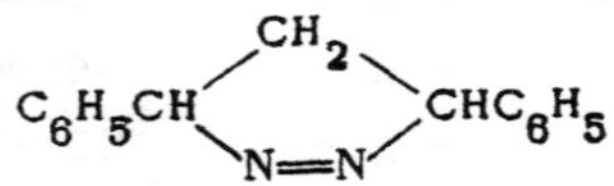		$CH_3C_6H_4CH_3$	25–75	$\bar{3}$,00	4,20	11,6	[149]
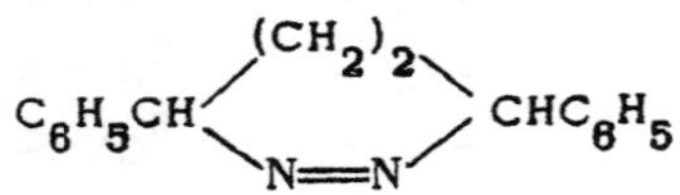		$C_6H_5C_2H_5$ + + $C_6H_5CH{=}CH_2$	80	$\bar{4}$,78			[150]
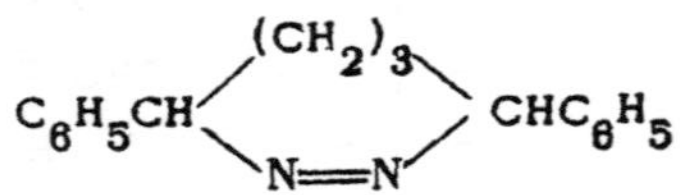		$CH_3C_6H_4CH_3$	61–100	$\bar{4}$,58	15,00	29,7	[151]

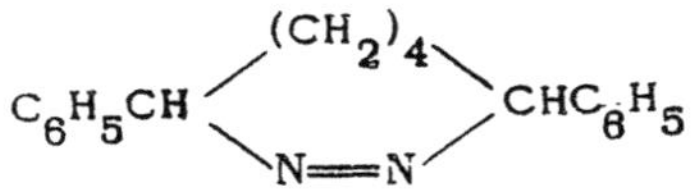	Tetralin	143–173	$\bar{8}$,48	15,24	36,7	[152]
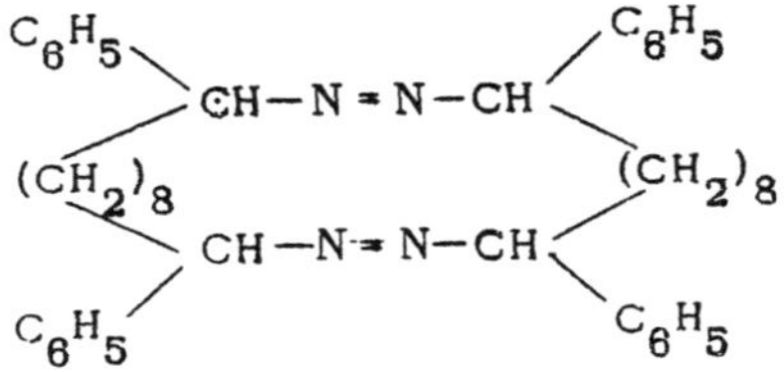	$C_6H_5C_2H_5$	113	$\bar{5}$,94			[153]
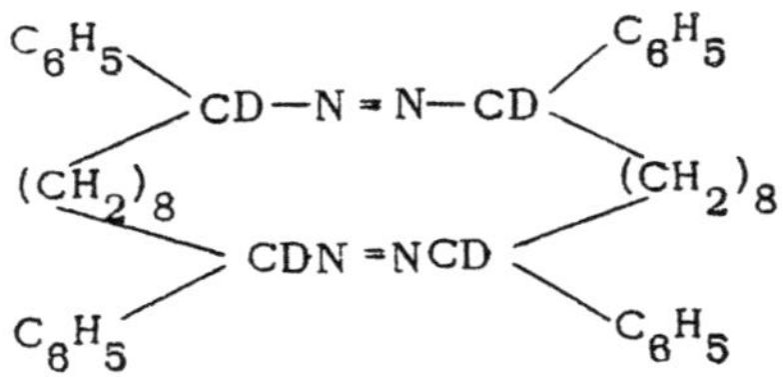	$C_6H_5C_2H_5$	113	$\bar{5}$,87			[153]

TABLE 15. Decomposition at C-C Bond (ICA Method)

Compound	Solvent	t°	log k (60°)	log A	E	Ref.
$(C_6H_5)_3C-C(C_6H_5)_3$	CCl_4	20–10	0,35	12,97	19,2	[154]
	$CHCl_3$	0	$\bar{3}$,545			
	Br CH_2CH_2Br	0	$\bar{3}$,52			
	$C_6H_5NH_2$	0	$\bar{3}$,505			
	C_2H_5OH	0	$\bar{3}$,46			
	$C_6H_5CH_3$	0	$\bar{3}$,40			
	$C_6H_5NO_2$	0	$\bar{3}$,40			
	$C_6H_5N(CH_3)_2$	0	$\bar{3}$,36			
	$ClCH_2CH_2OH$	0	$\bar{3}$,34			
	$CH_2(COOC_2H_5)_2$	0	$\bar{3}$,34			
	$HOCH_2CH_2OCH_3$	0	$\bar{3}$,34			
	C_5H_5N	0	$\bar{3}$,34			
	$CH_3CON(C_2H_5)_2$	0	$\bar{3}$,34			
	$(CH_3)_2C{=}CHCOCH_3$	0	$\bar{3}$,30			
	$HOC_6H_4COOCH_3$	0	$\bar{3}$,32			
	$NCCH_2COOC_2H_5$	0	$\bar{3}$,23			

$(C_6H_5)_2(CH_3)C-C(C_6H_5)_2(CH_3)$	C_6H_5Br	64,6	$\bar{4},57$			[155]
	$2,6-Cl_2C_6H_4$	65	$\bar{4},61$			[155]
	C_6H_5Cl, $C_6H_5NO_2$	60	$\bar{4},30$			[156]
	C_6H_5Cl	60–80	$\bar{4},37$	16,125	30,0	[155]
$C_6H_5(C_2H_5)_2C-C(C_2H_5)_2C_6H_5$	C_6H_5Cl	142,1	$\bar{4},58$			[155]
$C_6H_5(C_3H_7)_2C-C(C_3H_7)_2C_6H_5$	- " -	110–126	$\bar{8},70$	16,35	36,0	[155]
$C_6H_5(C_4H_9)_2C-C(C_4H_9)_2C_6H_5$	- " -	111-126	$\bar{7},57$	17,43	38,0	[155]
$CH_3CO-COCH_3$	$C_2H_5COCH_3$	110–145	$\overline{10},16$	13,46	35,4	[157]
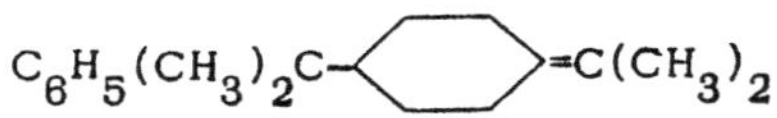	C_6H_{12}	26–40	$\bar{2},29$	15,79	26,6	[129]
$(C_6H_5)_2(C_2H_5)C-C(C_2H_5)(C_6H_5)_2$	$C_6H_5CH_3$	20–50	$\bar{3},58$	9,74	18,5	[158]
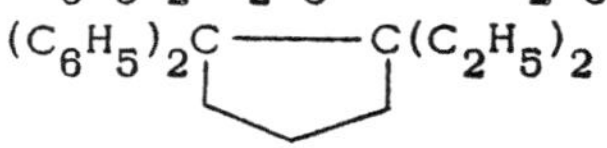	- " -	20–40	$\bar{2},16$	11,64	20,5	[158]
$p-IC_6H_4(C_3H_7)_2C-C(C_3H_7)_2-p-IC_6H_4$	C_6H_5Cl	100–120	$\bar{7},61$	13,91	31,0	[159]

TABLE 16. Decomposition at N-N Bond

Compound	Solvent	t°	log k (120°)	log A	E	Method	Ref.
$(C_6H_5)_2N-N(C_6H_5)_2$	C_6H_6	60–85	$\bar{3}$,96	13,76	28,4	ICA	[160]
$(C_6H_5)_2N-N(C_6H_5)_2$	o–ClC_6H_4Cl	75–100	$\bar{2}$,34	15,04	30,0	ICA	[161]
$C_6H_5N-N_2$	Tetralin	105–145	$\bar{6}$,30	12,40	32,5	KEPF	[162]
$C_6H_5N-N_2$	$C_6H_5NO_2$	105–145	$\bar{6}$,30	12,94	33,5	KEPF	[162]
p-$XC_6H_4(C_6H_5)_2CN-N_2$							
X=H	$C_6H_5NO_2$	170–190	$\bar{6}$,03	11,955	32,2	KEPF	
H	p–$C_{16}H_{34}$	170–190	$\bar{7}$,74	13,34	35,2	KEPF	
H	(p–$C_4H_9)_2$CHOH	170–190	$\bar{6}$,03	12,23	32,0	KEPF	
CH_3	– " –	170–190	$\bar{5}$,51	12,86	29,9	KEPF	[163]
CH_3O	– " –	170–190	$\bar{6}$,36	10,96	29,8	KEPF	
$(CH_3)_2N$	– " –	170–190	$\bar{6}$,75	9,40	26,3	KEPF	
Cl	– " –	170–190	$\bar{7}$,90	13,20	34,7	KEPF	
NO_2	– " –	170–190	$\bar{7}$,93	13,53	35,2	KEPF	
n–$C_{18}H_{37}OCON-N_2$	$C_6H_5OC_6H_5$	100–130	$\bar{4}$,34	13,88	31,5	KEPF	[164]

N_2-NCOO$(CH_2)_4$OCON– N_2	$C_6H_5OC_6H_5$	100–133	$\bar{4},36$	14,01	31,7	KEPF	[164]
p–N_2-NCOOC_6H_4OC$(CH_3)_2$OC_6H_4OCO– N_2-p	- " -	120	$\bar{4},54$			KEPF	[164]

$\left[\begin{matrix} p\text{-}R_1C_6H_4 \\ p\text{-}R_2C_6H_4 \end{matrix} \right\rangle N - N(COC_6H_5)\Big]_2$

R_1	R_2			log k (0°)				
CH_3	CH_3	CH_3COCH_3	35–40	$\bar{1},00$	11,01	15,0	ICA	[165]
CH_3	H	- " -	-20 to -40	$\bar{2},76$	11,435	15,8	ICA	[165]
H	H	- " -	-20 to -40	$\bar{2},57$	11,87	16,6	ICA	[165]
Br	H	- " -	-15 to -30	$\bar{2},47$	11,74	16,6	ICA	[165]
Br	Br	- " -	-15 to -40	$\bar{2},37$	12,15	17,2	ICA	[165]
NO_2	H	- " -	-20 to -30	$\bar{3},26$	12,73	19,3	ICA	[165]
NO_2	NO_2		0–30	$\bar{4},50$	13,795	21,6	ICA	[165]

TABLE 17. Decomposition of Compounds with Rupture of N-C or C-O Bond

Compound		Solvent	t°	log k (55,5°)	log A	E	Method	Ref.
$C_6H_5CH-N_2$		CH_3CN	85	$\bar{6},89$			CSM	[166]
$(C_6H_5)_2C-N_2$		CH_3CN	85	$\bar{4},20$			CSM	[167]
		CH_3CN	65–95	$\bar{6},68$	13,28	27,9	CSM	[168]
		C_6H_6	85	$\bar{4},00$			CSM	[168]
		$C_6H_5CH_3$	85	$\bar{5},92$			CSM	[168]
		$CH_3C_6H_4CH_3$	80–100	$\bar{6},53$	11,635	25,7	CSM	[169]
$C_6H_5COC(C_6H_5)-N_2$		$CH_3C_6H_4CH_3$	80–100	$\bar{5},61$	12,415	25,2	KEPF	[169]
$C_2H_5OCOCH-N_2$		Tetralin	150	$\bar{4},49$			KEPF	[170]
		$C_6H_5CH_2C$	150	$\bar{4},595$			KEPF	[170]
		$C_6H_5NO_2$	150	$\bar{4},87$			KEPF	[170]
$XC_6H_4N(NO)COOR$	X = H	C_6H_6	20–30	$\bar{3},68$	12,865	22,76	KEPF	[171]
	p-CH_3	- " -	20–30	$\bar{3},73$	13,23	23,22	KEPF	[171]

$XC_6H_4N(NO)COOR$								
	X = p-Cl	C_6H_6	20–30	$\bar{3}$,68	12,88	22,82	KEPF	[171]
	p-Br	- " -	20–30	$\bar{3}$,72	13,52	23,67	KEPF	
	p-NO_2	- " -	20–30	$\bar{3}$,23	13,55	23,55	KEPF	
$XC_6H_4-N=N-SC_6H_4Y$								
X = H	Y = p-$(CH_3)_3C$	C_6H_6	55,5	$\bar{5}$,92			CSM	[172]
p-CH_3	p-$(CH_3)_3C$		46–66	$\bar{5}$,78	20,23	36,67	CSM	
p-CH_3O	p-$(CH_3)_3C$		55,5	$\bar{5}$,83			CSM	
p-NO_2	p-$(CH_3)_3C$		55,5	$\bar{6}$,955			CSM	
2,4,6-$(CH_3)_3$	p-$(CH_3)_3C$		55,5	$\bar{3}$,06			CSM	
p-CH_3	H		55,5	$\bar{5}$,70			CSM	
p-CH_3	p-CH_3O		55,5	$\bar{4}$,82			CSM	
p-CH_3	p-NO_2		55,5	$\bar{6}$,48			CSM	
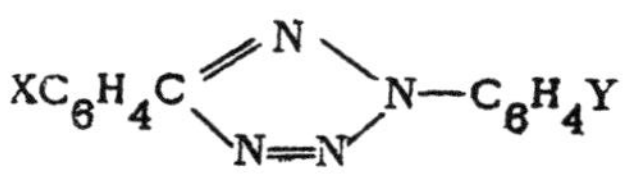		1-Cl-naphthalene						

Table 17 (continued)

Compound		Solvent	t^{o}	log k (55,5°)	log A	E	Method	Ref.
X = H	Y = H		168,5	$\bar{4}$,44	12,49	32,4	KEPF	[173]
p-CH_3O	H		168,5	$\bar{4}$,51			KEPF	[173]
p-CH_3	H		168,5	$\bar{4}$,48			KEPF	[173]
p-Cl	H		168,5	$\bar{4}$,31			KEPF	[173]
p-Br	H		168,5	$\bar{4}$,37			KEPF	[173]
p-CN	H		168,5	$\bar{4}$,28			KEPF	[173]
H	p-CH_3		168,5	$\bar{4}$,255			KEPF	[173]
H	m-CH_3		168,5	$\bar{4}$,52			KEPF	[173]
H	p-Cl		168,5	$\bar{4}$,74			KEPF	[173]
cyclo-C_6H_{10}(CN)N=C=C_6H_{10}-cyclo		C_6H_5Cl	80–100	$\bar{6}$,18	14,58	33,6	KEPF	[136]
$(CH_3)_2C$=C=N–C(CN)$(CH_3)_2$		CCl_4	62,5–81	$\bar{6}$,65	13,90	28,9	CSM	[174]
		C_6H_5Cl	62,5–81	$\bar{6}$,88	14,13	28,9	CSM	[174]
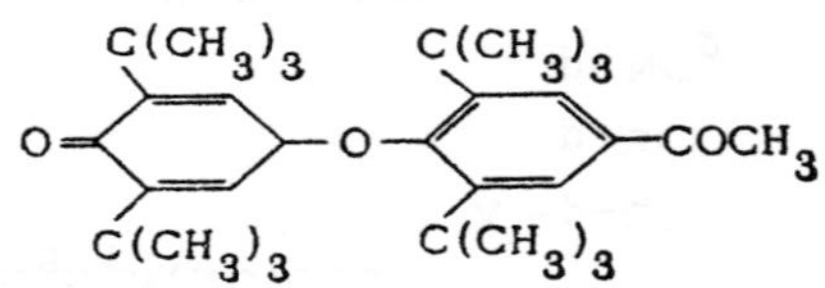		CH_3OH	-30 to +5	$\bar{3}$,13	9,53	9,6	NMR	[175]

TABLE 18. Decomposition of Nitrogen Pentoxide (KEPF Method)

Solvent	t°	log k (20°)	log A	E	Ref.
Gas phase	0–65	$\bar{5}$,22	13,67	24,7	[176]
CCl_4	25–55	$\bar{5}$,275	14,325	25,5	[177]
CCl_4	20–45	$\bar{5}$,33	13,41	24,2	[178]
$CHCl_3$	35–55	$\bar{5}$,44	13,64	24,4	[177]
CH_3CHCl_2	20–45	$\bar{5}$,51	14,11	24,9	
$C_2H_4Cl_2$	20–45	$\bar{5}$,38	13,58	24,4	
CCl_3CHCl_2	20–45	$\bar{5}$,34	14,01	25,0	[178]
CH_3NO_2	15–45	$\bar{5}$,19	13,49	24,5	
Br_2	15–25	$\bar{5}$,33	13,23	24,0	
N_2O_4	15–20	$\bar{5}$,54	14,87	25,0	

TABLE 19. Decomposition of Nitro Compounds

Nitro compound	Solvent	t°	log k (140°)	log A	E	Method	Ref.
$C(NO_2)_4$	CCl_4	85–100		16,7	39,7	KEPF	[180]
	Freon 114 B-2	120–135		17,7	41,3	KEPF	[180]
	Freon 113	100–135		16,8	39,8	KEPF	[180]
$ClC(NO_3)_2$	CCl_4	100–135		16,5	38,0	KEPF	[180]
$Br\ C(NO_2)_3$	CCl_4	100–135		16,6	37,6	KEPF	[180]
$ClC(NO_2)_2C(NO_2)_2Cl$	Freon 113	100–135		16,6	31,0	KEPF	[180]
$ClC(NO_2)_2C(NO_2)_2Cl$	cyclo-C_6H_{12}	100–135		16,3	30,7	KEPF	[180]
$(NO_2)_3CC(NO_2)_3$	CCl_4	60–100	$\bar{1}$,50	18,50	37,8	CSM	[181]
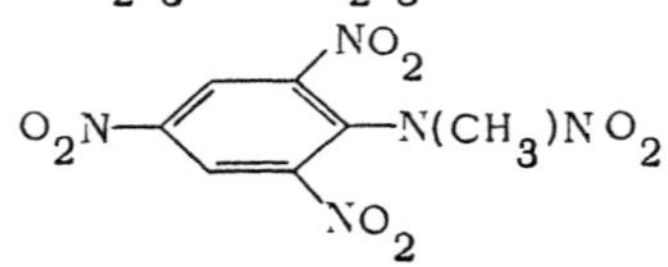	Liquid phase	211-260	$\bar{5}$,10	15,40	38,4	CSM	[182]
	m-$NO_2C_6H_4NO_2$	171-215	$\bar{7}$,30	14,30	39,7	CSM	[183]
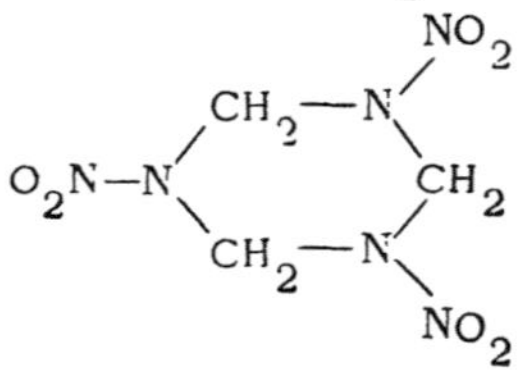	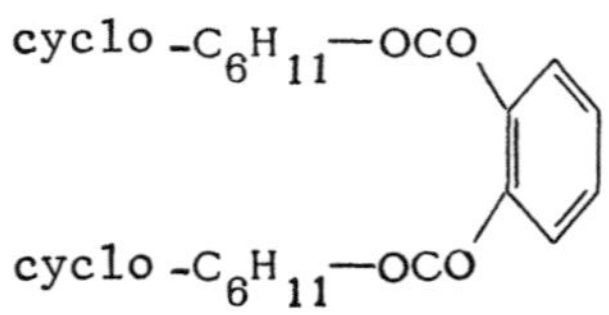	201-280	$\bar{6}$,76	15,46	41,0	CSM	[184]

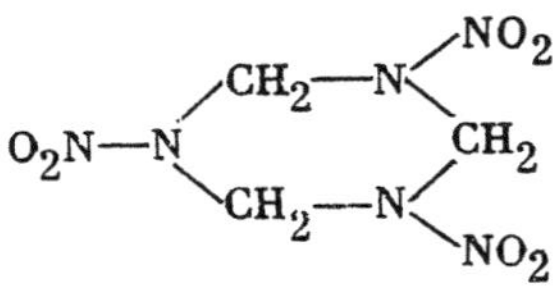	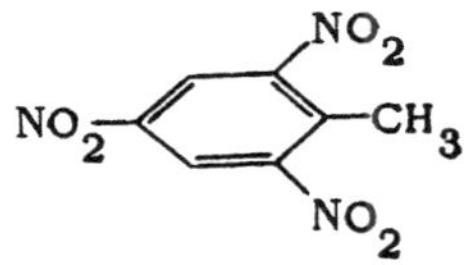	191–281	$\bar{7}$,55	15,505	41,5	CSM	[184]
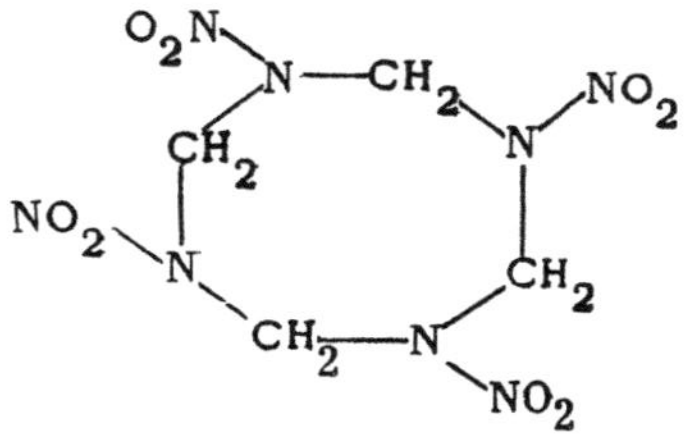	m-$NO_2C_6H_4NO_2$	171–215	$\bar{9}$,25	15,00	44,9	CSM	[183]
$C_2H_5ONO_2$	Liquid phase	70–100	$\bar{7}$,55	14,70	40,0	KEPF	[185]
$(CH_2ONO_2)_2$	- " -	80–140	$\bar{7}$,90	14,505	39,0	KEPF	[185]
$(O_2NNHCH_2)_2$	- " -	80–120	$\bar{4}$,68	12,80	30,5	KEPF	[182]
$O_2NOCH_2CH(ONO_2)CH_2ONO_2$	- " -	80–140	$\bar{7}$,80	14,40	39,0	KEPF	[186]
$C(CH_2ONO_2)_4$	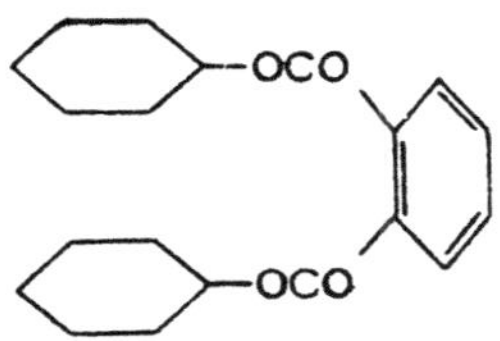	171–238	$\bar{5}$,18	16,08	39,5	KEPF	[187]

Table 19 (Continued)

Nitro compound	Solvent	t^{o}	log k (140°)	log A	E	Method	Ref.
$O_2NO(CH_2)_2O(CH_2)_2ONO_2$	Liquid phase	80–140	$\bar{6}$,30	16,505	42,0	KEPF	[188]
$(O_2NOCH_2CH_2)_2NNO_2$	- " -	80–140	$\bar{6}$,55	16,505	41,5	KEPF	[189]
$[CH(ONO_2)CH(ONO_2)CH_2ONO_2]_2$	- " -	80–140	$\bar{5}$,80	15,90	38,0	KEPF	[185]

TABLE 20. Decomposition at C-Metal Bonds

Compound	Solvent	t^{o}	log k (20°)	log A	E	Method	Ref.
$Ag-CH_3$	C_2H_5OH	-50 to -20	4,96	5,74	10,5	KEPF	[190]
$Hg(CH_2C_6H_5)_2$	$ClCH_2C_6H_5$	140–170	$\overline{13}$,25	16,25	38,8	-	[191]
$CH_3-TiCl_3\cdot 2\,cyclo\text{-}(CH_2)_4O$	$cyclo\text{-}(CH_2)_4O$	7–29	$\bar{5}$,37	11,00	19,6	CSM	[192]
$Pb(C_2H_5)_4$	$n\text{-}C_{11}H_{24}$	100–170	$\overline{15}$,99	18,79	44,0	CSM	[193]

TABLE 21. Decomposition of $XC_6H_4ICl_2$ (in CH_3COOH,CSM method)

X	t°	logk (25°)	logA	E	Ref.
H	25–45	$\bar{4}$,22	10,98	20,1	[194]
$p-NO_2$	25,2	$\bar{5}$,95			
$p-OC(CH_3)NH$	25,2	$\bar{4}$,58			
$p-CH_3$	25,2	$\bar{4}$,46			
$m-CH_3$	25,2	$\bar{4}$,31			[195]
$p-C_6H_5$	25–45	$\bar{4}$,22	11,37	20,6	
p-COOH	25–45	$\bar{4}$,16	9,46	18,1	
m-Cl	25–45	$\bar{4}$,16	10,26	19,2	
$p-CH_2OH$	25	$\bar{4}$,37			[196]
$p-CH_2OCH_3$	25–45	$\bar{4}$,12	12,52	22,3	
$p-CH_2OC_6H_5$	25–44	$\bar{4}$,21	7,46	15,3	
$p-CH_2COOCH_3$	25–45	$\bar{3}$,11	13,01	23,0	[197]
$p-COCH_3$	24–46	$\bar{4}$,00	9,45	18,3	
$p-COC_6H_5$	24–46	$\bar{4}$,05	9,75	18,8	
m-COOH	25	$\bar{4}$,245			
p-COOH	25	$\bar{4}$,145			
$p-COOCH_3$	25	$\bar{4}$,175			
$p-NHCOCH_3$	25	$\bar{4}$,58			[198]
$p-OCOCH_3$	25	$\bar{4}$,29			
$p-OCOC_6H_5$	25	$\bar{4}$,26			
$o-CH_2OH$	25	$\bar{2}$,11			[196]
$o-CH_2OCH_3$	25–35	$\bar{3}$,47	10,175	17,3	[197]
$o-CH_2OC_6H_5$	25–44	$\bar{4}$,22	8,49	16,7	[197]
$o-CH_2COOCH_3$	25–45	$\bar{4}$,99	13,89	23,0	
$o-COCH_3$	24,6	$\bar{2}$,07			[197]
$o-COC_6H_5$	24,4	$\bar{2}$,17			
$o-C_2H_5$	25	$\bar{2}$,19			
$o-CH_2OH$	25	$\bar{2}$,115			
$o-OCOCH_3$	25	$\bar{4}$,78			
$o-OCOC_6H_5$	25	$\bar{4}$,35			[196]
$o-NHCOC_6H_5$	25	$\bar{2}$,57			
$o-NHCOCH_3$	25	$\bar{2}$,505			

TABLE 22. Constants in Correlation Equations $\log k = \log k_o + \rho\sigma$

Compound class	Function	ρ	$\log k_o$	t^o	Solvent	Reference
$XC_6H_4COOOCOC_6H_4X$	σ	−0,38	$\bar{5}$,62	80	Dioxane	[15]
	σ	−0,38	$\bar{6}$,50	70	$C_6H_5CH{=}CH_2$	[10]
$XC_6H_4COOOCOCH_3$	σ	−0,88	$\bar{5}$,05	70	$C_6H_5CH{=}CH_2$	[11]
p-$XC_6H_4COOOC(CH_3)_3$	σ	−0,56	$\bar{5}$,95	120	$C_6H_5OC_6H_5$	[90]
	σ	−0,35		110	p-$C_{11}H_{24}$	[75]
$XC_6H_4CH_2COOOC(CH_3)_3$	σ^+	−1,09	$\bar{4}$,50	90	$C_6H_5CH_3$	[78]
	σ^+	−1,20	$\bar{6}$,73	56	$C_6H_5CH_3$	[78]
$(CH_3)_3COOCO(CH_2)COOOC(CH_3)_3$	σ	1,079		85	$C_6H_5C_2H_5$	[199]
$[(X_1C_6H_4)(X_2C_6H_4)N-N(COC_6H_5)]_2$	σ	−0,55	$\bar{4}$,92	30	CH_3COCH_3	[165]
$XC_6H_4N_2C(C_6H_5)_3$	σ	−0,88	$\bar{4}$,38	54	$C_6H_5CH_3$	[147]
$XC_6H_4(CH_3)CH-N_2-CH(CH_3)C_6H_4X$	σ	−0,55	$\bar{4}$,18	53,4	$C_6H_5CH_3$	[127]
$XC_6H_4ICl_2$	σ	−0,64	$\bar{4}$,23	25	CH_3COOH	[195–197]
$XC_6H_4N(NO)COOR$	σ	0,0	$\bar{4}$,16	25	C_6H_6	[171]

TABLE 23. Compensation Effect for Monomolecular Decomposition Reactions

Compound class	log A (relation between log A and E)	Isokinetic temperature	Range of E, kcal/mole	Table
RCOOOCOR, $RCOOOC(CH_3)_3$	0,3 + 0,47 E	466	21–40	5, 6, 9
$XC_6H_4COOOC(CH_3)_3$	0,3 + 0,44 E	500	36–41	9
$(p\text{-}XC_6H_4COO)_2$, $(m\text{-}XC_6H_4COO)_2$	5,6 + 0,29 E	755	28–31	3
$(o\text{-}XC_6H_4COO)_2$	5,0 + 0,33 E	654	27–30	3
ROOR	–6,2 + 0,59 E	372	32–40	7
$R_1R_2C(CN)N{=}N(CN)CR_1R_2$	–1,6 + 0,55 E	400	26–40	13
$p\text{-}XC_6H_4(CH_3)CH{-}N_2{-}CH(CH_3)(p\text{-}XC_6H_4)$	–3 + 0,64 E	220	24–30	13
$p\text{-}XC_6H_4N{=}NC(C_6H_5)_3$	–0,5 + 0,55 E	400	25–30	14
$p\text{-}XC_6H_4(C_6H_5)_2CN_3$	–2,6 + 0,46 E	476	26–35	16
$\left[(X_1C_6H_4)(X_2C_6H_4)N{-}N(COC_6H_5)\right]_2$	4,8 + 0,42 E	522	15–22	16
$XC_6H_4ICl_2$	–1 + 0,60 E	365	15–23	21
$(CH_3)_3COOC(CH_3)_3$ in various solvents	–	437	32–42	7

TABLE 24. Volume Change $\Delta V^{\neq}$ in Decomposition of Molecules

Substance	Solvent	t°	$\Delta V^{\neq}$, cm³/mole	Reference
$[(CH_3)_2C(CN)]_2N_2$	$C_6H_5CH_3$	62,5	+3,8	[201]
$(C_6H_5COO)_2$	CCl_4	60	+9,7	[200]
	$C_6H_5CH{=}CH_2$	30	+10	[202]
	$C_6H_5COCH_3$	80	+5	[203]
	$CH_3COOCH_2CH{=}CH_2$	80	+4,7	[203]
$[(CH_3)_3CO]_2$	C_6H_6	120	+12,6	[56]
	CCl_4	120	+13,3	[56]
	cyclo-C_6H_{12}	120	+6,7	[56]
	$C_6H_5CH_3$	120	+5,4	[56]
	n-C_7H_{16}	125	+13,3	[204]
	- " -	155	+10,0	[204]
$C_6H_5CH_2COOOC(CH_3)_3$	$C_6H_5CH(CH_3)_2$	79,55	+1,0	[204a]
	C_6H_5Cl	79,55	+1,5	[204a]
$[(CH_3)_2CHCOO]_2$	$CH_3C(CH_3)_2CH_2CH(CH_3)_2$	50	−5,1	[29]
$(C_6H_5)_3CCH(C_6H_5)_2$	$C_6H_5CH_3$	70	+13	[205]

TABLE 25. Range of Variation of k for Decomposition of Initiators in Various Solvents

Initiator	Range of variation	t^o	Number of solvents	Table
$(C_6H_5)_3C—C(C_6H_5)_3$	$2{,}0 - 4{,}0 \cdot 10^{-3}$	0	16	15
$[C_6H_5(CH_3)CH]_2N_2$	$3{,}2 - 4{,}1 \cdot 10^{-5}$	97,3	7	13
$[(CH_3)_2C(CN)]_2N_2$	$1{,}4 - 2{,}0 \cdot 10^{-4}$	80	24	12
$CH_3COOOCOCH_3$	$1{,}0 - 1{,}6 \cdot 10^{-5}$	85,2	15	4
$[(CH_3)_3CO]_2$	$8{,}5 - 15 \cdot 10^{-6}$	120	4	7
$[C_6H_5(CH_3)_2CO]_2$	$2{,}2 - 2{,}8 \cdot 10^{-4}$	140	5	7
$[C_6H_5COO]_2$	$(3{,}5 - 40) \cdot 10^{-6}$	80	40	2
$[(CH_3)_2CHCOO]_2$	$(3{,}1 - 58) \cdot 10^{-5}$	40	15	5

CHAPTER II

BIMOLECULAR AND TRIMOLECULAR REACTIONS

§1. Methods for Measuring Rate Constants of Bimolecular Reactions

The kinetics of a bimolecular reaction of the type A + B → reaction products is described by the following equation (where x is the concentration of starting material that has reacted):

$$-\frac{dA}{dt} = k[A][B] \quad \text{or} \quad \frac{dx}{dt} = k([A]_0 - x)([B]_0 - x),$$

whence

$$\ln\frac{[A]_0([B]_0 - x)}{[B]_0([A]_0 - x)} = k([B]_0 - [A]_0)t. \tag{II.1}$$

If one of the starting materials is taken in excess, so that $[B]_0 - x \approx [B]_0$, the reaction kinetics are described adequately by a first-order reaction law:

$$\ln\frac{[A]_0}{[A]_0 - x} = k\,B_0 t. \tag{II.2}$$

If the concentrations of the two starting materials are equal, i.e., if $[A]_0 = [B]_0$, then

$$\frac{x}{[A]_0 - x} = k[A]_0 t. \tag{II.3}$$

a. Consumption of Starting Material (CSM)

The kinetics of a bimolecular reaction can be followed according to the consumption of one of the starting materials. The rate constant is found from a plot using one of the formulas (II.1) - (II.3), depending on the initial conditions.

b. Kinetics of End Product Formation (KEPF)

The rate constant can be determined in an analogous manner by following the formation of the end product of the reaction.

c. Consumption of Radical Acceptor (ICA)

If a bimolecular reaction results in the formation of a free radical, the reaction rate can be measured according to the consumption of a free radical acceptor (see Chapter I, §1). Such a method permits the measurement of reaction velocity with insignificant consumption of starting materials; in this case, the rate constant for radical formation is found from the formula

$$k_i = \frac{W}{f[A]_o[B]_o},$$

where W is the rate of acceptor consumption. The only radicals reacting with the acceptor are those that have escaped into the bulk volume from a solvent cage. Hence, the constant k_i differs from the k measured from the consumption of starting material: $k_i = 2\,ek$.

d. Rate of Chain Initiated Reaction (RICR)

The rate of radical formation can be measured according to the velocity of a chain reaction W_ν (for example, a polymerization reaction) for which the ratio $k_p k_t^{-1/2}$ is known

$$W_\nu = k_p k_t^{-1/2} [RH] \sqrt{W_i/2},$$

where [RH] is the concentration of reacting substance (e.g., a monomer in a polymerization reaction) and W_i is the initiation rate. If the bimolecular reaction under investigation is the sole source of free radicals, then

$$k_i = W_i/[A]_o[B]_o, \quad k_i = 2\,ek.$$

§ 2. Diels-Alder Reaction

The dimerization of unsaturated compounds proceeds rapidly in various solvents without the presence of any ions or initiators. There is every reason to believe that these reactions proceed as the result of interaction of two molecules. Reaction rates for the dimerization of unsaturated compounds have been measured, as a rule, by the consumption of starting materials. The results of measurements are summarized in Table 26.

The rate constants for dimerization of unsaturated compounds are little affected by choice of solvent, and the rate constant for the gas-phase reaction will be the same as that in certain solvents. At 50°, k for the dimerization of cyclopentadiene has the following values (see Table 26): $2\cdot10^{-5}$(C_2H_5OH), $1\cdot10^{-5}$(CH_3COOH), $2\cdot10^{-5}$($C_6H_5NO_2$), $0.63\cdot10^{-5}$ (CS_2), $1\cdot10^{-5}$(C_6H_6), $1.3\cdot10^{-5}$(CCl_4), $2\cdot10^{-5}$(paraffin), $0.63\cdot10^{-5}$(C_6H_6), $0.63\cdot10^{-5}$(gas phase).

The activation energy and the preexponential coefficient change from solvent to solvent such that a compensation effect is observed: For the reaction of cyclopentadiene dimerization, $\log A = -5.9 + 0.74\ E$; and for the reaction of cyclopentadiene with benzoquinone, $\log A = -1.5 + 0.70\ E$ (with E in kcal/mole and A in liter/mole · sec). The dimerization reaction, as would be predicted, proceeds with a decrease in volume upon formation of an activated complex (see Table 27). For the reaction of XC_6H_4NO with dimethylbutadiene, a linear relation is maintained between log k and σ, with $\rho = +0.85$ (25°, CH_2Cl_2) [212].

§ 3. Bimolecular Reactions with Peroxide Participation

Reactions of Peroxides with Amines. Diacyl peroxides react rapidly with amines (aromatic or aliphatic). The reaction of benzoyl peroxide with tertiary alkylaromatic amines is used for the initiation of radical polymerization. In the case of primary and secondary amines, no initiation of polymerization is observed. The mechanism of the primary act in the reaction of diacyl peroxides with aromatic amines was studied in [216]. It was established by an EPR method that the primary product from the oxidation of the amine is a cation-radical (the same for tertiary as for secondary amines). The structure of this cation-radical and the properties of the medium govern the composition of the molecular products of reaction and the probability of escape of radicals into the bulk volume. The primary act in the reaction of peroxide with amine may be illustrated as follows:

$$ROOR + Ar\ddot{N}R_2 \longrightarrow RO\cdot + RO^- + Ar\dot{N}R_2^+ .$$

Hydroperoxides react with amines more slowly than do diacyl peroxides. The reaction apparently is preceded by the formation of a hydrogen bond between the hydroperoxide and amine (this has been demonstrated in the example of pyridine [217]). If this is true, then the measured bimolecular constant $k = k_1K$, where K relates to the equilibrium

$$ROOH + R_1R_2R_3N \rightleftharpoons ROOH \ldots NR_1R_2R_3 ,$$

and k_1 to the reaction

$$ROOH \ldots NR_1R_2R_3 \rightarrow \text{products}$$

and the experimental activation energy $E = E_1 - Q$, where Q is the heat of formation of the complex.

Bimolecular Reactions of Hydroperoxides. Hydroperoxides readily enter into many different reactions that form free radicals. These reactions have been studied by the use of oxidation inhibitors (phenols and aromatic amines) as free radical acceptors.

Hydroperoxides react with the C-H bonds of hydrocarbons according to the reaction [106]:

$$ROOH + R'H \longrightarrow RO\cdot + H_2O + R'\cdot .$$

In the case of unsaturated compounds, the hydroperoxide reacts with the double bond [235]:

$$ROOH + R_1CH{=}CHR_2 \longrightarrow RO\cdot + R_1CH(OH)\dot{C}HR_2 .$$

The bimolecular reaction often proceeds through preliminary formation of a complex between the reacting molecules:

$$ROOH + nM \overset{K}{\rightleftharpoons} \text{complex} \overset{k}{\longrightarrow} \text{free radicals} .$$

The formation of a complex proceeds by means of a hydrogen bond (in the reaction of a hydroperoxide with another hydroperoxide or with an alcohol, an acid, or pyridine) or by means of addition (in the reaction of a hydroperoxide with a carbonyl compound). The constant k is listed in sec^{-1} in all cases. If the reaction is performed under conditions such that the equilibrium is shifted to the left, in this case the experimental data are used to determine a pseudobimolecular constant equal to the product kK (see Table 30).

§4. Oxidation-Reduction Reactions with Participation of ClO_2, O_2, and Cl_2

The primary act in the oxidation of amines by such oxidants as ClO_2 and O_2 consists of the transfer of an electron from the amine to the oxidant, analogous to that observed in reaction with peroxides. Even primary and secondary amines are oxidized by oxygen not through abstraction of H, but by electron transfer, as evidenced by the lack of any isotope effect in the oxidation of amines with N-H and N-D bonds [251].

The reaction of oxygen with phenols and hydrocarbons apparently proceeds by the scheme [252]:

$$RH + O_2 \longrightarrow R\cdot + HO_2^{\cdot}$$

(in the experiment, only the rate of radical formation according to this reaction was measured). Styrene probably represents an exception, since the reaction with H abstraction must be extremely slow, and radical formation proceeds as a result of reaction at the double bond [253]:

$$C_6H_5CH{=}CH_2 + O_2 \longrightarrow C_6H_5\dot{C}H{-}CH_2OO.$$

Reaction of Chlorine Addition to Iodobenzene. The reaction

$$XC_6H_4I + Cl_2 \longrightarrow XC_6H_4ICl_2$$

is the reverse of the iodobenzene dichloride decomposition reaction (see Chapter I, §5). The rate constant of this reaction has been measured in acetic acid at 25° for various derivatives of iodobenzene (where X represents the substituent in the benzene ring): X = p-I, k = $6.3 \cdot 10^{-2}$; p-$CONH_2$, 0.35; p-NO_2, $6.0 \cdot 10^{-3}$; p-CH_3, 0.30; m-CH_3, 0.14; p-C_6H_5, 0.11; p-COOH, $9.2 \cdot 10^{-3}$; m-Cl, $1.13 \cdot 10^{-2}$ [195].

§5. Reactions with Rupture of Metal - Carbon Bond

Reactions between ethyllithium and various alkyl halides were studied in [256, 257]. The reactions were performed in decalin, and the kinetics were followed by LiX accumulation. A linear relation has been found between the activation energy and the heat of reaction q [258]: E = 13 - 0.33 q.

§6. Trimolecular Reactions

In a study of styrene polymerization in the absence of any initiator it was established that the polymerization rate $W_\nu \sim [\text{styrene}]^{5/2}$ [260]. Since $W_i \sim W_\nu^2/[M]^2$, it was concluded that chain generation involved a reaction among three molecules of styrene, for which the following mechanism was proposed:

$$3\ RCH{=}CH_2 \longrightarrow CH_3{-}\dot{C}HR + CH_3{-}CR{=}CH{-}\dot{C}HR.$$

The trimolecular reaction of chain generation

$$2\ RH + O_2 \longrightarrow R\cdot + H_2O_2 + R\cdot$$

was predicted theoretically [261] as energetically far more favorable than the bimolecular reaction

$$RH + O_2 \longrightarrow R\cdot + HO_2^{\cdot}$$

for compounds in which the C-H bond strength is less than 90 kcal/mole. The rate of the trimolecular reaction of chain generation was measured according to the consumption of inhibitor during the initial period of oxidation of the substance being investigated. A compensation effect was observed for this reaction in the case of hydrocarbons and alcohol:

$$\log A = -7.6 + 0.55\ E.$$

TABLE 26. Dimerization of Unsaturated Compounds (CSM Method)

Reaction	Solvent	t^{o}	log k (25°)	log A	E	Ref.
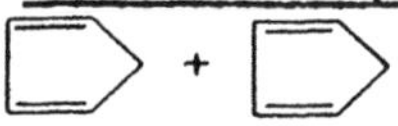	C_6H_6	12–55	$\bar{6}$,06	6,10	16,4	[209]
	CCl_4	0–55	$\bar{6}$,15	6,70	17,1	[209]
	CCl_4	0–22	$\bar{6}$,00	5,90	16,2	[210]
	Paraffin	0–172	$\bar{6}$,31	7,10	17,4	[209]
	CS_2	0–35	$\bar{7}$,80	6,20	16,9	[209]
	CS_2	0–22	$\bar{7}$,43	5,57	17,7	[210]
	Liquid phase	-2–35	$\bar{7}$,90	5,80	16,2	[209]
	- " -	0–58	$\bar{7}$,82	5,57	16,0	[210]
	C_2H_5OH	0–55	$\bar{6}$,36	6,40	16,4	[209]
	CH_3COOH	25–70	$\bar{6}$,20	5,00	14,7	[209]
	$C_6H_5NO_2$	0–55	$\bar{6}$,40	5,50	15,1	[209]
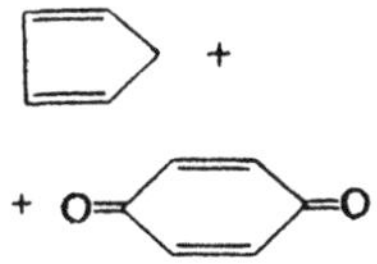	C_6H_6	10–50	$\bar{3}$,86	6,31	11,5	[211]
	C_6H_6	8–50	$\bar{2}$,00	6,505	11,6	[210]
	$n\text{-}C_6H_{14}$	20–40	$\bar{3}$,62	6,505	12,1	[210]
	CCl_4	10–50	$\bar{3}$,86	4,03	8,4	[211]
	CCl_4	2–49	$\bar{3}$,75	4,505	9,2	[210]
	$C_6H_5NO_2$	10–50	$\bar{2}$,54	4,99	8,8	[211]
	$C_6H_5NO_2$	21–42	$\bar{2}$,92	7,00	11,0	[210]
	C_6H_5CN	10–50	$\bar{2}$,63	4,50	8,0	[211]
	C_2H_5OH	2–39	$\bar{3}$,68	7,00	12,7	[210]
	CH_3COOH	18–30	$\bar{1}$,42	7,505	11,0	[210]
	CS_2	3–39	$\bar{3}$,75	4,00	8,5	[210]
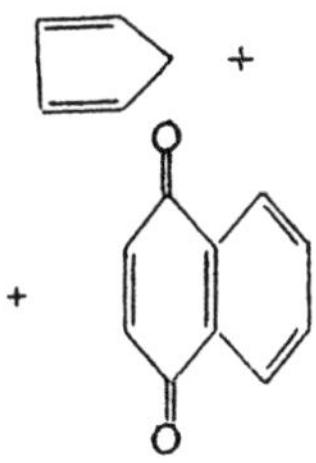	C_6H_6	11–51	$\bar{3}$,45	4,80	10,0	[210]
	CCl_4	12–54	$\bar{3}$,41	5,205	10,6	[210]

Table 26 (continued)

Reaction	Solvent	t°	log k (20°)	logA	E	Ref.
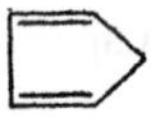 + $O{=}CH{-}CH{=}CH_2$	C_6H_6	6–77	$\bar{3}$,90	6,10	13,7	[210]
$CH_2{=}C(CH_3){-}C(CH_3){=}CH_2$ + XC_6H_4NO						
X = H	CH_2Cl_2	2–25	$\bar{3}$,32	7,74	14,2	[212]
p-Cl	- " -	2–25	$\bar{3}$,70	6,95	12,6	[212]
m-Cl	- " -	2–25	$\bar{3}$,885	6,775	12,1	[212]
p-Br	- " -	2–25	$\bar{3}$,88	6,98	12,4	[212]
m-Br	- " -	2–25	$\bar{3}$,90	6,65	11,9	[212]
p-I	- " -	2–25	$\bar{3}$,89	6,70	12,0	[212]
p-CH_3	- " -	2–25	$\bar{4}$,86	7,46	14,4	[212]

TABLE 27. Volume Change in Diels-Alder Reaction

Reaction	Solvent	t°	ΔV≠, cm^3/mole	Ref.
	Liquid phase	0	–20	[213]
		20	–30	[213]
		30	–31	[213]
		40	–34	[213]
	$n{-}C_4H_9Cl$	20	–20,2	[214]
		30	–22,3	[214]
		40	–22,5	[214]
2 $CH_2{=}C(CH_3)CH{=}CH_2$	Liquid phase	60	–24,3	[215]
		75	–25,6	[215]
$H_2C{=}C(CH_3){-}C(CH_3){=}CH_2$ + $CH_2{=}CH{-}COO(n{-}C_4H_9)$	C_6H_5Br	80	–22,3	[214]

TABLE 28. Reactions of Amines with Peroxides (CSM Method)

Peroxide	Amine	Solvent	t°	lg k (20°)	log A	E	Ref.
$(C_6H_5COO)_2$	$C_6H_5CH_2NH_2$	C_6H_6		$\bar{3},69$	6,79	12,2	[218]
	- " -	CCl_4		$\bar{2},255$	5,955	10,3	
	- " -	C_5H_5N		$\bar{2},32$	3,70	7,2	[219]
	- " -	$O{=}CHN(CH_3)_2$		$\bar{2},42$	3,79	7,2	
$(C_6H_5COO)_2$	$(C_2H_5)_2NH$	C_6H_6	0–20	$\bar{3},60$	4,62	9,4	
	- " -	$(C_2H_5)_2O$	0–20	$\bar{3},03$	4,80	10,4	
	- " -	CH_3COCH_3	20	$\bar{3},74$			[220]
	- " -	Dioxane	20	$\bar{3},755$			
	- " -	$CH_2{=}CHCOOCH_3$	20	$\bar{3},715$			
	- " -	$CH_2{=}CHC_6H_5$	20	$\bar{3},755$			[221]
	- " -	$CHON(CH_3)_2$	20	$\bar{3},98$			
	- " -	C_5H_5N	20	$\bar{2},46$			
$(C_6H_5COO)_2$	$(C_3H_7)_2NH$	$(C_2H_5)_2O$	20	$\bar{4},75$			
	$(n\text{-}C_4H_9)_2NH$	C_6H_6	0–20	$\bar{3},58$	4,44	9,2	[220]
	$[(CH_3)_2CHCH_2]_2NH$	$(CH_2H_5)_2O$	20	$\bar{4},05$			
	$(C_2H_5)_3N$	C_6H_6		$\bar{3},12$	7,12	14,12	[222]

Table 28 (continued)

Peroxide	Amine	Solvent	t°	$\log k$ (20°)	log A	E	Ref.
$(C_6H_5COO)_2$	$(CH_3)_2NCH_2CH_2OH$	C_6H_6		$\bar{2},09$	6,07	10,7	[223]
	- " -	$CH_2{=}CHCOOCH_3$		$\bar{3},96$	7,11	12,26	[223]
	$(C_2H_5)_2NCH_2CH_2OH$	C_6H_6		4,66	6,72	14,2	[222]
	$(CH_3)_2NCH_2C(CH_3)HOH$	- " -		$\bar{4},98$	6,08	12,2	
	$(CH_3)_2NCH_2C(CH_3)_2CH_2OH$	- " -		$\bar{4},335$	7,085	14,4	
	$(C_2H_5)_2NCH_2CH_2OH$	- " -		$\bar{3},055$	7,655	14,2	[223]
	$(C_2H_5)_2NCH_2CH(CH_3)OH$	- " -		$\bar{4},85$	6,77	13,3	
	- " -	$C_6H_5CH{=}CH_2$		$\bar{4},78$	4,76	10,7	
	$C_6H_5NH_2$	C_6H_6	20–70	$\bar{4},28$	4,27	10,7	[224]
	$C_6H_5N(CH_3)_2$	- " -		$\bar{2},18$	11,73	12,8	[225]
	- " -	- " -	5,7	$\bar{2},31$			[226]
	- " -	- " -	25	$\bar{3},36$			
	p-$CH_3C_6H_4N(CH_3)_2$	- " -	5–35	$\bar{3},90$	7,915	13,5	
	- " -	CCl_4	5–35	$\bar{3},48$	7,49	13,5	
	p-$CH_3C_6H_4N(CH_2CH{=}CH_2)_2$	C_6H_6	5–35	$\bar{3},085$	8,785	15,7	
	- " -	CCl_4	25	$\bar{4},905$			[227]
	p-$CH_3C_6H_4N$-(p-$C_3H_7)_2$	C_6H_6	25	$\bar{2},10$			
	- " -	CCl_4	25	$\bar{3},53$			
	p-$CH_3C_6H_4N(CH_3)(C_3H_7)$	C_6H_6	25	$\bar{2},07$			
	p-$CH_3C_6H_4N(CH_3)CH_2CH{=}CH_2$	- " -	25	$\bar{3},78$			

$(C_6H_5COO)_2$	$(C_6H_5)_2NH$	C_6H_6	15–30	$\bar{2},04$	7,96	13,3	[228]
	- " -	- " -	20–70	$\bar{3},85$	10,07	16,4	[224]
	- " -	- " -	20–40	$\bar{4},73^*$	6,33	12,8	[229]
	$CH_3N(C_6H_4CH_3\text{-}p)_2$	- " -	25	$\bar{3},11$			[227]
	$(p\text{-}CH_3C_6H_4)_3N$	- " -	25	$\bar{4},18$			[227]
	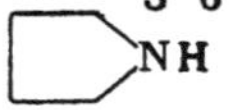	- " -	15–30	$\bar{1},00$	4,125	7,0	[228]
		- " -	15–30	$\bar{2},56$	5,50	9,3	[228]
$(p\text{-}CH_3OC_6H_4COO)_2$	$C_6H_5N(CH_3)_2$	C_6H_6	5,7	$\bar{3},67$			[226]
	$(C_6H_5)_2NH$	- " -	15–30	$\bar{3},55$	8,82	15,1	[228]
	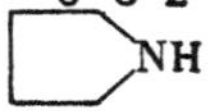	C_6H_6	15–30	$\bar{2},21$	7,09	11,9	[228]
$(p\text{-}CH_3C_6H_4COO)_2$	$C_6H_5N(CH_3)_2$	C_6H_6	5,7	$\bar{2},10$			[226]
	$(C_6H_5)_2NH$	- " -	15–30	$\bar{3},76$	8,56	14,5	[228]
	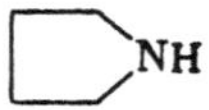	- " -	15–30	$\bar{2},57$	4,09	7,4	[228]
$(p\text{-}ClC_6H_4COO)_2$	$(C_6H_5)_2NH$	C_6H_6	15–30	$\bar{2},65$	8,11	12,7	[228]
	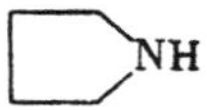	- " -	15–30	$\bar{1},31$	3,28	5,3	[228]
$(o\text{-}ClC_6H_4COO)_2$	$C_6H_5N(CH_3)_2$	C_6H_6	5,7	$\bar{2},70$			[228]

Table 28 (continued)

Peroxide	Amine	Solvent	t^{o}	log k (20°)	log A	E	Ref.
$(p\text{-}NO_2C_6H_4COO)_2$	$C_6H_5N(CH_3)_2$	C_6H_6	5,7	$\bar{2}$,63			[226]
	$(C_6H_5)_2NH$	- " -	15–30	$\bar{2}$,63	3,03	5,9	[228]
	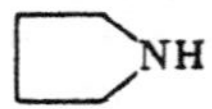	- " -	15–30	0,06	2,37	3,1	[228]
$(2,4\text{-}Cl_2C_6H_4COO)_2$	$C_6H_5N(CH_3)_2$	C_6H_6	5,7	$\bar{1}$,34			[226]
$(3,5\text{-}Br_2C_6H_4COO)_2$	- " -	- " -	5,7	$\bar{2}$,52			[226]
$[3,5\text{-}(NO_2)_2C_6H_4COO]_2$	- " -	- " -	5,7	$\bar{2}$,93			[226]
$(C_{11}H_{23}COO)_2$	$\alpha\text{-}C_{10}H_7NH_2$	$n\text{-}C_9H_{20}$	40–80	$\bar{5}$,37	18,92	31,5	[230]
H_2O_2	- " -	cyclo-$C_6H_{11}OH$	120–140	$\bar{7}$,73	3,26	12,9	[62]
$C_6H_5C(CH_3)_2OOH$	$(CH_2CH_2OH)_3N$	H_2O		$\bar{6}$,71	4,91	13,7	[231]
	$(C_2H_5)_3N$	- " -		$\bar{5}$,98	6,955	14,7	[232]
	$(C_2H_5)_2NCH_2CH_2OH$	- " -		$\bar{5}$,67	8,95	17,8	[232]
	$(C_2H_5)_2NH$	- " -		$\bar{5}$,48	8,16	17,0	[233]
	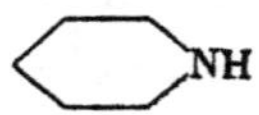	- " -		$\bar{4}$,37	10,17	18,5	[234]
		- " -		$\bar{5}$,30	8,94	18,3	[234]

*Measured by absorption of NO radicals formed in this reaction.

TABLE 29. Reaction ROOH + R'H → Free Radicals (ICA and RICR Methods)

ROOH	R'H	Solvent	t°	log k (130°)	log A	E	Ref.
sec-$C_{10}H_{21}OOH$	n-$C_{10}H_{22}$	C_6H_5Cl	130	$\bar{6}$,22			[106]
	iso-$C_{10}H_{22}$	- " -	130	$\bar{6}$,38			[106]
	$C_6H_5C_2H_5$	- " -	130	$\bar{6}$,61			[106]
	$CH_3(CH_2)_7CH{=}CH(CH_2)_7COOH$	- " -	130	$\bar{5}$,22			[106]
$C_6H_5C(CH_3)_2OOH$	$C_6H_5CH(CH_3)_2$	C_6H_5Cl	100–120	$\bar{7}$,58	7,70	26,0	[236]
	- " -	$C_6H_5CH(CH_3)_2$	80–95	$\bar{7}$,78	8,00	26,2	[237]
$(CH_3)_3COOH$	$C_6H_5CH{=}CH_2$	C_6H_5Cl	55–90	$\bar{6}$,73	4,08	17,2	[235]
	- " -	C_6H_6	70	$\bar{8}$,53			[238]
$C_6H_5C(CH_3)_2OOH$	$C_6H_5CH{=}CH_2$	C_6H_6	60–90	$\bar{5}$,38	5,80	19,2	[239]
	$C_6H_5C(CH_3){=}CH_2$	- " -	90	$\bar{7}$,85			[239]

TABLE 30. Reactions of Hydroperoxides 2 ROOH $\overset{K}{\rightleftarrows}$ Complex $\overset{k}{\rightarrow}$ Free Radicals (ICA Method)

Reaction	Solvent	t°	log k	log K	log (Kk)	Ref.
2 H_2O_2	cyclo-$C_6H_{11}OH$	120–140			9,84–28,1/θ	[62]
2$(CH_3)_3COOH$	n-C_7H_{16}	60–90	7,76–23,0/θ	1,90 (90°)	$\bar{7}$,79 (90°)	[240]
2 CH_2=CHCH(CH_3)OOH	CH_2=CHCH$_2$CH$_3$	65–80			11,70–26,0/θ	[241]
2 CH_3CH=CHCH$_2$OOH	CH_3CH=CHCH$_3$	65–80			12,70–28,0/θ	[241]
2 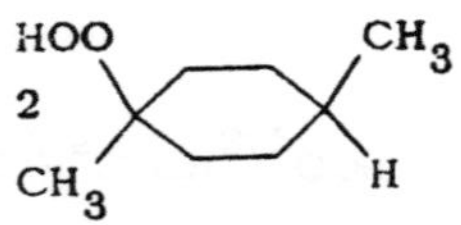	C_6H_6	60–80			11,40–28,6/θ	[88]
2 $C_6H_5C(CH_3)_2OOH$	C_6H_5Cl	100–120			2,38–16,4/θ	[236]
	$C_6H_5CH(CH_3)_2$	60–95	8,34–25,3/θ	4,52+6,0/θ	4,86–19,3/θ	[237]
	– " –	120			$\bar{6}$,28	[242]
	C_6H_5Cl	120			$\bar{7}$,78	[242]

$2(C_6H_5)_2C(CH_3)OOH$	$(C_6H_5)_2CHCH_3$	120	$\bar{5},42$	[242]
	C_6H_5Cl	120	$\bar{5},48$	
	$C_6H_5NO_2$	120	$\bar{5},65$	
$2\ CH_2{=}CH{-}CH(OOH)(CH_2)_5CH_3$	$CH_2{=}CH(CH_2)_6CH_3$	110	$\bar{6},82$	[242]
	C_6H_5Cl	110	$\bar{6},04$	
	$CH_3(CH_2)_3CH_2OH$	110	$\bar{6},40$	
$2\ C_6H_5CH(OOH){-}CH{=}CH_2$	$C_6H_5CH_2CH{=}CH_2$	55	$\bar{7},45$	[243]
$2\ CH_2{=}CHCH(OOH)(CH_2)_4CH_3$	$CH_2{=}CH(CH_2)_5CH_3$	55	$\bar{7},46$	[243]
$2CH_3(CH_2)_7CH{=}CH{-}CH(HOO)(CH_2)_8COOCH_3$	$CH_3(CH_2)_7CH{=}CH(CH_2)_7COOCH_3$	55	$\bar{7},65$	[243]
2 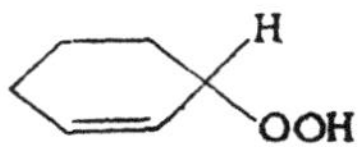	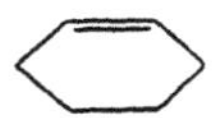	55	$\bar{7},73$	[243]

Table 30 (continued)

Reaction	Solvent	t°	logk	logK	log(Kk)	Ref.
2 $CH_3(CH_2)_4CH{=}CH{-}CH(OOH)$ $H_5C_2OOC(CH_2)_7HC{=}CH$	RH	55			$\bar{7}$,67	[243]
2 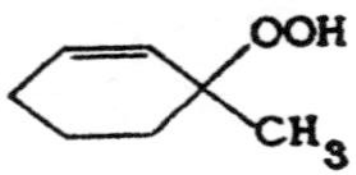		56			$\bar{6}$,06	[243]
2 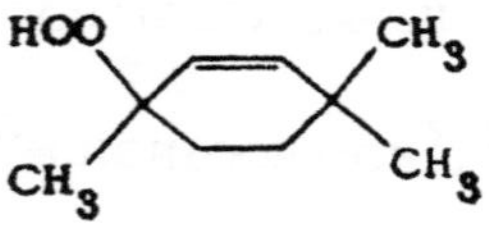	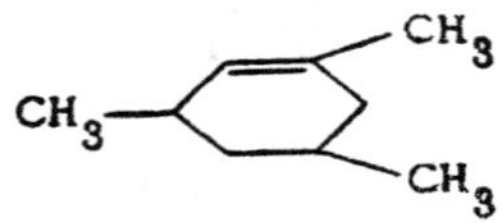	55			$\bar{6}$,51	[243]
$2CH_3CH_2CH{=}C(CH_3){-}CH(OOH)CH_2CH_3$	$C_2H_5CH{=}C(CH_3)C_3H_7$	55			$\bar{6}$,235	[243]
$2CH_3C(CH_3){=}CHCH(OOH)(CH_2)_3CH_3$	$CH_3C(CH_3){=}CH(CH_2)_4CH_3$	55			$\bar{7}$,70	[243]
2 (Squalene hydroperoxide)	Squalene	55			$\bar{6}$,47	[243]

TABLE 31. Reactions with Participation of Peroxide, $ROOH + M \underset{}{\overset{K}{\rightleftarrows}} Complex \xrightarrow{k} Free$ Radicals (ICA Method)

ROOH	M	Reaction	t°	log k, sec^{-1}	log K, liter/mole	log Kk	Ref.
$(CH_3)_3COOH$	n-C_4H_9OH	C_6H_5Cl	80–105	6,70–20,0/θ			[240]
$C_6H_5C(CH_3)_2OOH$	$C_6H_5CH_2OH$	$C_6H_5CH{=}CH_2$	90			$\bar{2}$,11	[239]
	- " -	$CH_2{=}C(CH_3)CO_2CH_3$	90			$\bar{2}$,165	[239]
$(CH_3)_3COOH$	cyclo-$C_6H_{11}O$	C_6H_5Cl	110–126	9,56–26,0/θ	$\bar{7}$,84+11,0/θ	3,40–15,0/θ	[244]
$C_6H_5C(CH_3)_2OOH$	- " -	- " -	120	$\bar{6}$,755	0,0	$\bar{3}$,755	[245]
H_2O_2	- " -	cyclo-$C_6H_{11}OH$	120–140	4,34–16,2/θ	$\bar{5}$,34+6,7/θ	$\bar{1}$,68–9,5/θ	[246]
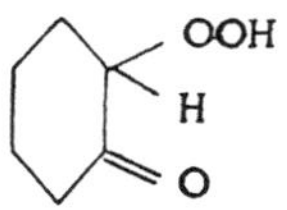	- " -	C_6H_5Cl	110–130	7,77–20,4/θ			[247]

Table 31 (continued)

ROOH	$\dot{M}$	Reaction	t°	log k, sec^{-1}	log k, liter/mole	log Kk	Ref.
$CH_3COCH(OOH)CH_3$	$CH_3COC_2H_5$	C_6H_6	30–70	10,06–23,0/θ	$\bar{1}$,90 (70°)		[105]
sec-$C_{10}H_{21}OOH$	2 $C_5H_{11}COOH$	$C_{10}H_{22}$		6,30–16,1/θ	$\bar{4}$,765+10,2/θ	3,065–5,9/θ	[248]
$C_6H_5(CH_3)_2OOH$	C_6H_5COOH	$C_6H_5CH(CH_3)_2$	90–110			9,85–20,6/θ	[249]
$(CH_3)_3COOH$	C_5H_5	C_6H_5Cl	60–85	7,28–21,0/θ	$\bar{4}$,255–8,3/θ	3,535–12,7/θ	[217]
$CH_3COCH(OOH)CH_3$	H_2O	$CH_3COC_2H_5$	30–70	8,97–20,2/θ	0,6* (70°)		
	$(CH_2)_3COH$	- " -	30–70	9,73–21,0/θ	0,65* (70°)		[250]
	CH_3COOH	- " -	30–70	8,86–20,5/θ	3,10* (70°)		

* K has the dimensions $liter^2/mole^2$.

TABLE 32. Reaction of Electron Transfer, $\geq N: + ClO_2 \longrightarrow \geq N^+ + ClO_2^-$
(CSM Method, Water as Solvent)

Amine	t°	log k	Ref.
$(CH_3)_3N$	25	5,00	[254]
$(C_2H_5)_3N$	26,5	5,30	[254]
$(CH_3)_2NC(CH_3)_3$	25	5,37	[255]
$[(CH_3)_2CH]_2NH$	25	2,56	[254]
$HOCH_2CH_2NH_2$	25	2,16	[255]
$C_6H_5CH_2NH_2$	25	$\bar{2}$,05	
$(C_6H_5CH_2)_2NH$	40,7	1,725	[254]
$C_6H_5CH_2NHC(CH_3)_3$	20–40	2,56 (25°)*	
$C_6H_5CH_2NHCH(CH_3)_2$	25	1,96	
$C_6H_5CH_2N(CH_3)_2$	25	4,44	
$p\text{-}CH_3OC_6H_4CH_2N(CH_3)_2$	25	4,695	
$p\text{-}CH_3C_6H_4CH_2N(CH_3)_2$	25	4,55	
$p\text{-}C_6H_4CH_2N(CH_3)_2$	25	4,30	
$m\text{-}NO_2C_6H_4CH_2N(CH_3)_2$	25	3,79	
$p\text{-}NO_2C_6H_4CH_2N(CH_3)_2$	25	3,54	
$m\text{-}ClC_6H_4CH_2N(CH_3)_2$	25	4,20	[255]
$p\text{-}ClC_6H_4CH_2N(CH_3)_2$	25	4,30	
$m\text{-}CH_3OC_6H_4CH_2N(CH_3)_2$	25	4,455	
$p\text{-}CH_3OC_6H_4CH_2NHCH_3$	25	2,43	
cyclo-$(CH_2)_5NH$	25	3,38	
cyclo-$(CH_2)_5NCH_3$	25	4,94	
cyclo-$(CH_2)_5NC(CH_3)_3$	25	6,10	

*log A = 7,55; E = 7,0.

TABLE 33. Reactions of Molecules with Oxygen

Compound	Solvent	t°	log k (180°)	log A	E	Method	Ref.
cyclo-C_6H_{12}	C_6H_6	110–140	$\bar{7},60$	12,90	40,0	ICA	[255]
o-$CH_3C_6H_4CH_3$	- " -	110–150	$\bar{7},30$	8,30	31,0	ICA	[255]
$C_6H_5CH(CH_3)_2$	C_6H_5Cl	101–130	$\bar{4},35$	9,545	27,4	ICA	[255]
$CH_3COCH_2CH_3$	C_6H_6	120–150	$\bar{5},00$	9,48	30,0	ICA	[254]
$C_6H_5CH{=}CH_2$	C_6H_5Cl	105–125	$\bar{3},05$	11,555	30,0	ICA	[253]
C_6H_5OH	C_6H_6	180–200	$\bar{5},95$	12,845	35,0	CSM	[252]
p-$CH_3C_6H_4OH$	- " -	180–203	$\bar{4},08$	11,48	32,0	CSM	[252]
p-$CH_3OC_6H_4OH$	- " -	160–200	$\bar{4},87$	9,875	27,0	CSM	[252]
β-$C_{10}H_7OH$	- " -	170–210	$\bar{4},65$	10,65	29,0	CSM	[252]
α-$C_{10}H_7OH$	- " -	140–195	$\bar{3},30$	9,34	25,0	CSM	[252]
α-$C_{10}H_7NHC_6H_5$	- " -	200–225	$\bar{5},23$	10,545	32,0	CSM	[251]
β-$C_{10}H_7NHC_6H_5$	- " -	190–220	$\bar{5},75$	10,46	31	CSM	[251]
α-$C_{10}H_7NH_2$	- " -	170–220	$\bar{5},78$	10,175	30	CSM	[251]
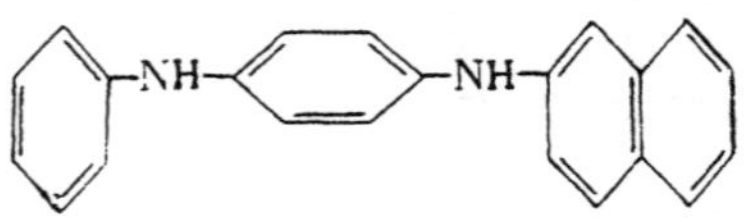	- " -	120–200	$\bar{2},145$	9,145	23	CSM	[251]

TABLE 34. Reaction $C_2H_5Li + RX \rightarrow C_2H_5^{\bullet} + LiX + R^{\bullet}$ (Decalin as Solvent, KEPF Method)

RX	t°	log k (25°)	log A	E	Ref.
C_2H_5I	3–23	$\bar{4},74$	6,74	13,6	[256]
C_2H_5Br	22–60	$\bar{5},10$	7,08	16,3	[256]
$n\text{-}C_4H_9Cl$	80–120	$\bar{7},15$	6,97	18,8	[257]
$C_6H_5CH_2Cl$	20–60		4,545	10,5	[257]
$CH_2{=}CH{-}CH_2Cl$	0–35	$\bar{3},21$	7,59	14,1	[257]
$(C_6H_5)_3CCl$	-80 to -44	0,06	4,32	5,8	[257]
C_2H_5Cl	90	$\bar{5},59$			[257]
$(CH_3)_2CHCl$	80	$\bar{4},40$			[257]
$(CH_3)_3CCl$	30	$\bar{2},75$			[257]

Reaction $C_5H_5VOCl_2 + D \longrightarrow C_5H_5^{\bullet} + DVOCl_2$ (toluene solvent)

D	t°	log k (25°)	log A	E	Ref.
Tetrahydrofuran	22–57	$\bar{6},62$	3,375	11,9	[259]
$(n\text{-}C_4H_9)_2O$	34–65	$\bar{6},06$	6,84	17,1	[259]
CH_3CN	19–55	$\bar{6},10$	5,415	15,4	[259]
$(C_6H_5)_3P$	11–40	$\bar{3},75$	5,255	10,2	[259]

TABLE 35. Trimolecular Reactions of Radical Formation

Reaction	Solvent	t°	log k (150°)	log A	E	Method	Ref.
3 CH_2=CHC_6H_5	C_6H_6	60–176	$\bar{9}$,16	6,12	28,9	RICR	[260]
2cyclo-$C_6H_{11}OH$ + O_2	C_6H_5Cl	111–130	$\bar{8}$,66	0,92	16,0	ICA	[262]
2 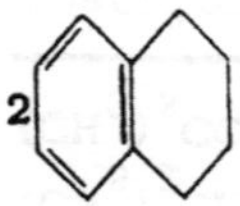 + O_2	n-$C_{10}H_{22}$	130–150	$\bar{8}$,84	3,545	20,7	ICA	[263]
	Decalin	105–124	$\bar{7}$,08	5,38	23,8	RICR	[264]
2 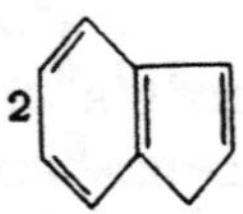 + O_2	Decalin	72–92	$\bar{7}$,89	3,59	18,8	RICR	[264]
2 n-$C_{10}H_{22}$ + O_2	p-ClC_6H_4Cl	150	$\bar{9}$,61			ICA	[265]

CHAPTER III

THE CAGE EFFECT

§1. Questions of Theory

Upon decomposition of a molecule in the gas phase, the decomposition products—radicals—immediately fly apart in different directions. The decomposition of a molecule in the liquid phase leads to the formation of two radicals that exist side by side for a certain time, surrounded by solvent molecules (in a cage). Since the rate constant for the interaction of two radicals is very high, some such radicals will have time to react with each other before the pair is separated by diffusion. The main principles of the cage-effect theory were formulated by Frank and Rabinowitch [266]. In the 1950's and 1960's, widespread experimental evidence was obtained to support this theory.

As a convincing example of the cage effect in a liquid, one may compare the products of photolysis of azomethane in the gas and liquid phases [267]. In the gas phase, the methyl radicals formed upon decomposition of an excited molecule of azomethane will fly apart immediately and enter into reactions of the type

$$CH_3^{\cdot} + RH \longrightarrow CH_4 + R^{\cdot},$$

and the probability of encounter and recombination of two methyl radicals is extremely small. The ratio $[C_2H_6]/[N_2] = 0.015$ for the gas phase. In the liquid phase, the ratio is 0.65, which is explained by recombination of methyl radicals in the solvent cage. Azomethane decomposition in solution, taking the cage effect into account, can be described by the following simple scheme:

$$CH_3\text{-}N{=}N\text{-}CH_3 \xrightarrow{k} \overline{CH_3^{\cdot} + N_2 + CH_3^{\cdot}}\ ,$$

$$\overline{CH_3^{\cdot} + N_2 + CH_3^{\cdot}} \xrightarrow{k'} C_2H_6 + N_2$$

$$\overline{CH_3\cdot + N_2 + CH_3\cdot} \xrightarrow{k_D} CH_3\cdot + CH_3\cdot + N_2 ,$$

where the overline indicates that the particles exist in a cage of solvent. The probability of radical escape into the bulk volume is $e = k_D/(k + k_D)$. The diffusion rate constant $k_D = 16\pi NDr$, where r is the particle radius and N is Avogadro's number. The diffusion coefficient D can be expressed in terms of the viscosity η by means of the Stokes-Einstein equation, $D = kT/6\pi r\eta$, where k is Boltzmann's constant; then $k_D = 8RT/(3000\eta)$ liter/mole·sec. Having substituted the expression for k_D into the formula for e, we obtain after simple transformations:

$$\frac{1-e}{e} = \frac{3000\, k'\eta}{8RT} .$$

A somewhat different expression was obtained by Bagdasar'yan [268]:

$$\frac{l}{1-e} = \frac{l}{l'} + \frac{l}{l'} \frac{8RT}{3\cdot 10^3 k'\eta} ,$$

where l and l' are (respectively) the distances between centers of radicals at the moment of formation of the pair and at the moment of reaction between them. Experimentally, some cases have been observed in which the relation between e and η matches the theoretical [268, 269], and some cases in which it does not [270-272]. The deviations from theory are related to the fact that the Stokes-Einstein equation is by no means universally applicable for describing the diffusion of molecules (for polar particles, solvation and hydrogen bond formation must be taken into account); moreover, in highly viscous media, it is necessary further to take into account the acceptor mechanism for escape of radicals into the bulk volume from a cage [270, 271].

In measuring the initiation rate and the magnitude of e by means of radical acceptors, the question arises whether the acceptor could not seize the radical from the cage and thus influence the magnitude of e (distort it). In [273], in the example of the dinitrile of azonaphthenic acid, it was shown that the yield of radical recombination products in the cage is not dependent on the concentration of radical acceptor (I_2, Br_2, diphenylpicrylhydrazyl) over a range from 10^{-5} to 10^{-1} mole/liter. The existence of a wide range of radical acceptor concentrations within which the acceptor does not seize the radicals from the cage can also be predicted from the cage-effect theory [268].

§2. Methods for Measuring Initiator Efficiency

a. Introduction of Initiator Fragments into Polymer (IIP)

When initiators are used in polymerization processes, the act of initiation occurs when an initiator radical that has escaped from a cage into the bulk volume adds to a monomer molecule, after which the chain growth begins. As a result, all initiator radicals that have escaped from a cage will enter into the macromolecule in the form of end groups. Having determined the number of initiator fragments that have entered the polymer, and knowing the number of molecules that were decomposed, one can determine

$$e = \frac{\text{number of initiator fragments in polymers}}{2(\text{number of initiator molecules decomposed})}.$$

The coefficient 2 accounts for the formation of two radicals from each molecule of initiator. If the initiator molecule contains several (m) groups that decompose to form radicals, the coefficient will be equal to 2m.

The number of inhibitor fragments in the polymer is determined by two general routes--by chemical methods, or by the use of radioactive tracers. When benzoyl peroxide is used as an initiator, the resulting radicals enter into the polymer in the form of terminal benzoate groups $C_6H_5COO^-$ The number of benzoate groups in the polymer is determined by saponifying and measuring the amount of caustic consumed in the saponification. The tracer method is used, in particular, for determining initiation efficiency for azobisisobutyronitrile (see Table 37), using an initiator labeled with C^{14}: $(CH_3)_2C^{14}(CN)\text{-}N{=}N\text{-}(CN)C^{14}(CH_3)_2$. The quantities measured in the experiment are the specific activity of the initiator, the specific activity of the polymer, and the amount of initiator that has been decomposed; from these data, the relative fraction of the fragments that have entered the polymer is calculated.

b. Consumption of Free Radical Acceptor (ICA)

If a free radical acceptor is introduced into the solution, the consumption of the acceptor InH can be used as a measure of the rate of radical formation: $W_i = fW_{InH}$, where f is the number of radicals reacting with one particle of acceptor. The initiator decomposition rate can be measured by one of the methods described in Chapter I. The initiator efficiency is

$$e = \frac{W_i}{2W} = \frac{fW_{InH}}{2W} .$$

The coefficient 2 accounts for the formation of two radicals by decomposition of one molecule (or one group) of the initiator. The acceptor concentration must be sufficiently large to suppress completely the reaction of initiator radicals with each other in the bulk volume; i.e., the minimum acceptor concentration must satisfy the requirement that

$$fk_{InH}[InH][R\cdot] \gg 2k_t[R\cdot]^2.$$

Since with this condition

$$W_i \approx fk_{InH}[InH][R\cdot],$$

or

$$[R\cdot] = \frac{W_i}{fk_{InH}[InH]} ,$$

then

$$[InH]_{min} \gg (2k_t W_i)^{1/2}/k_{InH}$$

If we set $2k_t = 10^8$, $k_{InH} = 10^6$, and $W_i = 10^{-7}$ mole/liter · sec, then $[InH]_{min} \gg 10^{-5}$. On the other hand, the acceptor must not react with the radicals in a solvent cage; i.e., its maximum concentration must meet the requirement that $k_{InH}[InH]_{max} \ll 10^{10}$, or $[InH]_{max} \ll 10^{10}/k_{InH}$. Thus, the concentration of radical acceptor must satisfy the following conditions: $(2k_tW_i)^{1/2} \ll k_{InH}[InH] \ll k_D$. If we set $2k_t = 10^9$, $W_i = 10^{-7}$ mole/liter · sec, and $k_D = 10^{10}$, then $10 \ll k_{InH}[InH] \ll 10^{10}$; i.e., there exists a wide range of acceptor concentrations and a large choice of acceptors suitable for measuring initiation rates.

The rate of acceptor consumption is usually determined by measuring its concentration at various time intervals.

When inhibitors of chain reactions are used, the acceptor consumption rate can be judged by the length of the retardation period [274].

In measuring W_i according to the consumption of a free radical acceptor, it is necessary to know the coefficient f. For such acceptors as stable free radicals, f = 1. For I_2, f = 2. For oxidation inhibitors (phenols and aromatic amines), values of f are shown in Table 36. These values were measured on the basis of the length of the retardation period in the inhibited oxidation of cumene at 65° [274, 275] or dihydroanthracene at 60° [276], or on the basis of inhibitor consumption in the inhibited oxidation of cyclohexanone at 75° [123] or cyclohexanol at 75° [122]. The reference substance was Ionol [2,6-di-tert-butyl-4-methylphenol], from which the following compound is formed in high yield in cumene oxidation:

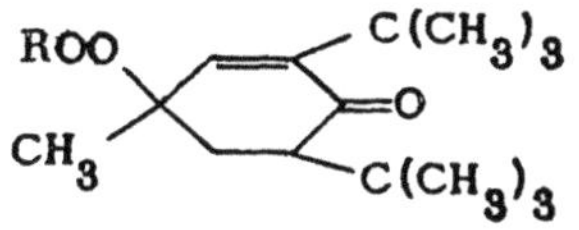

as a result of successive reactions of two peroxide radicals with the Ionol (f = 2).

c. Rate of Initiated Chain Reaction (RICR)

The method described in Chapter I consists of measuring the velocity of an initiated chain reaction W_ν and calculating the rate of initiation W_i. The rate of initiator consumption W is measured by an independent method. The initiation efficiency is $e = W_i/2W$. Another variant of this method is to continue the chain reaction until the initiator is completely consumed; the rate constant for initiator decomposition is determined from the kinetic curve of reaction-product accumulation (or starting material consumption), and the initiation efficiency is determined from the extent to which the reaction has proceeded (see Chapter I).

d. Products of Radical Recombination in Cage (RRC)

When decomposition occurs in a solvent consisting of molecules that react with the initiator radicals, or in the presence of a free radical acceptor, the initiator radicals will react with each other only in the cage. The quantity of products formed by interaction of radicals in cages can be used to determine the efficiency of the initiator

$$e = 1 - \frac{\text{reaction products of radicals in cage}}{\text{initiator decomposed}}$$

For example, in the photodecomposition of azomethane in n-heptane [277], ethane is formed as a result of recombination in the cage:

$$CH_3{-}N{=}N{-}CH_3 \longrightarrow \overline{CH_3^{\cdot} + N_2 + CH_3^{\cdot}}\ ,$$

$$\overline{CH_3^{\cdot} + CH_3^{\cdot}} \xrightarrow{k_p} C_2H_6\ ,$$

$$\overline{CH_3^{\cdot} + CH_3^{\cdot}} \xrightarrow{k_D} 2CH_3^{\cdot}$$

When the reaction is continued to the end,

$$e = 1 - \frac{[C_2H_6]}{[CH_3N_2CH_3]_0}.$$

In the decomposition of the dinitrile of azonaphthenic acid in chlorobenzene in the presence of acceptors [136], the following reactions are effected in the cage:

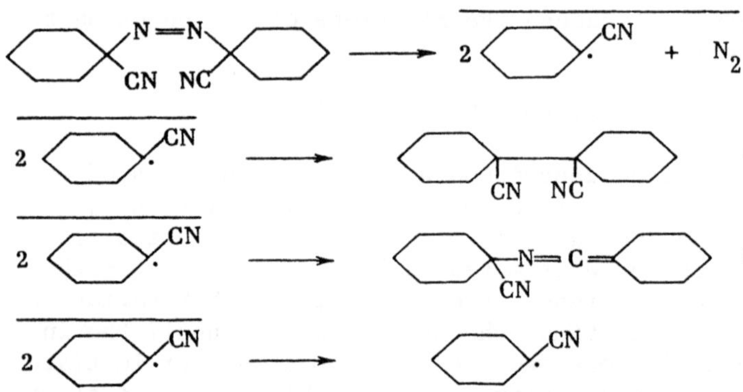

(I)

(II)

and the products I and II are formed as a result of radical recombination in the cage.

§3. Initiator Efficiency

Among the diverse initiators for which the probability of radical escape into the bulk volume $e = k_i/2k$ has been measured, a special place is occupied by azobisisobutyronitrile. The results of measurements of e for this compound are listed in Table 37.

TABLE 36. Stoichiometric Inhibition Coefficients
C = cumene CL = cyclohexanol
DHA = dihydroanthracene CN = cyclohexanone

Inhibitor	Compound oxidized	f	Ref.
C_6H_5OH	C	2,0	[274]
	C	2,1	[275]
p-$CH_3C_6H_4OH$	C	2,2	[274]
	C	2,5	[275]
o-$CH_3C_6H_4OH$	C	2,2	[274]
	C	2,1	[275]
p-$(CH_3)_3CC_6H_4OH$	C	2,3	
p-$CH_3OC_6H_4OH$	C	2,0	[275]
p-ClC_6H_4OH	C	1,7	
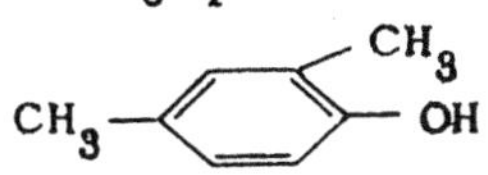	C	2,6	[275]
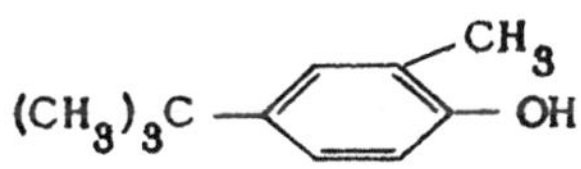	C	1,9	[275]
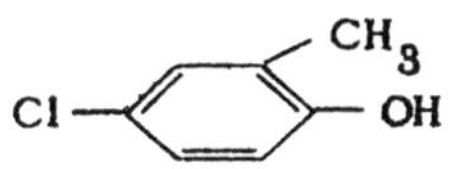	C	1,7	[275]
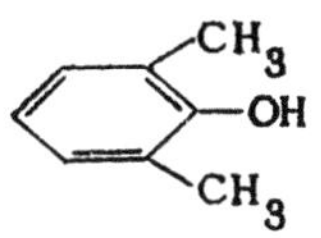	C	2,0	[275]
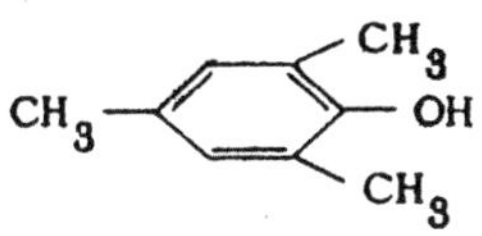	C	2,2	[275]
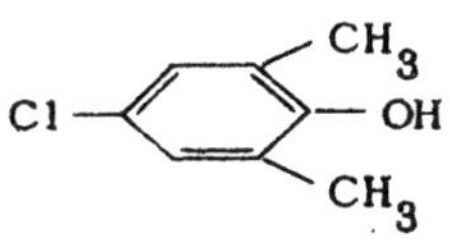	C	1,3	[275]

Table 36 (continued)

Inhibitor	Compound oxidized	f	Ref.
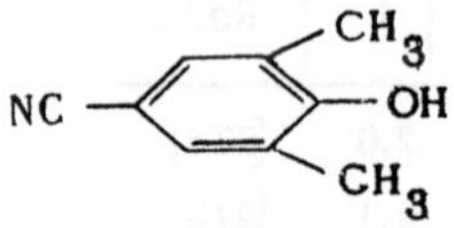	C	1,7	[275]
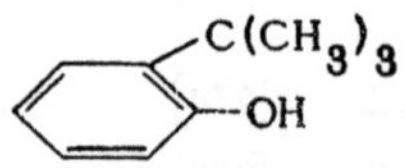	C	2,1	[275]
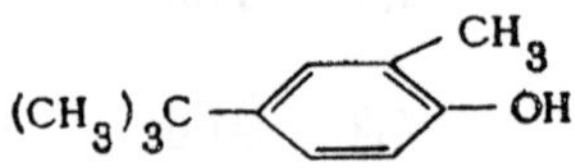	C	2,4	[275]
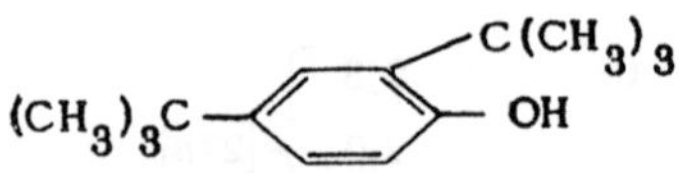	C	2,3	[275]
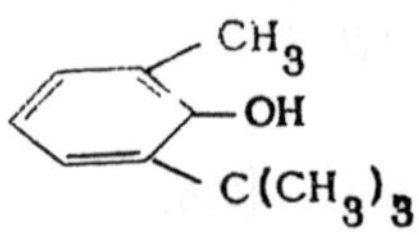	C	1,9	[275]
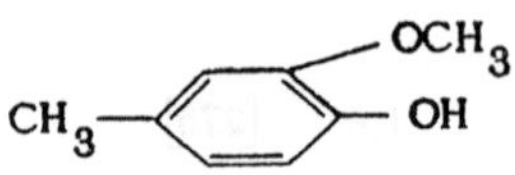	C	1,3	[275]
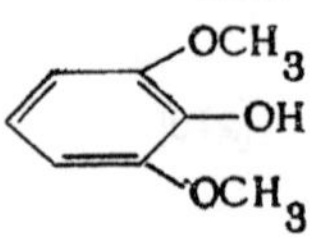	C	1,4	[275]
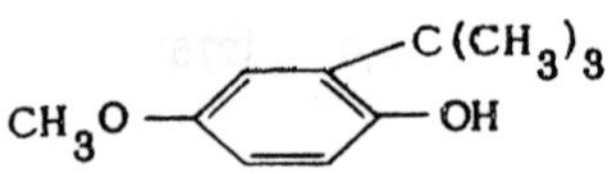	C	2,1	[275]
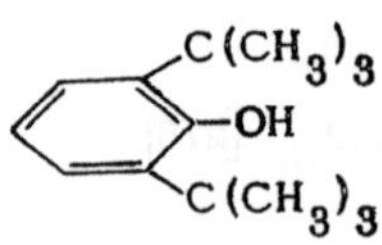	C	2,0	[275]
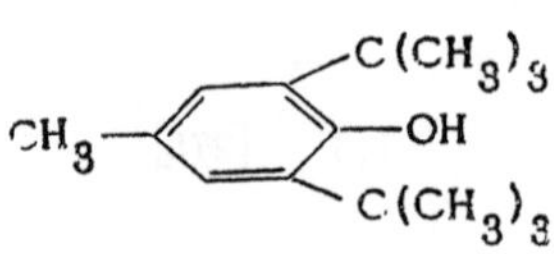	C	2,0	[275]
	C	2,0	[275]

Table 36 (continued)

Inhibitor	Compound oxidized	f	Ref.
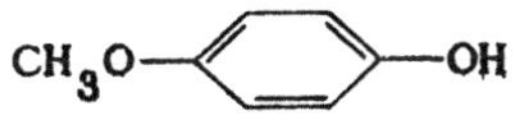	C	2,1	[275]
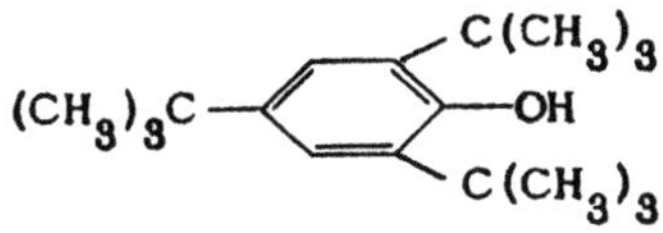	C	1,8	[275]
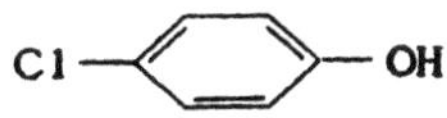	C	1,0	[275]
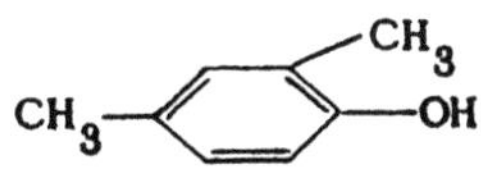	C	2,0	[275]
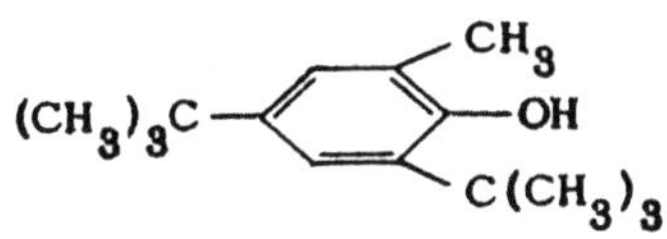	C	1,5	[275]
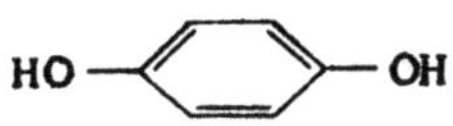	C DHA	1,8 1,75	[275] [276]
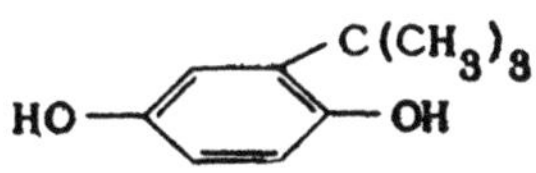	C	1,1	[275]
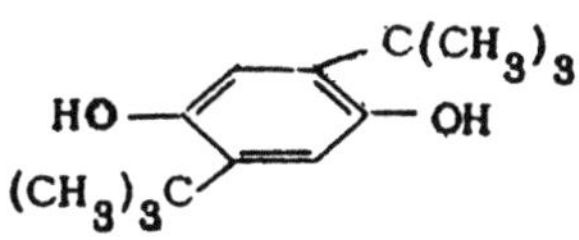	C	0,85	[274]
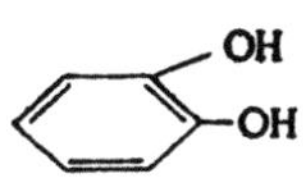	C	2,3	[275]
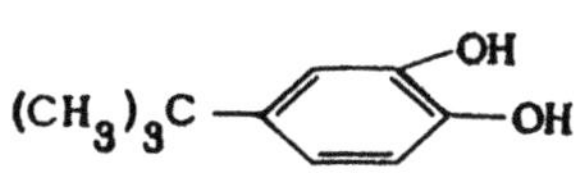	C	2,2	[275]
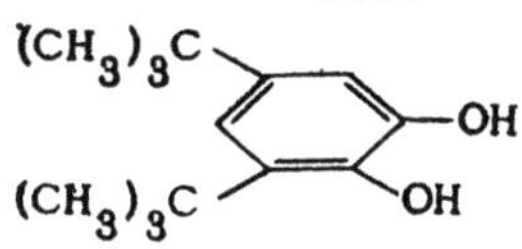	C	2,0	[275]

Table 36 (continued)

Inhibitor	Compound oxidized	f	Ref.
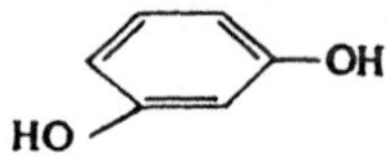	C	2,0	[275]
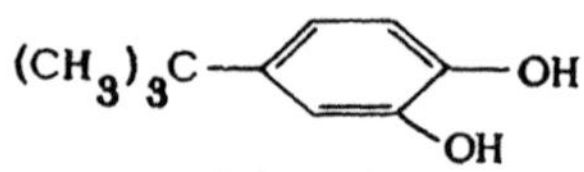	C	2,0	[274]
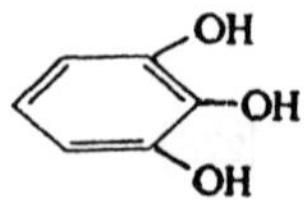	C	2,5	[275]
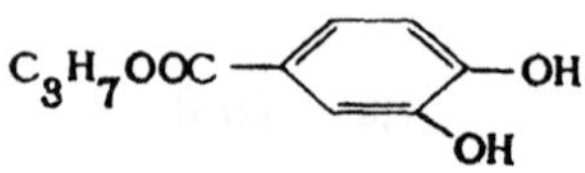	C	2,0	[275]
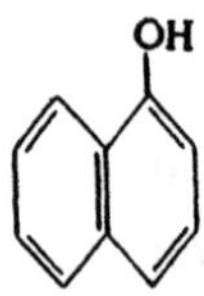	C	2,0	[275]
	DHA	2,08	[275]
	CN	2,0	[123]
	CL	2,7	[122]
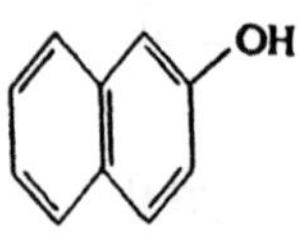	C	1,8	[275]
	C	2,1	[274]
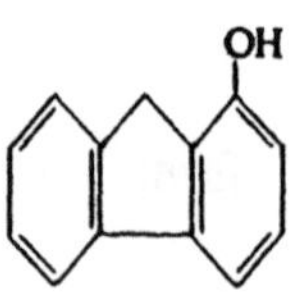	DHA	3,6	[276]
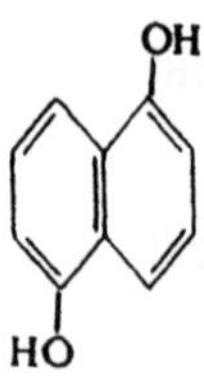	DHA	2,05	[276]

Table 36 (continued)

Inhibitor	Compound oxidized	f	Ref.
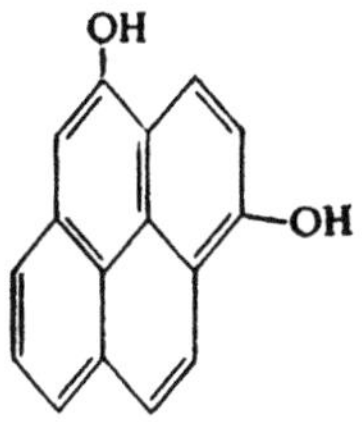	DHA	1,90	[276]
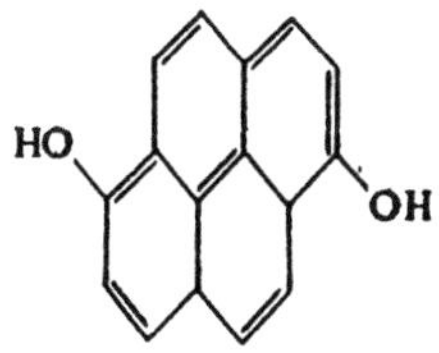	DHA	1,94	[276]
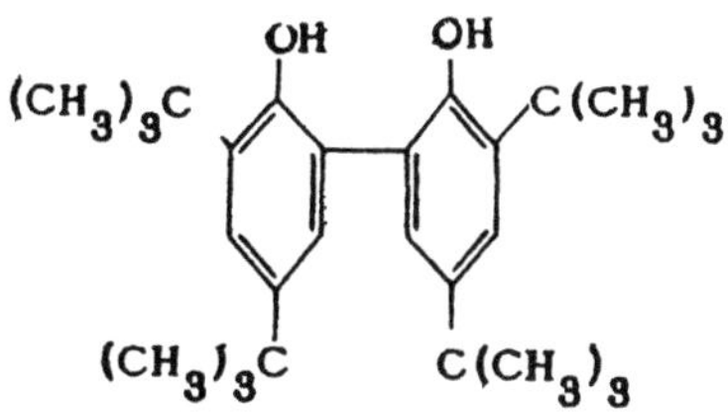	C	1,4	[275]
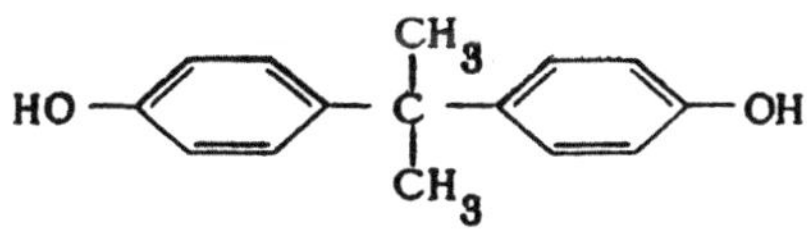	C	4,1	[274]
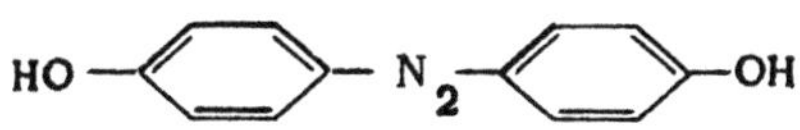	C	2,0	[274]
$(C_6H_5)_2NH$	C	2,8	[274]
$C_6H_5NHCH_3$	C	2,0	[274]
$(C_6H_5)_2NOCH_3$	C	3,3	[274]
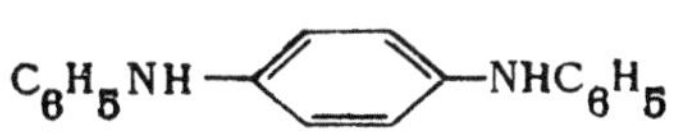	C	2,2	[274]
$HOC_6H_4NHC_6H_4OH$	C	2,1	[274]
α-$C_{10}H_7NH_2$	CN	2,0	[123]

Table 36 (continued)

Inhibitor	Compound oxidized	f	Ref.
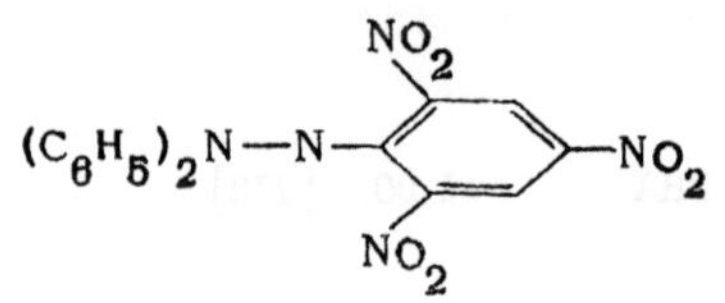	C	2,0	[274]
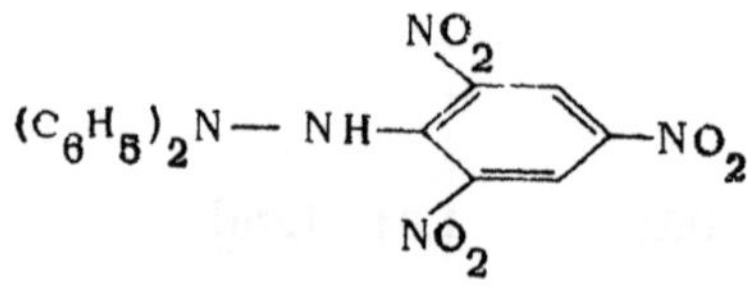	C	2,8	[274]
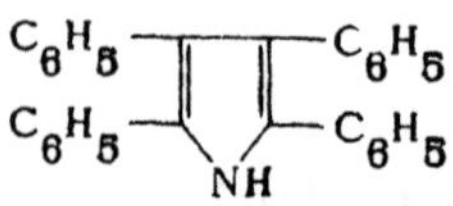	C	1,8	[274]

TABLE 37. Probability of Radical Escape into Bulk Volume (e) in Decomposition of Azobisisobutyronitrile

Solvent	t°	e	E_e	Method	Ref.
$C_6H_6 + C_6H_5CH{=}CH_2$	50	0,55		IIP	[278]
$C_6H_6 + CH_2{=}CH{-}CN$	60	0,50		IIP	[279]
$C_6H_6 + CH_2{=}CH{-}COOC_4H_9$	60	0,50		IIP	[280]
$C_6H_6 + CH_2{=}CH{-}OCOCH_3$	50	0,72		IIP	[281]
$C_6H_6 + CH_2{=}CHCl$	50	0,74		IIP	[281]
C_6H_6	62,5	0,51* 0,62[2*] 0,61[3*]		ICA	[113]
	30	0,44[1*]		ICA	[111][4*]
	65	0,44[3*]		ICA	[282]
C_6H_6 + 10% $CH_3COC_2H_5$	40	0,54[5*]	1,1	ICA	[272]
	70	0,62[5*]		ICA	
$C_6H_5CH{=}CH_2$	50	0,70		IIP	[281]

Table 37 (continued)

Solvent	t^{o}	e	E_e	Method	Ref.
$C_6H_5CH{=}CH_2$	50	0,52		IIP	[278]
	60	0,61		IIP	[283]
	60	0,48		IIP	[284]
	60	0,64		IIP	[285]
	60	0,70		IIP	[280]
	65	0,40[3*]		ICA	[282][4*]
$C_6H_5CH{=}CH_2$ + $CH_2{=}C(CH_3)COOCH_3$	60	0,46		IIP	[284]
$C_6H_5CH{=}CH_2$ + $CH_2{=}CH{-}COO(n{-}C_4H_8)$	60	0,52		IIP	[286]
	60	0,60		IIP	[280]
$C_6H_5CH_3$	65	0,44		ICA	[282][4*]
$C_6H_5C_2H_5$	60	0,60[5*]			
	80	0,72[5*]	2,1	ICA	[287]
	40	0,48[5*]			
	80	0,70[5*]	1,8	ICA	[288]

$C_6H_5CH(CH_3)_2$	40	0,50 [5x]			
	80	0,62 [5x]	1,2	ICA	[287]
$(C_6H_5)_2CHCH_3$	60	0,42 [5x]		ICA	[287]
		0,42 [1x]			
C_6H_5Cl	62,5	0,57 [3x]		ICA	[113]
		0,59 [2x]			
	65	0,40 [3x]		ICA	[282] [4x]
	75	0,54 [6x]		ICA	[122]
C_6H_5Br	50	0,50		IIP	[278]
$C_6H_5NO_2$	62,5	0,75 [2x]		ICA	[113]
		0,73 [3x]			
	65	0,49 [3x]		ICA	[282] [4x]
1,2-$Cl_2C_6H_4$	65	0,37 [3x]		ICA	[282] [4x]
$(CH_3)_3CCH_2CH(CH_3)_2$	60	0,45 [5x]			
	80	0,55 [5x]	2,4	ICA	[287]

Table 37 (continued)

Solvent	t^{o}	e	E_e	Method	Ref.
$(CH_3)_2CH(CH_2)_4CH(CH_3)_2$	60	0,50	4,3	ICA	[287]
	80	0,72			
$n\text{-}C_{10}H_{22}$	65	$0,36^{3*}$			[113]
	50	$0,28^{6*}$	1,4	ICA	[288]
	80	$0,35^{6*}$			
$CH_2{=}C(CH_3)COOC_2H_5$	45	0,50	1,5	RICR	[120]
	90	0,67			
$CH_2{=}C(CH_3)COOCH_2CH(CH_3)_2$	35	0,26	4,0	RICR	[120]
	80	0,60			
$CH_2{=}C(CH_3)COOC_9H_{19}$	35	0,14	4,2	RICR	[120]
	80	0,34			
$CH_2{=}C(CH_3)COOCH_3$	50	0,56		IIP	[281]
	60	0,52		IIP	[285]

	60	0,50	IIP	[289]
	60	0,43	IIP	[284]
CH_2=CH-CN	50	1,03	IIP	[281]
Dioxane	50	0,51	IIP	[278]
	50	0,56 0,63	IIP	[231]
	60	0,61	IIP	[281]
	60	0,43 0,46 0,48	IIP	[284]
	60	0,52 0,64	IIP	[285]
	60	0,60 0,70	IIP	[280]
	60	0,50	IIP	[289]
	60	0,52	IIP	[286]

Table 37 (continued)

Solvent	t^{o}	e	E_e	Method	Ref.
$CH_3COC_2H_5$	40	$0,32^{5*}$			
	70	$0,42^{5*}$	1,9	ICA	[272]
$CH_3COC_2H_5 + H_2O$ (1 : 1)	40	$0,20^{5*}$			
	70	$0,39^{4*}$	4,9	ICA	[272]
cyclo-$C_6H_{10}O$	55	0,48			
	80	0,66	3,1	ICA	[123]
$(CH_3)_3COH$	50	$0,28^{6*}$			
	70	$0,50^{6*}$	6,0	ICA	[288]
	65	$0,30^{3*}$		ICA	[282]
cyclo-$C_6H_{11}OH$	65	$0,27^{6*}$			
	90	$0,45^{6*}$	4,6	ICA	[122]
	50	$0,22^{6*}$			
	80	$0,38^{6*}$	6,2	ICA	[288]

CH_3COOH	65	0,51[3*]	ICA	[282][4*]
CH_3NO_2	65	0,40[3*]	ICA	[282][4*]
CH_3CN	65	0,67[3*]	ICA	[282][4*]
CCl_4		0,43[3*]		
	62,5	0,45[2*]	ICA	[113]
		0,39[1*]		
	65	0,36[3*]	ICA	[282][4*]

1* Measured by consumption of diphenylpicrylhydrazyl.

2* Measured by consumption of I_2.

3* Measured by induction period of inhibited oxidation in the presence of Ionol.

4* Almost all values of e, which were measured in this work by induction period in presence of inhibitor, are too low by a factor of about 1.5, because of a systematic error.

5* Measured by induction period of inhibited oxidation in the presence of hydroquinone.

6* Measured by consumption of nitroxy radical, $R_1R_2NO^{\bullet}$.

TABLE 38. Probability of Radical Escape into Bulk Volume in Decomposition of Various Initiators

Initiator	Solvent	t^o	e	Method	Ref.
$(C_6H_5COO)_2$	CCl_4	73	0,82	ICA	[29]
	$C_6H_5CH{=}CH_2$	90	0,72	RICR	
$(p\text{-}CH_3C_6H_4COO)_2$	- " -	90	0,73	RICR	
$(p\text{-}CH_3OC_6H_4COO)_2$	- " -	90	0,80	RICR	
$(p\text{-}ClC_6H_4COO)_2$	- " -	90	0,77	RICR	[290]
$(p\text{-}NO_2C_6H_4COO)_2$	- " -	90	0,87	RICR	
$(m\text{-}Br\,C_6H_4COO)_2$	- " -	90	0,47	RICR	
$(p\text{-}CNC_6H_4COO)_2$	- " -	90	0,34	RICR	
$(n\text{-}C_5H_{11}COO)_2$	$C_6H_5C_2H_5$	95	0,64	ICA	
$(n\text{-}C_6H_{13}COO)_2$	- " -	95	0,62	ICA	
$(n\text{-}C_7H_{15}COO)_2$	- " -	95	0,79	ICA	
$(n\text{-}C_8H_{17}COO)_2$	- " -	95	0,73	ICA	
$(n\text{-}C_9H_{19}COO)_2$	- " -	95	0,66	ICA	[28]
$(n\text{-}C_{11}H_{23}COO)_2$	- " -	95	0,72	ICA	
$(n\text{-}C_{15}H_{31}COO)_2$	- " -	95	0,57	ICA	
$(n\text{-}C_{17}H_{35}COO)_2$	- " -	95	0,64	ICA	

$[C_6H_5(CH_2)_3COO]_2$	$C_6H_5CH=CH_2$	80	0,5	ICA	[12]
$[C_6H_5(CH_2)_4COO]_2$	- " -	80	0,4	ICA	
$[CH_3(CH_2)COO]_2$	- " -	80	0,7	ICA	
$[CH_3(CH_2)_2COO]_2$	- " -	80	0,6	ICA	
$[CH_3(CH_2)_3COO]_2$	- " -	80	0,6	ICA	
$[CH_3(CH_2)_4COO]_2$	- " -	80	0,6	ICA	
$[CH_3(CH_2)_5COO]_2$	- " -	80	0,7	ICA	
$[CH_3(CH_2)_6COO]_2$	- " -	80	0,6	ICA	
$[CH_3(CH_2)_8COO]_2$	- " -	80	0,55	ICA	
$[CH_3(CH_2)_{10}COO]_2$	- " -	80	0,5	ICA	
$[CH_3(CH_2)_{14}COO]_2$	- " -	80	0,5	ICA	
$[CH_3(CH_2)_{16}COO]_2$	- " -	80	0,5	ICA	
$[(CH_3)_2CHCOO]_2$	- " -	40	0,25	ICA	
$(CH_3)_3COOCOC_6H_5$	- " -	100	0,9	ICA	
$(CH_3)_3COOCOCH_2C_6H_5$	- " -	100	0,5	ICA	
$(CH_3)_3COOCO(CH_2)_3C_6H_5$	- " -	100	0,4	ICA	
$(CH_3)_3COOCO(CH_2)_6CH_3$	- " -	100	0,9	ICA	
$(CH_3)_3COOCO(CH_2)_{10}CH_3$	- " -	100	0,85	ICA	
$(CH_3)_3COOCO(CH_2)_{14}CH_3$	- " -	100	0,70	ICA	
$(CH_3)_3COOCOCH(CH_3)_2$	- " -	100	0,70	ICA	

Table 38 (continued)

Initiator	Solvent	t^{o}	e	Method	Ref.
$(CH_3)_3COOCOCCl_3$	$C_6H_5CH{=}CH_2$	60	0,24	ICA	[12]
$[(CH_3)_2CHCOO]_2$	CCl_4	45	0,49	ICA	[219]
	- " -	80	0,80	ICA	[29]
(cyclo-$C_6H_{11}COO)_2$	C_6H_6	30	0,135	ICA	[38]
	- " -	85	0,23	ICA	[38]
(cyclo-$C_6H_{11}OCOO)_2$	$C_6H_5C_2H_5$	80	0,78	RICR	
	- " -	60	0,67	RICR	[41]
	C_6H_5Cl	75	0,22	RICR	
	n-C_7H_{16}	75	0,50	RICR	
	$CH_3COC_2H_5$	75	0,28	RICR	[2]
	$C_6H_{10}O$	75	0,34	RICR	[2]
$(CH_3)_3COOC(CH_3)_3$	n-$C_{10}H_{22}$	110	0,47	ICA	[60]
$[C_6H_5C(CH_3)_2O]_2$	- " -	110	0,51	ICA	[60]

$(CH_3)_3COOC(CH_3)_2C_6H_5$	$n\text{-}C_{10}H_{22}$	110	0,63	ICA	[60]
$(CH_3)_3COOC(CH_3)_2C_2H_5$	- " -	110	0,53	ICA	[60]
$[(CH_3)_3COO]_2$cyclo-C_6H_{10}	- " -	110	0,79	ICA	[60]
tert-Butylpinyl peroxide	- " -	110	0,92	ICA	[60]
$C_6H_5C(CH_3)_2OOC(C_6H_5)_2CH_3$	- " -	110	0,72	ICA	[60]
$(CH_3)_3COOCOCH_3$	$C_6H_5C_2H_5$	95	0,90	RICR	[28]
	$n\text{-}C_{11}H_{24}$	110	0,49	ICA	[75]
$(CH_3)_3COOCO(CH_2)_3CH_3$	$C_6H_5C_2H_5$	95	0,90	RICR	[28]
$(CH_3)_3COOCO(CH_2)_4CH_3$	$C_6H_5C_2H_5$	95	0,90	RICR	[28]
	$n\text{-}C_{11}H_{24}$	110	0,66	ICA	[75]
$(CH_3)_3COOCOCH(CH_3)_2$	$C_6H_5C_2H_5$	95	0,70	RICR	[28]
$(CH_3)_3COOCOC(CH_3)_3$	- " -	95	0,20	RICR	[28]
$o\text{-}C_6H_5SC_6H_4COOOC(CH_3)_3$	cyclo-C_6H_{12}	25	0,17	ICA	[94]
	CH_3COCH_3	25	0,16	ICA	[94]
	- " -	40	0,27	ICA	[94]
	$(CH_3)_3COH$	25	0,15	ICA	[94]

Table 38 (continued)

Initiator	Solvent	t°	e	Method	Ref.
o-$C_6H_5SC_6H_4COOOC(CH_3)_3$	CH_3CN	25	0,33	ICA	[94]
	$(CH_3)_2CHOH$	25	0,20	ICA	
	CH_3CH_2OH	25	0,26	ICA	
o-$ClC_6H_4COOOC(CH_3)_3$	n-$C_{11}H_{24}$	110	0,94	ICA	[94]
m-$ClC_6H_4COOOC(CH_3)_3$	n-$C_{11}H_{24}$	110	0,67	ICA	[75]
p-$ClC_6H_4COOOC(CH_3)_3$	- " -	110	0,57	ICA	[75]
2,4-$Cl_2C_6H_3COOOC(CH_3)_3$	- " -	110	0,88	ICA	
p-$NO_2C_6H_4COOOC(CH_3)_3$	- " -	110	0,95	ICA	
o-$CH_3C_6H_4COOOC(CH_3)_3$	- " -	110	0,73	ICA	
m-$CH_3C_6H_4COOOC(CH_3)_3$	- " -	110	0,62	ICA	
p-$CH_3C_6H_4COOOC(CH_3)_3$	- " -	110	0,61	ICA	

$o\text{-}(R_1R_2C{=}CH)C_6H_4\overset{O}{\overset{\|}{C}}OO\overset{O}{\overset{\|}{C}}R_3$

R_1	R_2	R_3					
H	C_6H_5	C_6H_5	C_6H_5Cl	70	0,11	ICA	[45]
H	$m\text{-}NO_2C_6H_4$	C_6H_5	- " -	70	0,15	ICA	
H	$p\text{-}NO_2C_6H_4$	C_6H_5	- " -	70	0,39	ICA	
H	$p\text{-}NO_2C_6H_4$	$p\text{-}NO_2C_6H_4$	- " -	70	0,34	ICA	
H	$p\text{-}NO_2C_6H_4$	$p\text{-}CH_3OC_6H_4$	- " -	70	0,41	ICA	
$p\text{-}NO_2C_6H_4$	$p\text{-}NO_2C_6H_4$	C_6H_5	- " -	70	0,13	ICA	
$CH_3\text{-}N{=}N\text{-}CH_3$			$n\text{-}C_7H_{16}$	-75 to +99	0,53–1,05/θ*	RRC	[277]
$CF_3\text{-}N{=}N\text{-}CF_3$			$(CH_3)_2C{=}C(CH_3)_2$	-30 to +95	$2{,}05 - \frac{2{,}3^*}{\theta}$	RRC	[291]

Table 38 (continued)

Initiator	Solvent	t^{o}	e	Method	Ref.
$[(CH_3)_2(COOC_2H_5)C]_2N_2$	C_6H_5Cl	70	0,64	ICA	[132]
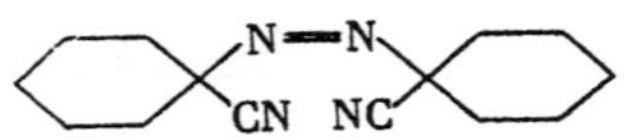	C_6H_5Cl	80	0,60 0,67	ICA	[136]
	- " -	100	0,70	ICA	[136]
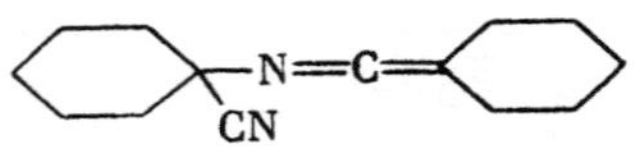	C_6H_5Cl	80	0,47	ICA	[136]
$(CH_3)_2C{=}C{=}N{-}C(CH_3)_2CN$	CCl_4	62,5	0,41	ICA	[174]

*Expressions are listed for $\log \frac{e}{1 - e}$.

LITERATURE CITED (PART I)

1. N. M. Émanuél, E. T. Denisov, and Z. K. Maizus, Chain Reactions of Liquid-Phase Hydrocarbon Oxidation. Moscow, "Nauka" (1965). [English translation: Liquid-Phase Oxidation of Hydrocarbons. New York, Plenum Press (1967).]

2. E. T. Denisov, Izv. Akad. Nauk SSSR, Ser. Khim., 1963, p. 2037.

3. E. T. Denisov, Izv. Akad. Nauk.SSSR, Otd. Khim. Nauk, 1960, p. 812.

4. V. Ya. Shlyapintokh, O. N. Karpukhin, L. M. Postnikov, I. V. Zakharov, A. A. Vichutinskii, and V. F. Tsepalov, Chemiluminescence Method for Studying Slow Chemical Processes. Moscow, "Nauka" (1966).

5. K. Nozaki and P. D. Bartlett, J. Am. Chem. Soc., 68:1686 (1946).

6. P. F. Hartman, H. G. Sellers, and D. Turnbull, J. Am. Chem. Soc., 69:2416 (1947).

7. C. E. H. Bawn and S. F. Mellish, Trans. Faraday Soc., 47:1216 (1951).

8. Kh. S. Bagdasar'yan and R. I. Milyutinskaya, Zh. Fiz. Khim., 27:420 (1953).

9. H. C. Bailey and G. W. Godin, Trans. Faraday Soc., 52:68 (1956).

10. W. Cooper, J. Chem. Soc., 1951, p. 3106.

11. Yu. A. Ol'dekop and G. S. Bylina, Vysokomolekul. Soedin., 6:1617 (1964).

12. V. I. Galibei, Candidate's Dissertation, Odessa State University (1965).

13. A. T. Blomquist and A. J. Buselli, J. Am. Chem. Soc., 73:3883 (1951).

14. K. Nozaki and D. P. Bartlett, J. Am. Chem. Soc., 68:2377 (1946).

15. C. G. Swein, W. H. Stockmayer, and J. T. Clarke, J. Am. Chem. Soc., 72:5426 (1950).

16. J. C. McGowan and T. Powell, J. Chem. Soc., 1960, p. 238.

17. W. R. Foster and G. H. Williams, J. Chem. Soc., 1962, p. 2862.

18. F. D. Greene, G. R. Van Norman, J. C. Cantrill, and R. D. Gilliom, J. Org. Chem., 25:1790 (1960).

19. A. V. Tobolsky and B. Baysal, J. Polymer Sci., 11:471 (1953).

20. J. E. Leffler, R. D. Faulkner, and C. C. Petropoulos, J. Am. Chem. Soc., 80:5435 (1958).

21. W. Honsberg and J. E. Leffler, J. Org. Chem., 26:733 (1961).

22. K. E. Russel, J. Am. Chem. Soc., 77:4814 (1955).

23. M. Levy, M. Steinberg, and M. Szwarc, J. Am. Chem. Soc., 76:5978 (1954).

24. H. J. Shine, J. A. Waters, and D. M. Hoffman, J. Am. Chem. Soc., 85:3613 (1963).

25. S. D. Ross and M. A. Fineman, J. Am. Chem. Soc., 73:2176 (1951).

26. O. J. Walker and G. L. E. Wold, J. Chem. Soc., 1937, p. 1132.

27. F. G. Edwards and F. R. Mayo, J. Am. Chem. Soc., 72:1265 (1950).

28. S. S. Ivanchev, Doctoral Dissertation, Odessa State University (1968).

29. R. C. Lamb, J. G. Pacifici, and P. W. Ayers, J. Am. Chem. Soc., 87:3928 (1956).

30. L. E. Redington, J. Polymer Sci., 3:503 (1948).

31. G. A. Razuvaev and L. M. Terman, Zh. Obshch. Khim., 30:2387 (1960).

32. J. E. Leffler and A. F. Wilson, J. Org. Chem., 25:424 (1960).

33. R. C. Lamb, F. F. Rogers, G. D. Dean, and F. W. Voight, J. Am. Chem. Soc., 84:2635 (1962).

34. N. Muramoto, T. Ochiai, O. Simamura, and M. Yoahida, Chem. Communications, 1968, p. 717.

35. V. I. Galibei, S. G. Erigova, S. S. Ivanchev, and A. I. Yurzhenko, Ukr. Khim. Zh., 33:191 (1967).

36. G. Greig and J. C. J. Thynne, Trans. Faraday Soc., 63:2196 (1967).

37. H. Hart and D. P. Wyman, J. Am. Chem. Soc., 81:4891 (1959).

38. R. C. Lamb and J. G. Pacifici, J. Am. Chem. Soc., 86:914 (1964).

39. H. H. Lau and H. Hart, J. Am. Chem. Soc., 81:4897 (1959).

40. F. Strain, W. E. Bissinger, W. R. Dial, H. Rudolf, B. J. DeWitt, H. C. Stevens, and J. H. Langston, J. Am. Chem. Soc., 72:1254 (1950).

41. Z. I. Kulitski, L. M. Terman, V. F. Tsepalov, and V. Ya. Shlyapintokh, Izv. Akad. Nauk SSSR, Ser. Khim., 1963, p. 253.

42. N. M. Kozyreva, R. I. Gramatikati, O. Ya. Fedotova, and G. S. Kolesnikov, Vysokomolekul. Soedin., 10B:132 (1968).

43. G. S. Bylina, A. P. El'nitskii, and Yu. A. Ol'dekop, Vysokomolekul. Soedin., 8:1386 (1966).

44. V. A. Dodonov and W. A. Waters, J. Chem. Soc., 1965, p. 2459.

45. T. W. Koenig and J. C. Martin, J. Org. Chem., 29:1520 (1964).

46. W. A. Pryor, D. M. Huston, T. R. Fiske, T. L. Rickering, and E. Ciuffarin, J. Am. Chem. Soc., 86:4237 (1964).

47. W. A. Pryor and E. P. Pultinas, J. Am. Chem. Soc., 85:133 (1963).

48. W. A. Pryor and G. L. Kaplan, J. Am. Chem. Soc., 86:4234 (1964).

49. W. A. Pryor, A. Lee, and C. E. Witt, J. Am. Chem. Soc., 86:4229 (1964).

50. J. H. Raley, F. F. Rust, and W. E. Vaughan, J. Am. Chem. Soc., 70:1336 (1948).

51. O. M. Mashnenko, A. E. Batog, N. I. Mironenko, and M. K. Romantsevich, Vysokomolekul. Soedin., 10B:444 (1968).

52. J. A. Offenbach and A. V. Tobolsky, J. Am. Chem. Soc., 79:278 (1957).

53. V. V. Zaitseva, A. I. Yurzhenko, and V. D. Enal'ev, Zh. Org. Khim., 4:1402 (1968).

54. W. A. Pryor, J. Phys. Chem., 67:519 (1963).

55. J. K. Allen and J. C. Bevington, Proc. Roy. Soc. (London), A262:271 (1961).

56. C. Walling and G. Metzger, J. Am. Chem. Soc., 81:5365 (1959).

57. S. M. Kavun and A. L. Buchachenko, Izv. Akad. Nauk SSSR, Ser. Khim., 1966, p. 1483.

58. E. S. Huyser and R. M. Van Scoy, J. Org. Chem., 33:3524 (1968).

59. I. A. Ivanchenko, V. V. Kharitonov, and E. T. Denisov, Vysokomolekul. Soedin., 11A:1622 (1969).

60. V. L. Antonovskii and L. D. Bezborodova, Zh. Fiz. Khim., 44:1224 (1970).

61. E. T. Denisov and V. M. Solyanikov, Neftekhimiya, 3:360 (1963).

62. E. T. Denisov and V. V. Kharitonov, Kinetika i Kataliz, 5:781 (1964).

63. M. S. Kharasch, A. Fono, and W. Nudenberg, J. Org. Chem., 16:105 (1951).

64. E. M. Dannenberg, M. E. Jordan, and H. M. Cole, J. Polymer Sci., 31:127 (1958).

65. B. M. E. Van der Hoff, Ind. Eng. Chem. Prod. Res. Develop., 2:273 (1963).

66. C. R. Parks and O. Lorenz, J. Polymer Sci., 50:287 (1961).

67. D. K. Thomas, J. Appl. Polymer Sci., 6:613 (1963).

68. K. Hummel, W. Sckeele, and K. H. Hillmer, Kautschuk Gummi, 14:171 (1961).

69. C. Walling and H. P. Waits, J. Phys. Chem., 71:2361 (1967).

70. R. R. Hiatt, Can. J. Chem., 42:985 (1964).

71. A. A. Turovskii, Candidate's Dissertation, L'vov (1968).

72. F. D. Greene, W. Adam, and G. A. Knudsen, J. Org. Chem., 31:2087 (1966).

73. G. P. Gerbert and V. P. Maslenikov, Zh. Obshch. Khim., 40:1105 (1970).

74. P. D. Bartlett and R. R. Hiatt, J. Am. Chem. Soc., 80:1398 (1958).

75. V. L. Antonovskii, L. D. Bezborodova, and M. É. Yasel'man, Zh. Fiz. Khim., 43:2286 (1969).

76. V. A. Puchin, T. I. Yurzhenko, O. E. Boisan, and L. M. Aparovich, Ukr. Khim. Zh., 33:478 (1967).

77. P. D. Bartlett and L. B. Gortler, J. Am. Chem. Soc., 85:1864 (1963).

78. P. D. Bartlett and D. M. Simons, J. Am. Chem. Soc., 82:1753 (1960).

79. J. P. Lorand, S. D. Chodroff, and R. W. Wallace, J. Am. Chem. Soc., 90:5266 (1968).

80. R. C. Fort and R. E. Franklin, J. Am. Chem. Soc., 90:5267 (1968).

81. M. Trachtman and J. G. Miller, J. Am. Chem. Soc., 84:4828 (1962).

82. V. A. Puchin, Doctoral Dissertation, Odessa State University (1969).

83. G. A. Nosaev and O. N. Romantsova, Vysokomolekul. Soedin., 8:16 (1966).

84. C. Ruchardt and H. Schwarzer, Chem. Ber., 99:1861, 1871, 1878 (1966).

85. D. R. Dixon and A. Pajaczkowski, Chem. Commun., 1966, p. 337.

86. C. Ruchardt, H. Boeck, and I. Ruthardt, Angew. Chem., 78:268 (1966).

87. P. D. Bartlett and H. Minato, J. Am. Chem. Soc., 85:1858 (1963).

88. A. V. Tobolsky and R. B. Mesrobian, Organic Peroxides, New York, Interscience (1954).

89. V. L. Antonovskii, L. D. Bezborodova, and M. É. Yasel'man, Zh. Fiz. Khim., 43:2281 (1969).

90. A. T. Blomquist and I. A. Berstein, J. Am. Chem. Soc., 73:5546 (1951).

91. H. Minato, H. Matsuzaki, and K. Miwa, Bull. Chem. Soc. Japan, 41:249 (1968).

92. T. H. Fisher and J. C. Martin, J. Am. Chem. Soc., 88:3382 (1966).

93. W. G. Bentrude and J. C. Martin, J. Am. Chem. Soc., 84:1561 (1962).

94. D. L. Tuleen, W. G. Bentrude, and J. C. Martin, J. Am. Chem. Soc., 85:1938 (1963).

95. C. Ruchardt and H. Boeck, Chem. Ber., 100:654 (1967).

95a. M. M. Martin, J. Am. Chem. Soc., 84:1986 (1962).

96. J. P. Lorand and P. D. Bartlett, J. Am. Chem. Soc., 88:3294 (1966).

97. P. D. Bartlett and R. E. Pinkock, J. Am. Chem. Soc., 82:1769 (1960).

98. P. D. Bartlett, E. P. Benzing, and R. E. Pincock, J. Am. Chem. Soc., 82:1762 (1960).

99. L. B. Gortler and M. D. Saltzman, J. Org. Chem., 31:3821 (1966).

100. L. M. Bobroff, L. B. Gortler, D. J. Sahn, and H. Wiland, J. Org. Chem., 31:2678 (1966).

101. V. D. Enal'ev, V. V. Zaitseva, et al. Vysokomolekul. Soedin., 7:275 (1965); V. V. Zaitseva, V. D. Enal'ev, and A. I. Yurzhenko, Ibid., 9A:1958 (1967).

102. G. P. Gerbert, V. P. Malennikov, and V. A. Shushunov, Zh. Obshch. Khim., 40:131 (1970).

102a. A. I. Prisyazhnyuk and S. S. Ivanchev, Zh. Org. Khim., 5:1172 (1969); E. G. Erigova, A. I. Prisyazhnyuk, and S. S. Ivanchev, Zh. Obshch. Khim., 38:2416 (1968).

103. R. R. Hiatt and W. M. J. Stachan, J. Org. Chem., 28:1893 (1963).

104. E. T. Denisov, Doctoral Dissertation, Moscow State University (1964).

105. G. E. Zaikov, Z. K. Maizus, and N. M. Émanuél', Izv. Akad., Nauk SSSR, Ser. Khim., 1968, p. 53.

106. Z. K. Maizus, I. P. Skibida, and N. M. Émanuél', Dokl. Akad., Nauk SSSR, 131:880 (1960).

107. N. V. Zolotova and E. T. Denisov, Vysokomolekul. Soedin., Vol. 12B, No. 9 (1970).

108. C. H. Bamford, W. G. Barb, A. D. Jenkins, and P. F. Onyon, The Kinetics of Vinyl Polymerization by Radical Mechanisms, London (1958).

109. J. W. Breitenbach and A. Shindler, Monatsh. Chem., 83:724 (1952).

110. J. P. Van Hook and A. V. Tobolsky, J. Am. Chem. Soc., 80:779 (1958).

111. C. E. H. Bawn and D. Verdin, Trans. Faraday Soc., 56:815 (1960).

112. D. C. Blackley and H. W. Melville, Makromol. Chem., 18/19:16 (1956).

113. G. S. Hammond, J. N. Sen, and C. E. Boozer, J. Am. Chem. Soc., 77:3244 (1955).

114. M. Talat-Erben and S. Bywater, J. Am. Chem. Soc., 77:3712 (1955).

115. C. G. Overberger, M. T. O'Shaugnessy, and H. Shalit, J. Am. Chem. Soc., 71:2661 (1949).

116. L. M. Arnett and J. H. Petersen, J. Am. Chem. Soc., 74:2027 (1952).

117. F. M. Lewis and M. S. Matheson, J. Am. Chem. Soc., 71:747 (1949).

118. R. C. Petersen, J. H. Markgraf, and S. D. Ross, J. Am. Chem. Soc., 83:3819 (1961).

119. K. Ziegler, V. Deparade, and W. Meye, Ann. Chem., 567:141 (1950).

120. S. N. Khanna, S. R. Chatterjee, U. S. Nandi, and S. R. Palit, Trans. Faraday Soc., 58:1827 (1962).

121. J. Betts and J. C. Robb, Trans. Faraday Soc., 64:2402 (1968).

122. A. L. Aleksandrov, T. I. Sapacheva, and V. F. Shuvalov, Izv. Akad. Nauk SSSR, Ser. Khim., 1969, p. 955.

123. A. L. Aleksandrov and E. T. Denisov, Kinetika i Kataliz, 10:912 (1969).

124. M. Prochazka, O. Ryba, and D. Lim, Collection Czech. Chem. Commun., 33:3387 (1968).

125. S. G. Cohen, S. J. Groszos, and D. B. Sparrow, J. Am. Chem. Soc., 72:3947 (1950).

126. S. Seltzer, J. Am. Chem. Soc., 83:2625 (1961).

127. S. G. Cohen and Chi Hua Wang, J. Am. Chem. Soc., 75:5504 (1953).

128. S. G. Cohen and Chi Hua Wang, J. Am. Chem. Soc., 77:2457 (1955).

129. S. F. Nelsen and P. D. Bartlett, J. Am. Chem. Soc., 88:137 (1966).

130. J. R. Shelton, Chin Kuo Liang, and P. Kovacic, J. Am. Chem. Soc., 90:354 (1968).

131. C. G. Overberger and H. Biletch, J. Am. Chem. Soc., 73:4880 (1951).

132. G. S. Hammond and J. R. Fox, J. Am. Chem. Soc., 86:1918 (1964).

133. C. G. Overberger, W. F. Hale, M. B. Berenbaum, and A. B. Finestone, J. Am. Chem. Soc., 76:6185 (1954).

134. C. G. Overberger, H. Biletch, A. B. Finestone, J. Lilker, and J. Herbert, J. Am. Chem. Soc., 75:2078 (1953).

135. A. F. Barton, J. C. Bevington, and A. Wahid, Makromol. Chem., 67:195 (1963).

136. Chin-Hua S. Wu, G. S. Hammond, and J. M. Wright, J. Am. Chem. Soc., 82:5386 (1960).

137. C. G. Overberger and A. Lebovits, J. Am. Chem. Soc., 76:2722 (1954).

138. E. A. S. Cavell and A. C. Meeks, Makromol. Chem., 108:304 (1967).

139. G. S. Hammond and R. C. Neuman, J. Am. Chem. Soc., 85:1501 (1963).

140. H. A. Ory, J. Phys. Chem., 65:571 (1961).

141. S. Seltzer and F. T. Dunne, J. Am. Chem. Soc., 87:2628 (1965).

142. C. G. Overberger and A. V. Diguilio, J. Am. Chem. Soc., 81:2154 (1959).

143. B. W. Brooks, F. S. Dainton, and K. J. Ivin, Trans. Faraday Soc., 61:1437 (1965).

144. S. G. Cohen and Chi Hua Wang, J. Am. Chem. Soc., 77:3628 (1955).

145. M. G. Adler and J. E. Leffler, J. Am. Chem. Soc., 76:1425 (1954).

146. G. L. Davies, D. H. Hey, and G. H. Williams, J. Chem. Soc., 1956, p. 4397.

147. S. Solomon, Chi Hua Wang, and S. G. Cohen, J. Am. Chem. Soc., 79:4104 (1957).

148. M. D. Cohen, J. E. Leffler, and L. M. Barbato, J. Am. Chem. Soc., 76:4169 (1954).

149. C. G. Overberger and J. P. Anselme, J. Am. Chem. Soc., 86:658 (1964).

150. S. G. Cohen, S. Hsiao, E. Saklad, and Chi Hua Wang, J. Am. Chem. Soc., 79:4400 (1957).

151. C. G. Overberger and J. G. Lombardino, J. Am. Chem. Soc., 80:2317 (1958).

152. C. G. Overberger and I. Tashlick, J. Am. Chem. Soc., 81:217 (1959).

153. S. G. Mylonakis and S. Seltzer, J. Am. Chem. Soc., 90:5487 (1968).

154. K. Ziegler, P. Orth, and K. Weber, Ann. Chem., 504:131 (1933).

155. K. Ziegler and W. Deparade, Ann. Chem. 567:123 (1950).

156. L. R. Mahoney, J. Am. Chem. Soc., 88:3035 (1966).

157. G. E. Zaikov, Z. K. Maizus, and N. M. Émanuél', Dokl. Akad. Nauk SSSR, 140:405 (1961).

158. E. Borsig, M. Lasar, and M. Čapla, Collection Czech. Chem. Commun., 32:4289 (1967).

159. K. Ziegler, W. Deparade, and H. Kuehlhorn, Ann. Chem., 567:151 (1950).

160. F. Tyudesh and M. Azori, Acta Chim. Acad. Sci. Hung., 34:61 (1962).

161. C. K. Cain and F. Y. Wiselogle, J. Am. Chem. Soc., 62:1163 (1940).

162. K. E. Russel, J. Am. Chem. Soc., 77:3487 (1955).

163. W. H. Saunders and J. C. Ware, J. Am. Chem. Soc., 80:3328 (1958).

164. D. S. Breslow, T. J. Prosser, A. F. Marcantonio, and C. A. Genge, J. Am. Chem. Soc., 89:2384 (1967).

165. W. K. Wilmarth and N. Schwartz, J. Am. Chem. Soc., 77:4551 (1955).

166. D. Bethell and D. Whittaker, J. Chem. Soc., B, 1966, p. 778.

167. D. Bethell, J. D. Callister, and D. Whittaker, Proc. Chem. Soc. (London), 1964, p. 192.

168. D. Bethell, D. Whittaker, and J. D. Callister, J. Chem. Soc., 1965, p. 2466.

169. J. Murgulesku and T. Oncescu, J. Chim. Phys., 58:508 (1961).

170. R. Huisgen, H. J. Sturm, and G. Binsch, Chem. Ber., 97:2864 (1964).

171. D. H. Hey, J. Stuart-Webb, and G. H. Williams, J. Chem. Soc., 1952, p. 4657.

172. H. Van Zwet and E. C. Kooyman, Rec. Trav. Chim., 86:1143 (1967).

173. S. Y. Hong and J. E. Baldwin, J. Org. Chem., 24:3787 (1959).

174. G. S. Hammond, O. D. Trapp, R. T. Keys, and D. L. Neff, J. Am. Chem. Soc., 81:4878 (1959).

175. D. J. Williams and R. Krelick, J. Am. Chem. Soc., 89:3408 (1967).

176. F. Daniels and E. H. Johnston, J. Am. Chem. Soc., 43:53 (1921).

177. R. H. Lueck, J. Am. Chem. Soc., 44:757 (1922).

178. H. Eyring and F. Daniels, J. Am. Chem. Soc., 52:1472 (1930).

179. G. M. Nazin, G. B. Manelis, and F. I. Dubovitskii, Usp. Khim., 37:1443 (1968).

180. H. P. Marshall, F. G. Borgardt, and P. Noble, J. Phys. Chem., 72:1513 (1968).

181. H. P. Marshall, F. G. Borgardt, and P. Noble, J. Phys. Chem., 69:25 (1965).

182. A. J. B. Robertson, Trans. Faraday Soc., 44:677 (1948).

183. Yu. Ya. Maksimov, Theory of Explosives. Tr. Mosk. Khim.-Tekh. Inst., No. 53, p. 73 (1967).

184. A. Robertson, Trans. Faraday Soc., 45:85 (1949).

185. B. A. Lur'e and B. S. Svetlov, Theory of Explosives. Tr. Mosk. Khim.-Tekh. Inst., No. 53, pp. 40, 51 (1967).

186. B. S. Svetlov, Kinetika i Kataliz, 2:38 (1961).

187. K. K. Andreev, Thermal Decomposition and Combustion of Explosives. Moscow, Gosénergoizdat (1957), p. 75.

188. B. A. Lur'e and B. S. Svetlov, in: Theory of Explosives. Oborongiz (1963), p. 281.

189. B. S. Svetlov and B. A. Lur'e, Zh. Fiz. Khim., 37:1979 (1963).

190. C. E. H. Bawn and F. J. Whitby, Discussions Faraday Soc., 2:228 (1947).

191. R. A. Jackson, J. Chem. Soc., 1963, p. 5284.

192. F. S. D'yachkovskii, N. E. Khrushch, and A. E. Shilov, Kinetika i Kataliz, 9:1006 (1968).

193. Yu. A. Aleksandrov and M. N. Spiridonova, Dokl. Akad. Nauk SSSR, 182:1319 (1968).

194. R. M. Keefer and L. J. Andrews, J. Am. Chem. Soc., 79:4348 (1957).

195. R. M. Keefer and L. J. Andrews, J. Am. Chem. Soc., 80:277 (1958).

196. R. M. Keefer and L. J. Andrews, J. Am. Chem. Soc., 81:5329 (1959).

197. L. J. Andrews, L. J. Spears, and R. M. Keefer, J. Am. Chem. Soc., 86:687 (1964).

198. L. J. Andrews and R. M. Keefer, J. Am. Chem. Soc., 81:4218 (1959).

199. S. S. Ivanchev and S. G. Erigova, Dokl. Akad. Nauk SSSR, 183:602 (1968).

200. A. E. Nicholson and R. G. Norrish, Discussions Faraday Soc., 22:97 (1956).

201. A. H. Ewald, Discussions Faraday Soc., 22:138 (1956).

202. A. F. Nicholson and R. G. W. Norrish, Discussions Faraday Soc., 22:104 (1956).

203. C. Walling and J. Pellon, J. Am. Chem. Soc., 79:4786 (1957).

204. A. S. Okunevich, S. M. Samoilov, and V. N. Monastyrskii, Zh. Fiz. Khim., 43:2265 (1969).

204a. R. E. Neuman and J. V. Behav, J. Am. Chem. Soc., 89:4549 (1967).

205. M. C. Ford and W. A. Waters, J. Chem. Soc., 1951, p. 1851.

206. J. C. Martin and S. A. Dombchik, Advances in Chemistry Series, 75:269 (1968).

207. F. P. Lossing and A. W. Tickner, J. Chem. Phys., 20:907 (1952).

208. M. Szwarc and J. S. Roberts, J. Chem. Phys., 18:561 (1950); 19:698 (1951).

209. A. Wasserman, Monatsh. Chem., 83:543 (1952).

210. A. Wasserman, Trans. Faraday Soc., 34:128 (1938).

211. R. A. Fairclough and C. N. Hinshelwood, J. Chem. Soc., 1938, p. 236.

212. M. Ahmed and J. Hamer, J. Org. Chem., 31:2829 (1966).

213. B. Raistrick, R. H. Sapiro, and D. M. Newitt, J. Chem. Soc., 1939, p. 1761.

214. C. Walling and H. J. Schugar, J. Am. Chem. Soc., 85:607 (1963).

215. C. Walling and J. Peisach, J. Am. Chem. Soc., 80:5819 (1958).

216. D. G. Pobedimskii, A. L. Buchachenko, and M. B. Neiman, Zh. Fiz. Khim., 42:1436 (1968).

217. N. V. Zolotova and E. T. Denisov, Izv. Akad. Nauk SSSR, Ser. Khim., 1966, p. 767.

218. N. M. Beileryan, F. O. Karapetyan, and O. A. Chaltykhan, Armen. Khim. Zh., 19:828 (1966).

219. N. M. Beileryan, F. O. Karapetyan, and O. A. Chaltykyan, Dokl. Akad. Nauk ArmSSR, 43:108 (1966).

220. O. A. Chaltykyan, E. N. Atanasyan, A. A. Varkisyan, G. A. Marmaryan, and D. S. Gaibakyan, Zh. Fiz. Khim., 32:2601 (1958).

221. O. A. Chaltykyan, N. M. Beileryan, and É. R. Sarukhanyan, Izv. Akad. Nauk ArmSSR, 17:21 (1964).

222. É. R. Sarukhanyan, N. M. Beileryan, and O. A. Chaltykyan, Dokl. Akad. Nauk ArmSSR, 37:285 (1964).

223. O. A. Chaltykyan, É. R. Sarukhanyan, and N. M. Beileryan, Dokl. Akad. Nauk ArmSSR, 44:213 (1967).

224. R. I. Milyutinskaya and Kh. S. Bagdasar'yan, Zh. Fiz. Khim., 34:405 (1960).

225. K. O. Driskol and A. Tobolsky, J. Colloid Sci., 11:244 (1956).

226. M. Imoto and S. Choe, J. Polymer Sci., 15:475 (1955).

227. F. Hrabak and N. Vaček, Collection Czech. Chem. Commun., 30:573 (1965).

228. S. Kashino, Y. Mugino, and S. Hasegawa, Bull. Chem. Soc., Japan, 40:2004 (1967).

229. O. A. Chaltykyan, E. N. Atanasyan, N. M. Beileryan, and G. A. Marmaryan, Zh. Fiz. Khim., 33:36 (1959).

230. V. L. Antonovskii and L. D. Bezborodova, Zh. Fiz. Khim., 42:351 (1968).

231. N. M. Beileryan, S. K. Grigoryan, and O. A. Chaltykyan, Izv. Akad. Nauk ArmSSR, 17:245 (1964).

232. O. A. Chaltykyan, S. K. Grigoryan, and N. M. Beileryan, Izv. Akad. Nauk ArmSSR, 18:133 (1965).

233. S. K. Grigoryan, Z. Z. Meliksetyan, and N. M. Beileryan, Arm. Khim. Zh., 20:333 (1967).

234. N. M. Beileryan, O. A. Chaltykyan, and G. A. Esayan, Arm. Khim. Zh., 21:7 (1968).

235. E. T. Denisov and L. N. Denisova, Dokl. Akad. Nauk SSSR, 157:907 (1964).

236. V. L. Antonovskii, E. T. Denisov, and L. V. Solntseva, Kinetika i Kataliz, 6:815 (1965).

237. N. I. Solomko, V. F. Tsepalov, and A. I. Yurzhenko, Kinetika i Kataliz, 9:766 (1968).

238. C. Walling and L. Heaton, J. Am. Chem. Soc., 87:38 (1965).

239. A. V. Tobolsky and L. R. Matlack, J. Polymer Sci., 55:49 (1961).

240. E. T. Denisov, Zh. Fiz. Khim., 38:2085 (1964).

241. A. Chauvel, G. Clement, and J. C. Balaceanu, Bull. Soc. Chim. France, 1963, p. 2025.

242. I. P. Shevchuk, Candidate's Dissertation, L'vov (1968).

243. L. Bateman, Quart. Rev. (London), 8:147 (1954).

244. E. T. Denisov, Dokl. Akad. Nauk SSSR, 146:394 (1962).

245. E. T. Denisov, Zh. Fiz. Khim., 37:1896 (1963).

246. E. T. Denisov, V. V. Kharitonov, and E. N. Raspopova, Kinetika i Kataliz, 5:981 (1964).

247. E. T. Denisov and L. N. Denisova, Izv. Akad. Nauk SSSR, Ser. Khim., 1963, p. 1731.

248. L. G. Privalova, Z. K. Maizus, and N. M. Émanuél', Dokl. Akad. Nauk SSSR, 161:1135 (1965).

249. V. L. Antonovskii, E. T. Denisov, and L. V. Solntseva, Kinetika i Kataliz, 7:409 (1966).

250. G. E. Zaikov, Z. K. Maizus, and N. M. Émanuél',Izv. Akad. Nauk SSSR, Ser. Khim., 1968, p. 256.

251. L. N. Denisova, E. T. Denisov, and D. L. Metelitsa, Zh. Fiz. Khim., 44:1670 (1970).

252. L. N. Denisova, E. T. Denisov, and D. I. Metelitsa, Izv. Akad. Nauk SSSR, Ser. Khim., 1969, p. 1657.

253. L. N. Denisova and E. T. Denisov, Izv. Akad. Nauk SSSR, 1965, p. 1702.

254. G. E. Zaikov, Z. K. Maizus, and N. M. Émanuél', Neftekhimiya, 4:91 (1964).

255. L. N. Denisova and E. T. Denisov, Kinetika i Kataliz, 10:1244 (1969).

256. F. S. D'yachkovskii and A. E. Shilov, Kinetika i Kataliz, 4:919 (1963).

257. F. S. D'yachkovskii, A. E. Shilov, and L. I. El'terman, Kinetika i Kataliz, 4:644 (1963).

258. F. S. D'yachkovskii and A. E. Shilov, Usp. Khim., 35:699 (1966).

259. F. S. D'yachkovskii, N. E. Khrushch, and A. E. Shilov, Kinetika i Kataliz, 8:1230 (1967).

260. F. R. Mayo, J. Am. Chem. Soc., 75:6133 (1953).

261. E. T. Denisov, Dokl. Akad. Nauk SSSR, 130:1055 (1960).

262. E. T. Denisov, Dokl. Akad. Nauk SSSR, 141:131 (1961).

263. E. T. Denisov, Kinetika i Kataliz, 4:53 (1963).

264. D. J. Carlsson and J. C. Robb, Trans. Faraday Soc., 62:3403 (1966).

265. Z. K. Maizus, N. M. Émanuél', and V. N. Yakovleva, Dokl. Akad. Nauk SSSR, 143:366 (1962).

266. J. Frank and E. Rabinowich, Trans. Faraday Soc., 30:120 (1934).

267. L. Herk, M. Feld, and M. Szwarc, J. Am. Chem. Soc., 83:2998 (1961).

268. Kh. S. Bagdasar'yan, Zh. Fiz. Khim., 41:1679 (1967).

269. N. V. Zolotova, E. T. Denisov, S. S. Ivanchev, and T. G. Podlipnaya, Vysokomolekul. Soedin., 11B:761 (1969).

270. G. P. Gladyshev and K. M. Gibov, Polymerization with High Degrees of Conversion and Methods of Studying This Process, Alma-Ata, "Nauka" (1968).

271. E. T. Denisov, S. S. Ivanchev, L. A. Zborshchik, and N. V. Zolotova, Izv. Akad. Nauk SSSR, Ser. Khim., 1968, p. 1500.

272. L. M. Andronov and G. E. Zaikov, Kinetika i Kataliz, 8:270 (1967).

273. H. P. Waits and G. S. Hammond, J. Am. Chem. Soc., 86:1911 (1964).

274. C. E. Boozer, G. S. Hammond, C. E. Hamilton, and J. N. Sen, J. Am. Chem. Soc., 77:3233 (1955).

275. E. C. Horswill, J. A. Howard, and K. U. Ingold, Can. J. Chem., 44:985 (1966).

276. L. R. Mahoney, J. Am. Chem. Soc., 88:3035 (1966).

277. S. Kodama, S. Fujita, J. Takeishi, and O. Toyama, Bull. Chem. Soc. Japan, 39:1009 (1966).

278. G. Henrici-Olive and S. Olive, Makromol. Chem., 58:188 (1962).

279. J. C. Bevington and D. E. Evans, Trans. Faraday Soc., 55:1777 (1959).

280. J. H. Bradbury and H. W. Melville, Proc. Roy. Soc., (London), A222:456 (1954).

281. L. M. Arnett and J. H. Peterson, J. Am. Chem. Soc., 74:2031 (1952).

282. J. A. Howard and K. U. Ingold, Can. J. Chem., 42:1044 (1964).

283. J. C. Bevington, Trans. Faraday Soc., 51:1392 (1955).

284. J. C. Bevington, H. W. Melville, and R. P. Taylor, J. Polymer Sci., 14:463 (1954).

285. J. C. Bevington, International Symposium on Macromolecular Chemistry (Milan) (1954), p. 3.

286. J. C. Bevington, J. H. Bradbury, and G. M. Burnett, J. Polymer Sci., 12:469 (1954).

287. V. F. Tsepalov and E. M. Kotsyuba, Izv. Akad. Nauk SSSR, Ser. Khim., 1967, p. 1220.

288. M. S. Khloplyankina, A. L. Buchachenko, A. G. Vasil'eva, and M. B. Neiman, Izv. Akad. Nauk SSSR, Ser. Khim., 1965, p. 1296.

289. G. Ayrey and C. G. Moore, J. Polymer Sci., 36:41 (1959).

290. K. F. O'Driscoll and P. J. White, J. Polymer Sci., A3:283 (1965).

291. G. E. Owen, J. M. Pearson, and M. Szwarc, Trans. Faraday Soc., 60:564 (1964).

PART TWO

REACTIONS OF FREE ATOMS AND RADICALS

CHAPTER IV

METHODS FOR MEASURING RATE CONSTANTS OF RADICAL REACTIONS

§1. Measurement of Relative Rate Constants of Radical Reactions

Atoms and free radicals are extremely active and enter into reaction very quickly with molecules and with each other (stable free radicals represent an exception). Therefore, the concentration of radicals in a system, when they are being formed continuously and reacting, is very small. Direct measurement of radical concentration, for instance by the EPR method, is possible only with specially selected systems and conditions. In the majority of cases, the concentrations of free radicals cannot be measured. This has led to a widespread use of methods that measure the relative rate constants of radical reactions. These methods are based on the ability of radicals to react in parallel, in two or in several directions; i.e., all these methods are, as a matter of fact, different variants of the method of competing reactions. The errors in measuring relative rate constants usually amount to 2 to 5%.

a. Measurement of Rate of Unbranched Chain Reaction (RUCR)

In the following paragraphs we cite examples of unbranched chain reactions in the liquid phase.

Polymerization reaction:

$$\text{Initiator} \longrightarrow r\cdot$$

$$r\cdot + CH_2{=}CHX \longrightarrow r{-}CH_2{-}\dot{C}HX$$

$$\sim CH_2{-}\dot{C}HX + CH_2{=}CHX \longrightarrow \sim CH_2{-}CHX{-}CH_2{-}\dot{C}HX$$

$$\sim CH_2{-}\dot{C}HX + \dot{C}HX{-}CH_2\sim \longrightarrow \sim CH_2{-}CHX{-}CHX{-}CH_2\sim$$

$$\searrow \sim CH{=}CHX + CH_2XCH_2\sim$$

Oxidation reaction:

$$r\cdot + RH \rightarrow rH + R\cdot$$
$$R\cdot + O_2 \rightarrow RO_2\cdot$$
$$RO_2\cdot + RH \rightarrow ROOH + R\cdot$$
$$RO_2\cdot + RO_2 \rightarrow ROOR + O_2$$

Halogenation with trichlorobromomethane:

$$r\cdot + CCl_3Br \rightarrow rBr + CCl_3\cdot$$
$$CCl_3\cdot + RH \rightarrow HCCl_3 + R\cdot$$
$$R\cdot + BrCCl_3 \rightarrow RBr + CCl_3\cdot$$
$$CCl_3\cdot + CCl_3\cdot \rightarrow CCl_3-CCl_3$$

In a liquid-phase chain reaction in the absence of any inhibitor, chain breaking occurs by reaction between two free radicals. With a constant rate of radical formation (from the starting materials, from an initiator, photochemical, or radiation-induced), in the stationary regime of the reaction, $W_i = 2\,k_t\,[R^\bullet]^2$, where $R^\bullet$ is the radical carrying the chain. The rate of the chain reaction is

$$W_\nu = k_p[M][R\cdot] = \frac{k_p}{\sqrt{2\,k_t}}\,[M]\sqrt{W_i}$$

where M is the reacting substance. The ratio of constants $k_p/\sqrt{2k_t}$ is found, measuring the rate of the chain reaction W_ν with a known initiation rate:

$$\frac{k_p}{\sqrt{2\,k_t}} = \frac{W_\nu}{[M]\sqrt{W_i}}\ .$$

For more details on this method, see [1-3]. The methods of measurement were described in Chapter I.

b. Measurement of Degree of Polymerization (DP)

In processes of radical polymerization, growth of the polymer chain proceeds as a result of addition of monomer to a polymer radical:

$$\sim CH_2-\dot{C}HX + CH_2{=}CHX \xrightarrow{k_p} \sim CH_2-\dot{C}HX.$$

A polymer radical will not only add to the monomer, but also will enter into other reactions--reactions of substitution with monomer, solvent, initiator, and the polymer being formed:

$$\sim\dot{C}H_2{-}CHX + RH \xrightarrow{k} \sim CH_2{-}CH_2X + R\cdot$$

$$R\cdot + CH_2{=}CHX \xrightarrow{k} \sim RCH_2{-}\dot{C}HX$$

As a result of such chain transfer reactions, the chain reaction is prolonged, but the growth of the polymer chain is terminated. The relative rate constant for the chain transfer reaction is determined by measuring the mean degree of polymerization $\bar{P}$ for various [RH]:

$$\frac{1}{\bar{P}} = \frac{1}{\bar{P}_0} + C_s \frac{[RH]}{[M]} \quad ,$$

where $C_s = k/k_p$ is the constant for chain transfer, [RH] is the concentration of the substance accomplishing the chain transfer, and $\bar{P}_0$ is the mean degree of polymerization with [RH] = 0.

If chain transfer is accomplished through the monomer, the formula for $\bar{P}$ is

$$\frac{1}{\bar{P}} = C_M + \frac{2k_t'' + k_t'}{k_p^2\,[M]^2} \quad ,$$

where $C_M = k/k_p$, and k_t' and k_t'' are the rate constants for recombination and disproportionation of the macroradicals. For more detail on the method, see [2, 3].

c. Method of Inhibited Chain Reaction (CINH)

Many compounds are inhibitors of chain reactions. For example, phenols and aromatic amines retard oxidation reactions, and quinones and nitro compounds retard radical polymerizations. The retarding action of inhibitors comes about by breaking chains through the reaction of free radicals with inhibitor molecules. Methods have been developed for determining the relative rate constants of radical reactions with inhibitor molecules [1, 2, 17].

The reactivity of an inhibitor with respect to a chain-carrying radical can be judged by the dependence of the chain reaction rate on the concentration of inhibitor introduced. If the inhibitor is added in sufficiently large concentration that the chain breaking takes place almost entirely through the reaction of radicals with the inhibitor, then $W_i = fk_{InH}[R^\bullet][InH]$, where k_{InH} and f are, respectively, the rate constant and the stoichiometric coefficient for the reaction of the inhibitor with the free radical $R^\bullet$; then, the rate of the chain reaction is

$$W_\nu = k_p [M][R\cdot] = \frac{k_p[M]\, W_i}{fk_{InH}\,[InH]} .$$

The ratio k_p/k_{InH} can be found from the relation between W_ν and $[InH]^{-1}$, provided that f and W_i are known. Since under such conditions the inhibitor is consumed at a constant rate, i.e.,

$$[InH] = [InH]_0 - \frac{W_i}{f} t ,$$

the ratio k_{InH}/k_p can be determined from the timewise variation in rate of the inhibited chain reaction:

$$\frac{1}{W_\nu} = \frac{k_{InH}}{k_p} \frac{f[InH]_0}{W_i[M]} - \frac{k_{InH}}{k_p} \frac{1}{[M]} t .$$

If, in the reaction under study, chains are broken both through reaction of radicals with inhibitor and through reaction of radicals with each other, the relation between W_ν and [InH] becomes more complex:

$$\frac{W_i}{W_\nu}\left(1 - \frac{W_\nu^2}{W_{\nu_0}^2}\right) = \frac{fk_{InH}[InH]}{k_p[M]} ,$$

where W_{ν_0} is the reaction rate in the absence of inhibitor.

The rate of a chain reaction may be measured by the consumption of starting material (e.g., oxygen absorption in oxidation), by the volume change of the system (dilatometrically in polymerization), by the rate of heat evolution, or by the intensity of chemiluminescence (in oxidation).

The ratio of constants k_{InH}/k_p can be determined if one measures both the inhibitor consumption and the starting material consumption (or accumulation of the reaction product

P); with sufficiently long chains, the following formula is obtained.

$$\ln\frac{[\mathrm{InH}]_0}{[\mathrm{InH}]} = \frac{k_{\mathrm{InH}}}{k_p}\,\frac{[P]}{[M]}\ .$$

The ratio of constants $k_{\mathrm{InH}}/\sqrt{2\,k_t}$ can be determined from the experimentally determined relation between inhibitor consumption rate and inhibitor concentration:

$$W_{\mathrm{InH}} = k_{\mathrm{InH}}\,[\mathrm{InH}][\mathrm{R}\cdot]$$

$$W_i = f\,k_{\mathrm{InH}}\,[\mathrm{InH}][\mathrm{R}\cdot] + 2\,k_t\,[\mathrm{R}\cdot]^2$$

$$W_{\mathrm{InH}} = \frac{f\,k^2_{\mathrm{InH}}\,[\mathrm{InH}]^2}{4\,k_t}\left(\sqrt{1 + 8\,W_i k_t / f^2 k^2_{\mathrm{InH}}\,[\mathrm{InH}]^2} - 1\right)$$

$$\frac{x}{\sqrt{1-x}} = \frac{f\,k_{\mathrm{InH}}\,[\mathrm{InH}]}{\sqrt{2\,k_t W_i}}\ ,$$

where $x = f\,W_{\mathrm{InH}}/W_i$.

If the inhibitor is introduced into the system in such a small concentration that $f\,k_{\mathrm{InH}} \ll 2\,k_t\,[\mathrm{R}^\bullet]$, the inhibitor will be consumed in accordance with first-order reaction kinetics:

$$-\frac{d\,[\mathrm{InH}]}{dt} = k_{\mathrm{InH}}\,[\mathrm{InH}]\,[\mathrm{R}\cdot] \cong \frac{k_{\mathrm{InH}}}{\sqrt{2\,k_t}}[\mathrm{InH}]\sqrt{W_i}$$

$$[\mathrm{InH}] = [\mathrm{InH}]_0\,e^{-kt}\ ,\ k \cong \frac{k_{\mathrm{InH}}}{\sqrt{2\,k_t}}\sqrt{W_i}\ .$$

d. Method of Competing Reactions (MCR)

As an example of the MCR, we may cite the method developed by Szwarc [4] for determining relative k's for reactions of methyl radicals. Methyl radicals may abstract an H atom from a molecule to form methane, or may add to a molecule with multiple bonds to form products other than methane:

$$CH_3\cdot + RH \xrightarrow{k_a} CH_4 + R\cdot$$

$$CH_3\cdot + M \xrightarrow{k_M} CH_3M\cdot$$

The method of determining k_M/k_a is as follows: In a solution of isooctane (RH) and the compound "M" under investigation (for example, benzene) acetyl peroxide is decomposed as a source of methyl radicals:

$$CH_3COOOCOCH_3 \longrightarrow \overline{2\ CH_3COO\cdot} \longrightarrow \overline{2\ CH_3\cdot} + 2\ CO_2$$

$$\overline{2\ CH_3\cdot} \longrightarrow C_2H_6$$

$$\overline{2\ CH_3\cdot} \longrightarrow CH_3\cdot + CH_3\cdot$$

The amounts of CO_2 and CH_4 that are formed are measured in the absence of M (subscript o) and in the presence of M (subscript M). The ratio of constants k_M/k_a is found from the relation

$$\frac{k_M}{k_a} = \left(\frac{([CH_4]/[CO_2])_o}{([CH_4]/[CO_2])_M} - 1\right)\frac{[RH]}{[M]}.$$

In the radiolysis of a hydrocarbon RH in the presence of a substance M, the following reactions take place [5]:

$$RH \rightsquigarrow H\cdot + R\cdot \ (G_i)$$

$$H\cdot + RH \xrightarrow{k_a} H_2 + R\cdot$$

$$H\cdot + M \xrightarrow{k_M} HM\cdot$$

$$H\cdot + M \xrightarrow{k_{Ma}} H_2 + M'\cdot$$

The radiochemical yield of H_2 is measured in the presence of M (G_M) and in the absence of M (G_o). The radiochemical yield of H· from RH is designated as G_i. The ratio of constants k_a/k_M is found from the relation

$$\frac{1}{G_o - G_M} = \frac{1}{G_i}\,\frac{k_a}{k_M}\,\frac{[RH]}{[M]} + \frac{1}{G_i}\left(\frac{k_{Ma}}{k_M} + 1\right).$$

A competing-reaction method using deuterium has been proposed by Bagdasar'yan and Milyutinskaya [6]. Benzoyl peroxide labeled with deuterium is decomposed in CCl_4 solution in the presence of an aromatic hydrocarbon such as benzene. Phenyl radicals in this system enter into the following reactions:

$$d\text{-}C_6H_5^{\cdot} + CCl_4 \xrightarrow{k_o} d\text{-}C_6H_5Cl$$

$$d\text{-}C_6H_5^{\cdot} + ArH \xrightarrow{k_a} d\text{-}C_6H_6 + Ar^{\cdot}$$

$$d\text{-}C_6H_5^{\cdot} + ArH \xrightarrow{k_M} d\text{-}C_6H_5ArH.$$

At the end of the reaction, by the method of isotopic dilution, C_6H_5Cl and C_6H_6 are recovered and their deuterium contents are determined. The ratio of constants is found from the relationships

$$\frac{\Delta[C_6H_5ArH]}{\Delta[C_6H_5Cl]} = \frac{k_M}{k_o}\frac{[ArH]}{[CCl_4]}$$

$$\frac{\Delta[C_6H_6]}{\Delta[C_6H_5Cl]} = \frac{k_a}{k_o}\frac{[ArH]}{[CCl_4]}$$

e. Tritium Isotope Method (TIM)

For measuring the relative reactivity of individual C-H bonds, Berezin [7, 8] developed a method using tritium. In a solution of the substance under investigation, labeled with tritium at one of the C-H bonds, acetyl peroxide is decomposed as a source of methyl radicals. The methyl radicals abstract H from all C-H bonds of the substance with an over-all rate constant $k_{\sigma H}$, and they abstract tritium from the C_j-T bond with a rate constant k_{jT}. Specific activities are measured for the starting material (I_{RH}) and for the methane that is formed (I_{CH_4}).

The ratio of constants is found from the relation

$$\frac{k_{jT}}{k_{\sigma H}} = \frac{I_{CH_4}}{I_{RH}}.$$

For measuring the relative rate constants for reactions of $CH_3^{\bullet}$ with the C-H bonds of various molecules, this method is used in a somewhat different variant. To the test substance RH, a standard hydrocarbon R_oT labeled with tritium (e.g., n-heptane-4T) is added in such an amount that the formation of CH_4 by reaction with C-H bonds of the standard hydrocarbon can be neglected.

Experiments are conducted in the presence of RH (measuring I_{CH_4}) and in the absence of RH (measuring $I^o_{CH_4}$). The ratio of rate constants k_{RH}/k_{R_oH} is found from the formula

$$\frac{k_{RH}}{k_{R_oH}} \quad \frac{I^o_{CH_4} - I_{CH_4}}{I_{CH_4}} \quad \frac{[R_oH]}{[RH]}.$$

§2. Measurement of Absolute Rate Constants of Reactions of Atoms and Radicals

At this point in time, chemical kinetics has at its disposal many different methods that make it possible to measure free radical concentrations or lifetimes, and to compute absolute rate constants of elementary reactions. These methods are described in detail in the literature, and hence only the essential features of each method will be set forth briefly in this section. The errors in measuring absolute reaction rate constants may reach 20 to 40%.

a. Electron Paramagnetic Resonance Method (EPR)

This method is based on the principle that a particle with an unpaired electron when placed in a magnetic field is capable of absorbing electromagnetic radiation. The absorption takes place at a resonance frequency ν that satisfies the condition $\nu = g\beta H/h$, where H is the intensity of the magnetic field, β is the Bohr magneton, h is the Planck constant, and g is a constant (g-factor).

The intensity of absorption of electromagnetic energy under otherwise equal conditions is directly proportional to the concentration of particles with unpaired electrons.

Hence, the concentrations of free radicals can be measured by the EPR method.

Knowing the rate of radical formation and having measured their concentration, one can determine the rate constant for the reaction of radicals with each other. Under stationary conditions,

$$W_i = 2k_t [R\cdot]^2,$$

whence

$$2k_t = W_i/[R\cdot]^2.$$

This sort of method has been used in particular to measure the concentration of peroxide radicals in the oxidation of certain hydrocarbons [9-11].

In the work of Buchachenko [12], an EPR method was applied successfully to the measurement of rate constants for the reaction of stable free radicals with molecules of organic compounds. If the substance with which the radical reacts is taken in excess, the radicals disappear in accordance with a first-order kinetic law:

$$-\frac{d[R\cdot]}{dt} = k[R\cdot][A] \cong k[A]_0[R\cdot]$$

and

$$\ln\frac{[R\cdot]_0}{[R\cdot]} = k[A]_0 t.$$

The velocity of fast reactions of electron transfer from one particle to another can be measured according to the broadening of the EPR spectral line when the reactant is added. If a reaction of the type $\dot{A}^- + A \rightleftharpoons A + \dot{A}^-$, is taking place in a solution, then the mean lifetime of the particle $\dot{A}^-$ is $\tau = 1/k[A]$. On the other hand, $\tau^{-1} = \pi\delta\nu$ where $\delta\nu$ is the line broadening. Having measured $\delta\nu$ with various [A], one can find k. For information on the EPR method, see [13-15].

b. Method of Nonstationary Kinetics of Radical Chain Reaction (NKR)

A fully developed unbranched chain reaction proceeds in a stationary state, such that $W_i = 2k_t[R^\bullet]^2$. The stationary regime of the reaction is established not all at once, but over an interval of time commensurate with the time of growth of a single chain. With quadratic chain-breaking, the mean time of chain growth is $(2\ k_t W_i)^{-1/2}$, varying over a range from 0.01 to 100 seconds. A study of the kinetics of a chain reaction in the nonstationary period, when $W_i \neq 2\ k_t[R^\cdot]^2$, makes it possible to determine the chain lifetime $(2\ k_t W_i)^{-1/2}$ or the ratio k_p/k_t and, with a known ratio $k_p/\sqrt{2\ k_t}$, to determine k_p and k_t.

Initial Nonstationary Kinetics [1, 2, 15, 17]. With the instantaneous creation of a constant rate of radical formation in a system, the radical concentration (with destruction by reaction with each other) will increase with time in accordance with the following kinetic law:

$$[R\cdot] = (W_i/2\ k_t)^{1/2} \tanh(t\sqrt{2\ k_t W_i}).$$

The chain-reaction rate will increase correspondingly, tending toward a constant value, and the chain-reaction kinetics (based on accumulation of reaction product, consumption of starting material, or heat evolution) will be described by the formula

$$\Delta[M] = \frac{k_p}{2\ k_t}[M] \ln \cosh \frac{t}{\tau}, \text{ and for } t \longrightarrow \infty$$

$$\Delta[M] \longrightarrow \frac{k_p[M]}{2\ k_t}\left(\frac{t}{\tau} - \ln 2\right), \quad \tau = (2\ k_t W_i)^{-1/2}.$$

By extrapolating $\Delta[M]$ to $t = 0$, we find $k_p/2\ k_t$. From the kinetics of the stationary chain reaction, we determine $k_p/\sqrt{2\ k_t}$ (see §1). Knowing $k_p/2\ k_t$ and $k_p/\sqrt{2\ k_t}$, we find k_p and k_t.

In studying the initial nonstationary kinetics of a chain reaction, it is important to create very rapidly a constant rate of free radical formation (within a time $t' \ll \tau$). Photochemical initiation of the chain reaction is most often used for this purpose. Also used is the technique of introducing the initiator into solution with very rapid stirring [16]. In oxidation reactions, a method is used in which inert gas is replaced rapidly with oxygen [17]. In this case, before introducing the oxygen, $[R^\cdot]_{st} = (W_i/2k_{tR\cdot})^{1/2}$; at the moment of oxygen introduction, all $R^\cdot$ radicals are converted to $RO_2^\cdot$ radicals and $[RO_2^\cdot]_0 \cong [R^\cdot]_{st}$, but in the course of the process the concentration of $RO_2^\cdot$

radicals changes, tending toward

$$[RO_2\cdot]_{st} = (W_i/2k_{t\ RO_2\cdot})^{1/2}.$$

Many different methods are used to measure the kinetics of nonstationary chain reactions. The kinetics may be followed by the adiabatic temperature increase during the time of measurement. Temperature changes are registered by change in the dielectric constant of the reaction mixture [18], by change in refractive index [19], or by a thermocouple [20, 21].

The degree of polymerization may be measured by volume increase using a dilatometric method [22] or viscometrically [23]. The kinetics of a chain oxidation can be followed through the oxygen absorption [24]. The use of chemiluminescence in studying oxidation reactions [17] makes it possible to follow the reaction rate according to the intensity of the chemiluminescence resulting from a reaction between two peroxy radicals:

$$2\ RO_2\cdot \longrightarrow P^* \begin{cases} \nearrow P + h\nu \\ \searrow \text{quenching} \end{cases}$$

A variant of the chemiluminescence method has been proposed in [25] for studying the nonstationary kinetics of chain oxidation.

Photochemical After-Effect [1, 2]. Another variant in the utilization of nonstationary kinetics of a chain reaction for the measurement of $k_p/2k_t$ has been realized in the photochemical after-effect method [24]. In this method, the stationary regime of a photoinitiated chain reaction is disrupted by instantaneous cutoff of the light. At this point, the rate of chain generation changes sharply from Wi_1 to Wi_2 ($Wi_1 \gg Wi_2$; Wi_2 is the dark rate of chain generation), and the rate of the chain reaction changes smoothly from W_1 to W_2, where W_1 is the rate of the stationary chain reaction. The quantity of starting material additionally reacted on account of the nonstationary period of the reaction is expressed by the formula

$$\Delta[M] = \frac{1}{2} k_p k_t^{-1} [M] \ln \left(\frac{2W_1}{W_1 + W_2} \right),$$

where W_1 and W_2 are the rates of the stationary reaction with illumination and in the dark.

From the experimental data (reaction kinetics in the nonstationary regime), $\Delta\Delta[M]_0$ is found by extrapolation $\Delta[M] \longrightarrow \Delta[M]_0$ (time t = 0 corresponds to moment of light cutoff).

c. Intermittent-Illumination Method (IIM)

In this method, free radicals are formed periodically by illuminating the reacting system for equal time intervals. Intermittent illumination of the reaction flask is accomplished as follows: A rotating disc with cutout slots is mounted in the light path between the lamp and the reactor. The frequency of alternation between light and dark is determined by the number of cutouts in the disc and the rotation rate of the disc. This method is often called the "sector" method. If $t' \gg \tau$ (t' is the period of illumination, τ is the time of chain growth), the chain reaction will proceed only during the period of illumination, and the mean reaction rate over an extended period will be

$$\overline{W}_M = \frac{k_p}{\sqrt{2k_t}}[M]\sqrt{\Phi I}\,(1+r)^{-1},$$

where Φ is the quantum yield of photoinitiation, I is the light intensity, and r is the ratio of dark to light period. With a very short period of illumination ($t' \ll \tau$), the mean rate of initiation

$$\overline{W}_i = \frac{\Phi I}{r+1},$$

and the mean rate of the chain reaction

$$\overline{W}_b = \frac{k_p}{\sqrt{2k_t}}[M]\left(\frac{\Phi I}{r+1}\right)^{1/2}.$$

The ratio $\overline{W}_b/\overline{W}_M = \sqrt{r+1}$; i.e., the mean rate of the chain reaction is changed by transition from the regime $t' \gg \tau$ to the regime $t' \ll \tau$. This transition is observed at a t' that is commensurate with τ. The time of chain growth, τ is found by comparing the experimental relation $\overline{W}/W_c = f(t')$ with that calculated theoretically ($\overline{W}$ is the rate with intermittent light, W_c the rate with continuous illumination). The theoretical relation has the following form:

$$\frac{\overline{W}}{W_c} = (r + 1)^{-1} \left\{ 1 + m^{-1} \ln \left(\frac{rm}{1 + A} + 1 \right) \right\}$$

$$A = \frac{2 (rm + \tanh m)}{rm \tanh m + \left[r^2 m^2 \tanh^2 m + 2 (rm + \tanh m) \right]^{1/2}}$$

$$= (2 k_t W_i)^{1/2} t' = t' / \tau .$$

The intermittent-illumination method (sector method) is described in [1, 2, 26].

d. Flash Method (FM)

In the system under investigation, a high concentration of radicals is created by irradiating the system with an intense flash of light lasting for some microseconds (10^{-6} to 10^{-4} sec). The changes in radical concentration are followed by means of high-speed spectrophotometry. The method is described in [15, 27-29].

e. Pulse Radiolysis (PR)

In the pulse radiolysis method [29-32], a high concentration of free radicals is created by a strong pulse of electrons or X-rays. The electron source may be a linear accelerator giving a 1 to 10 MeV pulse for 1 to 5 μsec. The same as in flash photolysis, the radical concentration is followed by high-speed spectrophotometric methods.

f. Jet Method (JM)

The principle of this method is as follows: Two solutions of substances that react rapidly with each other are mixed in a mixing chamber, and the resulting solution is passed through a tube where the degree of reaction is observed. If the solution passes through the tube with a velocity u, then a distance ℓ from the mixing chamber corresponds to a reaction time of ℓ/u. The degree of reaction is followed spectrophotometrically or by an EPR method. For a description of the jet method, see [15, 33, 34].

g. Emulsion Polymerization (EP)

When a polymerization is carried out in an emulsion, part of the monomer exists in the form of drops, and part is dissolved in the soap micelles. If a water-soluble initiator is used, the polymerization takes place both within the monomer drops, which become polymer-monomer particles, and within the soap micelles. The soap (emulsifier) gradually passes from the micelles to the adsorption layers of the polymer-monomer particles. When this process is complete, further increase in the number of polymer-monomer particles is ended, and the polymerization proceeds with a constant number of particles and a constant velocity. The velocity of an emulsion polymerization is expressed by a simple formula: $W = 0.5\, k_p[M]N$, where N is the number of polymer-monomer particles. For a description of the method, see [2, 35].

h. Nuclear Magnetic Resonance Method (NMR)

The NMR method is widely used to study the structure of molecules, to identify chemical compounds, and to study hydrogen bonds. In recent years, this method has also come into use for studying fast reactions of proton or hydrogen-atom transfer under equilibrium conditions, for example, in reactions such as

$$Ar_1OH + Ar_2O\cdot \rightleftharpoons Ar_1O\cdot + Ar_2OH$$

The use of NMR in studying fast exchange reactions of the type

$$R_1H + R_2^{\cdot} \rightleftharpoons R_1^{\cdot} + R_2H$$

or isomerization reactions is based on the changes in the NMR spectra of the hydrogen atoms in R_1H and R_2H when such an exchange is occurring. If the exchange is "slow" ($\tau\Delta\nu \gg 1$, where τ is the lifetime of R_1H and $\Delta\nu$ is the line broadening in the NMR spectrum for the proton in R_1H), then two peaks are observed in the spectrum (for R_1H and R_2H), each of them being broadened owing to the exchange reaction. If the broadening at half-height is denoted by $\delta'\nu$, then

$$k = \tau^{-1} = 2\pi\delta'\nu \quad .$$

If the exchange is "fast" ($\tau\Delta\nu \ll 1$), an "exchange narrowing" is observed; instead of two peaks (absorption bands of R_1H

and R_2H), there is a single line. The rate constant of the reaction

$$k = \tau^{-1} = 2^{4} \pi \Delta \nu_0^2 / \delta' \nu ,$$

where $\Delta\nu_0$ is the distance between the R_1H and R_2H absorption bands in the absence of exchange, and $\delta'\nu$ is the line broadening in comparison with the case for $\tau = 0$ (calculated). A description of this method is given in [15, 36, 37].

i. Photochemical Space-Discontinuity Method (PSD)

For measuring the rate constant of recombination of iodine atoms, Noyes [38] developed a photochemical method in which the reaction is carried out under conditions of spatial interruptions or discontinuities. The reaction mixture (I_2 + trans-1,2-diiodoethylene + solvent) is illuminated. A grid is placed in front of the reaction vessel such that some sections of the vessel are illuminated and others are in the dark. The reaction rate of iodine with diiodoethylene is measured. By the use of different grids, the width of the illuminated zones is varied, while retaining a constant total area of the illuminated zones. When the width of the illuminated zone l_{1t} is much greater than the mean distance to which an atom diffuses before the moment of recombination, the reaction takes place only in the illuminated zone, where

$$[I\cdot] = \sqrt{W_i / 2 k_t}$$

and the mean reaction rate for the entire volume of the vessel is

$$\overline{W} = k \, [I\cdot] \, [M] \, \frac{l_{1t}}{l_{1t} + l_{dk}} = \frac{l_{1t}}{l_{1t} + l_{dk}} \left(\frac{W_i}{2 k_t}\right)^{1/2} k [M].$$

When $l_{1t} \ll \lambda$ (where λ is the average distance to which an I atom diffuses before recombination), the iodine atoms are uniformly distributed throughout the entire volume, and their concentration is

$$\sqrt{\frac{W_i \, l_{1t}}{(l_{1t} + l_{dk}) \, 2 k_t}}$$

and the reaction rate is

$$\overline{W} = k[M]\sqrt{\frac{W_i\, l_{1t}}{(l_{1t} + l_{dk})\, 2\, k_t}}\ .$$

The ratio of reaction rates is

$$\frac{\overline{W}(l_{1t} \gg \lambda)}{\overline{W}(l_{1t} \ll \lambda)} = \sqrt{\frac{l_{1t}}{l_{1t} + l_{dk}}}\ .$$

As the grids are changed, the reaction rate varies from

$$\overline{W}(l_{1t} \gg \lambda) \text{ to } \overline{W}(l_{1t} \ll \lambda),$$

and the transition of the reaction from one regime to the other is determined by the grid geometry, the initiation rate, and the rate constant for recombination of atoms (radicals). This rate constant is found from a comparison of calculation with experiment.

CHAPTER V

ISOMERIZATION AND DECOMPOSITION OF FREE RADICALS

§1. Isomerization of Free Radicals

Voluminous experimental material has been accumulated on the isomerization of free radicals [39], but little quantitative data.

The isomerization of phenoxyl radicals

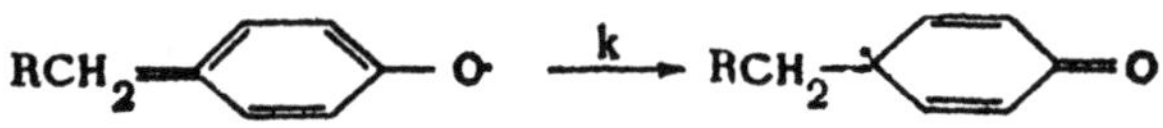

was studied quantitatively in [40] by means of EPR.

For  at 25° in benzene, $k = 1.33 \cdot 10^{-2}$; in methylcyclohexane, $k = 1.23 \cdot 10^{-2}$; $E = 11.4$, $\log A = 6.49$.

For  in benzene at 25°, $k = 7.5 \cdot 10^{-3}$.

The isomerization of alkyl radicals with shift of a phenyl group to the β-position was studied by Ruchardt [41-45] in the example of the radical $C_6H_5(CH_3)_2CCH_2$.

$$C_6H_5(CH_3)_2CCH_2^{\cdot} \longrightarrow (CH_3)_2\dot{C}CH_2C_6H_5$$

The radical $XC_6H_4\dot{C}(CH_3)_2CH_2$ (R•) was obtained in the peroxide-initiated decarbonylation of the aldehyde RCHO [42-43] and by the Kharasch reaction [44]:

$$RCl + CoCl \longrightarrow R^{\cdot} + CoCl_2$$

(The CoCl as an intermediate is obtained by the reaction 2 R'CoCl ⟶ R'R' + 2 CoCl). The ratio of rate constants for the reactions

$$C_6H_5(CH_3)CCH_2^{\cdot} \xrightarrow{k_1} (CH_3)_2CCH_2C_6H_5 \text{ and}$$

$$C_6H_5(CH)_2CCH_2^{\cdot} + C_6H_5(CH_3)_2CCH_2CHO \xrightarrow{k_2} C_6H_5(CH_3)_2CCH_3 + C_6H_5 + (CH_3)_2CCH_2\dot{C}O$$

$$\frac{k_1}{k_2} = 2.76 \text{ moles/liter } (129.7^{\circ}) \text{ and } 3.25 \ (144.5^{\circ})$$

in C_6H_5Cl [41].

§2. Decomposition of Free Radicals

The decomposition of alkoxy radicals has been studied quantitatively. The tert-butoxy radical is decomposed to acetone and a methyl radical with a rate constant [46] $\log k_d = 12.45 - 13.9/\theta$ [in CCl_4 at 10-55°, k_a for $(CH_3)_3CO^{\cdot} + CH_3C_6H_5$ was measured by an intermittent illumination method, and k_d was determined from the ratio k_a/k_d].

The relative rate constants for the decomposition of $(CH_3)_3CO^{\cdot}$ in various solvents are listed in Table 40, where k_a refers to the abstraction of H from cyclohexane.

The decomposition of $R_0R_1R_2CO^{\cdot}$ radicals with various alkyl substituents was studied by Kochi [48, 49].

The relative rates of splitting of various alkyl groups from trialkylmethoxy radicals were determined by a competing-reaction method, according to the composition of the products formed. In [49], it was shown that the relative splitting rate of a given alkyl group is little affected by the structure of the neighboring alkyl groups (in Table 42, the range of values of k is indicated for radicals with different alkyl substituents).

TABLE 39. Relative Rate Constants of Isomerization $XC_6H_4C(CH_3)_2CH_2\cdot \rightarrow (CH_3)_2\dot{C}CH_2C_6H_4X$

X	In C_6H_5Cl at 129.6° [45]	In C_6H_5Cl at 129.7° [42, 43]	In $C_2H_5OC_2H_5$ at 35° [44]
p-CH_3O	0,36	0,35	1,2
o-CH_3O	-	1,10	1,2
p-F	0,40	0,38	0,3
p-CH_3	0,76	0,65	-
o-Cl	0,90	0,90	-
H	1,00	1,00	1,0
m-Cl	1,33	1,55	-
m-Br	1,58	1,70	-
p-Br	1,90	1,79	-
p-Cl	1,88	1,82	1,8
p-CN	-	19	-
m-F	-	1,45	1,3

TABLE 40. Ratio of Constants k_d/k_a for $(CH_3)_3CO^{\cdot}$ (Various solvents, measured at 0-100° [47]).

Solvent	k_d/k_a (40°)	$E_d - E_a$	$\log \frac{A_d}{A_a}$
Gas phase	$1{,}7 \cdot 10^{-3}$	10,80	4,77
$C_2Cl_3F_3$	$1{,}8 \cdot 10^{-2}$	9,65	5,04
CH_3CN	$1{,}5 \cdot 10^{-1}$	9,54	5,73
C_6H_6	$4{,}0 \cdot 10^{-2}$	8,66	4,63
C_6H_5CN	$5{,}9 \cdot 10^{-2}$	8,28	4,58
C_6H_5CN	$6{,}1 \cdot 10^{-2}$	7,21	3,82
C_2Cl_4	$3{,}4 \cdot 10^{-2}$	8,72	4,49
trans-$C_2H_2Cl_2$	$7{,}0 \cdot 10^{-2}$	7,69	4,16
cis-$C_2H_5Cl_2$	$1{,}2 \cdot 10^{-1}$	7,04	3,92
CH_3COOH	$3{,}5 \cdot 10^{-1}$	5,80	3,64

TABLE 41. Relative Rate Constants for Decomposition of $R_0R_1R_2CO^{\cdot}$ Radicals

R_o	R_1	R_2	$(k_o+k_1+k_2)_{rel}$*	In CCl_4 at 0–40° [49]	
				E_2-E_1	$\log(A_2/A_1)$
CH_3	CH_3	CH_3	1,0		
CH_3	C_2H_5	n-C_3H_7	6,3	0,66	0,35
C_2H_5	C_2H_5	n-C_3H_7	7,3	0,55	0,28
CH_3	n-C_3H_7	n-C_4H_9	6,2	-	-
C_2H_5	n-C_3H_7	n-C_4H_9	5,5	-	-
CH_3	C_2H_5	iso-C_4H_9	4,3	1,10	0,52
CH_3	n-C_3H_7	iso-C_4H_9	3,8	-	-
n-C_3H_7	C_2H_5	n-C_4H_9	-	0,63	0,32
CH_3	C_2H_5	n-C_4H_9	-	0,79	0,34

Table 41 (continued)

R_o	R_1	R_2	$(k_o+k_1+k_2)_{rel}$*	In CCl_4 at 0–40° [49]	
				E_2-E_1	$\log(A_2/A_1)$
C_2H_5	C_2H_5	$n\text{-}C_4H_9$	-	0,90	0,50
CH_3	C_2H_5	$(CH_3)_3CCH_2$	-	0,63	0,86
CH_3	C_2H_5	$(CH_3)_3CCH_2$	-	-2,2	0,21
CH_3	$CH_2C(CH_3)_3$	$n\text{-}C_4H_9$	-	-0,50	$\bar{1}$,32
CH_3	$(CH_3)_3CCH_2$	$(CH_3)_3CCH_2$	-	0,35	0,20
CH_3	$(CH_3)_3CCH_2$	$(CH_3)_2CH(CH_2)_2CH_2$	-	-1,0	$\bar{1}$,52

*In H_2O_2 + $(CH_3)_2CHOH$ at 130° [48].

TABLE 42. Relative Rate of Splitting of Various Alkyl Groups from $R_1R_2R_3CO^\bullet$ Radicals

R	In CCl_4 at 25° [49]	In H_2O/CH_3CN at 15° [50]	In $H_2O/(CH_3)_2CHOH$ at 130° [48]
CH_3	< 0,005	0,01	0,003
C_2H_5	1,00	1,00	1,00
$n\text{-}C_3H_7$	0,63–0,74	0,34	0,65
$n\text{-}C_4H_9$	0,59–0,72	0,28	0,43
$(CH_3)_2CHCH_2$	0,47–0,57	0,17	-
$(CH_3)_3CCH_2$	0,42–0,45	-	-
$(CH_3)_3CCH_2CH_2$	0,92–0,94	-	-
$(CH_3)_2CH$	-	2,8	-
CH_3OCH_2	-	63,1	-
CH_3COO	-	0,22	-

CHAPTER VI

RADICAL SUBSTITUTION REACTIONS

§1. Reactions of Atoms

Hydrogen Atom. A large amount of data on reactions of the hydrogen atom has been obtained in radiochemical studies. In almost all cases, absolute constants have been calculated by a competing-reaction method. The calculation is based on the following value of the reaction rate constant:

$$H\cdot + O_2 \longrightarrow HO_2^{\cdot} \quad k = 2\cdot 10^{10} \quad \text{(at room temperature)}$$

The reactions of hydrogen atoms in hydrocarbon solutions have been studied by Hardwick. All constants were obtained by a competing-reaction method. Methyl methacrylate was used as the competing compound. Absolute values of the constants were calculated on the assumption that the rate constant for the addition of H to benzene in the liquid phase is the same as in the gas phase, where it is $1.25\cdot 10^8$ [67].

From a comparison of $k(H^{\cdot} + RH)$ with the structure of RH, partial rate constants were obtained for the reaction of $H^{\cdot}$ with the groups $-CH_3$, $>CH_2$ and $\geqslant CH$ [68]. The activation energies for these reactions were calculated on the assumption that, for a reaction in which $H^{\cdot}$ takes part, $k = z_0 e^{-E/RT}$, where z_0 is the number of collisions in solution ($z_0 = 3\pi\eta\sigma/2m$, η is viscosity, σ is diameter, m is mass of molecule).

Chlorine Atom. Relative rate constants for reactions of hydrogen abstraction from organic compounds by chlorine atoms were determined by a competing-reaction method. Absolute values of the reaction rate constants were calculated from the relative values on the assumption that the chlorine atom adds to butene-1 with a diffusion rate constant given by $\log k_D = 11.90 - 2.4/\theta$ and that the rate-constant ratio for addition of chlorine to butene vs. abstraction of hydrogen from cyclohexane by chlorine is equal to 11.7 at 9° (see Table 79). It was also assumed that this ratio of rate constants remains unchanged over a temperature range of 9 to 40°. For the gas phase, the reaction rate constants for abstraction of H by chlorine atoms are substantially higher. For example, the reaction rate constant for abstraction of H from a C-H bond of neopentane in the gas phase is $k_{C-H} = 2.9 \cdot 10^9$ (25°), E = 0.92 [72]. For the reaction of chlorine with a C-H bond of cyclopentane in the gas phase, $\log k = 10.50 - 0.6/\theta$; at 40°, log k = 10.08 [73]. In Table 47, both relative and absolute values of k are listed for reactions of H abstraction from various C-H bonds. All rate constants shown in Table 47 are partial, i.e., they relate to individual C-H bonds (C_1-H = primary, C_2-H = secondary, and C_3-H = tertiary C-H bond). The chlorine atom forms a complex with aromatic molecules (see Chapter X). Hence, in [74-76], in order to measure the true ratio of reaction rate constants in experiments with aromatic compounds, the obtained results were extrapolated to zero concentration of aromatic compound. In [77, 78], no such extrapolation was performed, and the obtained results relate to a chlorine atom that is to a significant degree bound in a complex with the aromatic molecules of the reactants.

<u>Bromine Atom.</u> One of the most widely used brominating agents is N-bromosuccinimide.

For bromination with N-bromosuccinimide (and with analogous compounds), the following mechanism has been substantiated experimentally:

$$Br\cdot + RH \longrightarrow BrH + R\cdot$$
$$R\cdot + Br_2 \longrightarrow RBr + Br\cdot$$

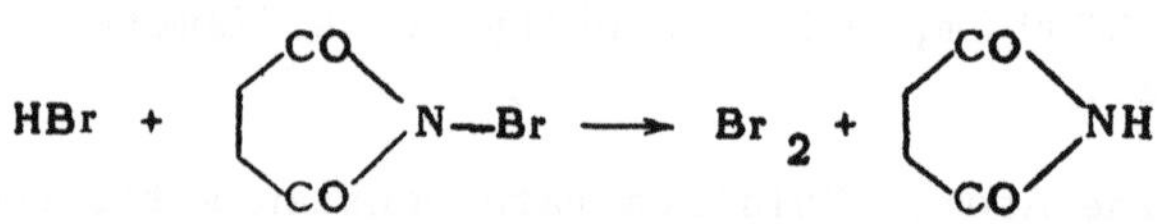

Hence, the relative rate of bromination with N-bromosuccinimide characterizes the relative reactivity of RH toward the bromine atom.

Absolute values of reaction rate constants for bromine atoms (Table 48) were calculated on the assumption that k ($Br\cdot + C_6H_5CH_3$) in CCl_4 is the same as in the gas phase, where log k = 10.52 - 7.2/Θ [79], or, calculated on one C-H bond, log k_{C-H} = 10.04 - 7.2/Θ.

§2. Reactions of Radicals Having Free Valence on Oxygen

Hydroxyl Radical. Using pulse radiolysis and competing-reaction methods, a large number of rate constants have been measured for reactions of hydroxyl with organic compounds in aqueous solution at room temperature. Absolute values of the constants have been computed from the relative constants on the basis of the following values of k:

$$k(OH + H_2O_2) = 4.5\cdot10^7,\ k(OH + I^-) = 1\cdot10^{10},$$
$$k(HO^\cdot + C_2H_5OH) = 1.1\cdot10^9,\ k(OH + Br^-) = 7.1\cdot10^8,$$
$$k(OH + CNS^-) = 6.6\cdot10^9,\ k(HO^\cdot + ONC_6H_4N(CH_3)_2) = 7.5\cdot10^9,$$
$$k(HO^\cdot + \text{thymine}) = 4.3\cdot10^9.$$

$(CH_3)_3CO^\cdot$ Radical. The rate constant for the reaction of $(CH_3)_3CO^\cdot$ with toluene, measured, by the IIM method [46] in CCl_4 at 10-55°, is equal to log k = 7.826 - 5.6/Θ . In a number of studies, measurements have been made of the ratio between the rate constant for abstraction of H from the solvent and the decomposition of the tert-butoxy radical. The rate constant for decomposition of $(CH_3)_3CO^\cdot$ is listed in Chapter V.

Partial rate constants for the abstraction of H from various C-H bonds by the $(CH_3)_3CO^\cdot$ radical, in CCl_4 at 40°, are listed in Table 53. These constants were calculated on the basis of the data of [121] (which were obtained by a competing-reaction method) and a value of $2.6\cdot10^3$ for k_{C-H} (toluene, 40°) [46].

Peroxy radicals are formed in the autoxidation of organic compounds. The reaction of H abstraction by peroxy radicals from the substance being oxidized is the limiting stage in prolonging the oxidation chain.

Rate constants for the reaction $RO_2^{\bullet}$ + RH have been measured many times for such hydrocarbons as cyclohexene, tetralin, and cumene. Averaging the experimental data leads to the following values of k ($RO_2^{\bullet}$ + RH):

For cyclohexene $\log k = 5.60 - 7.0/\theta$ $(\Delta \log k = 0.18)$

For tetralin $\log k = 5.58 - 6.4/\theta$ $(\Delta \log k = 0.05)$

For cumene $\log k = 6.51 - 10.1/\theta$ $(\Delta \log k = 0.10)$

In reactions of oxidation of individual substances, one can successfully measure k ($RO_2^{\bullet}$ + RH). In going from one compound to another, the structure of the attacking radical as well as that of the RH is changed. In order to compare the reactivities of various RH compounds, it is necessary to measure the rate constants for reaction of one particular peroxy radical with different RH compounds. Rate constants k ($R_1O_2^{\bullet}$ + R_2H) can be calculated on the basis of data on the initiated cooxidation of the two hydrocarbons R_1H and R_2H (see [1]). A more simple and reliable method for determining k ($R_1O_2^{\bullet}$ + R_2H) is to measure the rate of the initiated oxidation of R_2H in the presence of a hydroperoxide R_1OOH [154]. The hydroperoxide R_1OOH reacts rapidly with $R_2O_2^{\bullet}$ (rate constant for such exchange $\sim 10^3$ [154], leading to a complete replacement of $R_2O_2^{\bullet}$ by $R_1O_2^{\bullet}$ if an adequate concentration of the hydroperoxide is used. This method also permits a comparison of the reactivities of various peroxy radicals in reactions with a single hydrocarbon.

As is evident from Table 56, the most active is $HO_2^{\bullet}$, the least active the tertiary peroxy radical. The ratio of k's for $HO_2^{\bullet}$ vs. primary or secondary $RO_2^{\bullet}$ vs. tertiary $RO_2^{\bullet}$ is 3/1/0.25.

Phenoxyl radicals are formed when phenols are oxidized by various oxidants. Reactions of phenoxyl radicals play an important role in the oxidation of organic compounds inhibited with phenols.

Nitroxy radicals are quite stable and are used to measure radical formation rates. However, at high temperatures, they react with molecules, abstracting hydrogen.

§3. Reactions of Radicals Having Free Valence on Carbon

Methyl Radical. In the liquid phase, only relative rate constants have been measured for H abstraction by methyl radicals. The hydrocarbons used for comparison have been isooctane (Szwarc), n-heptane labeled with tritium in the 4-position (Berezin), and toluene labeled with tritium on the methyl group (Szwarc). Absolute values of the rate constants were calculated on the assumption that k ($CH_3^{\bullet}$ + toluene) in the liquid hydrocarbon is the same as in the gas phase [196]: $\log k_{toluene} = 7.35 - 7.30/\theta$. The ratio $k_{toluene}/k_{isooctane} = 0.37\ e^{1400/RT}$ [197], whence $\log k_{isooctane} = 7.79 - 8.7/\theta$. The ratio $k_{isooctane}/k_{C\text{-}H\text{-}heptane} = 0.3\ e^{1500/RT}$ [198, 199], whence we find $k_{C\text{-}H\text{-}heptane}$ = 8.31 - 10.2/θ. The isotope effect in the reaction $CH_3^{\bullet}$ + C-H-heptane when hydrogen is replaced by tritium in a CH_2 group is [200]: $\log(k_H/k_T) = \bar{1}.30 + 3.4/\theta$.

$F_3C^{\bullet}$ Radical. Absolute values of the rate constants for the reaction $F_3C^{\bullet} + RH \rightarrow CF_3H + R^{\bullet}$ have been calculated from the ratios of constants $k/k_{isooctane}$ on the assumption that in isooctane solution the value of $k(F_3C^{\bullet}$ + toluene) is the same as in the gas phase, where $\log k_{toluene} = 8.10 - 5.37/\theta$. [225].

Phenyl Radical. The C-H bond strength in benzene is high -- 102 kcal/mole. Therefore, the phenyl radical formed from benzene is extremely active and will rapidly abstract a hydrogen from hydrocarbons. Unfortunately, absolute constants for the abstraction of H by phenyl radicals are unknown. Therefore, relative rate constants for H abstraction are listed in Table 67.

§4. Chain Transfer in Radical Polymerization

The phenomenon of chain transfer in radical polymerization consists of the reaction of a macroradical with a molecule, forming a macromolecule and a new active radical. As a result of the act of chain transfer, the growth of the macroradical is stopped, but the new active radical that is formed will continue the chain polymerization. Chain transfer may take place through a substitution reaction, for example

$$\sim CH_2\dot{C}HX + RH \rightarrow \sim CH_2CH_2X + R\cdot$$
$$\sim CH_2\dot{C}HX + RCl \rightarrow \sim CH_2CHClX + R\cdot$$

through a reaction in which an H atom is transferred from the macroradical to a monomer molecule, for example

$$\sim CH_2\dot{C}HX + CH_2=CHX \rightarrow \sim CH=CHX + CH_3\dot{C}HX$$

or by a substitution reaction through a heteroatom, for example, oxygen or sulfur:

$$\sim CH_2\dot{C}HX + ROOR \rightarrow \sim CH_2CHXOR + RO\cdot$$
$$\sim CH_2\dot{C}HX + RSSR \rightarrow \sim CH_2CHXSR + RS\cdot$$

The chain transfer constant $C_S = k/k_p$ is determined from the dependence of the degree of polymerization on the concentration of the substance causing the chain transfer (see Chapter IV, §1). The absolute value of k can be found by using a known value of k_p.

§5. Reactions of Radicals Having Free Valence on Nitrogen

See Table 76.

§6. Correlation Equations for Radical Substitution Reactions

For reactions of atoms and free radicals with aromatic compounds, good linear correlations are observed between the logarithm of the reaction rate constant and the function σ (see Part One, Chapter I, §6). The coefficient σ in the linear relation $\log(k_X/k_H) = \rho\sigma$ is always negative; the less active the radical, the larger the absolute value of ρ. Values of ρ are listed in Table 77.

TABLE 43. Hydrogen-Atom Reactions in Aqueous Solution at Room Temperature (23°)

Reaction	pH	Method	log k	Ref.
$H\cdot + Br_2 \rightarrow HBr + Br\cdot$	7	PR	10	[51]
$H + H_2O_2 \rightarrow H_2 + HO_2\cdot$	<7	MCR with O_2	7,60	[52]
	<7	PR	7,60	[53]
	0,4–3	PR	8,20	[54]
	2	PR	7,96	[55]
$H + N_2O \rightarrow N_2 + OH$	0–2	MCR with Fe^{2+}	5,34	[56]
	3,5–11	Photo-chemical	4,10	[57]
$H + CH_3OH \rightarrow H_2 + \dot{C}H_2OH$	1	MCR with Fe^{3+}	6,20	[58]
	7	MCR with $Fe(CN)_6^{3-}$	6,23	[59]
	7	MCR with NO_2^-	6,20	[60]
	7	MCR with $DCOO^-$	6,22	[61]
$H + CD_3OH \rightarrow HD + \dot{C}D_2OH$	7	MCR with CH_3OH	4,90	[62]
$H + CH_3CH_2OH \rightarrow H_2 + CH_3\dot{C}H(OH)$	1	MCR with Fe^{3+}	7,18	[58]
$H\cdot + CH_3CH_2OH \rightarrow H_2 + CH_3\dot{C}HOH$	1–3	MCR with $Fe(CN)_6^{3-}$	7,22	[59]
	7	MCR with DCO_2^-	7,20	[61]
	7	MCR with CH_3OH	7,20	[63]
$\cdot H + (CH_3)_2CHOH \rightarrow H_2 + (CH_3)_2\dot{C}OH$	1	MCR with $Fe(CN)_6^{3-}$	7,70	[59]
	7	MCR with $DCOO^-$	7,70	[61]

Table 43 (continued)

Reaction	pH	Method	log k	Ref.
$\cdot H+(CH_3)_2CHOH \longrightarrow H_2 + (CH_3)_2\dot{C}OH$	7	MCR with CH_3OH	7,70	[63]
$\cdot H+(CH_3)_2CDOH \longrightarrow HD + (CH_3)_2\dot{C}OH$	7	MCR with $(CH_3)_2CHOH$	6,93	[62]
$\cdot H+(CH_3)_3COH \longrightarrow H_2 + \cdot CH_2(CH_3)_2COH$	7	MCR with DCO_2^-	5,00	[61]
$\cdot H+HOCH_2CH_2OH \longrightarrow H_2 + R\cdot$	7	- " -	6,98	[61]
$\cdot H+HOCH_2CH(OH)CH_2OH \longrightarrow H_2 + R\cdot$	1	MCR with $Fe(CN)_6^{3-}$	7,30	[59]
$\cdot H$ + 2-desoxy-D-ribose	7	MCR with $DCOO^-$	7,34	[64]
$H^\cdot$ + saccharose	1	MCR with Fe^{3+}	7,57	[58]
$\cdot H + CH_2O \longrightarrow H_2 + CHO$	1	- " -	6,70	[58]
$\cdot H+CH_3COCH_3 \longrightarrow H_2 + \dot{C}H_2COCH_3$	1	MCR with Fe^{3+}	5,54	[58]
	11–13	MCR with OH^-	6,30	[65]
	7	MCR with DCO_2^-	5,78	[61]
$H+CH_3COOC_2H_5 \longrightarrow H_2 + CH_3COO\dot{C}H(CH_3)$	1	MCR with Fe^{3+}	5,78	[58]
$\cdot H + HCOOH \longrightarrow H_2 + \dot{C}OOH$	1	MCR with Fe^{3+}	6,04	[58]
	1	MCR with $Fe(CN)_6^{3-}$	6,04	[59]
$\cdot H+ DCOOH \longrightarrow HD + \dot{C}OOH$	1	MCR with Fe^{3+}	5,18	[58]
$\cdot H+CH_3COOH \longrightarrow H_2 + \dot{C}H_2COOH$	1	- " -	5,00	[58]
$\cdot H+Cl_3COOH \longrightarrow H_2 + CCl_3COO\cdot$	<7	MCR with $H^\cdot$	5,26	[66]
$\cdot H+CCl_3COOH \longrightarrow HCl + \dot{C}Cl_2COOH$	<7	- " -	4,90	[66]

TABLE 44. Hydrogen-Atom Reactions of the Type $RH + H^{\bullet} \rightarrow H_2 + R^{\bullet}$ in Hydrocarbon Solution at 23° (MCR method)

RH	log k	Ref.
$n\text{-}C_5H_{12}$	6,67	[68]
$cyclo\text{-}C_5H_{10}$	6,78	
$n\text{-}C_6H_{14}$	6,69	
$(CH_3)_2CHCH_2CH_2CH_3$	6,97	
$CH_3CH_2(CH_3)CH\ CH_2CH_3$	6,945	
$(CH_3)_3CCH_2CH_3$	7,38	
$(CH_3)_2CHCH(CH_3)_2$	7,28	
$cyclo\text{-}C_5H_9CH_3$	7,18	
$n\text{-}C_7H_{16}$	6,82	
$(CH_3)_2CHCH_2CH(CH_3)_2$	7,32	
$cyclo\text{-}C_6H_{11}CH_3$	7,175	
$n\text{-}C_8H_{18}$	6,94	
$(CH_3)_3CCH_2CH(CH_3)_2$	7,04	
$n\text{-}C_9H_{20}$	6,995	
$cyclo\text{-}C_5H_8$	8,125	[69]
$cyclo\text{-}C_6H_{10}$	8,05	
$CH_3\text{-}CH\text{=}CH\text{-}CH_2\text{-}CH_3$	8,09	
$CH_3\text{-}CH\text{=}CH\text{-}CH_2CH_2CH_3$	8,10	
$CH_3\text{-}CH\text{=}CH\text{-}(CH_2)_3CH_3$	8,13	
$CH_2\text{=}CH(CH_2)_2CH_3$	8,36	
$CH_2\text{=}CH(CH_2)_3CH_3$	8,38	
$CH_2\text{=}CH(CH_2)_4CH_3$	8,46	
$CH_2\text{=}CH(CH_2)_5CH_3$	8,30	
$CH_2\text{=}CH(CH_2)_7CH_3$	8,40	
$CH_2\text{=}C(CH_3)CH_2C(CH_3)_2CH_3$	8,53	
$CH_2\text{=}C(CH_3)CH_2CH_3$	8,53	
$CH_2\text{=}C(CH_3)CH_2CH_2CH_3$	8,57	
$CH_2\text{=}C(CH_3)(CH_2)_4CH_3$	8,49	
$CH_2\text{=}C(C_2H_5)CH_2CH_3$	8,57	

Table 44 (continued)

RH	log k	Ref.
—CH=CH_2	8,32	
$(CH_3)_2$=$CHCH_3$	7,78	[69]
$(CH_3)_2C$=$CHCH_2CH_3$	8,08	
CH_3CH=$C(CH_2CH_3)_2$	8,255	
$C_6H_5CH_2CH_3$	8,04	
$C_6H_5CH_2CH_2CH_3$	7,78	
$C_6H_5(CH_2)_3CH_3$	7,70	
$C_6H_5(CH_3)_2CH$	7,30	
$C_6H_5CH(CH_3)CH_2CH_3$	7,90	
o-$C_2H_5C_6H_4C_2H_5$	7,78	
m-C_2H_5-$C_6H_4C_2H_5$	8,00	
p-$C_2H_5C_6H_4C_2H_5$	8,28	
$(C_2H_5)_6C_6$	8,34	
p-$(CH_3)_2CHC_6H_4CH(CH_3)_2$	7,90	
1,3,5-$[(CH_3)_2CH]_3C_6H_3$	7,30	
CH_3COCH_3	8,20	
$CH_3COC_2H_5$	8,18	
$C_2H_5COC_2H_5$	8,23	
$CH_3COC_3H_7$	7,875	[70]
$CH_3COCH(CH_3)_2$	7,78	
$CH_3COCH_2CH(CH_3)_2$	8,12	
$(CH_3)_2CHCOCH(CH_3)_2$	7,845	
cyclo-$C_6H_{10}O$	8,38	
CH_3COOH	8,18	
CH_3CH_2COOH	8,20	
$CH_3(CH_2)_2COOH$	8,20	
$CH_3(CH_2)_3COOH$	8,20	
$(CH_3)_2CH$ CCOH	8,20	
F_3CCOOH	8,64	
C_6H_5COOH	8,92	

Table 44 (continued)

RH	log k	Ref.
CH_2Cl_2	8,60	[71]
$CHCl_3$	8,34	
CH_2Br_2	8,935	
$CHBr_3$	8,415	
CH_3I	8,845	
Cl_2CHCH_3	8,23	
$Cl_2CHCH_2CH_2CH_3$	8,23	
$CH_3Cl_2CH_2CH_3$	8,36	
$ClCH_2CH_2Cl$	8,87	
Cl_2CHCH_2Cl	8,18	
Cl_3CCH_3	8,34	
$Cl_2CHCHCl_2$	8,15	
$CH_2Br\,CH_2Br$	8,65	
$CHBr_2CH_2Br$	8,23	
$CHBr_2CHBr_2$	7,70	
$Br\,CH_2CHBr\,CH_3$	8,61	
$Br\,CH_2CH_2CH_2Br$	8,64	
$Br\,CH_2CH_2CH_2CH_2Br$	8,08	
$Br\,CH_2(CH_2)_3CH_2Br$	8,57	
$Br_2CHCH_2CH_2CH_3$	8,32	
$CH_3CBr_2CH_2CH_3$	8,63	
$CH_3CHBr\,CHBr\,CH_3$	9,03	
$CH_2Br\,CHBr\,CH_2Br$	8,59	

TABLE 45. Reactivities of Different Groups in Reactions with H$^{\cdot}$

Reacting group	k (23°)	E(calculated)
$-CH_3$	$0{,}25\cdot10^6$	8,7
$>CH_2$	$1{,}10\cdot10^6$	7,5
$>CH$	$7{,}75\cdot10^6$	6,0

TABLE 46. Abstraction of Halogen Atom by Hydrogen ($H^{\cdot} + RX \rightarrow HX + R^{\cdot}$) in n-Hexane Solution at 23° (MCR method [71])

RX	log k	RX	log k
n-C_3H_7Cl	8,15	$CH_2Br\,CHBr\,CH_3$	9,32
n-C_4H_9Cl	8,15	$CH_2Br\,CH_2CH_2Br$	9,32
$CH_3CHClCH_2CH_3$	8,15	$CH_2Br\,CH_2CH_2CH_2Br$	9,29
$CHCl_2CH_3$	8,935	$CH_2Br\,(CH_2)_3CH_2Br$	9,24
$CHCl_2CH_2CH_2CH_3$	8,93	$CHBr_2CH_2CH_2CH_3$	9,20
$CH_3CCl_2CH_2CH_3$	9,06	$CH_3CHBr\,CHBr\,CH_3$	9,51
CH_2ClCH_2Cl	9,19	$CH_3CBr_2CH_2CH_3$	9,32
$CHCl_2CH_2Cl$	9,14	$CH_2Br\,CHBr\,CH_2Br$	9,48
$CHCl_2CHCl_2$	9,52	CH_2Br_2	9,36
CCl_3CHCl_2	9,56	$CHBr_3$	9,71
CCl_3CCl_3	9,255	CBr_4	10,00
CH_2Cl_2	9,17	CH_2I_2	10,14
$CHCl_3$	9,21	CHI_3	10,27
CCl_4	9,24	CI_4	10,32
CCl_3CH_3	9,15	$ClCH_2COOCH_3$	9,34
n-C_3H_7Br	8,08	$CH_3CHClCOOCH_3$	9,165
n-C_4H_9Br	8,08	$Br\,CH_2COOCH_3$	9,59
$CH_3CHBr\,CH_2CH_3$	9,22	ICH_2COOCH_3	9,965
$(CH_3)_3CBr$	9,18	CH_3I	9,78
cyclo-$C_6H_{11}Br$	9,25	n-C_3H_7I	9,765
$CH_2Br\,CH_2Br$	9,40	n-C_4H_9I	9,88
$CHBr_2CH_2Br$	9,40	$CH_3CHICH_2CH_3$	9,88
$CHBr_2CHBr_2$	9,53	cyclo-$C_6H_{11}I$	9,88
CBr_3CHBr_2	9,69	$(CH_3)_3CI$	9,83

TABLE 47. Partial Reaction Rate Constants for Abstraction of H from C-H Bonds of Hydrocarbons by Chlorine Atoms

Compound	Bond type	Solvent	t °	Relative k_{C-H}	log k_{C-H}	Ref.
$(CH_3)_2CHCH(CH_3)_2$	C_1–H	Reactants	40	1,00*	7,63	[75]
	C_3–H	- " -	40	3,9*	8,24	[75]
cyclo-C_5H_{10}	C_2–H	Reactants	40	2,8*	8,08	[75]
cyclo-C_6H_{12}	C_2–H	- " -	40	2,7*	8,06	[75]
cyclo-C_7H_{14}	C_2–H	- " -	40	3,0*	8,11	[75]
cyclo-C_8H_{16}	C_2–H	- " -	40	4,3*	8,28	[75]
$C_6H_5CH_3$	C_1–H	$C_6H_5NO_2$	40	1,4*	7,77	[75]
$C_6H_5CH_2CH_3$	C_2–H	$C_6H_5NO_2$	40	3,5*	8,18	[75]
	C_3–H	- " -	40	1,8*	7,88	[75]
$C_6H_5CH(CH_3)_2$	C_1–H	$C_6H_5NO_2$	40	2,2*	7,98	[75]
	C_3–H	- " -	40	7,8*	8,52	[75]
Indan	α-C_2–H	$C_6H_5NO_2$	40	6,1*	8,40	[75]
	β-C_2–H	- " -	40	4,7*	8,31	[75]

Table 47 (continued)

Compound	Bond type	Solvent	t^{o}	Relative k_{C-H}	$\log k_{C-H}$	Ref.
Tetralin	α-C_2-H	$C_6H_5NO_2$	40	$5,8^{*}$	8,40	[75]
	β-C_2-H	- " -	40	$5,5^{*}$	8,38	[75]
$(C_6H_5)_2CH_2$	C_2-H	$C_6H_5NO_2$	40	$2,7^{*}$	8,06	[75]
$(C_6H_5)_3CH$	C_3-H	- " -	40	10^{*}	8,65	[75]
cyclo-C_5H_9-C_6H_5	C_3-H	$C_6H_5NO_2$	40	13^{*}	8,745	[75]
cyclo-C_6H_{11}-C_6H_5	C_3-H	- " -	40	$9,9^{*}$	8,63	[75]
$C_6H_5(CH_3)_3$	C_1-H	CCl_4	40	$0,63^{*}$	7,43	[75]
$C_6H_5(CH_2)_3CH_3$	α-C_2-H	CCl_4	40	$5,9^{*}$	8,40	[75]
	β-C_2-H	- " -	40	$2,7^{*}$	8,06	[75]
$C_6H_5(CH_2)_3CH_3$	γ-C_2-H	CCl_4	40	$4,0^{*}$	8,23	[75]
	C_1-H	- " -	40	$1,0^{*}$	7,63	[75]
$(CH_3)_3CCN$	C_1-H	Reactants	40	$0,12^{*}$	6,72	[75]
CH_2Cl_2	C_1-H	- " -	40	$0,011^{*}$	5,68	[75]
$CHCl_3$	C_1-H	- " -	40	$0,0051^{*}$	5,345	[75]

CH_3CN	C_1-H	- " -	40	$0{,}0040^{*}$	5,24	[75]
p-$CH_3C_6H_4CH_3$	C_1-H	Reactants	70	$1{,}62^{2*}$	-	[77]
	C_1-H	CCl_4	40	$1{,}57^{2*}$	7,965	[74]
p-$C_6H_5C_6H_4CH_3$	C_1-H	Reactants	70	$1{,}59^{2*}$	-	[77]
	C_1-H	CCl_4	40	$1{,}55^{2*}$	7,96	[74]
m-$CH_3C_6H_4CH_3$	C_1-H	Reactants	70	$1{,}33^{2*}$	-	[77]
	C_1-H	CCl_4	40	$1{,}30^{2*}$	7,86	[74]
m-$C_6H_5C_6H_4CH_3$	C_1-H	Reactants	70	$1{,}12^{2*}$	-	[77]
p-$C_6H_5OC_6H_4CH_3$	C_1-H	CCl_4	40	$2{,}50^{2*}$	8,17	[74]
m-$C_6H_5OC_6H_4CH_3$	C_1-H	- " -	40	$0{,}86^{2*}$	7,70	[74]
p-$ClC_6H_4CH_3$	C_1-H	Reactants	70	$0{,}72^{2*}$	-	
	C_1-H	C_6H_6	70	$0{,}69^{2*}$	-	[77]
	C_1-H	CCl_4	70	$0{,}47^{2*}$	-	
	C_1-H	- " -	40	$0{,}79^{2*}$	7,67	[74]
m-$ClC_6H_4CH_3$	C_1-H	Reactants	70	$0{,}54^{2*}$	-	[77]

Table 47 (continued)

Compound	Bond type	Solvent	t°	Relative k_{C-H}	$\log k_{C-H}$	Ref.
$m\text{-}ClC_6H_4CH_3$	C_1-H	CCl_4	40	$0,60^{2*}$	7,55	[74]
$p\text{-}CNC_6H_4CH_3$	C_1-H	Reactants	70	$0,38^{2*}$	-	[77]
$m\text{-}CNC_6H_4CH_3$	C_1-H	- " -	70	$0,36^{2*}$	-	[77]
$p\text{-}NO_2C_6H_4CH_3$	C_1-H	CCl_4	40	$0,32^{2*}$	7,21	[74]
$C_6H_5C(CH_3)_3$	C_1-H	Reactants	80	$0,22^{2*}$	-	[78]
$C_6H_5CH(CH_3)_2$	C_3-H	Reactants	80	$4,0^{2*}$	-	[78]
	C_1-H	- " -	80	$0,3^{2*}$	-	[78]

* k_{C-H}/k_{C_I-H} (in 2,3-dimethylbutane)

2* k_{C-H}/k_{C_1-H} (in toluene)

Table 48. Partial Rate Constants for Bromination ($Br^{\bullet} + RH \rightarrow BrH + R^{\bullet}$) (MCR method, toluene as standard hydrocarbon)

RH	Br• source	Solvent	t°	k_{C-H} (rel.)	log k_{C-H}	Ref.
$C_6H_5CH_2CH_3$	$(CH_2-CO)_2NBr$	CCl_4	80	22,8	6,94	[80]
	- " - + hν	CH_2Cl_2	40	21	6,32	[81]
	- " - + hν	CCl_4	80	23,6	6,95	[82]
	Br_2 + hν	CCl_4	40	17,2	6,235	[83]
$C_6H_5CH(CH_3)_2$	$(CH_2-CO)_2NBr$ + hν	CH_2Cl_2	40	45	6,65	[81]
	- " -	CCl_4	80	50	7,28	[82]
	Br_2 + hν	CCl_4	40	37	6,57	[83]
$C_6H_5CH_2CH_2CH_3$	$(CH_2-CO)_2NBr$ + hν	CCl_4	80	28,2	7,03	[82]
$C_6H_5CH(CH_3)(C_2H_5)$	- " -	CCl_4	80	61	7,365	[82]
$C_6H_5(CH_2)_3CH_3$	$(CH_2-CO)_2NBr$ + hν	CCl_4	80	28,5	6,035	[82]
$C_6H_5CH_2CH=CH_2$	- " -	CH_2Cl_2	40	28	6,45	[81]

Table 48 (continued)

RH	Br· source	Solvent	t°	k_{C-H} (rel.)	$\log k_{C-H}$	Ref.
Tetralin	$(CH_2CO)_2NBr$ + hν	CCl_4	80	25,3	6,98	[82]
	- " -	CH_2Cl_2	40	36	8,56	[81]
Indan	$(CH_2CO)_2NBr$ + hν	CH_2Cl_2	40	36	6,58	[81]
$C_6H_5CH(OCH_3)_2$	- " -	CCl_4	80	50,6	7,285	[80]
$C_6H_5CH_2OCH_3$	- " -	- " -	80	52,6	7,30	[80]
$C_6H_5CH(OCH_3)CO_2CH_3$	- " -	- " -	80	52,9	7,35	[80]
$C_6H_5CH_2OCH_2C_6H_5$	- " -	- " -	80	77,5	7,47	[80]
$(C_6H_5)_2CHOCH_3$	- " -	- " -	80	109	7,62	[80]
$(C_6H_5)_2CH_2$	$(CH_2CO)_2NBr$ + hν	CCl_4	80	10	6,58	[81]
	- " -	- " -	40	9,6	5,98	[83]
$(C_6H_5)_3CH$	Br_2 + hν	CCl_4	40	17,8	6,25	[83]
$(C_6H_5)_2CHCH_3$	- " -	- " -	40	42	6,62	[83]
$(CH_3)_2CHCH(CH_3)_2$	Br_2 + hν	CCl_4	40	0,3	4,48	[83]
	$(CH_2CO)_2NBr$	- " -	80	0,105	4,60	[82]

n-C_6H_{14}	CH_2-CO CH_2-CO >NBr	CCl_4	80	0,071	4,43	
n-C_8H_{18}	- " -	- " -	80	0,0054	3,31	
cyclo-C_6H_{12}	- " -	- " -	80	0,0028	3,03	
cyclo-$C_6H_{11}CH_3$	- " -	- " -	80	0,071	4,43	
$(CH_3)_2C=CH-CH_3$	- " -	CCl_4	80	76	7,46	[82]
$CH_2=C(CH_3)CH=CH_2$	- " -	- " -	80	48	7,26	
$CH_2=CH(CH_2)_2CH_3$	- " -	- " -	80	35,6	7,13	
$CH_2=CH(CH_2)_3CH_3$	- " -	- " -	80	35	7,125	
$CH_2=CH(CH_2)_5CH_3$	- " -	- " -	80	40	7,18	
cyclo-C_5H_8	CH_2-CO CH_2-CO >NBr	CCl_4	80	99	7,575	[82]
	- " -	CH_2Cl_2	40	600	7,78	[81]
cyclo-C_6H_{10}	CH_2-CO CH_2-CO >NBr	CCl_4	80	129	7,69	[82]
	- " -	CH_2Cl_2	40	160	7,20	[81]
cyclo-C_7H_{12}	CH_2-CO CH_2-CO >NBr	CH_2Cl_2	40	825	7,92	[81]
cyclo-C_8H_{14}	- " -	- " -	40	440	7,64	[81]

TABLE 49. Relative Rate Constants for Bromination of Compounds XC_6H_4R (MCR method, toluene as standard hydrocarbon)

XC_6H_4R	Solvent	t°	k/k_T*, X = H	Substituent									Ref.
				p-OCH_3	p-CH_3	m-CH_3	p-Cl	p-Br	p-F	m-Cl	m-Br	p-CN	
$XC_6H_4CH_3$	CCl_4	80	1,00	11,7	2,56	1,52	0,80	0,58	0,54		0,24		[84]
	- " -	80	1,00	12,6	2,33	1,80	0,77	0,69	0,55				[84]
	- " -	80	1,00	6,75	1,59	1,18	0,68	0,46	0,77		0,15		[84]
	- " -	80	1,00	5,7	1,34	1,06	0,46	0,38	0,62		0,19		[84]
	- " -	80	1,00	8,2	1,21		0,72	0,49	0,64		0,38		[84]
	- " -	80	1,00	12,4	2,6		0,71				0,24	0,11	[85]
	- " -	19	1,00	7,4	3,2		0,76				0,16		[85]
	- " -	80	1,00	9,0							0,24	0,14	[86]
	- " -	80	1,00	6,0							0,37	0,22	[86]
$XC_6H_4C_2H_5$	CCl_4	80	22,8	69	40			18		11,4			[87]
$XC_6H_4CH{=}CH_2$	C_6H_5Cl	69,5	26,2	85	50	34	20		28	13			[88]
$XC_6H_4CH_2OCH_3$	CCl_4	80	52,6	100	66		49			38			[89]
$XC_6H_4CH(OCH_3)_2$	- " -	80	50,6	91	67		42			34			[89]

*$k_T = k_{toluene}$.

Table 49 (continued)

XC_6H_4R	Solvent	t°	k/k_T;* X = H	Substituent								Ref.
				p-$C(CH_3)_3$	m-OCH_3	m-CN	p-NO_2	m-NO_2	p-$N(CH_3)_2$	p-CF_3	m-CF_3	
$XC_6H_4CH_3$	CCl_4	80	1,00									
	- " -	80	1,00									
	- " -	80	1,00									[84]
	- " -	80	1,00									
	- " -	80	1,00									
	- " -	80	1,00	2,8								[85]
	- " -	19	1,00									[85]
	- " -	80	1,00	2,0	0,75	0,11	0,05					[86]
	- " -	80	1,00	1,5			0,11					[86]
$XC_6H_4C_2H_5$	CCl_4	80	22,8	36								[87]
$XC_6H_4CH{-}CH_2$	C_6H_5Cl	69,5	26,2		27				550	9,5	10,5	[88]
$XC_6H_4CH_2OCH_3$	CCl_4	80	52,6	60								[89]
$XC_6H_4CH(OCH_3)_2$	- " -	80	50,6	62				27				[89]

TABLE 50. Reactions of H Abstraction by Hydroxyl in Aqueous Solutions at 23°
(In case of MCR method, competing compound is shown)

Compound	pH	Method	log k	Ref.
H_2	13	MCR, $Fe(CN)_6^{4-}$	8,205	[90]
	7	MCR, H_2O_2	7,62	[91]
	7	- " -	7,65	[91]
D_2	1	MCR, Fe^{2+}	7,205	[92]
H_2O_2	2,5–7	MCR, H_2	7,65	[91]
	7	MCR, $HO^{\cdot}$	7,65	[93]
	0,4–3	PR	7,08	[54]
	7	MCR, I^-	7,34	[94]
		Mean	7,43	
CH_4	9	MCR, p-$ONC_6H_4N(CH_3)_2$	8,15	[95]
$CHCl_3$	9	- " -	6,93	[95]
CH_3OH	1	MCR, Fe^{2+}	8,87	[96]

	1	MCR	9,255	[97]
	4	MCR, $(NH_2CH_2)_2$	8,48	[90]
	2	MCR, thymine	8,84	[98]
	10,7	MCR, CO_3^{2-}	8,66	
	7	MCR, CNS^-	8,67	[99]
	7	MCR, SeO_3^{2-}	8,70	
	6	MCR, Br^-	8,71	[95]
	9	MCR, $p\text{-}ONC_6H_4N(CH_3)_2$	8,785	[95]
	7	MCR, I^-	8,67	[94]
	10,[illegible]	MCR, C_6H_5COOH	8,825	[100]
	9	MCR, $p\text{-}ONC_6H_4N(CH_3)_2$	8,81	[101]
		Mean	8,75	
CD_3OH	6	MCR, Br^-	8,40	[95]
CH_3CH_2OH	1	MCR, Fe^{2+}	9,08	[102]
	6–10,6	MCR, $Fe(CN)_6^{4-}$	8,35	[103]
	7	MCR, CO_3^{2-}	8,46	[104]
	7	MCR	9,48	[97]
	2	MCR, thymine	9,20	[98]
	7	MCR, CNS^-	9,04	[99]

Table 50 (continued)

Compound	pH	Method	log k	Ref.
CH_3CH_2OH	6; 9	MCR, Br^-	9,04	[95]
	7	MCR, I^-	8,86	[94]
	9	MCR, p-$ONC_6H_4N(CH_3)_2$	9,04	[101]
		Mean	8,95	
CD_3CD_2OH	6	MCR, Br^-	8,83	[95]
$CH_3CH_2CH_2OH$	1	MCR, Fe^{2+}	8,92	[102]
	7	MCR, CNS^-, SeO_3^{2-}	9,175	[98]
	9	MCR, p-$ONC_6H_4N(CH_3)_2$	9,22	[95]
	2	MCR, thymine	29,415	[98]
$(CH_3)_2CHOH$	1	MCR, Fe^{2+}	8,98	[102]
	9	MCR, $ONC_6H_4N(CH_3)_2$	9,11	[101]
	7	MCR, I^-	9,23	[94]
	6	MCR, Br^-	9,11	[95]
	2	MCR, thymine	9,255	[98]
$(CH_3)_2CDOH$	6	MCR, Br^-	8,935	[95]

$CH_3(CH_2)_3OH$	7	MCR, CNS^-	9,34	[99]
	9	MCR, p-$ONC_6H_4N(CH_3)_2$	9,34	[95]
	2	MCR, thymine	9,57	[98]
$C_2H_5CH(CH_3)OH$	9	MCR, p-$ONC_6H_4N(CH_3)_2$	9,145	[95]
	2	MCR, thymine	9,34	[98]
$(CH_3)_2CHCH_2OH$	9	MCR, p-$ONC_6H_4N(CH_3)_2$	9,32	[95]
	2	MCR, thymine	9,30	[98]
$(CH_3)_3COH$	9	MCR, Br^-	8,45	[95]
	1	MCR, Fe^{2+}	8,32	[102]
$(CH_3)_3COH$	2	MCR, thymine	8,765	[98]
$C_5H_{11}OH$	2	- " -	9,61	[98]
$C_6H_{13}OH$	2	- " -	9,67	[98]
$C_8H_{17}OH$	2	- " -	9,72	[98]
$CH_3(CH_2)_3CH_2OH$	9	MCR, p-$ONC_6H_4N(CH_3)_2$	9,45	[95]
$C_2H_5C(CH_3)_2OH$	9	- " -	9,04	[95]

Table 50 (continued)

Compound	pH	Method	log k	Ref.
$HOCH_2CH_2OH$	9	MCR, p-$ONC_6H_4N(CH_3)_2$	8,955	[95]
	6	MCR, Br^-	8,83	[95]
	1	MCR, Fe^{2+}	8,06	[96]
	7	MCR, CNS^-, SeO_3^{2-}	8,935	[99]
	2	MCR, thymine	9,15	[98]
$CH_3CH(OH)CH_2OH$	6	MCR, Br^-	8,97	[95]
	9	MCR, p-$ONC_6H_4N(CH_3)_2$	8,995	[95]
	1	MCR, Fe^{2+}	6,505	[96]
	2	MCR, thymine	9,18	[98]
$C_2H_5CH(OH)CH_2OH$	4	MCR, Fe^{2+}	6,78	[96]
$[(CH_3)_2COH]_2$	1	MCR, Fe^{2+}	8,60	[96]
	9	MCR, p-$ONC_6H_4N(CH_3)_2$	8,505	[95]
$HO(CH_2)_3OH$	9	MCR, p-$ONC_6H_4N(CH_3)_2$	9,15	[95]
$CH_3CH(OH)CH_2CH_2OH$	9	- " -	9,15	[95]

$HO(CH_2)_4OH$	9	- " -	9,255	[95]
$HO(CH_2)_5OH$	9	- " -	9,32	[95]
$HO(CH_2)_6OH$	9	- " -	9,43	[95]
$HOCH_2CH(OH)CH_2OH$	9	MCR, p-$ONC_6H_4N(CH_3)_2$	9,04	[95]
	9	- " -	9,09	[101]
	7	MCR, CO_3^{2-}	8,62	[104]
	7	MCR, thymine	9,18	[98]
	7	MCR, CNS^-	8,98	[99]
$HOCH_2CH(OH)CH(OH)CH_2(OH)$	9	MCR, p-$ONC_6H_4N(CH_3)_2$	9,08	[95]
$C(CH_2OH)_4$	9	- " -	9,28	[95]
CH_2O	1	MCR, Fe^{2+}	8,98	[96]
CH_2O	1	- " -	9,30	[97]
CH_3CHO	1	- " -	8,845	[102]
CH_3COCH_3	9	MCR, p-$ONC_6H_4N(CH_3)_2$	7,63	[95]
	7	MCR, I^-	7,89	[94]
	10; 7	MCR, C_6H_5COOH	8,80	[100]
	6–7	MCR, CNS^-	7,76	[105]
	2	MCR, thymine	7,78	[98]
		Mean	7,78	

Table 50 (continued)

Compound	pH	Method	log k	Ref.
$CH_3COCH_2CH_3$	6–7	MCR, CNS^-, SeO_3^{2-}	8,73	
$C_2H_5COC_2H_5$	6–7	- " -	8,91	
$CH_3COC_3H_7$	6–7	- " -	9,06	[105]
$CH_3COCHOHCH_3$	2	- " -	8,71	
$CH_3CH_2CH_2CHO$	2	- " -	9,36	
$C_2H_5OC_2H_5$	1	MCR, Fe^{2+}	9,30	[102]
	9	MCR, p-$ONC_6H_4N(CH_3)_2$	8,91	[95]
	7	MCR, I^-	10,59	[94]
	2	MCR, thymine	9,34	[98]
$CH_3OCH_2OCH_3$	9	MCR, p-$ONC_6H_4N(CH_3)_2$	8,53	
$C_2H_5OCH_2OC_2H_5$	9	- " -	8,96	
$CH_3OCH_2CH_2OH$	9	- " -	8,89	
$CH_3OCH_2CH_2OCH_3$	9	- " -	8,97	[95]
$C_2H_5OCH_2CH_2OH$	9	- " -	9,00	
$C_2H_5OCH_2CH_2OC_2H_5$	9	- " -	9,115	
$(HOCH_2CH_2)_2O$	9	- " -	9,08	
$(CH_3CH_2OCH_2CH_2)_2O$	9	- " -	9,28	

Tetrahydrofuran	9	MCR, p-$ONC_6H_4N(CH_3)_2$	9,205	[95]
	1	MCR, Fe^{2+}	9,30	[96]
Dioxane	9	MCR, $ONC_6H_4N(CH_3)_2$	9,04	[95]
	2	MCR, thymine	9,205	[98]
	7	MCR, I^-	9,37	[94]
	4	MCR, $(NH_2CH_2)_2$	8,65	[90]
$\overline{CH_2OCH_2OCH_2O}$	9	MCR, $ONC_6H_4N(CH_3)_2$	8,46	[95]
$\overline{CH(CH_3)OCH(CH_3)OCH(CH_3)O}$	9	- " -	8,78	[95]
$\overline{CH_2CH_2O}$	9	- " -	7,59	[95]
$CH_3\overline{CHCH_2O}$	9	- " -	8,15	[95]
$C_2H_5\overline{CHCH_2O}$	9	- " -	8,65	[95]
$HOCH_2\overline{CHCH_2O}$	9	- " -	8,43	[95]
$\overline{CH_2(CH_2)_4O}$	1	MCR, Fe^{2+}	9,15	[96]
HCOOH	3	MCR, H_2O_2	8,45	[106]
	0	- " -	7,70	[107]
	3	Photochemical	8,40	[108]

Table 50 (continued)

Compound	pH	Method	log k	Ref.
HCOOH	0,1	MCR, Ce^{3+}	8,08	[109]
	2,5–4	MCR, $Fe(CN)_6^{4-}$	8,40	[103]
	3	–	8,48	[97]
	0,4	–	7,62	[97]
	1	MCR, I^-	8,15	[94]
	1	MCR, CNS^-	7,875	[105]
	2	MCR, thymine	8,15	[98]
		Mean	8,13	
CH_3COOH	1	MCR, thymine	7,15	[105]
	1	MCR, I^-	7,15	[94]
	2	MCR, thymine	7,28	[98]
		Mean	7,19	
CH_3CH_2COOH	1	MCR, Fe^{2+}	8,43	[96]
	2	MCR, thymine	8,62	[98]
$ClCH_2COOH$	1	MCR, CNS^-	7,415	[105]

$CH_3(CH_2)_2COOH$	1	MCR, Fe^{2+}	8,36	[96]
	2	MCR, thymine	9,18	[98]
$CH_3CH_2CH(CH_3)COOH$	1	MCR, Fe^{2+}	9,00	[96]
$HOOCCOOH$	2	MCR, thymine	7,49	[98]
$HOOCCH_2COOH$	6–7	MCR, CNS^-	8,255	[105]
	2	MCR, thymine	7,205	[98]
$HOOCCH_2CH_2COOH$	1	MCR, Fe^{2+}	6,98	[96]
	2	MCR, thymine	8,00	[98]
$HOOCCH_2CH_2CH_2COOH$	2	MCR, thymine	8,75	[98]
$HOOC(CH_2)_4COOH$	1	MCR, Fe^{2+}	8,60	[96]
	2	MCR, thymine	9,15	[98]
$HOCH_2COOH$	1	MCR, Fe^{2+}	8,81	[96]
	2	MCR, thymine	8,57	[98]
$CH_3CH(OH)COOH$	1	MCR, Fe^{2+}	8,74	[96]
	1	MCR, CNS^-	8,415	[105]
$CH_3CH(OH)COOH$	2	MCR, thymine	8,715	[105]

Table 50 (continued)

Compound	pH	Method	log k	Ref.
$CH_3CH_2CH(OH)COOH$	1	MCR, Fe^{2+}	8,98	[98]
$HOOC(CH_2)_5COOH$	2	MCR, thymine	9,38	[98]
$HOOC(CH_2)_6COOH$	2	- " -	9,505	[98]
$HOOC(CH_2)_7COOH$	2	- " -	9,57	[98]
$HOOC(CH_2)_8COOH$	2	- " -	9,63	[98]
$HOOCCH(OH)CH_2COOH$	2	- " -	8,63	[98]
$HOOCCH(OH)-CH(OH)COOH$	2	- " -	8,67	[98]
$(HOOCCH_2)_2C(OH)COOH$	2	- " -	8,65	[98]
$ClCH_2COOH$	2	- " -	7,81	[98]
$O=C-C(OH)=C(OH)-CHCH(OH)CH_2OH$ (ring closed through O between first C and CH)	1	MCR, CNS^-	8,83	[105]
$CH_2(COOH)C(COOH)=CHCOOH$	1	- " -	7,48	[105]
$HCOOC_2H_5$	6-7	MCR, CNS^-	8,36	[105]
CH_3COOCH_3	1	MCR, Fe^{2+}	8,43	[98]
	2	MCR, CNS^-	7,90	[105]
	8	MCR, $p\text{-}ONC_6H_4N(CH_3)_2$	7,81	[95]
	6-7	MCR, CNS^-	7,845	[105]

$CH_3COOC_2H_5$	1	MCR, Fe^{2+}	8,54	[102]
	2	MCR, thymine	8,34	[98]
	6–7	MCR, CNS^-	8,38	[105]
$CH_3COOCH(CH_3)_2$	6–7	MCR, CNS^-	8,43	[105]
$CH_3COOCH(CH_3)_2$	1	MCR, Fe^{2+}	8,60	[96]
$CH_3COOCH_2CH_2CH_3$	6–7	MCR, CNS^-	8,91	[105]
$CH_3CH_2COOCH_3$	6–7	- " -	8,43	[105]
$CH_3CH_2COOC_2H_5$	6–7	- " -	8,40	[105]
$CH_3(CH_2)_2COOCH_3$	6–7	- " -	9,00	[105]
$CH_3(CH_2)_2COOC_2H_5$	6–7	- " -	8,98	[105]
$HSCH_2COOH$	1	MCR, Fe^{2+}	8,935	[96]
$HSCH_2CH(NH_2)COOH$	1	MCR, CNS^-	9,90	[105]
	2	MCR, thymine	9,67	[98]
$HOOCCH_2-S-CH_2COOH$	1	MCR, CNS^-	9,555	[105]
NH_2CH_2COOH	1	MCR, Fe^{2+}	7,04	[96]
	6	MCR, $Fe(CN)_6^{4-}$	6,43	[103]
	9,5–10,5	- " -	9,115	[103]
	10,5	MCR, C_6H_5COOH	9,255	[100]
	1	MCR, CNS^-	7,00	[105]

Table 50 (continued)

Compound	pH	Method	log k	Ref.
NH_2CH_2COOH	2,8–3,0	MCR, CNS^-	6,78	[110]
	5,8–6,0	- " -	7,00	[110]
	9,5–9,7	- " -	9,[illegible]4	[110]
$CH_3CH(NH_2)COOH$	1	MCR, Fe^{2+}	7,58	[96]
	6	MCR, $Fe(CN)_6^{3-}$	7,34	[111]
	9,8	- " -	8,80	[111]
	5,5–6,0	MCR, CNS^-	6,66	[110]
	2	MCR, thymine	7,58	[98]
$CH_2(OH)CH(NH_2)COOH$	2,0–2,2	MCR, CNS^-	8,175	[110]
	5,5–6,0	- " -	8,28	[110]
$CH_2(SCH)_3CH_2CH(NH_2)COOH$	5,5–5,7	MCR, CNS^-	9,69	[110]
$CH_3SCH_2CH_2CH(NH_2)COOH$	6–7	- " -	9,71	[105]
$(CH_3)_2CHCH_2CH(NH_2)COOH$	2,0–2,2	MCR, CNS^-	8,97	[110]
	5,5–6,0	- " -	8,99	[110]
	9,7–9,9	- " -	9,34	[110]

$HOOCCH(NH_2)CH_2COOH$	6,8–7,0	MCR, CNS^-	7,65	[110]
	2	MCR, thymine	7,415	[98]
$H_2NC(=NH)NH(CH_2)_3CH(NH_2)COOH$	2,0–2,2	MCR, CNS^-	8,60	[110]
	6,5–7,5	- " -	9,32	[110]
$CH_3CH_2CH(NH_2)COOH$	2	MCR, thymine	8,48	[98]
$CH_3CH(NH_2)CH_2COOH$	2	- " -	7,80	
$NH_2(CH_2)_3COOH$	2	- " -	8,23	
$HOOC(CH_2)_2CH(NH_2)COOH$	2	- " -	8,115	
$CH_3(CH_2)_3CH(NH_2)COOH$	2	- " -	9,38	
$H_2N(CH_2)_3CH(NH_2)COOH$	2	- " -	8,15	
$C_6H_5CH_2CH(NH_2)COOH$	2	- " -	9,78	
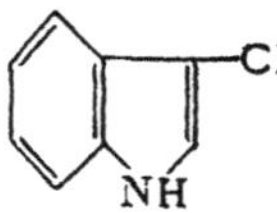 $CH_2CH(NH_2)COOH$	2	- " -	9,79	
$CH_3(CH_2)_2CH(NH_2)COOH$	2	- " -	9,08	
$(CH_3CH_2OCO)_2CH_2$	6–7	MCR, CNS^-	8,59	[105]
$C_2H_5OCO(CH_2)_2COOC_2H_5$	6–7	- " -	8,67	[105]
CH_3NO_2	9	MCR, p-$ONC_6H_4N(CH_3)_2$	8,255	[95]

Table 50 (continued)

Compound	pH	Method	log k	Ref.
CH_3NH_2	12	MCR, $p\text{-}ONC_6H_4N(CH_3)_2$	9,15	[95]
$NH_2CH_2CH_4NH_2$	5	MCR, $(CH_3)_2CHOH$	7,175	[112]
	4	MCR, C_2H_5OH	6,845	[90]
CH_3CONH_2	9	MCR, $ONC_6H_4N(CH_3)_2$	6,89	[95]
CH_3CN	9	- " -	6,32	[95]
CH_3CH_2SH	6–7	MCR, CNS^-	9,71	[105]
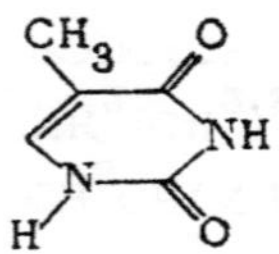	1	- " -	9,49	[105]
$HOOCCH(NH_2)(CH_2)_2CONHCH(CH_2SH)CONHCH_2COOH$	1	- " -	9,945	[105]
$HOOCCH_2NHCOCH_2NH_2$	6–7	- " -	8,20	[105]
Glucose	2	MCR, thymine	9,23	[98]
Saccharose	2	- " -	9,34	[98]

TABLE 51. Partial Rate Constants for Reaction of Hydroxyl Radicals with C-H Bond in Aqueous Solution at 23° and pH 9 (MCR method)

p-$ONC_6H_4N(CH_3)_2$

$k(HO^{\cdot} + ONC_6H_4N(CH_3)_2) = 7{,}5 \cdot 10^9$ [95]

Compound	k_{C-H}	Compound	k_{C-H}
CH_4	$3{,}6 \cdot 10^7$	CH_3OR	$6 \cdot 10^7$
RCH_3	$3{,}6 \cdot 10^7$	RCH_2OR	$1{,}4 \cdot 10^8$
R_2CH_2	$3{,}0 \cdot 10^8$	$(RO)_2CH_2$	$5 \cdot 10^7$
R_3CH	$8{,}4 \cdot 10^8$	$(RO)_2CHR$	$9 \cdot 10^7$
CH_3COOR	$5{,}0 \cdot 10^6$	CH_3OCOR	$2 \cdot 10^7$
RCH_2COOR	$6{,}5 \cdot 10^7$	RCH_2OCOR	$5 \cdot 10^7$
$CH_2(COOR)_2$	$<1 \cdot 10^6$	$ROCR_2CH_2OH$	$1{,}6 \cdot 10^8$
CH_3OH	$1{,}8 \cdot 10^8$	$HOCR_2CH_2OH$	$2{,}2 \cdot 10^8$
RCH_2OH	$5{,}0 \cdot 10^8$	$HOCR_2CH_2CR_2OH$	$2{,}2 \cdot 10^8$
R_2CHOH	$8{,}8 \cdot 10^8$	$HOCR_2CH(OH)R$	$4{,}2 \cdot 10^8$

TABLE 52. Rate Constants for Reaction $(CH_3)_3CO^{\bullet} + RH \rightarrow (CH_3)_3COH + R^{\bullet}$ (Note: k_d relates to reaction $(CH_3)_3CO^{\bullet} \rightarrow CH_3COCH_3 + CH_3^{\bullet}$)

RH	Solvent	t^{o}	k/k_d	$\frac{k}{k_{toluene}}$	logk	Ref.
$C_6H_5CH_3$	CCl_4	24			3,635	
	$C_2Cl_3F_3$	24			3,68	
	C_6H_6	24			4,03	[46]
	C_6H_5CN	24			3,32	
	$C_6H_5CH_3$	135	0,22			[113]
	$C_2Cl_3F_3$	45	2,19			[114]
	C_6H_5Cl	45	1,31			[114]
	- " -	130	0,094			[115]
	CH_3CN	45	0,52			[114]
$p\text{-}CH_3C_6H_4CH_3$	CCl_4	24			3,845	[46]
	- " -	40		1,47	4,08	[119]
	RH	135	0,48			[113]
	$C_2Cl_3F_3$	24			4,04	[46]
	- " -	45	2,88			[114]
	C_6H_5Cl	45	1,73			[114]

	CH_3CN	45	0,663			[114]
	C_6H_6	139,6		1,52		[116]
	RH	165	0,142			[117]
	RH	40		1,53		[118]
$m\text{-}CH_3C_6H_4CH_3$	$C_2Cl_3F_3$	45	2,35			[114]
	C_6H_5Cl	45	1,32			[114]
	CH_3CN	45	0,555			[114]
	CCl_4	40		1,07	3,93	[119]
	RH	40		1,17		[118]
	C_6H_6	39,6		1,14		[116]
$p\text{-}C_6H_5C_6H_4CH_3$	C_6H_6	39,6		1,64		[116]
$p\text{-}C_6H_5OC_6H_4CH_3$	$C_2Cl_3F_3$	45	3,10		3,37	[114]
	C_6H_5Cl	45	1,82		3,14	[114]
	CH_3CN	45	0,873		2,82	[114]
	CCl_4	40		2,13	4,23	[119]
	C_6H_6	39,6		1,42		[116]
$p\text{-}CH_3OC_6H_4CH_3$	CCl_4	40		3,2	4,41	[119]
	C_6H_5Cl	130	0,0174			[115]

Table 52 (continued)

RH	Solvent	t^{o}	k/k_d	$\frac{k}{k_{toluene}}$	log k	Ref.
$C_6H_5OCH_3$	C_6H_5Cl	130	0,0136			
p-$ClC_6H_4OCH_3$	- " -	130	0,0112			
m-$ClC_6H_4OCH_3$	- " -	130	0,0100			[115]
p-$CNC_6H_4OCH_3$	- " -	130	0,0078			
m-$CH_3OC_6H_4OCH_3$	- " -	130	0,0113			
p-$ClC_6H_4CH_3$	$C_2Cl_3F_3$	45	2,19			
	C_6H_5Cl	45	1,10			[114]
	CH_3CN	45	0,462			
	CCl_4	40		0,89	3,85	[119]
	RH	40		0,71		[118]
	C_6H_6	39,6		0,67		[116]
	$C_6H_5CH_3$	110		0,90		[120]
m-$ClC_6H_4CH_3$	$C_2Cl_3F_3$	25			3,255	[46]
	$C_2Cl_3F_3$	45	1,68		3,11	
	C_6H_5Cl	45	0,89			[114]
	CH_3CN	45	0,36			

	$C_6H_5CH_3$	110		0,70		[120]
	C_6H_6	39,6		0,52		[116]
	CCl_4	40		0,59	3,67	[119]
	RH	40		0,52		[118]
o–$ClC_6H_4CH_3$	$C_6H_5CH_3$	110		0,73		[120]
m–$CNC_6H_4CH_3$	$C_2Cl_3F_3$	45	1,37		3,02	
	C_6H_5Cl	45	0,87			[114]
	CH_3CN	45	0,31			
p–$CNC_6H_4CH_3$	$C_2Cl_3F_3$	45	1,24		2,97	
	C_6H_5Cl	45	0,82			[114]
	CH_3CN	45	0,31			
	C_6H_6	39,6		0,29		[116]
	CCl_4	40		0,36	3,46	[119]
p–Br $C_6H_4CH_3$	C_6H_6	39,6		0,71		[116]
p–$NO_2C_6H_4CH_3$	CCl_4	40		0,25	3,30	[119]
m–$NO_2C_6H_4CH_3$	CCl_4	40		0,29	3,37	[119]
$C_6H_5CH_2CH_3$	RH	40		2,1		[121]
$C_6H_5CH_2CH_3$	RH	135	0,59			[113]

Table 52 (continued)

RH	Solvent	t^{o}	k/k_d	$\frac{k}{k_{toluene}}$	log k	Ref.
$C_6H_5CH_2CH_3$	$C_6H_5CH_3$	110		2,6		[120]
	RH	135	0,32			[122]
$C_6H_5(CH_3)_2CH$	RH	135	0,47			[113]
	RH	40		2,3		[121]
	$C_6H_5CH_3$	110		3,1		[120]
	RH	135	0,50			[122]
$n\text{-}C_4H_9C_6H_5$	RH	135	0,67			[113]
$C_6H_5C(CH_3)_3$	RH	40		0,275		[121]
$C_6H_5(CH_3)_3$	$C_2Cl_3F_3$	24			3,04	[46]
$1,3,5\text{-}(CH_3)_3C_6H_3$	RH	135	0,12			[122]
$(C_6H_5)_2CH_2$	RH	135	0,61			[113]
$(C_6H_5)_2CH_2$	RH	165	0,228			[117]
$(C_6H_5)_3CH$	CCl_4	24			3,505	[46]
$C_6H_5CH_2CH{=}CH_2$	RH	165	0,465			[114]
$p\text{-}CH_3C_6H_4CH_2CH{=}CH_2$	RH	165	0,529			[114]

p-$CH_3OC_6H_4CH_2CH{=}CH_2$	RH	165	1,21		[117]
p-$(CH_3)_2CHC_6H_4CH(CH_3)_2$	RH	165	0,523		[117]
p-$(CH_3)_2CHC_6H_4CH_3$	$C_6H_5CH_3$			3,2	[120]
cyclo-C_5H_9-C_6H_5	RH	135	1,49		[113]
cyclo-C_6H_4-C_6H_5	RH	135	1,02		[113]
	RH	135	1,25		[123]
Tetralin	RH	135	3,27		[113]
	RH	135	3,0		[122]
α-Methylnaphthalene	RH	135	0,51		[122]
β-Methylnaphthalene	RH	135	0,3		[122]
p-$CH_3C_6H_4C_6H_5$	$C_6H_5CH_3$	110		1,75	[120]
β-Methylpyridine	- " -	110		0,44	[120]
γ-Methylpyridine	- " -	110		0,33	[120]
C_6H_6	RH	135	$3{,}2 \cdot 10^{-2}$		[124]
	RH	135	$1{,}8 \cdot 10^{-3}$		[122]
	RH	135	$2{,}2 \cdot 10^{-2}$		[125]
Pyridine	RH	135	0,105		[125]
IC_6H_5	RH	135	$8{,}5 \cdot 10^{-2}$		[125]

Table 52 (continued)

RH	Solvent	t°	k/k_d	$\frac{k}{k_{toluene}}$	log k	Ref.
C_6H_5CN	RH	135	$7{,}1\cdot10^{-2}$			[125]
	RH	165	$9{,}6\cdot10^{-3}$			[117]
$C_6H_5NO_2$	RH	135	$7{,}0\cdot10^{-2}$			[125]
$(C_6H_5)_3N$	RH	135	0,173			[125]
C_6H_5F	RH	135	$5{,}0\cdot10^{-2}$			[125]
C_6H_5Cl	RH	135	$4{,}5\cdot10^{-2}$			[125]
C_6H_5Br	RH	135	$2{,}0\cdot10^{-2}$			[125]
$C_6H_5C_6H_5$	RH	135	$5{,}0\cdot10^{-2}$			[125]
	RH	165	$1{,}3\cdot10^{-2}$			[117]
$C_6H_5OC_6H_5$	RH	135	0,085			[125]
$C_6H_5COC_6H_5$	RH	135	0,094			[125]
	RH	165	0,0305			[117]
$n\text{-}C_4H_{10}$	RH	40		1,6		[121]
$(CH_3)_2CHCH(CH_3)_2$	RH	135	0,83			[113]

n-C_7H_{16}	RH	135	0,70		
$(CH_3)_2CH(CH_2)_2CH_3$	RH	135	0,59		
$C_2H_5CH(CH_3)C_2H_5$	RH	135	0,69		[113]
$(CH_3)_2CHCH_2CH(CH_3)_2$	RH	135	0,49		
$(CH_3)_3CCH_2CH(CH_3)_2$	RH	135	0,26		
$(CH_3)_2CH(CH_2)_2CH(CH_3)_2$	RH	135	0,89		
n-$C_{16}H_{34}$	RH	135	2,2		[122]
$CH_2{=}CHCH_2CH_3$	RH	40		4,0	
$CH_2{=}C(CH_3)_2$	RH	40		2,4	
cis-$CH_3CH{=}CHCH_3$	RH	40		4,2	
trans-$CH_3CH{=}CHCH_3$	RH	40		4,0	
$(CH_3)_2C{=}C(CH_3)_2$	RH	40		17	
$CH_2{=}CHCH(CH_3)_2$	RH	40		5,8	[121]
cis-$CH_3CH{=}CHC_2H_5$	RH	40		6,1	
trans-$CH_3CH{=}CHC_2H_5$	RH	40		6,3	
cyclo-C_5H_8	RH	40		50,5	
cyclo-C_6H_{10}	RH	40		48,6	
cyclo-C_5H_{10}	RH	40		4,6	
cyclo-C_5H_9-C_6H_5	RH	135	0,60		[113]
cyclo-C_5H_9-cyclo-C_5H_9	RH	135	3,22		[113]

Table 52 (continued)

RH	Solvent	t°	k/k_d	$\frac{k}{k_{toluene}}$	log k	Ref.
cyclo-C_6H_{12}	RH	135	1,17			[113]
	$C_2Cl_3F_3$	24			4,415	[46]
	CH_3COOH	24			4,28	[46]
	CS_2	24			2,78	[46]
	RH	135	0,20			[122]
	RH	40		6,0		[118]
	RH	40		6,0		[121]
	C_6H_6	135	1,59			[123]
cyclo-$C_6H_{11}OCH_3$	C_6H_6	135	2,03			[123]
cyclo-$C_6H_{11}C_2H_5$	C_6H_6	135	1,61			[123]
	RH	135	1,72			[113]
cyclo-$C_6H_{11}CH_3$	RH	135	1,55			[113]
	RH	135	1,44			[123]

cyclo $-C_6H_{11}COCl$	RH	135	1,25	[123]
cyclo $-C_6H_{11}OCOCH_3$	RH	135	1,01	
cyclo $-C_6H_{11}OH$	RH	135	0,84	
cyclo $-C_6H_{11}Cl$	RH	135	0,718	
cyclo $-C_6H_{11}COOCH_3$	RH	135	0,973	
cyclo $-C_6H_{11}CN$	RH	135	0,573	
cyclo $-C_6H_{11}COOH$	RH	135	0,293	
cyclo $-C_6H_{11}NO_2$	RH	135	0,282	
cyclo $-C_6H_{11}NH_2$	RH	135	0,770	
cyclo $-C_6H_{11}NHCH_3$	RH	135	5,75	
cyclo-$C_6H_{11}N(CH_3)_2$	RH	135	6,03	
(cyclo $-C_6H_{11})_2$	RH	135	2,17	[113]
(cyclo $-C_6H_{11})_2CH_2$	RH	135	1,94	[123]
cyclo-$C_6H_{11}-CH_2CH_2$cyclo-C_6H_{11}	RH	135	2,72	
cis-Tetrahydroindan	RH	135	2,38	
trans-Tetrahydroindan	RH	135	2,12	
cis-Decalin	RH	135	2,85	
trans-Decalin	RH	135	2,38	[113]
$\overline{CH_2(CH_2)_2O}$	RH	135	0,045	[124]
$\overline{CH_2(CH_2)_3O}$	RH	135	0,47	
$\overline{CH_2(CH_2)_4O}$	RH	135	0,45	

Table 52 (continued)

RH	Solvent	t^{o}	k/k_d	$\frac{k}{k_{toluene}}$	log k	Ref.
$C_6H_5OCH_2\overset{O}{\widehat{CH-CH_2}}$	RH	135	0,18			[124]
$C_2H_5\underset{O}{\widetilde{CH-CH_2}}$	RH	135	0,08			[124]
$n\text{-}C_6H_{13}\overset{O}{\widehat{CH-CH_2}}$	RH	135	0,21			[124]
$CH_3\text{-}\overset{O}{\overset{\Vert}{C}}\text{-}\overset{O}{\widehat{CH-C}}(CH_3)\text{-}CH_3$	RH	135	0,10			[124]
$C_6H_5\overset{O}{\widehat{CH-CH_2}}$	RH	135	0,34			[124]
$CH_3\text{-}\overset{O}{\widehat{CH-CH_2}}$	RH	135	0,035			[124]
CH_3COCH_3	RH	135	0,023			[126]
cyclo $-C_5H_8O$	RH	40		1,53		[118]
cyclo $-C_6H_{10}O$	RH	135	0,467			[126]
$C_2H_5COC_2H_5$	RH	135	0,215			[126]
$n\text{-}C_3H_7OCOCH_3$	RH	40		0,64		[118]

$C_6H_5CH_2OCOC_6H_5$	RH	165	0,097			[117]
$C_6H_5CH_2COC_6H_5$	RH	165	0,094			[117]
n-C_4H_9Cl	RH	40		1,48		
$C_6H_5CH_2Cl$	RH	40		0,48		[118]
n-C_3H_7CN	RH	40		0,18		
C_6H_5OH	CCl_4	122	540		7,47	
m-$CH_3C_6H_4OH$	CCl_4	122	380		7,12	
p-$CH_3C_6H_4OH$	- " -	122	1460		7,905	
m-$CH_3OC_6H_4OH$	- " -	122	400		7,34	
p-ClC_6H_4OH	- " -	122	380		7,12	[127]
m-FC_6H_4OH	- " -	122	192		7,02	
m-ClC_6H_4OH	- " -	122	192		7,02	
p-CNC_6H_4OH	- " -	122	88		6,685	
p-$(CH_3)_3CC_6H_4OH$	- " -	122	1080		7,77	
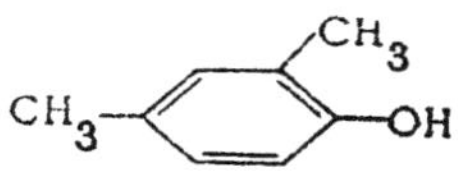	CCl_4	122	1020		7,75	[127]
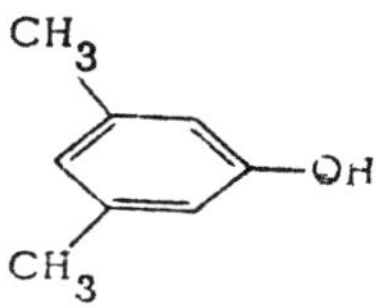	CCl_4	122	740		7,61	[127]

Table 52 (continued)

RH	Solvent	t°	k/k_d	$\frac{k}{k_{toluene}}$	log k	Ref.
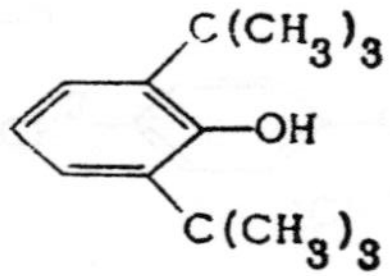	CCl_4	122	27		6,17	[127]
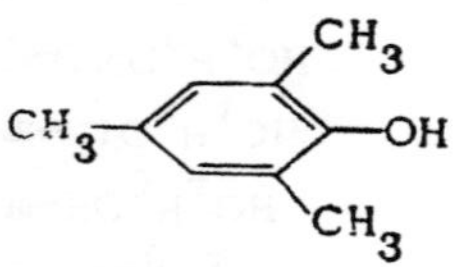	- " -	122	1340		7,87	[127]
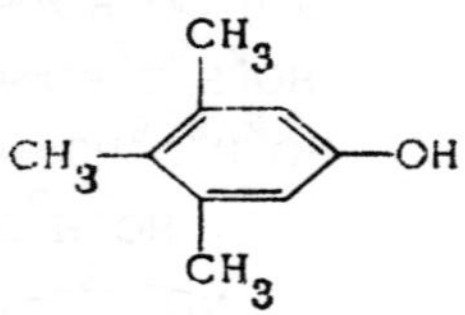	- " -	122	1400		7,885	[127]

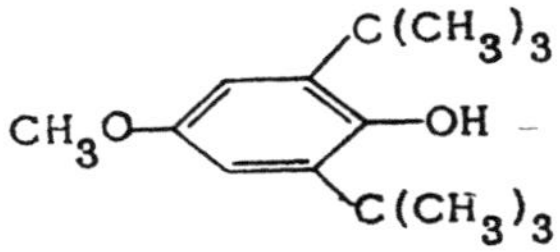	- " -	122	180	6,995	[127]
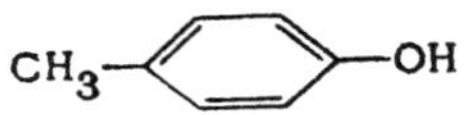	- " -	122	60	6,52	[127]
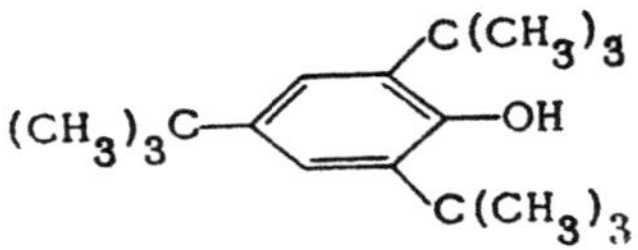	- " -	122	56	6,49	[127]
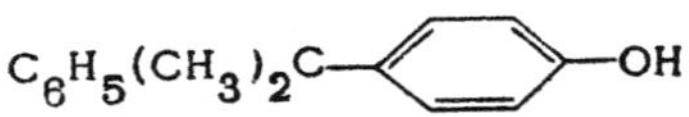	- " -	122	56	6,49	[127]
	- " -	122	36	6,295	[127]
	- " -	122	9,6	5,72	[127]

Table 52 (continued)

RH	Solvent	t°	k/k_d	$\frac{k}{k_{toluene}}$	log k	Ref.
	CCl_4	122	3,6		5,295	[127]
 $R = CH(CH_3)C_2H_5$	- " -	122	620		7,53	[127]
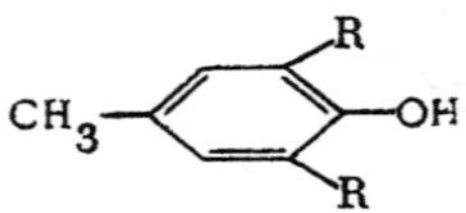						
$R = C(CH_3)_2C_2H_5$	CCl_4	122	46		6,70	[127]
$(CH_3)_2CH$	- " -	122	380		7,685	
$(CH_3)_3CCH_2(CH_3)_2C$	- " -	122	46		6,70	

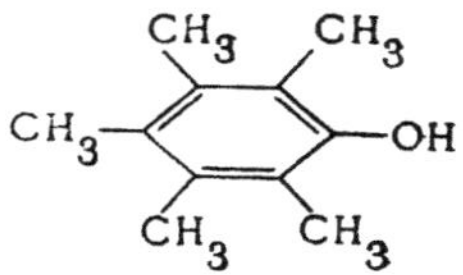	- " -	122	800	7,64	[127]
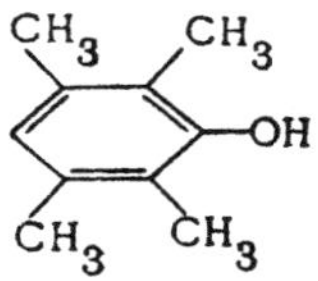	- " -	122	680	7,57	[127]
$C_6H_5N(CH_3)_2$	C_6H_5Cl	130	7,74		[115]
p-$(CH_3)_2NC_6H_4N(CH_3)_2$	- " -	130	17,2		[115]
p-$C_6H_5OC_6H_4N(CH_3)_2$	- " -	130	8,84		
p-$ClC_6H_4N(CH_3)_2$	- " -	130	6,52		
m-$ClC_6H_4N(CH_3)_2$	- " -	130	5,81		
p-$CNC_6H_4N(CH_3)_2$	- " -	130	3,66		
$C_6H_5SCH_3$	- " -	130	0,20		
p-$C_6H_5OC_6H_4SCH_3$	- " -	130	0,23		

Table 52 (continued)

RH	Solvent	t°	k/k_d	$\frac{k}{k_{toluene}}$	log k	Ref.
p-$CH_3C_6H_4SCH_3$	C_6H_5Cl	130	0,24			
p-$ClC_6H_4SCH_3$	- " -	130	0,19			
m-$ClC_6H_4SCH_3$	- " -	130	0,18			
p-$CNC_6H_4SCH_3$	- " -	130	0,17			
$C_6H_5SO_2Cl$	- " -	130	$4{,}8\cdot10^{-3}$			
p-$C_6H_5OC_6H_4SO_2Cl$	- " -	130	$5{,}3\cdot10^{-3}$			[115]
p-$ClC_6H_4SO_2Cl$	- " -	130	$4{,}7\cdot10^{-3}$			
p-$CNC_6H_4SO_2Cl$	- " -	130	$4{,}0\cdot10^{-3}$			
$(C_6H_5S)_2CH_2$	- " -	130	0,94			
$(C_6H_5)_3CH$	- " -	130	1,31			
$(C_6H_5O)_2CH_2$	- " -	130	0,39			
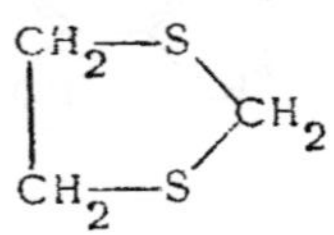	- " -	130	4,6			
$(CH_3C{\equiv}C)_2S$	- " -	130	0,67			

TABLE 53. Partial Rate Constants for Reaction $(CH_3)_3CO^{\bullet} + H\text{-}C\Leftarrow \rightarrow (CH_3)_3COH + {}^{\bullet}C\Leftarrow$ (40°)

Bond type	k_{C-H}	Bond type	k_{C-H}	Bond type	k_{C-H}
RCH_2-H	$2,6\cdot10^2$	RAr CH–H	$8,3\cdot10^3$	$>C{=}C(CH_3)CHR$–H	$1,6\cdot10^4$
R_1R_2CH–H	$3,2\cdot10^3$	R_1R_2Ar C–H	$1,76\cdot10^4$	$>C{=}C(CH_2R)CHR$–H	$2,4\cdot10^4$
$R_1R_2R_3$C–H	$1,14\cdot10^4$	$>C{=}C(CH_3)CH_2$–H	$3,1\cdot10^3$	$(CH_2)_n$ CH=CHCH–H	$8,75\cdot10^4$
Ar CH_2–H	$2,6\cdot10^3$	$>C{=}C(CH_2R)CH_2$–H	$5,2\cdot10^3$	$>C{=}CC(CH_3)_2$–H	$4,6\cdot10^4$

TABLE 54. Abstraction of H by Cumyloxy Radicals

RH	Solvent	t°	k/k_d	$k/k_{C_6H_5OH}$	Method	Ref.
$C_6H_5CH_3$	RH	140	$7,2\cdot10^{-2}$		MCR	[128]
$C_6H_5C_2H_5$	RH	140	0,10		MCR	[128]
$C_6H_5CH(CH_3)_2$	RH	140	$8,7\cdot10^{-2}$		MCR	[128]
$(C_6H_5)_2CH_2$	RH	140	$8,2\cdot10^{-2}$		MCR	[128]
p-$CH_3C_6H_4CH_3$	RH	140	0,12		MCR	[128]
cyclo-$C_6H_{11}C_6H_5$	RH	140	0,28		MCR	[128]
Tetralin	RH	140	0,73		MCR	[128]
o-$CH_3C_6H_4OH$	n-$C_{10}H_{22}$	25		1,8	CINH	[129]

Table 54 (continued)

RH	Solvent	t^{o}	k/k_d	$k/k_{C_6H_5OH}$	Method	Ref.
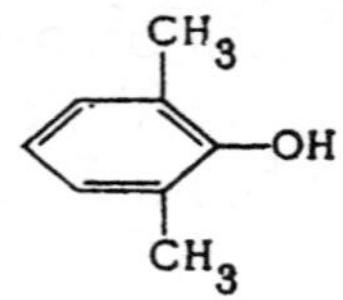	$n-C_{10}H_{22}$	25		3,8	CINH	[129]
$p-HOC_6H_4OH$	$n-C_{10}H_{22}$	25		2,6	CINH	[129]
$m-HOC_6H_4OH$	- " -	25		1,01	CINH	
$o-HOC_6H_4OH$	- " -	25		3,9	CINH	
α-Naphthol	- " -	25		13,2	CINH	
β-Naphthol	- " -	25		2,2	CINH	
β-Thionaphthol	- " -	25		10,7	CINH	
$C_6H_5NHCH_3$	- " -	25		2,7	CINH	
α-Naphthylamine	- " -	25		3,8	CINH	
Phenyl-β-naphthylamine	- " -	25		2,5	CINH	
$(C_6H_5)_2\dot{N}H$	- " -	25		1,1	CINH	
Naphthyl-p-phenylenediamine	" -	25		3,9	CINH	
$CO(NH_2)_2$	- " -	25		0,08	CINH	
CH_3CSNH_2	- " -	25		2,3	CINH	
$CH_3CSNHC_6H_5$	- " -	25		6,2	CINH	
$C_6H_5CSNHC_6H_5$	- " -	25		8,4	CINH	

TABLE 55. Rate Constants for Reaction $RO_2^{\bullet} + RH \rightarrow ROOH + R^{\bullet}$ in RH Medium (Note: If reaction was studied in solvent, the solvent is listed in parentheses)

RH	t°	log k (30°)	log A	E	Method	Ref.
$(CH_3)_2CHCH(CH_3)CH(CH_3)_2$	40–80	$\bar{2}$,72	5,28	9,1	IIM	[130]
cyclo-C_6H_{12}	60	$\bar{1}$,725			NKR	[131]
$CH_2{=}CH(CH_2)_5CH_3$	25–40	$\bar{2}$,570	3,12	6,3	NKR	[132]
	30	0,00			IIM	[133]
$(CH_3)_2C{=}CHCH_2CH_3$	40–60	0,05	6,40	8,8	NKR	[131]
cyclo-C_6H_{10}	15	0,04			IIM	[132]
	20–50	0,42	5,47	7,0	NKR	[134]
	56	0,845			NKR	[136]
	15	$\bar{1}$,81			IIM	[137]
	30	0,825			NKR	[16]

Table 55 (continued)

RH	t°	log k (30°)	log A	E	Method	Ref.
cyclo-C_6H_{10} (C_6H_5Cl)	30	0,785			IIM	[138]
1-Methylcyclohexene	15	0,04			IIM	[132]
	15–50	0,60	4,94	6,0	NKR	[135]
4-Methylcyclohexene	15–50	0,64	7,49	9,5	NKR	[135]
4,5-Dimethylcyclohexene	15–50	1,06	7,06	8,3	NKR	[135]
$CH_3(CH_2)_4CH{=}CH{-}CH_2{-}CH{=}CH(CH_2)_7COOC_2H_5$	11	0,755			IIM	[132]
	25	0,82			NKR	[132]
$(CH_3)_2C{=}CH(CH_2)_2C(CH_3){=}CHCH_3$	0–25	$\bar{1}$,415 (25°)			NKR	[132]
	15	$\bar{1}$,60			IIM	[137]
$CH_2{=}\overset{\overset{CH_3}{\vert}}{C}(CH_2)_3\overset{\overset{CH_3}{\vert}}{C}{=}CH(CH_2)_2CH{=}\overset{\overset{CH_3}{\vert}}{C}(CH_2)_3\overset{\overset{CH_3}{\vert}}{C}{=}CH_2$	25–40	$\bar{1}$,875			NKR	[132]
$CH_3(CH_2)_3C{=}C(CH_2)_3CH_3$	30	0,45			IIM	[133]
$(CH_3)_2C=C(CH_3)_2$	30	0,23			IIM	[139]
$CH_3CH_2CH{=}CHCH_2CH_2CH_3$	30	0,145			IIM	[139]
$CH_2{=}CH(CH_2)_5CH_3$	30	0,00			IIM	[139]

$CH_3(CH_2)_7CH{=}CH(CH_2)_7COOCH_3$	30	$\bar{1}$,945			IIM	
$CH_3(CH_2)_4CH{=}CHCH_2CH{=}CH(CH_2)_7COOCH_3$	30	1,79			IIM	
$CH_3(CH_2CH{=}CH)_3(CH_2)_7COOH$	30	2,37			IIM	
$(CH_3)_2CHCH{=}CHCH(CH_3)_2$	30	0,38			IIM	
cyclo -C_5H_8	30	0,83			IIM	
Cyclopentadiene	30	1,145			IIM	
1,3- Cyclohexadiene	30	2,34			IIM	[139]
1,4- Cyclohexadiene	30	3,17			IIM	
$CH_3(CH_2)_3C{\equiv}C(CH_2)_3CH_3$	30	0,45			IIM	
$C_6H_5CH_3$	30	$\bar{1}$,38			IIM	
m -$CH_3C_6H_4CH_3$	30–65	$\bar{1}$,38	6,89	10,0	IIM	
o-$CH_3C_6H_4CH_3$	30	$\bar{1}$,32			IIM	
p -$CH_3C_6H_4CH_3$	30	$\bar{1}$,62			IIM	
$C_6H_5CH_2CH_3$	50–90	$\bar{1}$,84	5,98	3,5	IIM	[140]
	30	0,04			IIM	[138]
	30	0,115			IIM	[133]
$C_6H_5(CH_3)_2CH$	50	$\bar{1}$,49			IIM	[141]
	65	$\bar{1}$,75			IIM	[141]
	56	0,04			NKR	[136]
	25–65	$\bar{1}$,19	6,51	10,13	IIM	[142]
	40	$\bar{1}$,61			NKR	[16]
	30	$\bar{1}$,255			IIM	[139]

Table 55 (continued)

RH	t°	log k (30°)	log A	E	Method	Ref.
p-$CH_3C_6H_4CH(CH_3)_2$	56	$\bar{1}$,70	3,70	6,0	NKR	[136]
p-$CH_3OC_6H_4CH(CH_3)_2$	30	$\bar{1}$,60			IIM	
m-$CH_3OC_6H_4CH(CH_3)_2$	30	$\bar{1}$,23			IIM	[133]
p-$CH_3OCOC_6H_4CH(CH_3)_2$	30	$\bar{1}$,175			IIM	
$C_6H_5(CH_2)_3CH_3$	30	$\bar{1}$,75			IIM	[139]
$C_6H_5CH(CH_3)C_2H_5$	30	$\bar{2}$,88			IIM	[143]
$C_6H_5CH(CH_3)C_3H_7$	30	$\bar{2}$,845			IIM	[139]
$C_6H_5CH_2CH_2C_6H_5$	30	$\bar{1}$,75			IIM	[139]
$(C_6H_5)_2CH_2$	30	0,30			IIM	[133]
	30	0,68			IIM	[138]
$(C_6H_5)_2CD_2$	30	$\bar{1}$,975			IIM	[138]
$(C_6H_5)_2CHCH_3$	50–88	$\bar{1}$,30	8,57	12,1		[144]
	30	$\bar{1}$,53			IIM	[139]
cyclo-C_6H_{11}-C_6H_5	30	$\bar{2}$,78			IIM	[139]
Indan	30	0,68			IIM	[139]
Tetralin	25	1,125			IIM	[145]
	13–50	0,75	6,75	8,3	IIM	[138]

	40	0,965			NKR	[16]
	30	0,81			IIM	[133]
Tetralin (n-$C_{10}H_{22}$)	30	0,66			IIM	[138]
(C_6H_5Cl)	30	0,805			IIM	[138]
(CH_3CN)	30	1,00			IIM	[138]
$(CH_3)_2SO$	30	0,82			IIM	[138]
9,10-Dihydrophenanthrene	30	1,75			IIM	[139]
9,10-Dihydroanthracene	30	2,245			IIM	[139]
$C_6H_5CH_2CH{=}CH_2$	30	1,00			IIM	[138]
	65	1,60			IIM	[146]
$C_6H_5CH_2CH{=}CHCH_3$	30	0,91			IIM	[139]
Indene (C_6H_5Cl)	30	1,145				[147]
cyclo-$C_6H_{11}OH$	75–90	$\bar{2}$,74	7,43	12,0	NKR	[148]
cyclo-$C_6H_{10}O$	55–80	$\bar{2}$,685	7,375	12,0	NKR	[149]
$CH_3COCH_2CH_3$	35–75	$\bar{1}$,04	5,10	8,4	NKR	[150]
$CH_3COCH_2CH_2CH_3$	50–90	$\bar{2}$,85	5,92	9,8	NKR	[151]
$CH_3COCH(CH_3)_2$	40–80	$\bar{1}$,56	4,76	7,2	NKR	[151]
n-$C_9H_{19}CHO$	5–20	2,86 (5°)	6,16	4,2	IIM	[152]
C_6H_5CHO	5–20	3,28 (5°)	4,70	1,8	IIM	[153]

TABLE 56. Reactivities of Various Peroxy Radicals [154] (Note: Values listed are for $k(R_iO_2^{\bullet} + R_jH)$ at 30°)

R_jH	$HO_2^{\bullet}$*	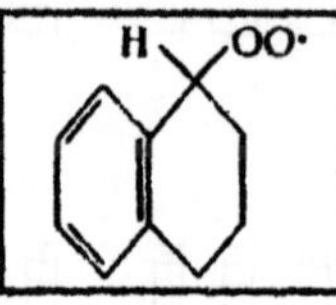	$C_6H_5C(CH_3)_2OO^{\bullet}$
$C_6H_5CH_3$ [2*]	0,7	0,26	0,03
$C_6H_5CH_2CH_3$	1,6	0,50	0,20
Tetralin	13	6,4	1,65
$C_6H_5C(CH_3)_2H$	1,6	0,77	0,18

*$HO_2^{\bullet}$ radical formed by oxidation of dihydroanthracene [155].

2*For reaction $C_6H_5CH_2OO^{\bullet} + CH_3C_6H_5$, constant k = 0.24 at 30°.

TABLE 57. Peroxy-Radical Reactions of the Type $ROO^{\cdot} + R'H \rightarrow ROOH + R'^{\cdot}$
(Note: If reaction was studied not in RH medium but in solvent, the solvent is listed in parentheses. $C_{10}H_{11}OO^{\cdot}$ is the tetralylperoxy radical)

$RO_2^{\cdot}$	R'H	t°	log k	Ref.
$HO_2^{\cdot}$ *	1,4-Cyclohexadiene ($n\text{-}C_{10}H_{22}$)	30	2,53	[155]
	- " - (CCl_4)	30	3,58	[155]
	- " - (C_6H_5Cl)	30	3,15	[155]
	- " - (CH_3CN)	30	2,54	[155]
$HO_2^{\cdot}$ 2*	1,4-Dihydronaphthalene (C_6H_5Cl)	30	2,95	[155]
$HO_2^{\cdot}$	$HO_2^{\cdot} + H_2O_2 \longrightarrow HO^{\cdot} + O_2 + H_2O$ (H_2O)	0	0,04	[156]
	- " -	25	0,57	[156]
	$C_6H_5CH_3$	30	$\bar{1}$,845	[154]
	$C_6H_5CH_2CH_3$	30	0,205	[154]
	$C_6H_5(CH_3)_2CH$	30	0,206	[154]
	Tetralin	30	1,115	[154]
$CH_3(CH_2)_2CH_2OO^{\cdot}$	$C_6H_5CH_3$	30	$\bar{1}$,00	[157]
	$C_6H_5CH_2CH_3$	30	$\bar{1}$,74	[157]

Table 57 (continued)

$RO_2^{\cdot}$	R'H	t°	log k	Ref.
$CH_3(CH_2)_2CH_2OO^{\cdot}$	$C_6H_5CH(CH_3)_2$	30	1,875	[157]
	Tetralin	30	0,81	
	9,10-Dihydroanthracene	30	2,49	
$CH_3CH_2CH(CH_3)OO^{\cdot}$	$C_6H_5CH_3$	30	1,00	[157]
	$C_6H_5CH_2CH_3$	30	1,70	
	$C_6H_5CH(CH_3)_2$	30	1,60	
	Tetralin	30	0,62	
	9,10-Dihydroanthracene	30	2,145	
cyclo-$C_6H_{11}OO^{\cdot}$	$C_6H_5CH_3$	30	1,00	[157]
	$C_6H_5C_2H_5$	30	1,70	
	$C_6H_5CH(CH_3)_2$	30	1,70	
	Tetralin	30	0,80	
	9,10-Dihydroanthracene	30	2,38	
$C_6H_5CH(CH_3)OO^{\cdot}$	$C_6H_5CH(CH_3)_2$	60–80	6,23–8,9/θ	[158]
	$C_6H_5CH_2CH_3$	50–90	5,98–8,5/θ	[140]

$(C_6H_5)_2CHOO^\cdot$	$C_6H_5CH_2CH_3$	30	$\bar{1}$,905	[157]
	$C_6H_5CH(CH_3)_2$	30	$\bar{1}$,70	
	Tetralin	30	0,82	
	9,10-Dihydroanthracene	30	2,40	
$C_6H_5CH(OO)CH{=}CH_2$	o-$CH_3C_6H_4CH_3$	65	0,60	[146]
	$C_6H_5(CH_3)_2CH$	65	0,81	
	$C_6H_5CH_2CH_3$	65	0,79	
	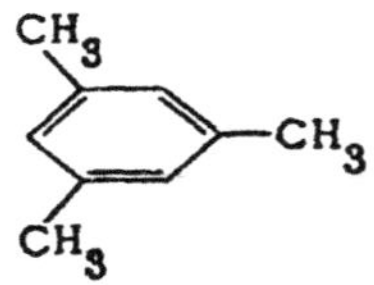	65	0,84	
	$C_6H_5CH_2CH{=}CH_2$	65	1,60	
	Tetralin	65	1,64	
$C_6H_5CH(OO)CH{=}CH_2$	$C_2H_5(CH_3)CH{=}CH(CH_3)C_2H_5$	65	1,83	[146]
	cyclo-C_6H_{10}	65	2,00	[146]
$C_{10}H_{11}OO^\cdot$	$C_6H_5CH_3$	30	$\bar{1}$,00	[157]
	$C_6H_5CH_2CH_3$	30	$\bar{1}$,70	[157]

Table 57 (continued)

$RO_2^{\cdot}$	$R'H$	t^{o}	log k	Ref.
$C_{10}H_{11}OO^{\cdot}$	$C_6H_5CH_2CH_3$	65	0,59	[146]
	- " -	30	$\bar{1}$,70	[157]
	- " -	90	0,505	[159]
	- " -	65	0,415	[146]
	o-$CH_3C_6H_4CH_3$	65	0,30	[146]
	1,3,5-$(CH_3)_3C_6H_3$	65	0,34	[146]
	$C_6H_5CH_2CH{=}CH_2$	65	0,28	[146]
	$C_2H_5(CH_3)CH{=}CH(CH_3)C_2H_5$	65	0,48	[146]
	cyclo-C_6H_{10}	65	0,71	[146]
	Tetralin	65	1,38	[146]
	- " -	30	0,805	[154]
	9,10-Dihydroanthracene	30	2,38	[157]
	$C_6H_5(CH_3)_2COOH$	30	3,40	[157]
	- " -	56	3,45	[154]
$\sim CH_2{-}C(OO^{\cdot})(CH_3)C_6H_5$	o-$CH_3C_6H_4CH_3$	65	1,955	[146]
	$C_6H_5(CH_3)_2CH$	65	0,00	[146]
	$C_6H_5CH_2CH_3$	65	0,145	[146]
	1,3,5-$(CH_3)_3C_6H_3$	65	0,115	[146]

	$C_6H_5CH_2CH{=}CH_2$	65	0,965	
	Tetralin	65	1,08	[146]
	$C_2H_5(CH_3)CH{=}CH(CH_3)C_2H_5$	65	1,175	
	cyclo $-C_6H_{10}$	65	1,43	
	$(CH_3)_3COOH$	60	2,81	[160]
$C_6H_5COOO^{\cdot}$	$n{-}C_9H_{19}CHO$	5	3,98	[161]
$n{-}C_9H_{19}COOO^{\cdot}$	C_6H_5CHO	5	3,49	[161]
$(CH_3)_3COO^{\cdot}$	$C_6H_5CH_3$	30	$\bar{2}$,70	
	$C_6H_5C_2H_5$	30	$\bar{1}$,30	
	$C_6H_5CH(CH_3)_2$	30	$\bar{1}$,34	[143]
	Tetralin	30	0,30	
$(CH_3)_3CCH_2C(CH_3)_2OO^{\cdot}$	$C_6H_5CH_3$	30	$\bar{2}$,60	
	$C_6H_5C_2H_5$	30	$\bar{1}$,505	
	$C_6H_5CH(CH_3)_2$	30	$\bar{1}$,15	[143]
	Tetralin	30	0,30	
$C_6H_5(CH_3)(C_2H_5)COO^{\cdot}$	$C_6H_5CH_3$	30	$\bar{2}$,70	
	$C_6H_5C_2H_5$	30	$\bar{1}$,34	
	$C_6H_5CH(CH_3)_2$	30	$\bar{1}$,18	[143]
	Tetralin	30	0,115	

Table 57 (continued)

RO_2	R′H	t°	log k	Ref.
$(C_6H_5)_2(CH_3)COO^{\cdot}$	$C_6H_5CH(CH_3)_2$	30	$\bar{1}$,255	[143]
$C_6H_5C(CH_3)_2OO^{\cdot}$	$(CH_3)_2C{=}C(CH_3)_2$	30	0,23	[133]
	$CH_2{=}CH(CH_2)_5CH_3$	30	$\bar{1}$,43	[133]
	$CH_3(CH_2)_3C{=}C(CH_2)_3CH_3$	30	$\bar{1}$,78	[133]
	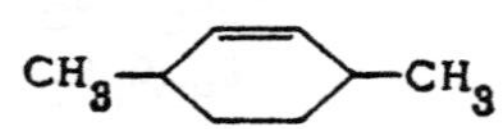	30	0,32	[133]
	$CH_3C_6H_5$	30	$\bar{2}$,48	[143]
	$CH_3C_6H_5$	30	$\bar{2}$,53	[133]
	$C_2H_5C_6H_5$	60–80	4,83–7,0/θ	[158]
	$C_2H_5C_6H_5$	30	$\bar{1}$,32	[133]
	$C_2H_5C_6H_5$	30	$\bar{1}$,26	[143]
	$C_6H_5CH(CH_3)C_3H_7$	30	$\bar{2}$,845	[133]
	$C_6H_5CH(CH_3)_2$	30	$\bar{1}$,255	[133, 143]
	$C_6H_5CD(CH_3)_2$	30	$\bar{2}$,30	[133]
	$C_6H_5CD(CD_3)_2$	30	$\bar{2}$,115	[133]
	$C_6H_5CH(CD_3)_2$	30	$\bar{1}$,04	[133]

$C_6H_5C(CH_3)_2OO^{\cdot}$	p-$CH_3OC_6H_4CH(CH_3)_2$	30	$\bar{1}$,52	[133]
	p-$CH_3C_6H_4CH(CH_3)_2$	30	$\bar{1}$,34	[133]
	p-$(CH_3)_2CHC_6H_4CH(CH_3)_2$	30	$\bar{1}$,38	[133]
	m-$CH_3OC_6H_4CH(CH_3)_2$	30	$\bar{1}$,255	[133]
	p-$ClC_6H_4CH(CH_3)_2$	30	$\bar{1}$,18	[133]
	m-Br $C_6H_4CH(CH_3)_2$	30	$\bar{1}$,08	[133]
	p-$CH_3OCOC_6H_4CH(CH_3)_2$	30	$\bar{1}$,115	[133]
	p-$CNC_6H_4CH(CH_3)_2$	30	$\bar{1}$,115	[133]
	m-$NO_2C_6H_4CH(CH_3)_2$	30	$\bar{1}$,08	[133]
	p-$NO_2C_6H_4CH(CH_3)_2$	30	$\bar{1}$,08	[133]
	$C_6H_5CH_2CH_2C_6H_5$	30	$\bar{1}$,30	[133]
	$(C_6H_5)_2CH_2$	30	$\bar{1}$,70	[133]
	Tetralin	30	0,22	[133, 143]
	- " -	80	0,115	[159]
	Indan	30	0,23	[133]
	9,10-Dihydroanthracene	30	1,845	[133]
	$C_{10}H_{11}OOH$	30	2,78	[154]
	- " -	56	3,04	[154]
	- " -	57	2,48	[162]

Table 57 (continued)

RO_2	$R'H$	t^o	log k	Ref.
$C_6H_5C(CH_3)_2OO^{\bullet}$	$CH_3(CH_2)_8COOH$	120	$\bar{1},205$	[163]
	$HOOC(CH_2)_5COOH$	120	$\bar{1},40$	
	$HOOC(CH_2)_7COOH$	120	$\bar{1},64$	
	$HOOC(CH_2)_8COOH$	120	$\bar{1},76$	
	HOOCCOOH	75–95	7,87–14,4/θ	[163]
	$HOOC(CH_2)_8COOH$	120–140	6,58–12,0/θ	[163]

* $HO_2^{\bullet}$ formation assumed to occur by the reaction

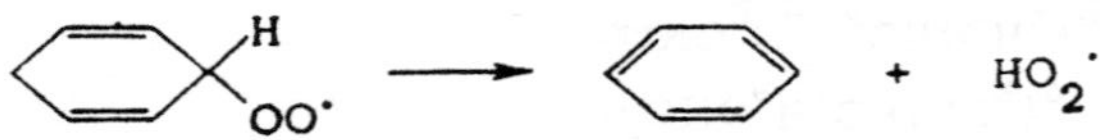

2* $HO_2^{\bullet}$ formation assumed to occur by the reaction

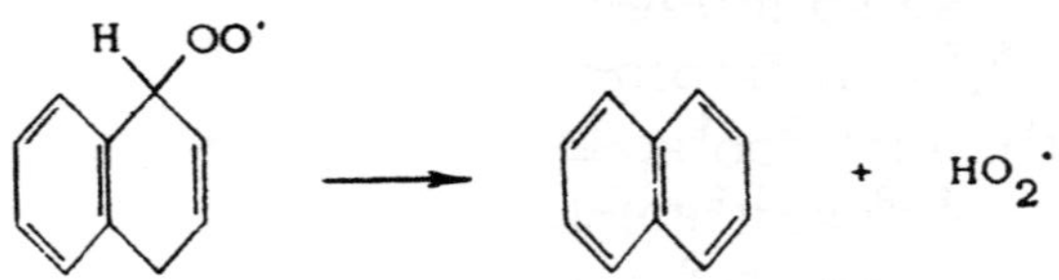

TABLE 58. Reactions of Peroxy Radicals with Inhibitors (Phenols, Amines)
(Note: $C_{10}H_{11}OO^{\bullet}$ is tetralylperoxy radical; $C_{10}H_7$ is naphthyl.)

Inhibitor	Radical	Solvent	t°	log k	log A	E	Ref.
C_6H_5OH	$^{\bullet}OOCH(CH_3)C_6H_5$	$C_6H_5C_2H_5$	60	3,48			[164, 165]
	$^{\bullet}OOCH(C_6H_5)CH_2$~	$C_6H_5CH{=}CH_2$	65	3,805			[166]
	$C_{10}H_{11}OO^{\bullet}$	$C_{10}H_{12}$	50	3,05			[167]
p-$CH_3OC_6H_4OH$	$^{\bullet}OOCH(C_6H_5)CH_2$~	$C_6H_5CH{=}CH_2$	65	4,68			[166]
p-$CH_3C_6H_4OH$	$^{\bullet}OOCH(C_6H_5)CH_2$~	$C_6H_5CH{=}CH_2$	65	4,30			[166]
	$C_{10}H_{11}OO^{\bullet}$	$C_{10}H_{12}$	50	3,59			[167]
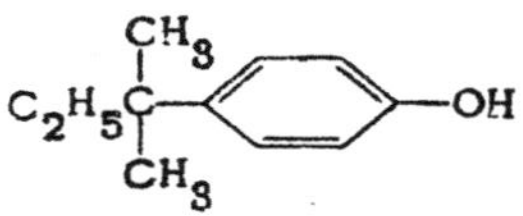	$^{\bullet}OOCH(C_6H_5)CH_2$~	$C_6H_5CH{=}CH_2$	65	4,205			[166]
p-$(CH_3)_3C{-}C_6H_4OH$	$C_{10}H_{11}OO^{\bullet}$	$C_{10}H_{12}$	50	3,79			[167]
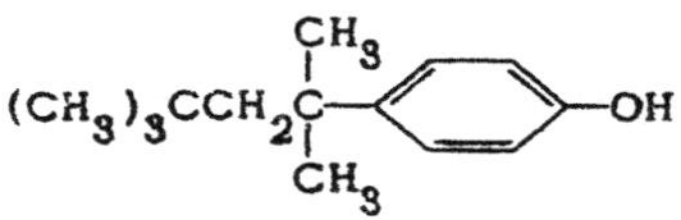	$^{\bullet}OOCH(C_6H_5)CH_2$~	$C_6H_5CH{=}CH_2$	65	4,28			[166]

Table 58 (continued)

Inhibitor	Radical	Solvent	t°	log k	log A	E	Ref.
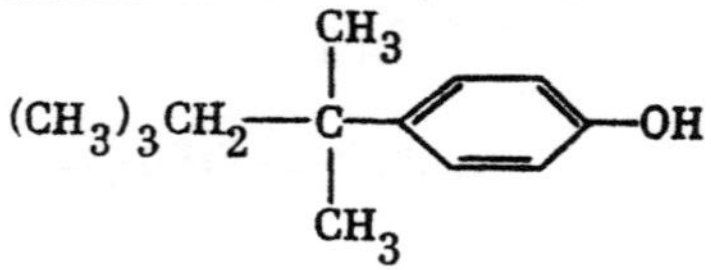	$\cdot OOCH(CH_3)C_6H_5$	$C_6H_5C_2H_5$	60	4,58			[165]
p-ClC_6H_4OH	$\cdot OOCH(C_6H_5)CH_2$~	$C_6H_5CH{=}CH_2$	65	3,67			[166]
p-CNC_6H_4OH	- " -	- " -	65	3,81			[166]
p-$NO_2C_6H_4OH$	$C_{10}H_{11}OO\cdot$	$C_{10}H_{12}$	50	2,845			[167]
m-$CH_3OC_6H_4OH$	$\cdot OOCH(C_6H_5)CH_2$~	$C_6H_5CH{=}CH_2$	65	3,755			[166]
m-ClC_6H_4OH	- " -	- " -	65	4,09			[166]
m-$NO_2C_6H_4OH$	$\cdot OOC(CH_3)_2C_6H_5$	$C_6H_5CH(CH_3)_2$	60	2,67			
m-$NO_2C_6H_4OH$	$\cdot OOCH(CH_3)C_6H_5$	$C_6H_5C_2H_5$	60	2,23			[164]
o-$CH_3C_6H_4OH$	$OOCH(CH_3)C_6H_5$	- " -	60	4,40			
o-$CH_3C_6H_4OH$	$C_{10}H_{11}OO\cdot$	$C_{10}H_{12}$	50	3,85			[167]
o-$(CH_3)_3CC_6H_4OH$	$\cdot OOCH(C_6H_5)CH_2$~	$C_6H_5CH{=}CH_2$	65	4,22			[168]
	$C_{10}H_{11}OO\cdot$	$C_{10}H_{12}$	50	4,02			[167]
	$\cdot OOCH(CH_3)C_6H_5$	$C_6H_5C_2H_5$	60	4,80			[165]
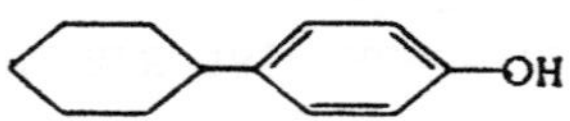	$\cdot OOCH(CH_3)C_6H_5$	$C_6H_5C_2H_5$	60	4,58			[165]

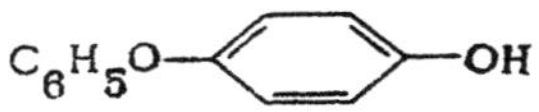	$C_{10}H_{11}OO^{\cdot}$	$C_{10}H_{12}$	50	4,365	[167]
	$^{\cdot}OOCH(C_6H_5)CH_2$~	$C_6H_5CH{=}CH_2$	65	5,37	[168]
	$OOCH(C_6H_5)CH_2$~ $C_{10}H_{11}OO^{\cdot}$	$C_6H_5CH{=}CH_2$ $C_{10}H_{12}$	65 50	4,725 4,35	[168] [167]
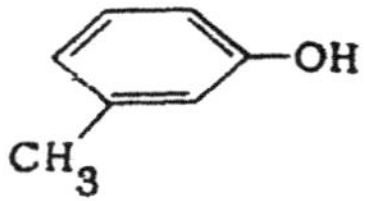	$^{\cdot}OOCH(C_6H_5)CH_2$~	$C_6H_5CH{=}CH_2$	65	4,38	[168]
	- " -	- " -	65	3,28	[168]
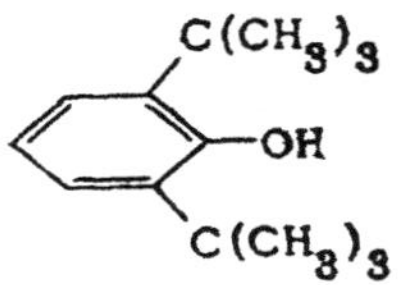	$OOCH(C_6H_5)CH_2$~	$C_6H_5CH{=}CH_2$	65	3,70	[168]

Table 58 (continued)

Inhibitor	Radical	Solvent	t°	log k	log A	E	Ref.
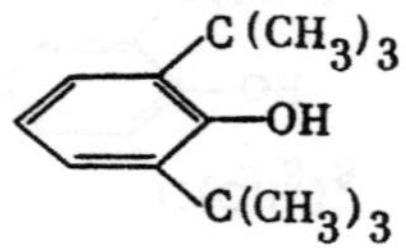	$\cdot OOCH(CH_3)COCH_3$	$CH_3CHC_2H_5$	60	3,38*			[169, 170]
	$\cdot OOCH(CH_3)C_6H_5$	$C_6H_5CH_2CH_3$	60	3,98			[165]
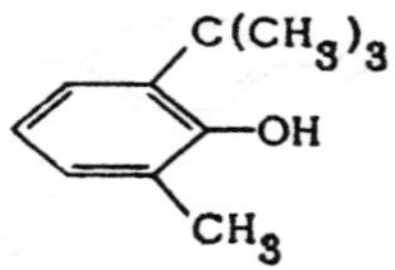	$\cdot OOCH(CH_3)C_6H_5$	$C_6H_5CH_2CH_3$	60	4,83			[165]
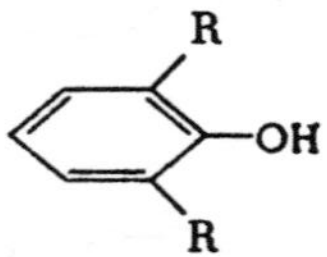 R-cyclo-C_6H_{11}	$\cdot OOCH(CH_3)C_6H_5$	$C_6H_5CH_2CH_3$	60	4,935			[165]
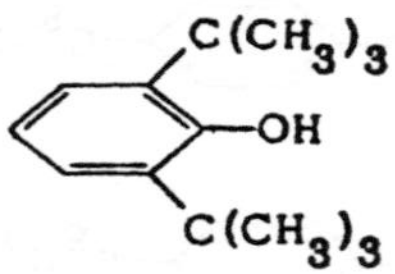	$C_{10}H_{11}OO\cdot$	$C_{10}H_{12}$	50	3,115			[167]

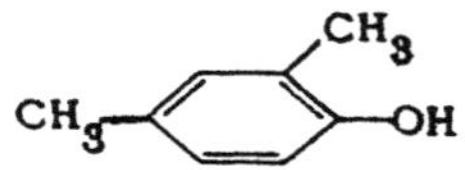	$\cdot OOCH(C_6H_5)CH_2\sim$	$C_6H_5CH{=}CH_2$	65	4,61	[168]
	$C_{10}H_{11}OO\cdot$	$C_{10}H_{12}$	50	4,25	[167]
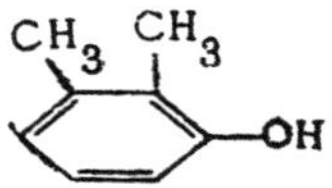	$\cdot OOCH(C_6H_5)CH_2\sim$	$C_6H_5CH{=}CH_2$	65	4,30	[168]
	$C_{10}H_{11}OO\cdot$	$C_{10}H_{12}$	50	3,90	[167]
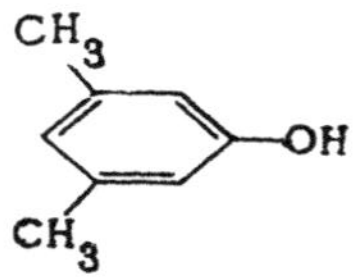	$C_{10}H_{11}OO\cdot$	$C_{10}H_{12}$	50	3,865	[167]
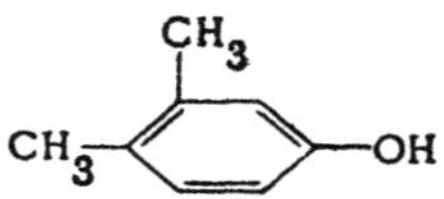	- " -	- " -	50	3,97	[167]
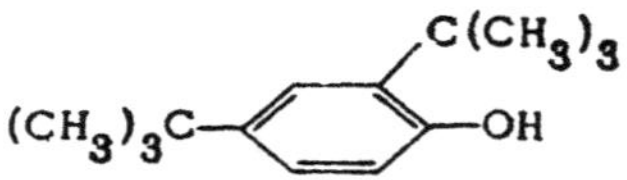	- " -	- " -	50	4,28	[167]

Table 58 (continued)

Inhibitor	Radical	Solvent	t^{o}	log k	log A	E	Ref.
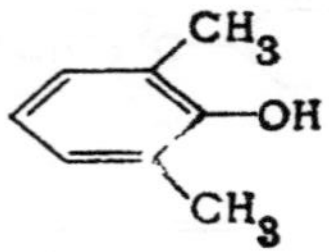	$C_{10}H_{11}OO^{\cdot}$	$C_{10}H_{12}$	50	4,04			[167]
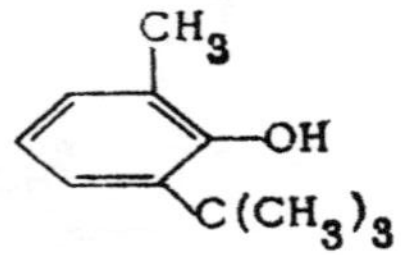	- " -	- " -	50	4,28			[167]
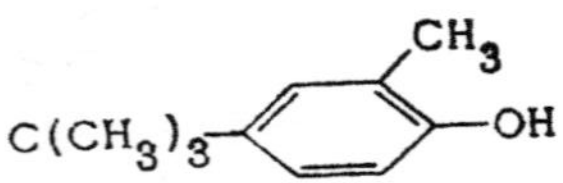	- " -	- " -	50	4,33			[167]
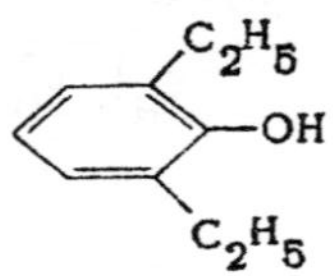	$^{\cdot}OOCH(C_6H_5)CH_2$~	$C_6H_5CH{=}CH_2$	65	4,335			[168]

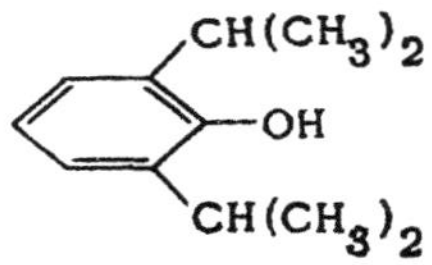	- " -	- " -	65	4,32	[168]
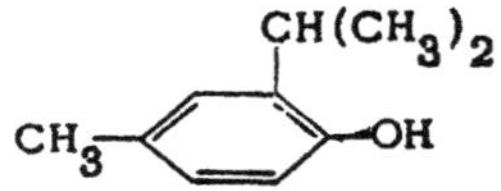	- " -	- " -	65	4,59	[168]
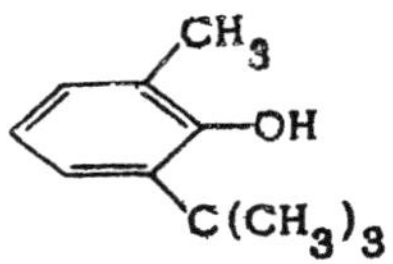	- " -	- " -	65	4,20	[168]
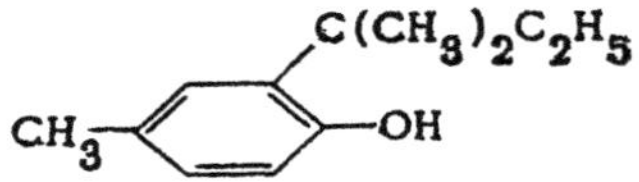	- " -	- " -	65	4,64	[168]
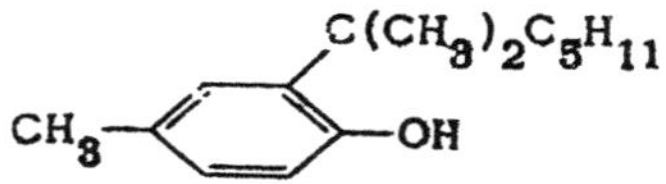	$\cdot OOCH(C_6H_5)CH_2$~	$C_6H_5CH{=}CH_2$	65	4,56	[168]

Table 58 (continued)

Inhibitor	Radical	Solvent	t^{o}	log k	log A	E	Ref.
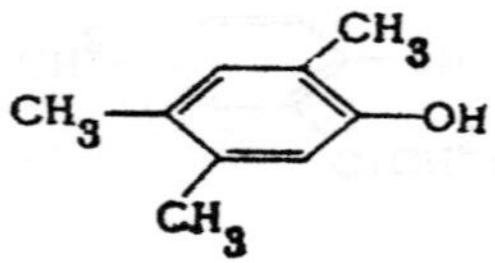	$^{\cdot}OOCH(C_6H_5)CH_2\sim$	$C_6H_5CH=CH_2$	65	4,675			[168]
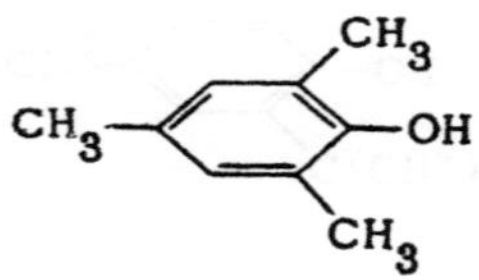	$^{\cdot}OOCH(CH_3)C_6H_5$ $C_{10}H_{11}OO^{\cdot}$	$C_6H_5C_2H_5$ $C_{10}H_{12}$	60 50	5,28 4,62			[165] [167]
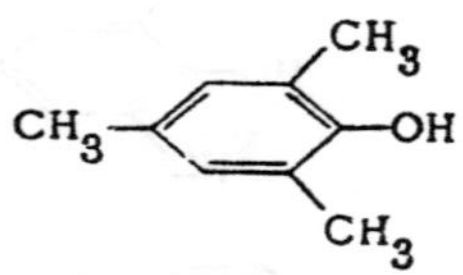	$^{\cdot}OOCH(CH_3)COCH_3$	$CH_3COC_2H_5$	60	4,555*			[168, 170]
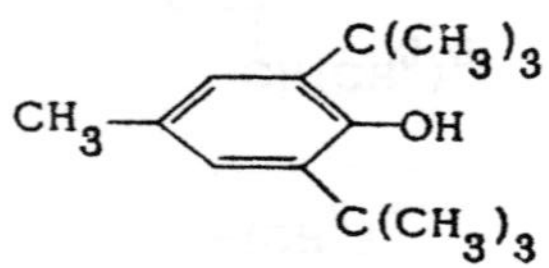	- " -	- " -	60	4,16			[169, 170]

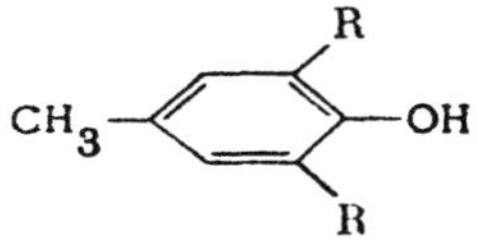 $R = CH(CH_3)_2$	$\cdot OOCH(C_6H_5)CH_2\sim$	$C_6H_5CH{=}CH_2$	65	4,755			[168]
	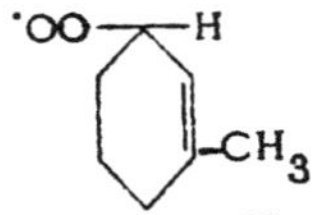	1–CH_3–cyclo–C_6H_9	40	4,10	7,60	5,0	[35]
	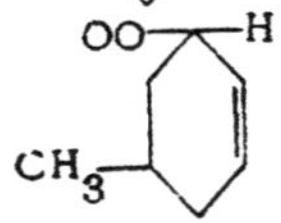	4–CH_3–cyclo–C_6H_9	40	4,07	7,57	5,0	[35]
	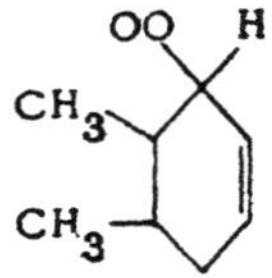	4,5–$(CH_3)_2$–cyclo–C_6H_8	40	4,46	8,31	5,5	[35]
	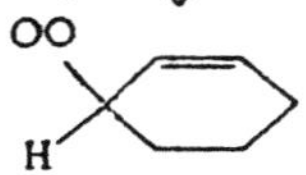	cyclo–C_6H_{10}	40	3,88	7,02	4,5	[134]
	– " –	– " –	56	4,30	7,30	4,5	[136]
	– " –	– " –	30	4,48	7,73	4,5	[136]

Table 58 (continued)

Inhibitor	Radical	Solvent	t°	log k	log A	E	Ref.
	$\cdot OOC(CH_3)_2C_6H_5$	$C_6H_5CH(CH_3)_2$	56	4,115	6,775	4,0	[136]
	- " -	- " -	40	4,08			[16]
	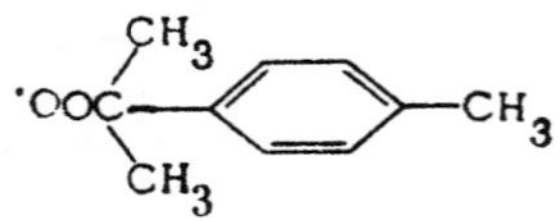	$p\text{-}CH_3C_6H_4CH(CH_3)_2$	56	3,60	5,60	3,0	[136]
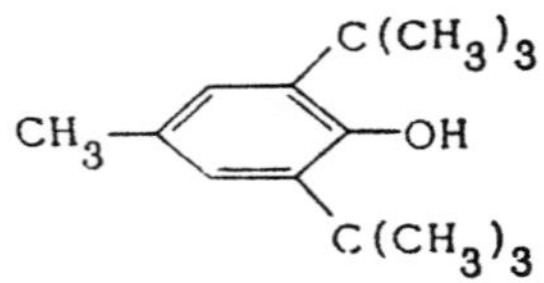	$\cdot OOCH(C_6H_5)CH_2\sim$	$C_6H_5CH{=}CH_2$	13–50	4,00 (40°)	7,924	5,6	[171]
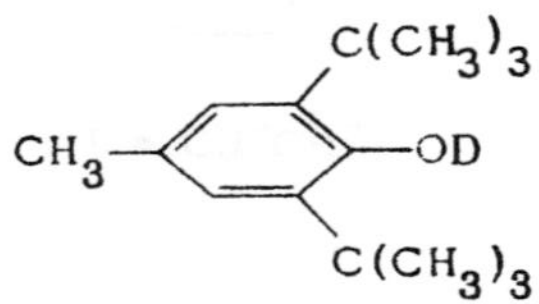	- " -	- " -	45–70	3,12 (40°)	7,81	6,7	[172]

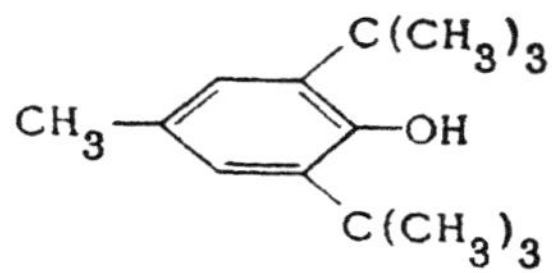	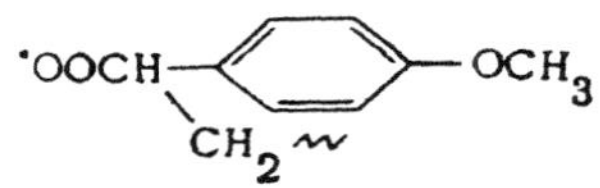	$CH_3OC_6H_4CH{=}CH_2$	40	4,93	[173]
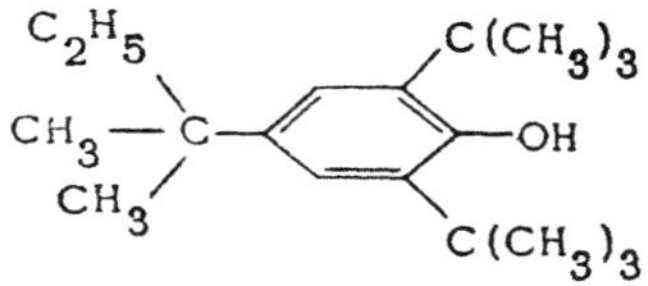	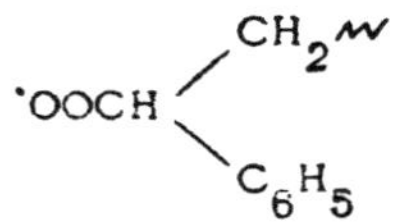	$C_6H_5CH{=}CH_2$	65	4,20	[168]
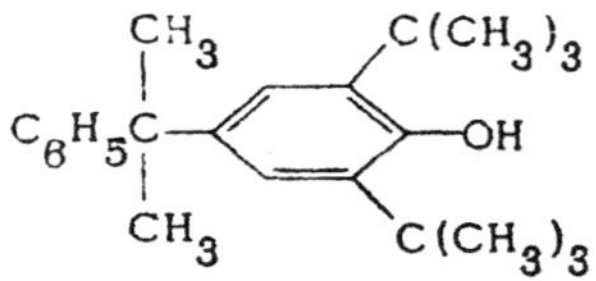	- " -	- " -	65	4,13	[168]
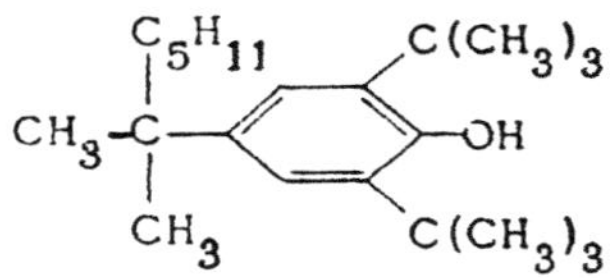	- " -	- " -	65	4,155	[168]
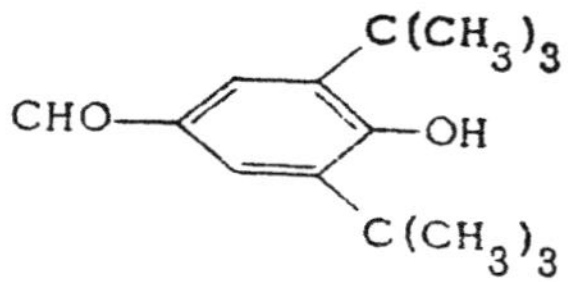	- " -	- " -	65	3,245	[168]

Table 58 (continued)

Inhibitor	Radical	Solvent	t°	log k	log A	E	Ref.
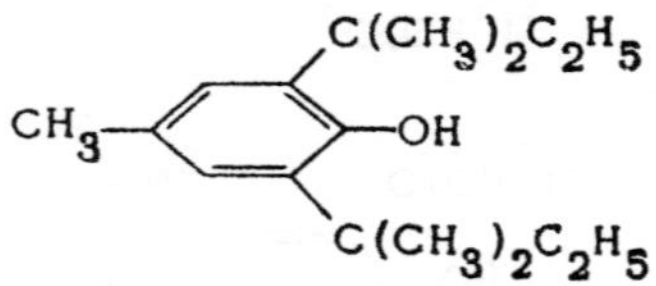	$\cdot OOCH(C_6H_5)CH_2\sim$	$C_6H_5CH{=}CH_2$	65	4,20			[168]
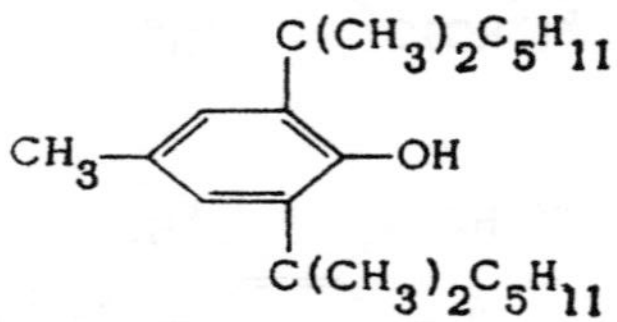	- " -	- " -	65	4,56			[168]
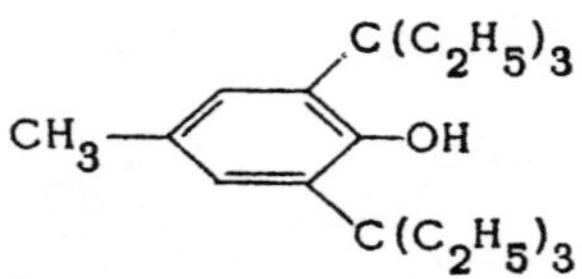	- " -	- " -	65	4,20			[168]
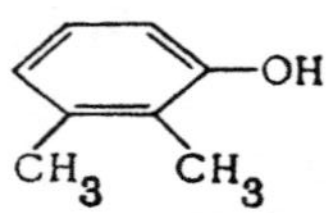	- " -	- " -	65	4,55			[168]

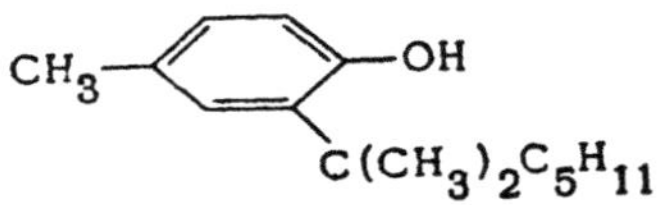	- " -	- " -	65	4,16	[168]
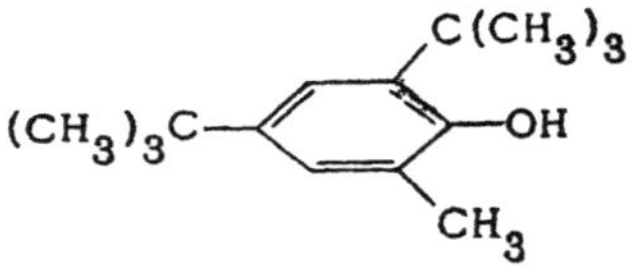	$C_{10}H_{11}OO^{\cdot}$	$C_{10}H_{12}$	50	4,64	[167]
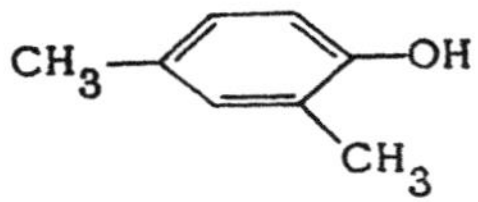	- " -	- " -	50	4,65	[167]
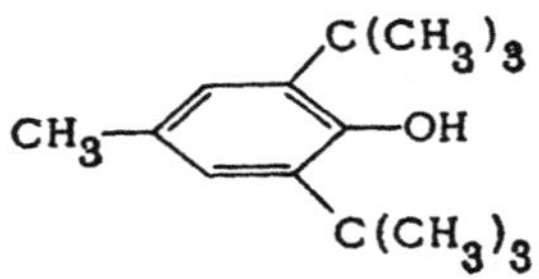	$^{\cdot}OOCH(m\text{-}C_6H_4Cl)CH_2\sim$	$ClC_6H_4CH{=}CH_2$	40	4,17	[173]
	$^{\cdot}OOC(CH_3)(C_6H_5)CH_2\sim$	$C_6H_5C(CH_3){=}CH_2$	65	3,905	[147]
	$^{\cdot}OOCH(CH_3)C_6H_5$	$C_6H_5C_2H_5$	60	4,40	[164]

Table 58 (continued)

Inhibitor	Radical	Solvent	t°	log k	logA	E	Ref.
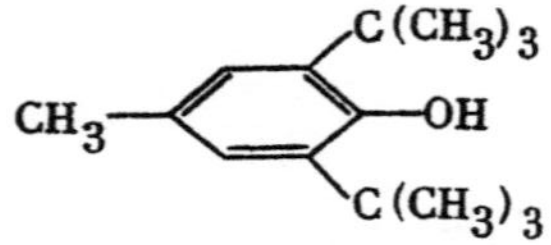	$\cdot OOCH(C_6H_5)CH_2\sim$	$C_6H_5CH{=}CH_2$	65	4,245			[168]
	$C_{10}H_{11}OO\cdot$	$C_{10}H_{12}$	50	4,43			[167]
	- " -	- " -	40	4,34			[16]
	- " -	- " -	65	4,38			[174]
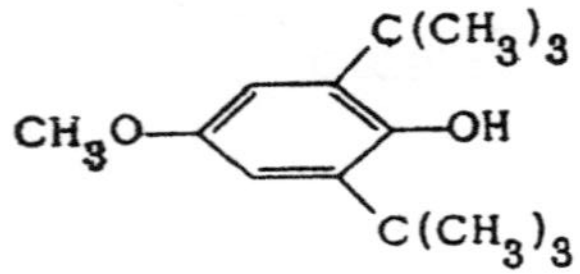	$\cdot OOC(C_6H_5)CH_2\sim$	$C_6H_5CH{=}CH_2$	65	4,89			[174]
	$C_{10}H_{11}OO\cdot$	$C_{10}H_{12}$	65	4,95			[174]
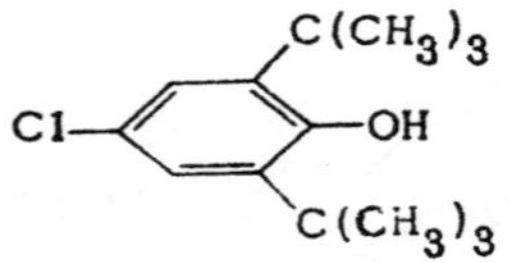	$\cdot OOC(C_6H_5)CH_2\sim$	$C_6H_5CH{=}CH_2$	65	3,83			[174]
	$C_{10}H_{11}OO\cdot$	$C_{10}H_{12}$	65	3,93			[174]

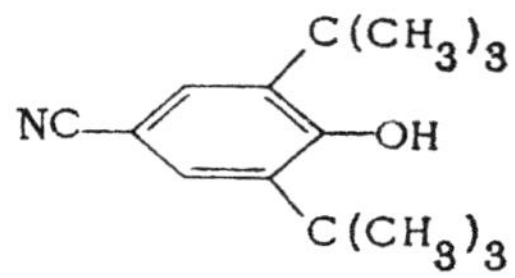	$^{\cdot}OOC(C_6H_5)CH_2$〰	$C_6H_5CH{=}CH_2$	65	3,23		[174]
	$C_{10}H_{11}OO^{\cdot}$	$C_{10}H_{12}$	65	3,20		[174]
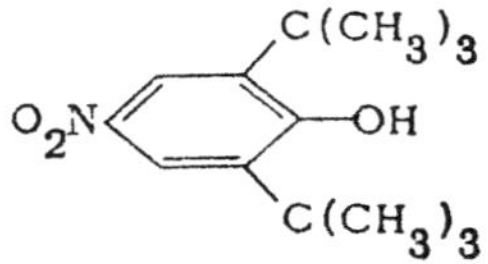	$^{\cdot}OOC(C_6H_5)CH_2$〰	$C_6H_5CH{=}CH_2$	65	3,00		[174]
	$C_{10}H_{11}OO^{\cdot}$	$C_{10}H_{12}$	65	2,78		[174]
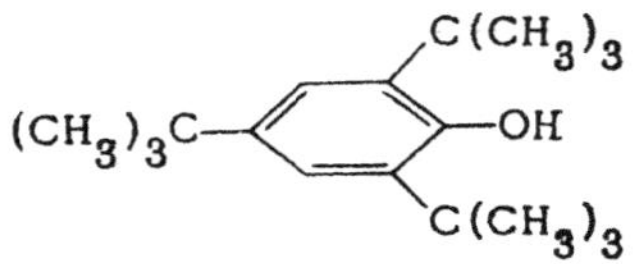	$^{\cdot}OOCH(CH_3)COCH_3$	$CH_3COC_2H_5$	60	3,98*		[169]
	$^{\cdot}OOC(C_6H_5)CH_2$〰	$C_6H_5CH{=}CH_2$	65	4,205		[174]
	$C_{10}H_{11}OO^{\cdot}$	$C_{10}H_{12}$	65	4,45		[174]
	- " -	- " -	50	4,285		[167]
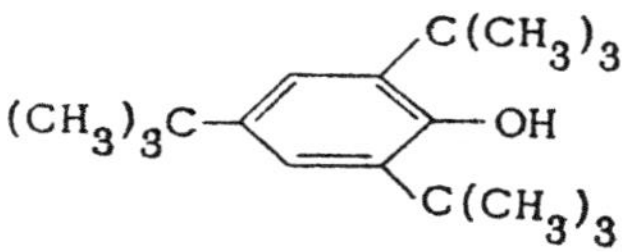	$^{\cdot}OOCH(CH_3)C_6H_5$	$C_6H_5C_2H_5$	40–80	4,34 (60°)	6,58 3,4	[165]
	- " -	- " -	60	4,115		[164]

Table 58 (continued)

Inhibitor	Radical	Solvent	t^{o}	log k	logA	E	Ref.
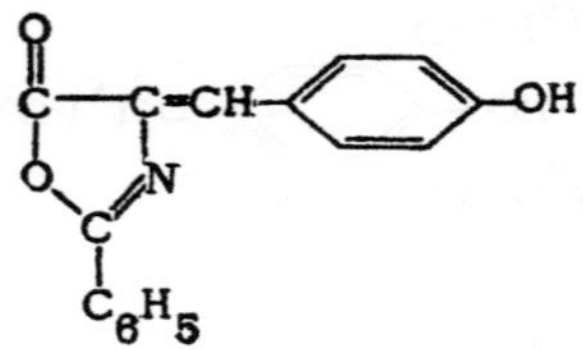	$^{\cdot}OOCH(CH_3)COCH_3$	$CH_3COC_2H_5$	60	4,08*			[169]
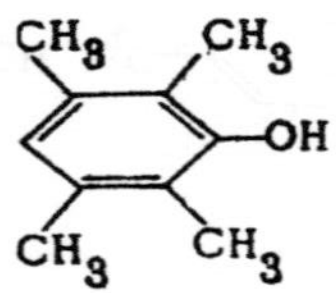	$^{\cdot}OOCH(C_6H_5)CH_2\sim$	$C_6H_5CH{=}CH_2$	65	4,47			[168]
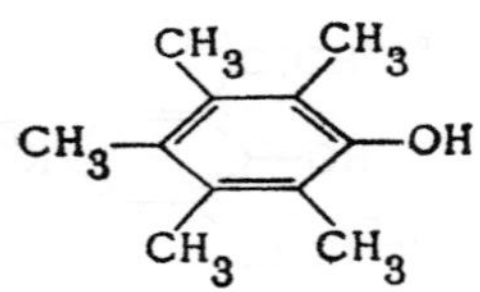	- " -	- " -	65	4,93			[168]
p-HOC_6H_4OH	$^{\cdot}OOCH(CH_3)COCH_3$	$CH_3COC_2H_5$	60	5,13*			[169]
	$^{\cdot}OOCOC_9H_{19}$	$C_9H_{19}CHO$	5	5,20			[152]

	$OOCOC_6H_5$	C_6H_5CHO	5	5,59	[153]
	$^{\bullet}OOCH(C_6H_5)CH_2$~	$C_6H_5CH{=}CH_2$	65	5,31	[166]
	$RO_2^{\bullet}$ from ethyl linoleate	Ethyl linoleate	45	6,04	[175]
	$C_{10}H_{11}OO^{\bullet}$	$C_{10}H_{12}$	50	4,875	[167]
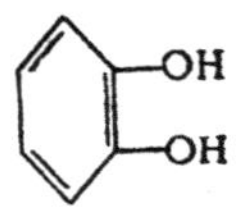	$RO_2^{\bullet}$ from ethyl linoleate	Ethyl linoleate	45	5,845	[175]
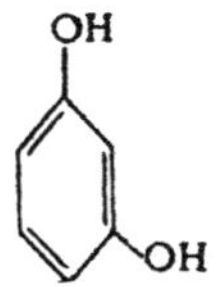	- " -	- " -	45	4,245	[175]
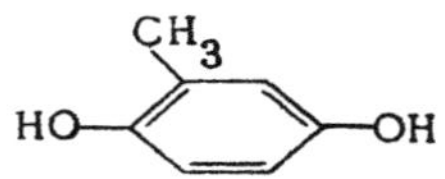	- " -	- " -	45	6,22	[175]
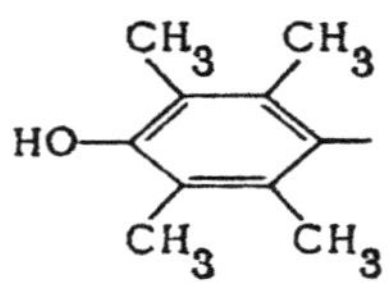	- " -	- " -	45	6,80	[175]

Table 58 (continued)

Inhibitor	Radical	Solvent	t^{o}	log k	logA	E	Ref.
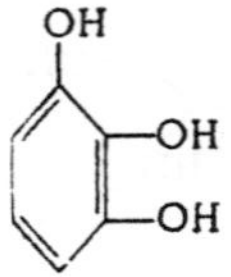	$RO_2^{\bullet}$ from ethyl linoleate	Ethyl linoleate	45	6,52			[175]
α-Naphthol	- " -	- " -	45	5,79			[175]
α-Naphthol	cyclo-$C_6H_{11}OO^{\cdot}$	cyclo-C_6H_{12}	75	5,30			[176]
	$^{\cdot}OOCH_2C_6H_5$	$C_6H_5CH_3$	60	5,505			[164]
α-$C_{10}H_7OH$	$^{\cdot}OOC(CH_3)_2C_6H_5$	$C_6H_5CH(CH_3)_2$	60	5,15			[164]
	$^{\cdot}OOCH(CH_3)COCH_3$	$CH_3COC_2H_5$	60	4,49*			[169]
	$^{\cdot}OOCH_2(CH_3)C_6H_5$	$C_6H_5C_2H_5$	60	5,65	8,15	3,8	[164]
	$C_{10}H_{11}OO^{\cdot}$	$C_{10}H_{12}$	150	4,86			[177]
	- " -	- " -	50	4,73			[167]
β-$C_{10}H_7OH$	$C_{10}H_{11}OO^{\cdot}$	$C_{10}H_{12}$	50	3,05			[167]
	$RO_2^{\bullet}$ from ethyl linoleate	Ethyl linoleate	45	4,93			[175]
	cyclo-$C_6H_{11}OO^{\cdot}$	cyclo-C_6H_{12}	75	3,98			[178]
	$^{\cdot}OOCH_3(CH_3)C_6H_5$	$C_6H_5C_2H_5$	60	4,955			[164]

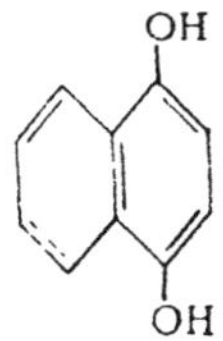	$RO_2^\bullet$ from ethyl linoleate	Ethyl linoleate	45	7,645	[175]
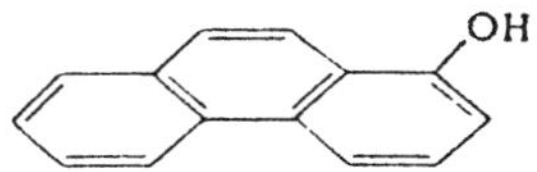	$C_{10}H_{11}OO^\bullet$	$C_{10}H_{12}$	50	4,74	[167]
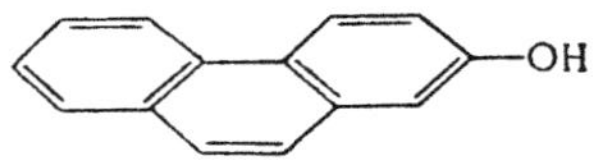	- " -	- " -	50	3,77	[167]
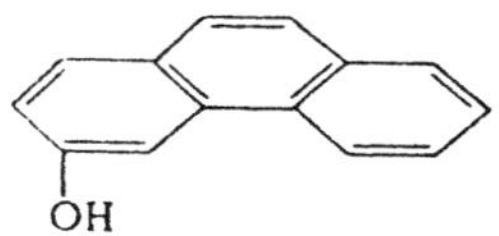	$C_{10}H_{11}OO^\bullet$	$C_{10}H_{12}$	50	4,13	[167]
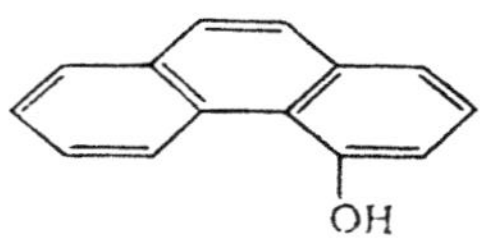	- " -	- " -	50	4,62	[167]

Table 58 (continued)

Inhibitor	Radical	Solvent	t°	log k	log A	E	Ref.
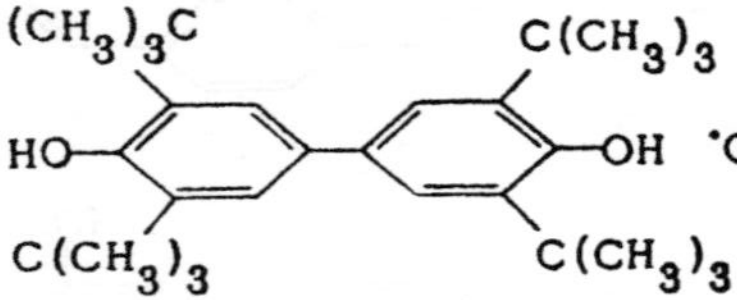	$\cdot OOCH(C_6H_5)CH_2\sim$	$C_6H_5CH=CH_2$	65	4,62			[168]
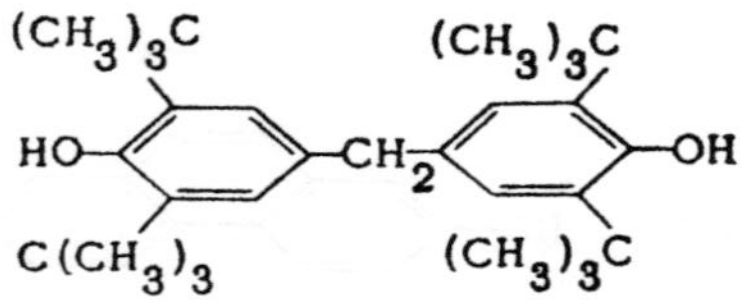	- " -	- " -	65	4,48			[168]
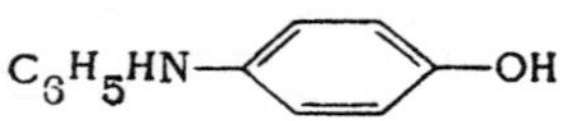	$\cdot OOCH(CH_3)C_6H_5$	$C_6H_5C_2H_5$	60	6,0			[179]
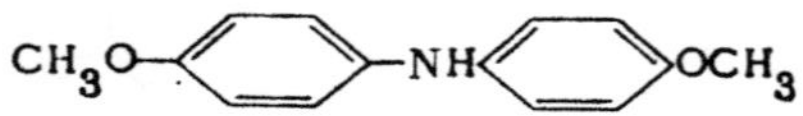	- " -	- " -	60	6,58			[179]

$m\text{-}NH_2C_6H_4NH_2$	- " -	- " -	60	2,95			[179]
p-$(CH_3)CHNH$ $(CH_3)_2CHNH$–C_6H_4	- " -	- " -	60	6,65			[179]
	- " -	- " -	60	6,23			[179]
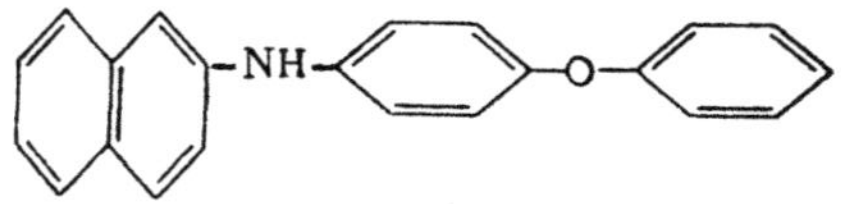		- " -	60	5,68			[179]
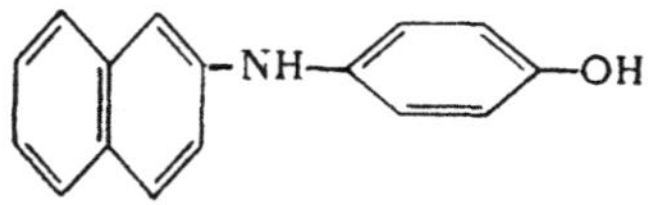	- " -	- " -	60	6,20			[179]
$(C_6H_5)_2NH$	$(CH_3)_2C(OO^\cdot)$–$CH_2CH(CH_3)$ $(CH_3)_2CH_2$	RH	42–78	2,48 (60°)	4,78	3,5	[180]
	$^\cdot OOCH_2(CH_3)C_6H_5$	$C_6H_5C_2H_5$	60	4,64			[164]
	$^\cdot OOCH(C_6H_5)CH_2$~	$C_6H_5CH{=}CH_2$	65	4,30			[181]
	- " -	o–ClC_6H_4Cl	65	4,60			[182]

Table 58 (continued)

Inhibitor	Radical	Solvent	t°	log k	log A	E	Ref.
p-$CH_3C_6H_4NHCH_3$	$\cdot OOCH(C_6H_5)CH_2$~	o-ClC_6H_4Cl	65	4,08			
m-$CH_3C_6H_4NHCH_3$	- " -	- " -	65	3,70			
$C_6H_5NHCH_3$	- " -	- " -	65	3,60			
$C_6H_5NDCH_3$	- " -	- " -	65	3,12			
p-$CH_3OCOC_6H_4NHCH_3$	- " -	- " -	65	2,955			
(p-$CH_3OC_6H_4)_2NH$	- " -	- " -	65	5,52			[182]
p-$CH_3OC_6H_4NHC_6H_5$	- " -	- " -	65	5,30			
(p-$CH_3C_6H_4)_2NH$	- " -	- " -	65	5,00			
m-$ClC_6H_4NHC_6H_5$	- " -	- " -	65	4,255			
p-$NO_2C_6H_4NHC_6H_5$	- " -	- " -	65	3,78			
(p-$NO_2C_6H_4)_2NH_2$	- " -	- " -	65	3,205			
α-$C_{10}H_7NH_2$	cyclo-$C_6H_{11}OO\cdot$	cyclo-C_6H_{12}	75	4,20			[176]
	$\cdot OOCH(CH_3)C_6H_5$	$C_6H_5C_2H_5$	60	4,115			[179]
α-$C_{10}H_7NHC_6H_5$	$\cdot OOCH(C_6H_5)CH_2$~	o-ClC_6H_4Cl	65	5,15			[181]
		$C_6H_5CH{=}CH_2$	65	4,845			[180]

α–$C_{10}H_7NDC_6H_5$	$^{\bullet}OOCH(C_6H_5)CH_2$~	$C_6H_5CH{=}CH_2$	65	4,255	[180]
	- " -	o–ClC_6H_4Cl	65	4,55	[181]
α–$C_{10}H_7NHC_6H_5$	$^{\bullet}OOCH(CH_3)C_6H_5$	$C_6H_5C_2H_5$	60	5,00	[164]
(α–$C_{10}H_7)_2NH$	- " -	- " -	60	6,00	[178]
β–$C_{10}H_7NHC_6H_5$	$^{\bullet}OOCH(CH_3)C_6H_5$	$C_6H_5C_2H_5$	60	5,115	[164]
	- " -	o–ClC_6H_5Cl	65	5,00	[181]
	$^{\bullet}OOC(CH_3)_2C_6H_5$	$C_6H_5CH(CH_3)_2$	60	4,545	[164]
	$^{\bullet}OOCH(C_6H_5)CH_2$~	$C_6H_5CH{=}CH_2$	65	4,70	[180]
β–$C_{10}H_7NDC_6H_5$	$OOCH(C_6H_5)CH_2$~	$C_6H_5CH{=}CH_2$	65	4,23	[181]
	- " -	o–ClC_6H_4Cl	65	4,52	[182]
(β–$C_{10}H_7)_2NH$	$^{\bullet}OOCH(C_6H_5)CH_2$~	o–ClC_6H_4Cl	65	5,255	[182]

Table 58 (continued)

Inhibitor	Radical	Solvent	t°	log k	log A	E	Ref.
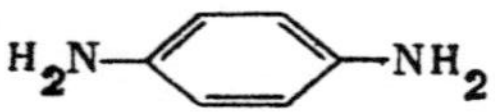	$\cdot OOCH(C_6H_5)CH_2$~	$o\text{-}ClC_6H_4Cl$	65	5,875			[182]
	$\cdot OOCH(CH_3)C_6H_5$	$C_6H_5C_2H_5$	60	6,00			[179]
$(p\text{-}CH_3OC_6H_4)_2NOH$	$\cdot OOCH_2(CH_3)C_6H_5$	$o\text{-}ClC_6H_4Cl$	65	5,70			[183]
$(p\text{-}CH_3OC_6H_4)_2NOD$	- " -	- " -	65	5,40			[183]
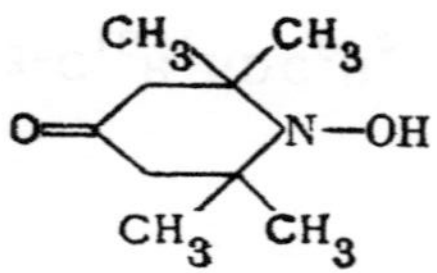	- " -	- " -	65	5,70			[183]
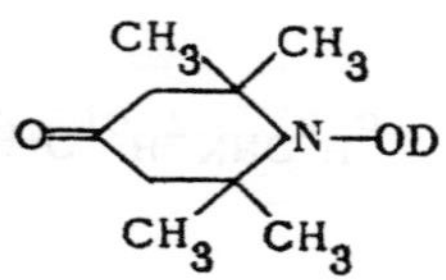	- " -	- " -	65	5,22			[183]

*These constants computed for reaction of $RO_2^{\bullet}$ with inhibitor molecules that are not hydrogen-bonded to the ketone.

TABLE 59. Rate Constants for Reactions of H Abstraction by Phenoxyl Radicals

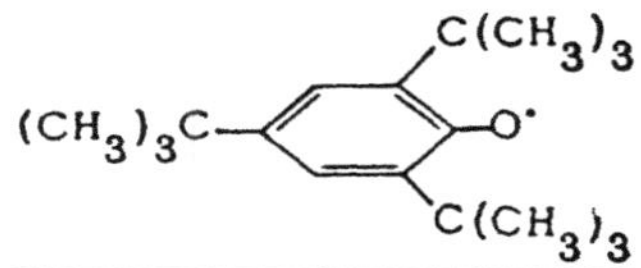

RH	Solvent	t°	log k (24°)	log A	E	Method	Ref.
$n\text{-}C_{10}H_{22}$	$n\text{-}C_{10}H_{22}$	120-150	$\overline{10},18$	10,415	27,5	EPR	[184]
$(CH_3)_2CH(CH_2)_7CH_3$	$(CH_3)_2CH(CH_2)_7CH_3$	90-150	$\overline{10},51$	9,280	25,5	EPR	
$C_6H_5CH_3$	$C_6H_5CH_3$	100-150	$\bar{9},86$	8,780	23,0	EPR	
$C_6H_5CH_2CH_3$	$C_6H_5CH_2CH_3$	70-100	$\bar{8},91$	5,430	17,0	EPR	
$C_6H_5CH(CH_3)_2$	$C_6H_5CH(CH_3)_2$	100-160	$\bar{8},55$	4,690	16,5	EPR	
$(C_6H_5)_2CH_2$	$(C_6H_5)_2CH_2$	80-150	$\bar{6},40$	5,080	14,5	EPR	

Table 59 (continued)

RH	Solvent	t^o	logk (24°)	logA	E	Method	Ref.
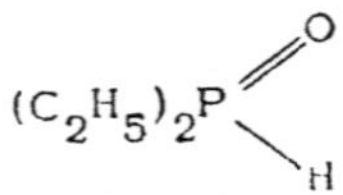	C_6H_6	20	$\bar{5}$,86			EPR	[185]
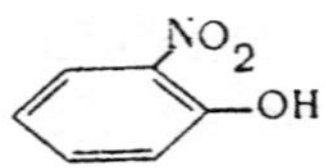	C_6H_6	20–90	$\bar{5}$,88	4,780	12,1	EPR	[186]
C_6H_5OH	C_6H_6 C_6H_5Cl	25–60 60	0,79 1,21	4,85	5,5	Kinetics of oxidation with phenols	[187] [187]
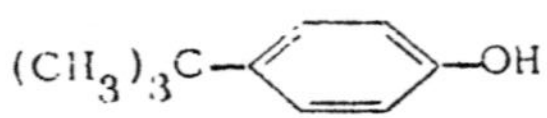	C_6H_6	25–60	1,97	5,88	5,3	Kinetics of oxidation with phenols	[187]
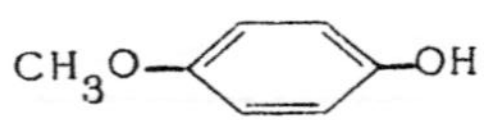	C_6H_6	25	3,78				[187]

p-$CH_3OC_6H_4OH$	C_6H_6	24	3,48				
p-$C_6H_5C_6H_4OH$	C_6H_6	24	2,10				
p-$(CH_3)_3CC_6H_4OH$	C_6H_6	24	1,66				
p-Br C_6H_4OH	C_6H_6	24	0,64				[188]
p-$CH_3OCOC_6H_4OH$	C_6H_6	24	$\bar{1}$,32				
p-CNC_6H_4OH	C_6H_6	24	$\bar{1}$,88				
C_6H_5OH	C_6H_6	24	0,49				[189]
m-$(CH_3)_3CC_6H_4OH$	C_6H_6	24	1,02				[188]
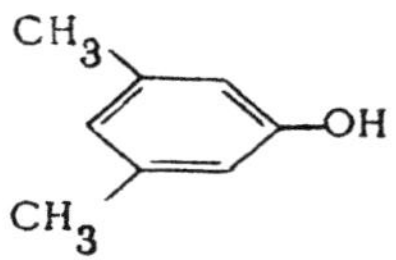	C_6H_6	24	1,19				
m-$C_2H_5OCOC_6H_4OH$	C_6H_6	24	$\bar{1}$,78				[189]
m-CNC_6H_4OH	C_6H_6	24	$\bar{2}$,90				
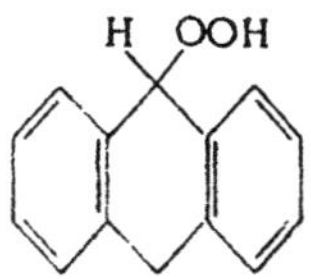	C_6H_6	24	$\bar{1}$,30			Kinetics of oxidation with phenols	[187]
	C_6H_5Cl	24–60	$\bar{1}$,255	5,60	8,6		[187]

Table 59 (continued)

RH	Solvent	t°	log k (24°)	log A	E	Method	Ref.
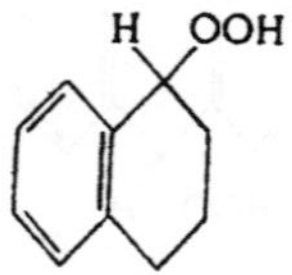	C_6H_5Cl	24–60	$\bar{1}$,84	4,11	5,8	Kinetics of oxidation with phenols	[187]
$(CH_3)_3COOH$	CCl_4	0–21	0,21	6,33	8,3	Kinetics	[189]
m-$ClC_6H_4NH_2$	CCl_4	23,5	0,60			- " -	[189]
β-$C_{10}H_7$-NHC_6H_5	$C_6H_5C_2H_5$	–30–23	1,65 (23)	4,61	4,0	EPR	[179]
	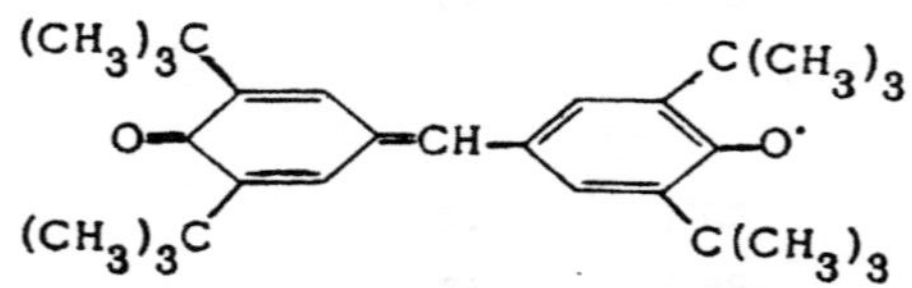						
$C_6H_5C_2H_5$	$C_6H_5C_2H_5$	60–140	$\bar{7}$,55	3,255	14,0	EPR	[190]

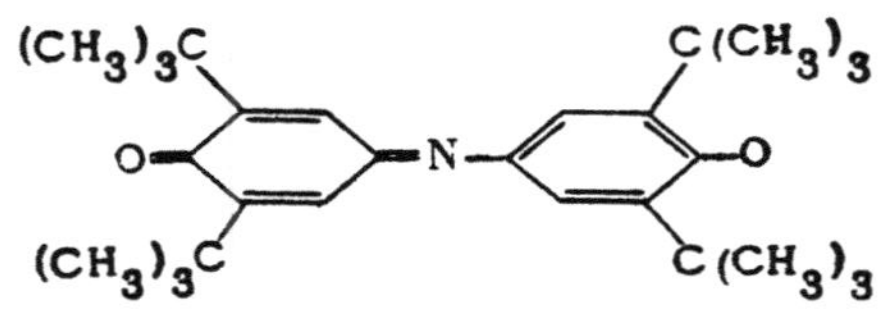

$C_6H_5C_2H_5$	$C_6H_5C_2H_5$	80–130	$\bar{7},58$	3,48	14,8	EPR	[190]
$(C_6H_5)_2NH$	o–$CH_3C_6H_4CH_3$	30–50	$\bar{4},01$	1,30	6,4		
$C_6H_5NH_2$	- " -	30–50	$\bar{2},48$	3,85	7,3		
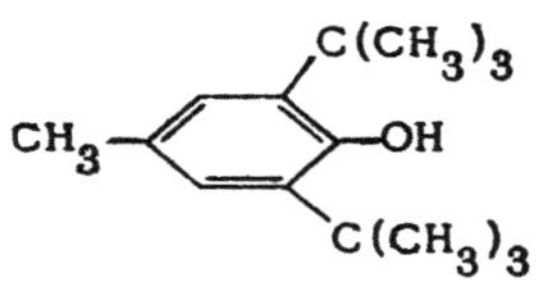	- " -	30–50	$\bar{1},79$	3,84	5,5		[191]
C_6H_5OH	- " -	30–50	$\bar{2},75$	6,49	10,5	EPR	
$(p\text{-}NO_2C_6H_4)_2NH$	- " -	50	$\bar{3},38$				
$(m\text{-}NO_2C_6H_4)_2NH$	- " -	50	$\bar{3},32$				
$(o\text{-}NO_2C_6H_4)_2NH$	o–$CH_3C_6H_4CH_3$	50	$\bar{3},19$			EPR	
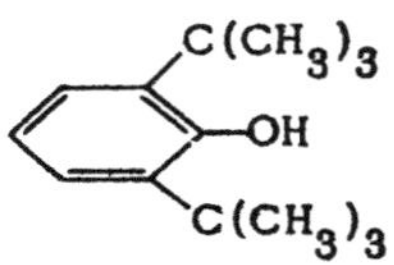	o–$CH_3C_6H_4CH_3$	30	0,49			EPR	[191]

Table 59 (continued)

RH	Solvent	t^o	log k (24°)	log A	E	Method	Ref.
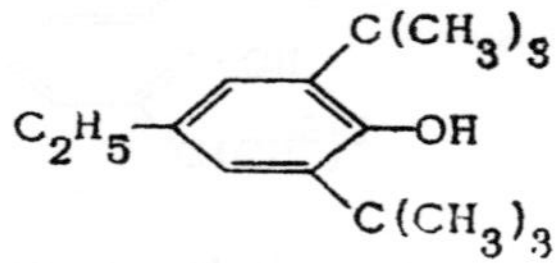	o-$CH_3C_6H_4CH_3$	30	$\bar{1}$,74			EPR	[191]
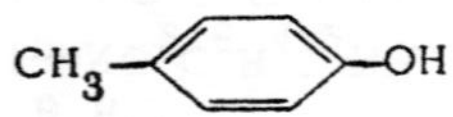	o-$CH_3C_6H_4CH_3$	30	$\bar{2}$,90			EPR	[191]
p-$NO_2C_6H_4OH$	- " -	30	$\bar{2}$,90			EPR	[191]
	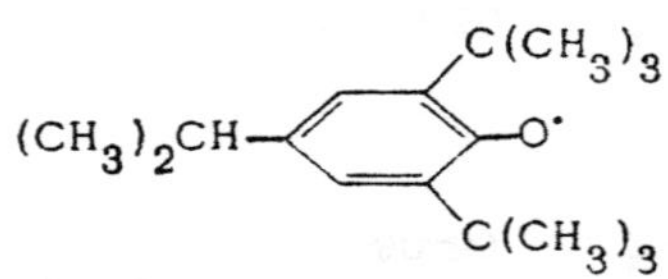						
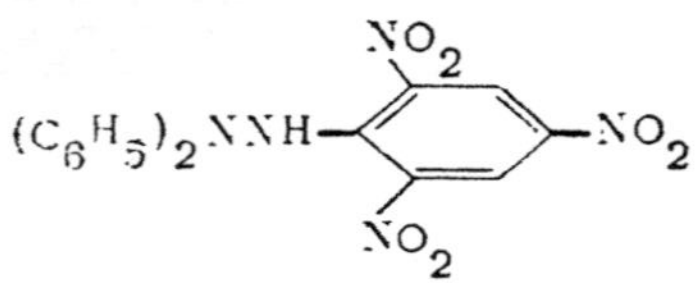	C_6H_6	20–60	2,7	4,62	6,0	EPR	[192]

TABLE 60. Reaction of Hydrogen Exchange between Phenoxyl Radical and Phenol (NMR method)

Phenol	Solvent	t°	log k	Ref.
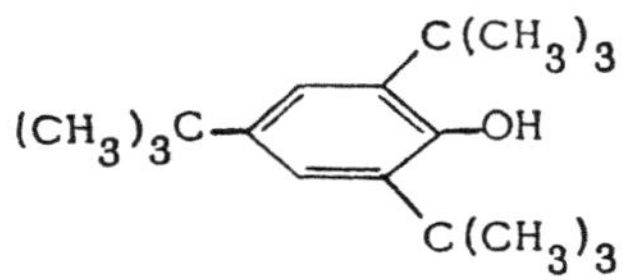	CCl_4	30	2,48	[193]
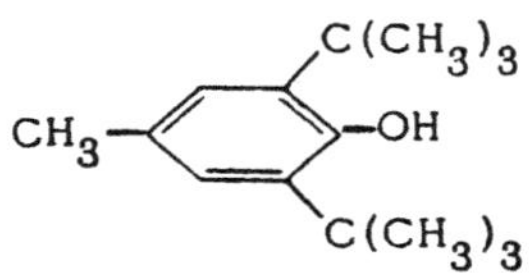	CCl_4	27	2,52	[194]
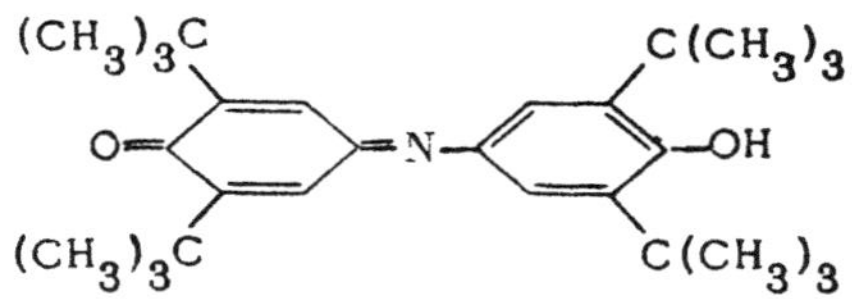	CCl_4	27	2,85	[194]
	C_6H_{12}	27	3,04	[194]

Table 60 (continued)

Phenol	Solvent	t°	log k	Ref.
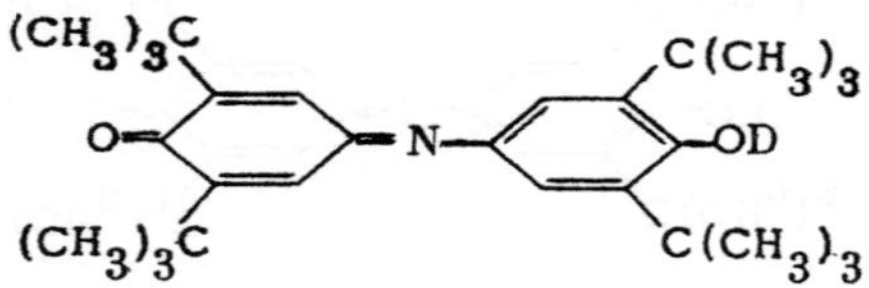	CCl_4	27	2,555	[194]
$(C_6H_5)_2NOH$	CCl_4	27	7	[194]
$[(CH_3)_3C]_2NOH$	CCl_4	27	2,20	[194]
	C_6H_{12}	27	2,08	[194]
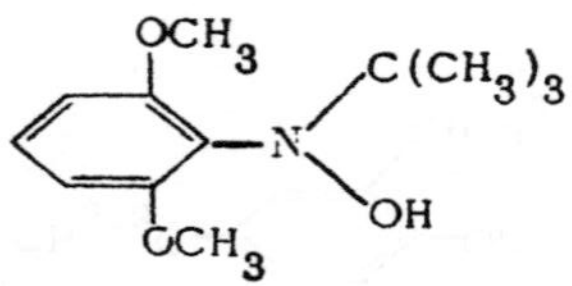	CCl_4	27	3,00	[194]
	C_6H_5Cl	27	2,415	[194]
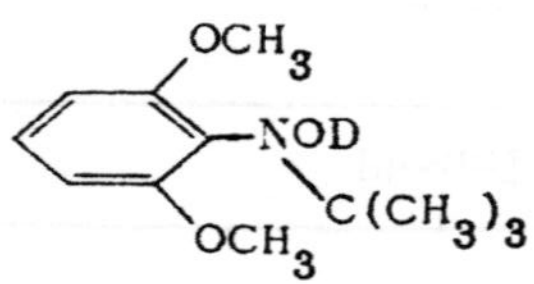	CCl_4	27	2,81	[194]

TABLE 61. Reactions of H Abstraction from Hydrocarbons (RH) by Nitroxy Radicals
(Note: EPR method [195, 190]; hydrocarbon with which radical was reacted also served as solvent).

Radical	RH	t°	logk (190°)	logA	E
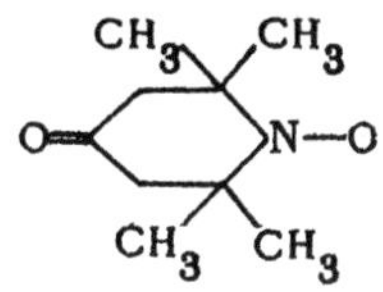	$C_6H_5C_2H_5$	155–205	$\bar{5}$,545	2,89	15,5
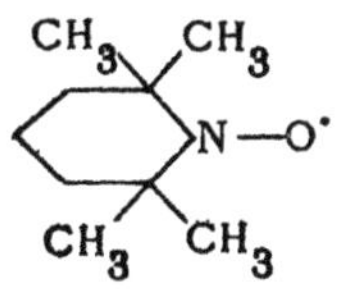	$C_6H_5C_2H_5$	130–20	$\bar{5}$,77	3,11	15,5
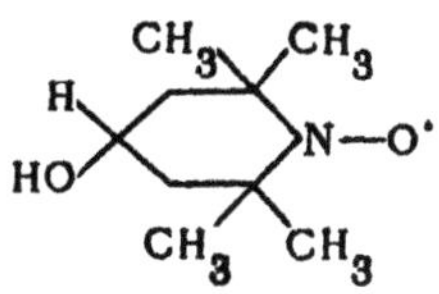	$C_6H_5C_2H_5$	140–190	$\bar{5}$,79	3,08	15,4
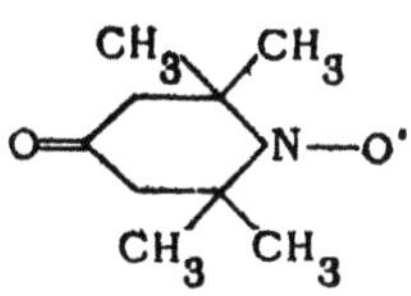	n-$C_{100}H_{202}$	180–220	$\bar{7}$,71	10,73	36,0
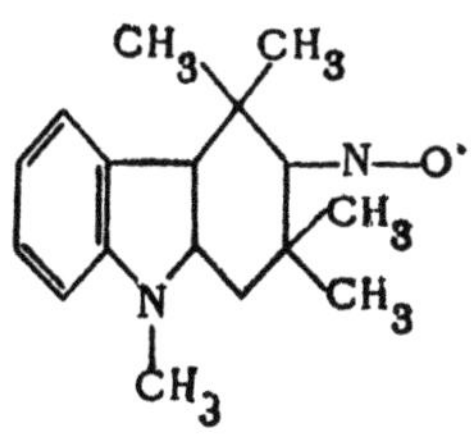	n-$C_{100}H_{202}$	190–215	$\bar{7}$,145	11,11	38,0

Table 61 (continued)

Radical	RH	t^{o}	log k (190°)	log A	E
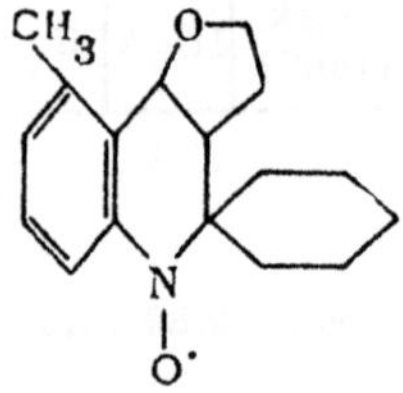	n-$C_{100}H_{202}$	160–195	$\bar{6}$,34	5,23	23,0
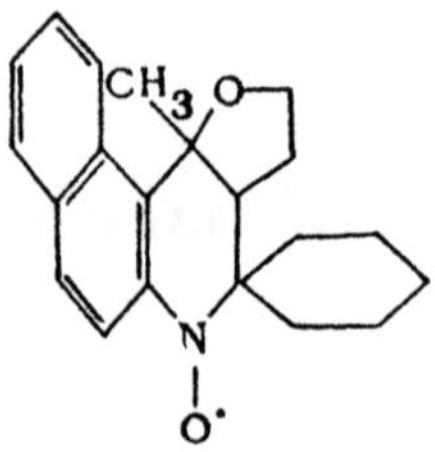	n-$C_{100}H_{202}$	150–190	$\bar{6}$,28	5,15	23,0

TABLE 62. Abstraction of H by Methyl Radicals

RH	Solvent	t°	log k (65°)	log A	E	Method	Ref.
CH_4	H_2O	65	0,52			MCR, $CH_3C_6H_5$	[201]
C_2H_6	- " -	65	2,44			- " -	
C_3H_8	- " -	65	3,125			- " -	
$(CH_3)_3CH$	- " -	65	3,89			- " -	
cyclo-C_5H_{10}	n-C_7H_{16}	80	2,90	11,265	12,90	TIM, 4-T-C_7H_{15}	[198]
cyclo-C_6H_{12}	- " -	60–90	2,70	10,06	11,33	- " -	
cyclo-$C_5H_9CH_3$	- " -	60–90	3,095	9,095	9,25	- " -	
cyclo-$C_6H_{11}CH_3$	- " -	60–90	2,935	9,865	10,68	- "	
cis-Decalin	- " -	60–90	3,51	8,89	8,30	- " -	
trans-Decalin	- " -	60–90	3,19	7,54	6,70	- " -	

Table 62 (continued)

RH	Solvent	t°	log k (65°)	log A	E	Method	Ref.
cis-$CH_3CH{=}CHCH_3$	$(CH_3)_3CCH_2CH(CH_3)_2$	65	2,13			MCR, isooctane	
trans-$CH_3CH{=}CHCH_3$	- " -	65	1,995			- " -	[202]
$(CH_3)_2C{=}CH_2$	- " -	65	0,93			- " -	
$CH_2{=}CHCH_2OCOCH_3$	$(CH_3)_3CCH_2CH(CH_3)_2$	65	2,43			- " -	[202]
	- " -	85	2,68			- " -	[202]
cyclo-C_5H_8	$(CH_3)_3CCH_2CH(CH_3)_2$	65	2,66			- " -	[203]
3-CH_3-cyclo-C_5H_7	n-C_7H_{16}	60–90	3,95	9,40	8,40	TIM, 4-T-C_7H_{15}	[198]
cyclo-C_6H_{10}	n-C_7H_{16}	60–90	3,90	9,265	8,27	TIM, 4-T-C_7H_{15}	[198]
	$(CH_3)_3CCH_2CH(CH_3)_2$	65	1,925			MCR, isooctane	[203]
$(CH_2)_5$ / $CH{=}CH$ (ring)	$(CH_3)_3CCH_2CH(CH_3)_2$	65	2,54			- " -	[203]
$(CH_3)_6$ / $CH{=}CH$ (ring)	- " -	65	2,35			- " -	[203]

Cyclopentadiene	- " -	65	3,93			- " -	[203]
Cyclohexadiene-1,3	$(CH_3)_3CCH_2CH(CH_3)_2$	65	4,46			- " -	[203]
	n-C_7H_{16}	60–90	4,21	12,10	12,15	TIM, 4-T-C_7H_{15}	[198]
Cycloheptatriene-1,4,6	$(CH_3)_3CCH_2CH(CH_3)_2$	65	3,68			MCR, isooctane	[203]
C_6H_6	n-C_7H_{16}	80	1,57			TIM, 4-T-C_7H_{15}	[198]
o-$CH_3C_6H_4CH_3$	$C_6H_5CH_3$	65	2,92			MCR, $C_6H_5CH_3$	[204]
	o-$CH_3C_6H_4CH_3$	0–95	2,98	7,65	7,2	TIM, $C_6H_5CH_2T$	[205]
m-$CH_3C_6H_4CH_3$	$C_6H_5CH_3$	65	2,92			MCR, $C_6H_5CH_3$	[204]
	m-$CH_3C_6H_4CH_3$	0–95	2,91	7,65	7,3	TIM, $C_6H_5CH_2T$	[205]*
p-$CH_3C_6H_4CH_3$	$C_6H_5CH_3$	65	3,00			MCR, $C_6H_5CH_3$	[204]
	p-$CH_3C_6H_5CH_3$	0–95	2,98	7,65	7,2	TIM, $C_6H_5CH_2T$	[205]*
o-$ClC_6H_4CH_3$	$C_6H_5CH_3$	100	2,76			MCR, $C_6H_5CH_3$	[206]
p-$ClC_6H_4CH_3$	- " -	100	3,06			- " -	[206]
p-Br $C_6H_4CH_3$	- " -	100	3,015			- " -	[206]

Table 62 (continued)

RH	Solvent	t°	log k (65°)	log A	E	Method	Ref.
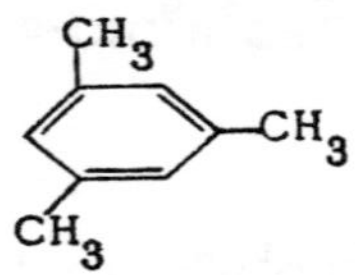	$C_6H_5CH_3$	65	3,07			MCR, $C_6H_5CH_3$	[204]
	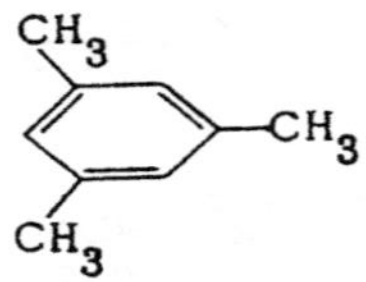	0–95	3,08	7,625	7,0	TIM, $C_6H_5CH_2T$	[205]*
$C_6H_5C_2H_5$	$C_6H_5CH_3$	65	3,06			MCR, $C_6H_5CH_3$	[204]
	- " -	100	3,65			- " -	[206]
	$C_6H_5C_2H_5$	0–95	3,04	7,45	6,8	TIM, $C_6H_5CH_2T$	[205]*
$C_6H_5CH(CH_3)_2$	$C_6H_5CH_3$	65	3,25			MCR, $C_6H_5CH_3$	[204]
	$C_6H_5CH_3$	100	3,71			- " -	[206]
	$C_6H_5CH(CH_3)_2$	0–95	3,25	7,47	6,5	TIM, $C_6H_5CH_2T$	[205]*

p-$(CH_3)_3CC_6H_4CH(CH_3)_2$	p-$(CH_3)_3CC_6H_4CH(CH_3)_2$	0–95	3,25	7,40	6,4	TIM, $C_6H_5CH_2T$	[205]*
Indan	$C_6H_5CH_3$	65	4,14			MCR, $C_6H_5CH_3$	[204]
	Indan	0–95	3,63	7,65	6,2	TIM, $C_6H_5CH_2T$	[205]*
Tetralin	$C_6H_5CH_3$	65	4,58			MCR, $C_6H_5CH_3$	[204]
	Tetralin	0–95	4,11	7,61	5,4	TIM, $C_6H_5CH_2T$	[205]*
cyclo-C_6H_{11}-C_6H_5	cyclo-C_6H_{11}-C_6H_5	0–95	3,27	7,29	6,2	TIM, $C_6H_5CH_2T$	[205]*
$(C_6H_5)_2CH_2$	$(C_6H_5)_2CH_2$	0–95	3,54	7,75	6,5	- " -	
$(C_6H_5)_3CH$	$(C_6H_5)_3CH$	0–95	4,00	7,56	5,5	- " -	
α-$C_{10}H_7CH_3$	$(CH_3)_3CCH_2CH(CH_3)_2$	65	2,55			MCR, isooctane	[207]
β-$C_{10}H_7CH_3$	- " -	65	2,94			- " -	
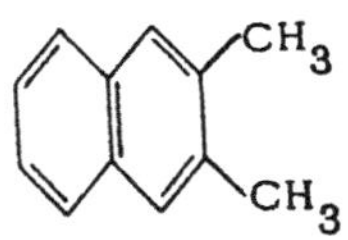	- " -	65	2,955			- " -	

Table 62 (continued)

RH	Solvent	t°	log k (65°)	log A	E	Method	Ref.
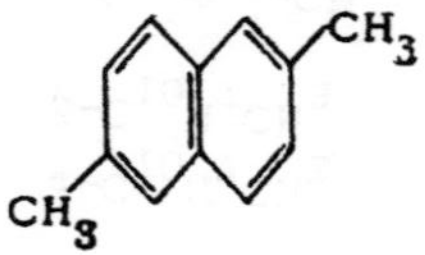	$(CH_3)_3CCH_2CH(CH_3)_2$	65	2,96			MCR, isooctane	[207]
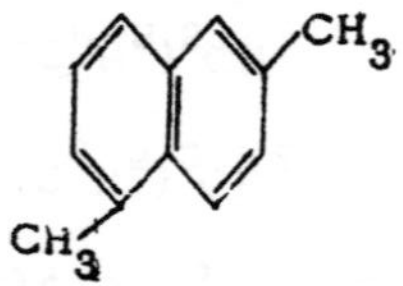	- " -	65	2,965			- " -	[207]
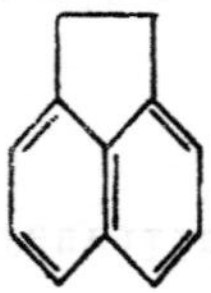	- " -	65	3,465			- " -	[207]
α-$C_{10}H_7CH_2CH_3$	- " -	65	3,08			- " -	[207]
β-$C_{10}H_7C_2H_5$	- " -	65	3,08			- " -	[207]
CH_2ClBr	$C_6H_5CH_3$	48,5-77	2,89	11,52	13,3	MCR, toluene	[201]

$CHCl_2Br$	- " -	25,5-65	4,09	8,83	7,3	- " -	[201]
CH_3OH	H_2O			23	2,34	PR	
C_2H_5OH	- " -			23	2,77	- " -	[63]
$(CH_3)_2CHOH$	- " -			23	3,53	- " -	
CH_3O-T	$n-C_7H_{16}$	60-90	0,48	7,61	11,0	TIM, CH_3OT	[208]
$(CH_3)_2CHO-T$	- " -	80	0,725			TIM, $(CH_3)_2CHOT$	[208]
$(CH_3)_3CO-T$	- " -	60-90	0,66	9,93	14,3	TIM, $(CH_3)_3COT$	[209]
$(CH_3)_3COO-H$	$n-C_7H_{16}$	60-90	4,74	6,79	3,16	TIM, $4-T-C_7H_{15}$	[210]
	H_2O[2*]	60-90	2,93	10,26	11,3	- " -	[210]
$C_2H_5COC_2H_5$	$n-C_7H_{16}$	60-90	3,535	8,135	7,1	TIM, $4-T-C_7H_{15}$	
$C_2H_5CH_2COCH_2C_2H_5$	- " -	80	3,82			- " -	
$(n-C_4H_9)_2CO$	- " -	60-90	3,45	8,76	8,2	- " -	[211]
cyclo-$C_6H_{10}O$	- " -	60-90	3,54	8,40	7,5	- " -	

Table 62 (continued)

R-X	Solvent	t°	log k (65°)	log A	E	Method	Ref.
cyclo-C_5H_8O	n-C_7H_{16}	80	3,26			TIM, 4-T-C_7H_{15}	[211]
$(CH_3)_2CHCOCH(CH_3)_2$	- " -	60–90	3,59	7,41	5,9	- " -	
$C_6H_5COC_2H_5$	- " -	80	3,67			- " -	
$C_6H_5COCH(CH_3)_2$	- " -	80	3,27			- " -	
CH_3COOH	- " -	60–90	1,95	9,91	12,28	- " -	[212]
CH_3CH_2COOH	- " -	60–90	3,12	8,42	8,18	- " -	
$CH_3(CH_2)_2COOH$	- " -	60–90	3,00	8,72	8,83	- " -	
$CH_3(CH_2)_3COOH$	- " -	60–90	3,01	9,01	9,23	- " -	
$CH_3(CH_2)_4COOH$	- " -	80	3,115			- " -	
$CH_3(CH_2)_5COOH$	- " -	80	3,185			- " -	
$(CH_3)_2CHCOOH$	- " -	60–90	3,46	9,60	9,48	- " -	
cyclo-$C_6H_{11}COOH$	- " -	60–90	3,33	9,13	8,94	- " -	

C_6H_5OH	- " -	60–90	4,35	9,02	7,2	TIM, 4-T–C_7H_{15}	[213]
o–$CH_3C_6H_4OH$	- " -	80	4,56			- " -	
p–$CH_3C_6H_4OH$	- " -	80	4,56			- " -	
o–$(CH_3)_3CC_6H_4OH$	- " -	80	4,56			- " -	
o–$NO_2C_6H_4O$–T	- " -	80	2,06			TIM, ArOT	[214, 215]
p–$NO_2C_6H_4O$–T	- " -	80	2,45			TIM, ArOT	
o–$CHOC_6H_4O$–T	- " -	80	1,35			TIM, ArOT	
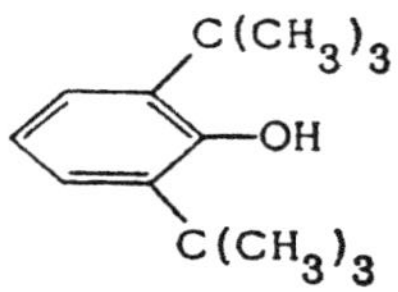	- " -	60–90	3,80	6,46	4,1	TIM, 4-T–C_7H_{15}	[216]
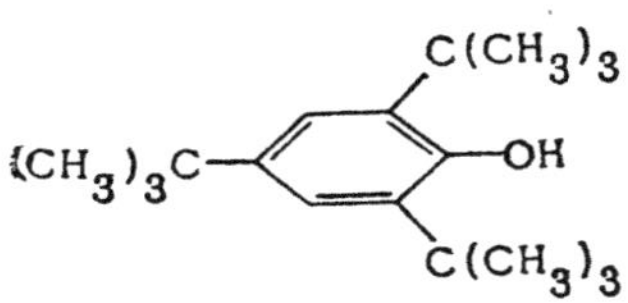	- " -	60–75	3,82	7,92	6,32	- " -	[216]

Table 62 (continued)

RH˙	Solvent	t°	log k (65°)	log A	E	Method	Ref.
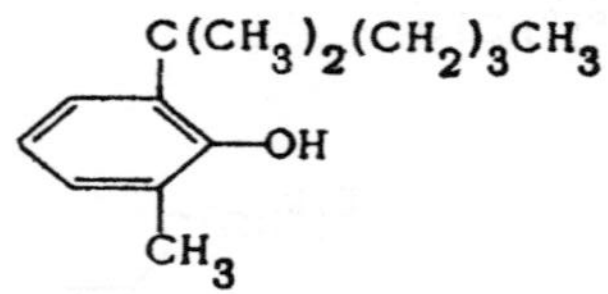	$n\text{-}C_7H_{16}$	80–95	3,07	5,79	4,2	- " -	[216]
$n\text{-}C_4H_9SnH$	cyclo $\text{-}C_6H_{12}$	25	6,76			IIM	[217]

* Absolute values of constants calculated on assumption that rate constant for abstraction of tritium from $C_6H_5CH_2T$ by methyl radical is given by log k = 6.97 - 7.5/θ.

2* In water, a considerable part of the O-H groups of the hydroperoxide are hydrogen-bonded; hence, the rate constants listed in the table are apparent rate constants.

TABLE 63. Partial Rate Constants for Reactions $X\text{-}H + CH_3^{\bullet} \rightarrow CH_4 + X^{\bullet}$ (65°)

Bond type	k_{C-H}	Bond type	k_{C-H}
$RCH_2 - H$	5,0	$ArC(CH_3)_2 - H$	1800
$R_2CH - H$	50	$-COCH(CH_3) - H$	850
$R_3C - H$	500	$-COC(CH_3)_2 - H$	1950
$>C{=}CHCH_2 - H$	18	$HOOCCH_2 - H$	30
$>C{=}CHCHR - H$	120	$HOOCCRH - H$	660
$ArCH_2 - H$	420	$ArO - H$	22000
$ArCH(CH_3) - H$	560	$(CH_3)_3COO - H$	55000

TABLE 64. Abstraction of Halogen Atoms by Methyl Radicals $R\text{-}X + CH_3^{\bullet} \rightarrow CH_3X + R^{\bullet}$

R-X	Solvent	t°	log k (65°)	logA	E	Method	Ref.
I-I	$(CH_3)_3CCH_2CH(CH_3)_2$	25	8,48			MCR, I_2^{131}	[218]
	H_2O	23	9,78			NR	[63]
CH_3-I	$(CH_3)_3CCH_2CH(CH_3)_2$	55–85	4,25	7,85	5,5	MCR, $C^{14}H_3I$	[197]
	$C_6H_5CH_3$	55–85	4,22	7,82	5,5	- " -	[197]
C_2H_5-I	$C_6H_5CH_3$	45–85	4,93	8,53	5,5	MCR, $C_6H_5CH_3$	[201]
$(CH_3)_2CHI$	- " -	45–85	6,53	8,08	3,9	- " -	[201]
CF_3-I	- " -	65	6,93			- " -	[201]
CH_2Cl-I	- " -	65	3,425			- " -	[201]
$C_6H_5CH_2-I$	- " -	65	5,50			- " -	[201]
$(CH_3)_3C-I$	- " -	65	2,89			- " -	[201]
CH_3-Br	$C_6H_5CH_3$	70	0,635			MCR, $C^{14}H_3Br$	[197]
	- " -	80,4	0,98			- " -	[197]
$CHCl_2-Br$	$C_6H_5CH_3$	0–65	4,73	7,81	4,7	MCR, $C_6H_5CH_3$	[201]
CCl_3-Br	- " -	65	6,49			- " -	[201]
$C_6H_5CH_2-Br$	- " -	65	3,48			- " -	[201]
CCl_3-Cl	- " -	65	3,24			- " -	[201]

TABLE 65. Abstraction of H and Cl Atoms by $Cl_3C^{\bullet}$ Radicals
(Note: MCR method, source of $Cl_3C^{\bullet}$ was CCl_3Br in majority of the studies)

RH	Solvent	t°	$k/k_{toluene}$	Ref.
C_6H_6	$C_6H_5CH_3$	91	0,16	[219]
p-$CH_3C_6H_4CH_3$	$C_6H_5CH_3$	50	5,52	[220]
	- " -	91	2,2	[219]
m-$CH_3C_6H_4CH_3$	$C_6H_5CH_3$	50	3,16	[220]
	- " -	91	2,0	[219]
p-$CH_3OC_6H_4CH_3$	$C_6H_5CH_3$	50	14,2	[220]
m-$CH_3OC_6H_4CH_3$	- " -	50	0,87	[220]
p-$ClC_6H_4CH_3$	- " -	50	0,73	[220]
p-Br $C_6H_4CH_3$	- " -	50	0,70	[220]
m-Br $C_6H_4CH_3$	- " -	50	0,25	[220]
$C_6H_5CH_2CH_3$	$C_6H_5CH_3$	91	3,0	[219]
	CCl_4	80	16	[80]
$C_6H_5CH_2OCH_3$	CCl_4	80	35	[80]

Table 65 (continued)

RH	Solvent	t^{o}	$k/k_{toluene}$	Ref.
$p\text{-}CH_3OC_6H_4CH_2OCH_3$	CCl_4	80	68	[89]
$p\text{-}CH_3C_6H_4CH_2OCH_3$	- " -	80	44	
$p\text{-}CH_3(CH_2)_3C_6H_4CH_2OCH_3$	- " -	80	41	
$p\text{-}ClC_6H_4CH_2OCH_3$	- " -	80	32	
$m\text{-}ClC_6H_4CH_2OCH_3$	- " -	80	26	
$C_6H_5CH_2OCH_2C_6H_5$	- " -	80	103	[80]
$(C_6H_5)_2CHOCH_3$	- " -	80	36	
$C_6H_5CH(OCH_3)_2$	- " -	80	17	
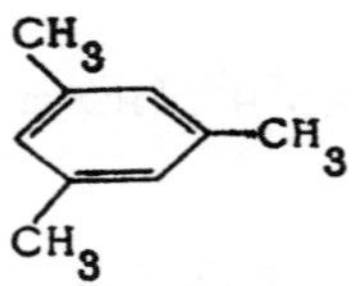	$C_6H_5CH_3$	91	3,0	[219]
$(C_6H_5)_2CH_2$	- " -	91	8,0	
$(C_6H_5)_3CH$	- " -	91	17	
$(C_6H_5CH_2)_2$	- " -	91	,3,1	
$C_6H_5C(CH_3)_3$	- " -	91	0,24	

$C_6H_5CH(CH_3)_2$	$C_6H_5CH_3$	91	4,2	[219]
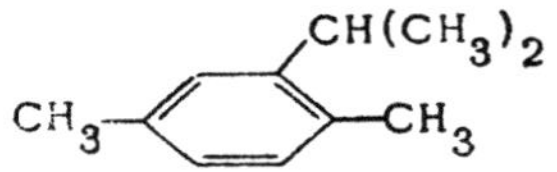	$C_6H_5CH_3$	91	1,85	[219]
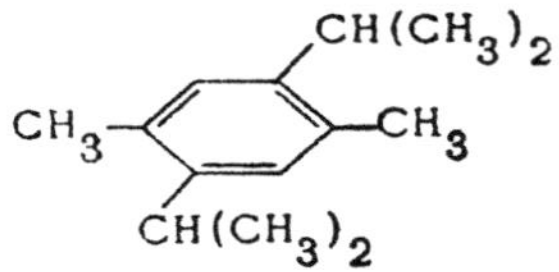	$C_6H_5CH_3$	91	1,76	[219]
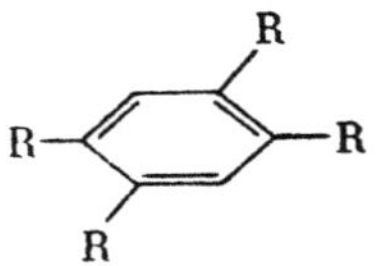 R = $CH(CH_3)_2$	$C_6H_5CH_3$	91	1,14	[219]
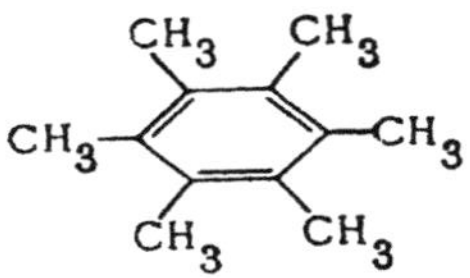	$C_6H_5CH_3$	91	7,4	[219]

Table 65 (continued)

RH	Solvent	t°	$k/k_{toluene}$	Ref.
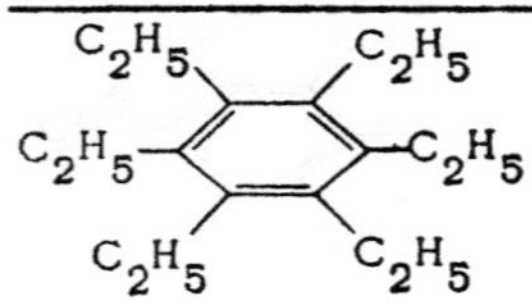	$C_6H_5CH_3$	91	1,6	[219]
$p\text{-}CH_3C_6H_4CH(CH_3)_2$	$C_6H_5CH_3$	91	6,0	[219]
$m\text{-}CH_3C_6H_4CH(CH_3)_2$	- " -	91	4,8	
$o\text{-}CH_3C_6H_4CH(CH_3)_2$	- " -	91	1,05	
cyclo-C_5H_9-C_6H_5	- " -	91	13,3	
cyclo-C_6H_{11}-C_6H_5	- " -	91	0,83	
Fluorene	- " -	91	112	
Indan	- " -	91	9,6	
9,10-Dihydroanthracene	- " -	91	7,7	
$C_{10}H_{12}$		91	17	
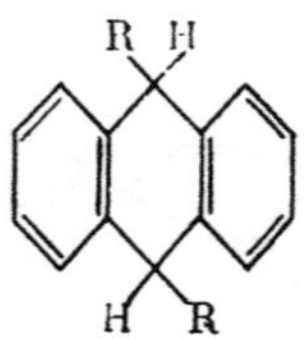	$C_6H_5CH_3$	91	14,7	[219]

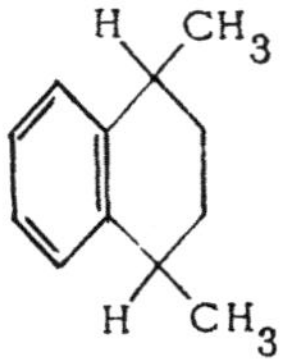	$C_6H_5CH_3$	91	43	[219]
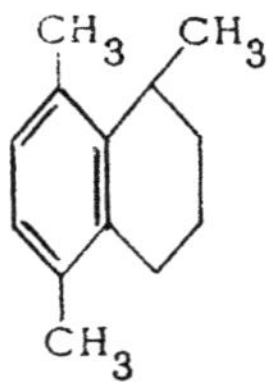	$C_6H_5CH_3$	91	32	[219]
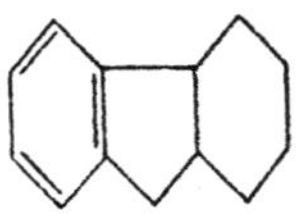	$C_6H_5CH_3$	91	95	[219]
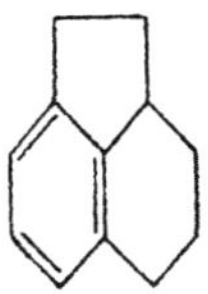	$C_6H_5CH_3$	91	178	[219]

Table 65 (continued)

RH	Solvent	t^o	$k/k_{toluene}$	Ref.
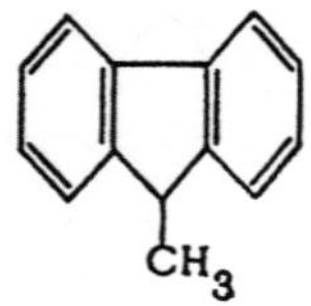	$C_6H_5CH_3$	91	214	[219]
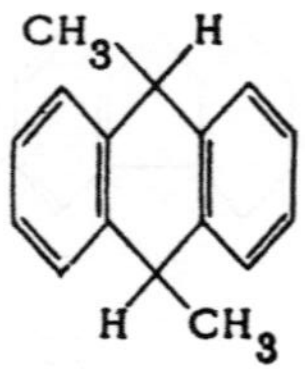	$C_6H_5CH_3$	91	52,5	[219]
n-$C_{13}H_{27}$-CH_2-$CH{=}CH_2$	$C_6H_5CH_3$	91	4,3	[219]
$CH_3CH_2CH{=}CHCH_2C_2H_5$	- " -	91	11,5	[219]
cyclo -C_6H_{10}	- " -	91	26	[219]
$C_6H_5CH_2CH{=}CH_2$	- " -	91	29	[219]
p-$CH_3OC_6H_4CH_2CH{=}CH_2$	- " -	69,5	42	[221]
p-$CH_3C_6H_4CH_2CH{=}CH_2$	- " -	69,5	41	[221]
m-$CH_3C_6H_4CH_2CH{=}CH_2$	- " -	69,5	33	[221]
p-$C_6H_5C_6H_4CH_2CH{=}CH_2$	- " -	69,5	29,5	[221]

$m\text{-}CH_3OC_6H_4CH_2CH{=}CH_2$	- " -	69,5	24	[221]*
$p\text{-}FC_6H_4CH_2CH{=}CH_2$	- " -	69,5	27	[221]*
$p\text{-}ClC_6H_4CH_2CH{=}CH_2$	- " -	69,5	23	[221]*
$m\text{-}ClC_6H_4CH_2CH{=}CH_2$	- " -	69,5	17,4	[221]*
$m\text{-}CF_3C_6H_4CH_2CH{=}CH_2$	- " -	69,5	15,6	[221]*
$p\text{-}CF_3C_6H_4CH_2CH{=}CH_2$	- " -	69,5	13,6	[221]*
$C_6H_5(CH_2)_n\,CH{=}CH_2$				
n = 2	$C_6H_5CH_3$	69,5	8,7	[222]*
n = 3	- " -	69,5	8,4	[222]*
n = 4	- " -	69,5	7,0	[222]*
n = 5	- " -	69,5	3,8	[222]*
n = 6	- " -	69,5	2,3	[222]*
$CHCl_3$	$CHCl_3$	25	3,38[2*]	[223]
$(CH_3)_3CO\text{-}Cl$	- " -	24	2,08[2*]	[224]

* Values of $k/k_{toluene}$ were calculated from the ratios of constants $k/k_{allylbenzene}$ on the assumption that $k_{allylbenzene}/k_{toluene}$ = 29 (the same as at 91°).

2* Values listed are for log k; k measured by IIM method.

TABLE 66. Abstraction of H Atoms by $F_3C^{\bullet}$ Radicals (MCR method, 65°, isooctane)

RH	log k	Ref.
$(CH_3)_3CCH_2CH(CH_3)_2$	5,05	[226]
cis-$CH_3CH{=}CHCH_3$	7,31	[226]
trans-$CH_3CH{=}CHCH_3$	7,34	[226]
$CH_3CH{=}C{=}CHCH_3$	6,75	[226]
$CH_2{=}CHCH_2CH_3$	7,215	[226]
$(CH_3)_2CHCH{=}CH_2$	6,89	[226]
$C_6H_5CH_2CH_3$	5,27	[227]
$C_6H_5CH(CH_3)_2$	5,525	[227]
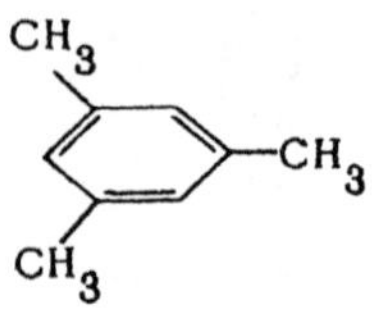	5,15	[227]

TABLE 67. Relative Rate Constants for Reactions of Phenyl Radical in CCl_4 at 60° (Note: MCR method; $k_{rel} = k(C_6H_5^{\bullet} + RH)/k(C_6H_5^{\bullet} + CCl_4)$)

RH	k_{rel}	Ref.
n-C_5H_{12}	0,60	[228]
$CH_3C(CH_3)_2CH_3$	0,14	[228]
n-C_6H_{14}	0,80	[228]

Table 67 (continued)

RH	k_{rel}	Ref.
$n\text{-}C_6H_{14}$	0,80	[229]
$(CH_3)_3CCH_2CH_3$	0,31	[228]
$(CH_3)_2CHCH(CH_3)_2$	1,19	[228]
$CH_3CH_2CH(CH_3)CH_2CH_3$	0,90	[228]
$(CH_3)_2CHCH_2CH_2CH_3$	0,86	[228]
$n\text{-}C_7H_{16}$	0,97	[228]
$C_2H_5CH(CH_3)CH_2CH_2CH_3$	1,07	[228]
$(CH_3)_2CHCH_2CH(CH_3)CH_3$	0,67	[228]
$(CH_3)_3CCH(CH_3)_2$	0,71	[228]
$n\text{-}C_8H_{18}$	1,17	[228]
$(CH_3)_2CH(CH_2)_2CH(CH_3)_2$	1,27	[228]
$(CH_3)_3CCH_2CH(CH_3)_2$	0,34	[228]
$(CH_3)_3CC(CH_3)_3$	0,21	[228]
$(CH_3)_2CH(CH_3)CH(CH_3)_2$	0,97	[228]
$(CH_3)_3CCH_2CH_2CH(CH_3)_2$	0,81	[228]
$n\text{-}C_{16}H_{34}$	2,5	[228]
cyclo-C_5H_{10}	1,04	[228]
cyclo-C_6H_{12}	1,08	[228]
cyclo-C_7H_{14}	2,3	[228]
cyclo-C_8H_{16}	2,9	[228]
cyclo-$C_5H_9CH_3$	1,56	[228]
cyclo-$C_6H_{11}CH_3$	1,55	[228]
$CH_2{=}CH{-}CH_2CH_3$	0,66	[228]
cis-$CH_3CH{=}CHCH_3$	0,88	[228]
trans-$CH_3CH{=}CHCH_3$	0,63	[228]
$CH_2{=}CHCH_2CH_2CH_3$	0,73	[228]
$CH_3CH{=}CH{-}CH_2CH_3$	1,29	[228]
$CH_3C(CH_3){=}CHCH_3$	1,24	[228]
$CH_2{=}C(CH_3)CH_2CH_3$	0,93	[228]
$CH_2{=}CHCH(CH_3)_2$	1,50	[228]
$CH_3{-}C(CH_3){=}C(CH_3)_2$	2,3	[228]

Table 67 (continued)

RH		k_{rel}	Ref.
$CH_3CH{=}CHCH(CH_3)_2$		1,40	
$CH_2{=}CH(CH_2)_5CH_3$		1,38	
$CH_3CH{=}CH(CH_2)_4CH_3$		1,60	
cyclo-C_5H_8		3,7	
cyclo-C_6H_{10}		4,4	[228]
$(CH_3)_2C{=}CH{-}CH{=}C(CH_3)_2$		2,9	
$H_2C{=}CH{-}CH_2{-}C_6H_5$		1,82	
$CH{\equiv}C{-}CH_3$		0,3	
$CH{\equiv}C{-}C_2H_5$		1,02	
$CH_3{-}C{\equiv}C{-}CH_3$		0,87	
$XC_6H_4CH_3$			
	X = H	0,27	[228]
	H	0,36	[229]
	H	0,32*	[230]*
	p-CH_3	0,79	[228]
	p-CH_3	1,02	[229]
	p-CH_3	2,9 *	[230]*
	m-CH_3	2,15*	[230]
	p-CH_3O	0,57	[229]
	p-Cl	0,29	[228]
	p-Cl	0,80*	[230]*
	m-Cl	0,24	
	p-C_6H_5O	0,28	[228]
	p-NO_2	0,22	
	p-CN	0,89*	[230]*
	p-C_2H_5OCO	0,72*	[230]
	m-C_2H_5OCO	0,86*	[230]
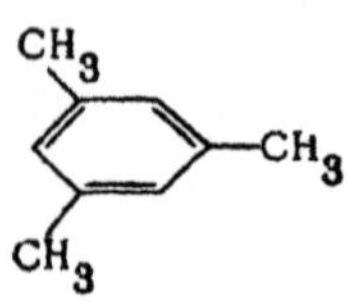		0,85	[228]
		1,14	[229]

Table 67 (continued)

RH	k_{rel}	Ref.
$C_6H_5C_2H_5$	0,84	
$C_6H_5C(CH_3)_3$	0,12	
$C_6H_5CH(CH_3)_2$	0,93	
$(C_6H_5)_2CH_2$	1,4	[228]
$(C_6H_5)_3CH$	3,5	
$C_6H_5CD_3$	0,06	
Indan	3,1	
Tetralin	4,8	
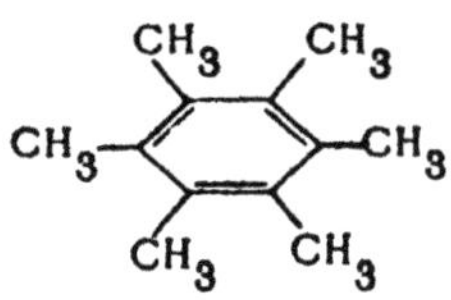	3,1	[228]
$CHCl_3$	3,2	
CH_2Cl_2	0,48	[228]
CH_3Cl	0,80	
$N(CH_3)_3$	3,3	
$C_6H_5NH_2$	1,7	[231]
$(C_6H_5)_2NH$	2,3	[228]
$C_6H_5(CH_3)_2CNH_2$	0,38	[231]
$C_6H_5N(CH_3)_2$	2,7	[228]
CH_3OH	0,13	[228]
C_6H_5OH	8,7	
p-$NO_2C_6H_4OH$	0,6	[231]
p-$CH_3OC_6H_4OH$	52	
$C_6H_5(CH_3)_2COH$	0,14	[231]
CH_3OCH_3	0,28	[228]
$C_6H_5OCH_3$	0,094	[228]
$C_6H_5(CH_3)_2COCH_3$	0,20	[232]
CH_3CHO	2,4	[231]
C_6H_5CHO	2,3	[231]

Table 67 (continued)

RH	k_{rel}	Ref.
m-$NO_2C_6H_4CHO$	3,6	[231]
p-$NO_2C_6H_4CHO$	4,5	[231]
$C_6H_5(CH_3)_2CCHO$	8,6	[231]
CH_3COCH_3	0,17	[228]
CH_3COCH_3	0,17	[229]
CD_3COCD_3	0,041	[228]
$C_2H_5COC_2H_5$	1,30	[228]
$(CH_3)_2CHCOCH(CH_3)_2$	1,30	[228]
$C_6H_5COCH_3$	0,17	[231]
$C_6H_5(CH_3)_2CCOCH_3$	0,49	[232]
CH_3COOH	0,086	[228]
CH_3COOCH_3	0,086	[228]
CH_3COOCH_3	0,086	[229]
$C_6H_5COOCH_3$	0,01	[228]
$CH_3OCOCH_2COOCH_3$	0,18	[228]
$CH_3COCH_2COCH_3$	0,48	[228]
$(CH_3)_2CHCN$	0,73	[228]
CH_3CH_2CN	0,37	[228]
CH_3CN	0,09	[228]
$(CH_3)_4Si$	0,29	[228]
$C_6H_5Si(CH_3)_3$	0,21	[228]
CH_3NO_2	0,047	[228]
$C_2H_5NO_2$	0,19	[228]
$(CH_3)_2CHNO_2$	0,25	[228]
$(CH_3)_2SO$	0,039	[228]
$(C_6H_5)_2SiH_2$	7,7	[228]
$(C_6H_5)_3SiH$	4,8	[228]
$(C_6H_5)_2PH$	60	[228]
3-Methylpyridine	0,12	[228]
4-Methylpyridine	0,12	[228]

Table 67 (continued)

RH	k_{rel}	Ref.
2,4,6-Trimethylpyridine	0,37	[228]
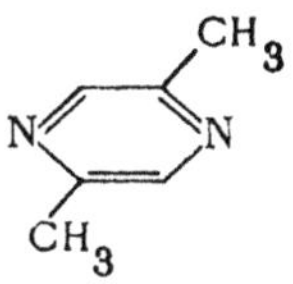	0,33	[228]
α-Methylfuran	0,25	
β-Methylthiophene	0,24	
α-Methylthiophene	0,44	[228]
N-Methylpyrrole	0,3	
α,α'-Dimethylpyrrole	1,43	
$C_6H_5SCH_3$	0,39	
C_6H_5SH	518	[229]
$C_6H_5COON(CH_3)_2$	1,05	[232]
$C_6H_5C(CH_3)_2OCH_3$	0,20	[232]
$C_6H_5C(CH_3)_2COCH_3$	0,49	[232]

* At 100°C.

TABLE 68. Competition of Two Substitution Reactions:

$$C_6H_5^{\cdot} + CCl_4 \xrightarrow{k_3*} C_6H_5Cl + CCl_3^{\cdot}$$

$$C_6H_5^{\cdot} + RSSR \xrightarrow{k_3*} C_6H_5SR + RS^{\cdot}$$

$$C_6H_5^{\cdot} + RSSR \xrightarrow{k_1} C_6H_6 + RSSR^{\cdot}$$

(in CCl_4 at 60°) [233, 234]

R	k_1/k_2	k_1/k_3	R	k_1/k_2	k_1/k_3
CH_3	$1,7\cdot10^{-2}$	0,57	$n\text{-}C_4H_9$	0,14	-
C_2H_5	$7,5\cdot10^{-2}$	1,3	$C_2H_5(CH_3)CH$	0,61	-
$n\text{-}C_3H_7$	0,112	1,8	$(CH_3)_3C$	1,05	0,25
$(CH_3)_2CH$	0,58	1,9			

*As in original

TABLE 69. Relative Rate Constants for Reactions of p-$XC_6H_4^{\cdot}$ Radicals in CCl_4 at 60° [229]

$$k_{rel} = \frac{k(XC_6H_4^{\cdot} + RH)}{k(XC_6H_4^{\cdot} + CCl_4)}$$

RH	k_{rel} of p-XC_6H_4 radicals			
	X=H	CH_3	Br	NO_2
n-C_6H_{14}	0,80	0,89	2,76	7,5
n-C_8H_{18}	–	–	5,1	11,0
$(CH_3)_2CHCH_2CH(CH_3)_2$	–	–	4,1	9,7
$(CH_3)_3CCH_2CH(CH_3)_2$	–	–	0,82	2,1
$(CH_3)_2CHCH_2CH_3$	–	–	0,64	2,2
cyclo-C_6H_{12}	–	1,11	3,5	9,3
CH_3COCH_3	0,17	0,14	0,35	0,61
C_6H_5SH	518	438	569	870
cyclo-C_6H_{10}	–	7,1	14,9	31,8
$C_6H_5CH_3$	0,36	–	0,79	1,67
$C_6H_5CH_2CH_3$	–	–	2,49	7,0
$C_6H_5(CH_3)_2CH$	–	–	3,75	14,6
$C_6H_5C(CH_3)_3$	–	–	0,21	0,19
p-$CH_3OC_6H_4CH_3$	0,57	0,35	1,28	2,98
p-$CH_3C_6H_4CH_3$	1,02	–	2,16	4,86
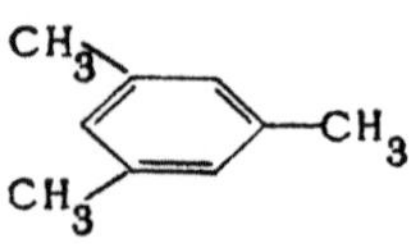	1,14	–	2,84	5,90
p-$FC_6H_4CH_3$	–	0,30	0,61	1,24
p-$NO_2C_6H_4CH_3$	–	0,26	0,54	0,68
CH_3OCOCH_3	0,086	0,075	0,22	

TABLE 70. Reactions of H Abstraction by Various Alkyl Radicals

RH	Solvent	t°	k_{rel}	log k	Method	Ref.
	Radical $n\text{-}C_4H_9$		$\frac{k(C_4H_9^{\cdot} + RH)}{k(C_4H_9^{\cdot} + Cu^{2+})}$			
$C_6H_5CH_2CHO$	67% CH_3COOH + 33% H_2O	57	$6{,}7 \cdot 10^{-4}$		MCR	[235]
$(CH_3)_2CHCHO$	- " -	57	$4{,}6 \cdot 10^{-4}$		MCR	
$CH_3CH_2CH(C_2H_5)CHO$	- " -	57	$3{,}1 \cdot 10^{-4}$		MCR	
$CH_3CH_2CH_2CHO$	- " -	57	$1{,}7 \cdot 10^{-4}$		MCR	
$C_6H_5CH_2OH$	- " -	57	$3 \cdot 10^{-5}$		MCR	
$C_6H_5CH_2OCH_2C_6H_5$	- " -	57	$3 \cdot 10^{-4}$		MCR	
$(CH_2{=}CH{-}CH_2)_2O$	- " -	57	$9 \cdot 10^{-5}$		MCR	
CH_3COOH	- " -	57	$2{,}3 \cdot 10^{-6}$		MCR	
$ClCH_2COOH$	- " -	57	$3{,}1 \cdot 10^{-4}$		MCR	
$CHCl_3$	- " -	57	$7 \cdot 10^{-4}$		MCR	
	Radical $(CH_3)_3C^{\cdot}$					
$(CH_3)_3SnH$	cyclo-C_6H_{12}	25		5,46	IIM	[236]
$(C_6H_5)_3SnH$	- " -	25		6,49	IIM	
$(n\text{-}C_4H_9)_3SnH$	- " -	25		5,87	IIM	
$(n\text{-}C_4H_4)_3SnD$	- " -	25		5,43	IIM	

Table 70 (continued)

RH	Solvent	t^{o}	k_{rel}	log k	Method	Ref.
		Radical $(CH_3)_2\dot{C}CN$				
$(C_2H_5)_3N$	$n-C_6H_{14}$	80-100		6,15-8,0/Θ	MCR	[237]
		Radical $n-C_6H_{13}$	$\frac{k(C_6H_{13}^{\cdot} + RH)}{k(C_6H_{13}^{\cdot} + C_6H_5CH_3)}$			
$(CH_3)_3CH_2CH(CH_3)_2$	CCl_4	74	2,2		MCR CCl_4	[238]
$CH_3CH{=}CH(CH_2)_4CH_3$	- " -	74	8,5		- " -	
$CH_2{=}CH(CH_2)_5CH_3$	- " -	74	10,8		- " -	
cyclo-C_6H_{10}	- " -	74	16,7		- " -	
cyclo-C_6H_{12}	- " -	74	3,3		- " -	
CH_3OH	- " -	74	3,3		- " -	
$(CH_3)_3COH$	- " -	74	0,7		- " -	
$(CH_3)_2CHOH$	- " -	74	25		- " -	
$CH_2{=}CHCH_2OH$		74	4,0			
CH_3COCH_3	- " -	74	1,7		- " -	
$(CH_3)_2CHOCH(CH_3)_2$	- " -	74	108		- " -	
Tetrahydrofuran	- " -	74	1,7		- " -	
Dioxane	- " -	74	5,0		- " -	
CH_3OCOCH_3	- " -	74	0,3		- " -	

$C_2H_5OCOCH_3$	- " -	74	0,3		- " -	[238]
$CH_3OCOC_6H_5$	- " -	74	0,3		- " -	
$CHCl_3$	- " -	74	67		- " -	
$n\text{-}C_4H_9Cl$	- " -	74	1,0		- " -	
cyclo-$C_6H_{11}Cl$	- " -	74	1,3		- " -	
$C_6H_5CH_2Cl$	- " -	74	4,6		- " -	
$(n\text{-}C_4H_9)_3Sn\text{-}H$	cyclo-C_6H_{12}	25		6,00	IIM	[236]
		Radical	cyclo-C_6H_{11}			
$(n\text{-}C_4H_9)_3Sn\text{-}H$	cyclo-C_6H_{12}	25		6,08	IIM	[236]
$(n\text{-}C_4H_9)_3Sn\text{-}D$	- " -	25		5,64	- " -	[236]
	Radical		k/k(p-ClC_6H_4SnH)			
HSC_6H_4X						
X-p-CH_3O	$C_6H_5CH_3$	110	1,75		MCR	[239]
p-$(CH_3)_3C$	- " -	110	1,40		MCR	
p-C_2H_5	- " -	110	1,35		MCR	
H	- " -	110	1,20		MCR	
p-NO_2	- " -	110	0,45		MCR	
$HSC_6H_3X_2$						
X_2=3,4-$(C_6H_5CH_2)_2$	- " -	110	1,15		MCR	
2,3-$(C_6H_5CH_2)_2$	- " -	110	0,80		MCR	

TABLE 71. Abstraction of Halogen Atoms by Various Radicals

Reaction	Solvent	t°	logk	Method	Ref.
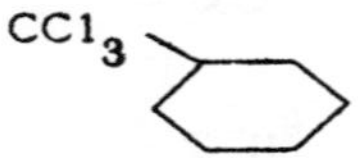 + CClBr	CCl_3Br	30	1,805	IIM	[240]
$CCl_3CH_2\dot{C}HOCOCH_3 + CCl_3Br$	- " -	30	3,44	IIM	[240]
$CCl_3CH_2\dot{C}H(CH_2)_4CH_3 + CCl_3Br$	- " -	15–40	3,24 (30°) (log A=9,4; E=8,5)	NKR	[241]
$CCl_3CH_2\dot{C}HC_6H_5 + CCl_3Br$	- " -	30	1,30	IIM	
$CCl_3CH_2\dot{C}(CH_3)C_6H_5 + CCl_3Br$	- " -	30	0,90	IIM	[240]
$CCl_3CH_2\dot{C}HCH_2Cl + CCl_3Br$	- " -	30	1,20	IIM	
$(CH_3)_3Sn^{\cdot} + (CH_3)_3CCl$	cyclo-C_6H_{12}	25	3,77	IIM	
$(CH_3)_3Sn^{\cdot}$ + cyclo-$C_6H_{11}Cl$	- " -	25	2,69	IIM	
$(n\text{-}C_4H_9)_3Sn^{\cdot} + ClCH_2(CH_2)_3CH_3$	- " -	25	2,93	IIM	
$+ ClC(CH_3)_3$	- " -	25	4,20	IIM	[236]
+cyclo-$C_6H_{11}Cl$	- " -	25	3,30	IIM	
$+ ClCH_2C_6H_5$	- " -	25	5,81	IIM	
$+ Br\,CH_2(CH_2)_4CH_3$	- " -	25	7,28	IIM	

$(n\text{-}C_4H_9)_3Sn^{\cdot}$ + 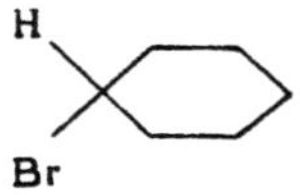	- " -	25	7,34	IIM	[236]
+ $BrC(CH_3)_3$	- " -	25	7,93	IIM	
+ ICH_3	- " -	25	9,40	IIM	
$(C_6H_5)_3Sn^{\cdot} + ClC(CH_3)_3$	- " -	25	4,30	IIM	
$(n\text{-}C_4H_9)_2ClSn^{\cdot} + ClC(CH_3)_3$	- " -	25	3,59	IIM	
$(n\text{-}C_4H_9)_2HSn^{\cdot} + ClC(CH_3)_3$	- " -	25	3,67	IIM	
C_2H_5N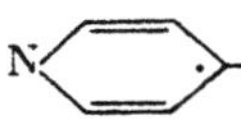$COOCH_3 + CH_2Br_2$	CH_3CN	25	$\bar{4}$,30	Kinetics	[242]
+ CH_2Br_2	CH_2Cl_2	25	$\bar{4}$,00	- " -	[243]
+ CH_2Br_2	$(CH_3)_2CHOH$	25	$\bar{5}$,78	- " -	
+ CH_2Br_2	C_2H_5OH	25	$\bar{5}$,845	- " -	
+ CH_2BrCl	CH_3CN	25	$\bar{4}$.00	- " -	
+ CH_2Cl_2	- " -	25–75	$\bar{8}$,73 (25) (log A=6,51; E=18,8)	- " -	

Table 71 (continued)

Reaction	Solvent	t°	log k	Method	Ref.
C_2H_5N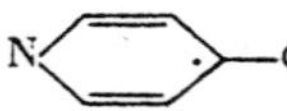$COOCH_3$ + $C_6H_5CH_2Cl$	CH_3CN	25	$\bar{4},52$	Kinetics	[243]
+ $p\text{-}CH_3OC_6H_4CH_2Cl$	- " -	25	$\bar{3},05$	- " -	
+ $p\text{-}CH_3C_6H_4CH_2Cl$	- " -	25	$\bar{4},585$	- " -	
+ $p\text{-}ClC_6H_4CH_2Cl$	- " -	25	$\bar{4},815$	- " -	
C_2H_5—N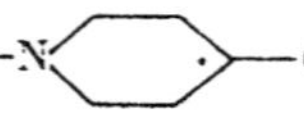$COOCH_3$ + $p\text{-}NO_2C_6H_4CH_2Cl$	CH_3CN	25	4,38	Kinetics	
	$CHON(CH_3)_2$	25	4,08	- " -	
	CH_3COCH_3	25	2,65	- " -	
	CH_2Cl_2	25	1,88	- " -	
	$(CH_3OCH_2)_2$	25	0,92	- " -	
	α-Methyl-tetra-hydrofuran	25	0,21	- " -	

Relative k for H abstraction by ketones $(XC_6H_4)_2CO$ from hydroxyalkyl radical $CH_3C(OH)(C_2H_5)$ [244] (MCR method, 125°)

Substituent X	$m\text{-}CF_3$	$p\text{-}Cl$	$m\text{-}CH_3$	$p\text{-}CH_3$	$p\text{-}OCH_3$	$2,4\text{-}(CH_3)_2$	$2,4,6\text{-}(CH_3)_3$
k_X/k_H	4,85	3,01	0,85	0,59	0,17	0,18	0,05

TABLE 72. Chain Transfer in Bulk Polymerization of Styrene ($\log k_p = 7.095 - 7.2/\theta$)

Compound	t°	C_S	log k	log A	E	Ref.
C_6H_6	60	$1,8 \cdot 10^{-6}$	$\bar{4},505$	10,975	22,0	[245]
	60	$6,0 \cdot 10^{-6}$	$\bar{3},02$			[246]
$C_6H_5CH_3$	60	$1,25 \cdot 10^{-5}$	$\bar{3},34$	8,73	17,3	[245]
	60	$1,6 \cdot 10^{-5}$	$\bar{3},44$			[246]
$C_6H_5C_2H_5$	60	$6,7 \cdot 10^{-5}$	$\bar{2},08$	6,43	12,7	[245]
$C_6H_5CH(CH_3)_2$	60	$8,2 \cdot 10^{-5}$	$\bar{2},15$	6,50	12,7	[245]
	60	$1,8 \cdot 10^{-4}$	$\bar{2},30$			[246]
	60	$3,9 \cdot 10^{-4}$	$\bar{2},83$			[247]
$(C_6H_5)_2CH_2$	60	$2,3 \cdot 10^{-4}$	$\bar{2},60$	5,77	10,9	[245]
	60	$2,8 \cdot 10^{-4}$	$\bar{2},69$			[246]
$(C_6H_5)_3CH$	60	$3,5 \cdot 10^{-4}$	$\bar{2},785$	6,885	12,3	[245]
	60	$6,5 \cdot 10^{-5}$	$\bar{2},06$			[246]
$C_6H_5C(CH_3)_3$	60	$6,0 \cdot 10^{-6}$	$\bar{3},02$	10,77	20,9	[245]
$(C_6H_5)_3CCH(C_6H_5)_2$	60	2,0	2,545			[245]
	60	2,0	2,545			[246]

Table 72 (continued)

Compound	t°	C_S	log k	log A	E	Ref.
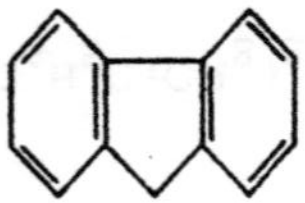	60	$7{,}5 \cdot 10^{-3}$	0,115	6,895	10,3	[245]
Naphthalene	60	$1{,}1 \cdot 10^{-3}$	$\bar{1},285$			[248]
p-$CH_3OC_6H_4CH(CH_3)_2$	60	$3{,}23 \cdot 10^{-4}$	$\bar{2},755$			[247]
p-$(CH_3)_3CC_6H_4CH(CH_3)_2$	60	$3{,}46 \cdot 10^{-4}$	$\bar{2},785$			[247]
p-$(CH_3)_2CHC_6H_4CH(CH_3)_2$	60	$6{,}60 \cdot 10^{-4}$	$\bar{1},065$			[247]
p-$ClC_6H_4CH(CH_3)_2$	60	$6{,}90 \cdot 10^{-4}$	1,085			[247]
p-Br $C_6H_4CH(CH_3)_2$	60	$5{,}57 \cdot 10^{-4}$	$\bar{2},99$			[247]
p-$CNC_6H_4CH(CH_3)_2$	60	$1{,}86 \cdot 10^{-3}$	$\bar{1},515$			[247]
m-$CH_3OC_6H_4CH(CH_3)_2$	60	$5{,}23 \cdot 10^{-4}$	$\bar{2},965$			[247]
m-Br $C_6H_4CH(CH_3)_2$	60	$8{,}29 \cdot 10^{-4}$	$\bar{1},165$			[247]
cyclo-C_6H_{12}	60	$1{,}8 \cdot 10^{-6}$	$\bar{4},62$	10,17	20,6	[245]
	60	$4{,}0 \cdot 10^{-6}$	$\bar{4},84$			[246]
Decalin	60	$3{,}1 \cdot 10^{-5}$	$\bar{3},845$			[245]

$C_6H_5CH{=}CH_2$	0–25	$1,07\cdot10^{-5}$	$\bar{4},83$ (25°)	7,26	14,2	[249]
	0–25	$1,0\cdot10^{-5}$	$\bar{4},80$ (25°)	7,38	14,4	[250]
	60	$4,7\cdot10^{-5}$	$\bar{2},025$			[251]
	60	$8,0\cdot10^{-5}$	$\bar{2},255$			[246]
	30	$1,07\cdot10^{-5}$	$\bar{4},92$			[252]
$(CH_3)_2CHOH$	100–130	$1,7\cdot10^{-4}$	$\bar{1},10$ (100°)	6,10	11,9	[253]
	60	$8\cdot10^{-5}$	$\bar{2},15$			[246]
$(CH_3)_2CHOD$	100	$1,6\cdot10^{-4}$	$\bar{1},07$			
$(CH_3)_2CDOD$	100	$7,8\cdot10^{-5}$	$\bar{2},76$			
$(CH_3)_3COH$	100–130	$5,5\cdot10^{-5}$	$\bar{2},61$ (100°)	6,36	13,2	[253]
$(CH_3)_3COD$	100–130	$2,9\cdot10^{-5}$	$\bar{2},33$ (100°)	8,33	17,0	
$CH_3CH_2CH_2CH_2OH$	60	$5,1\cdot10^{-6}$	$\bar{3},06$			[254]
$CH_3CH_2CH_2CHO$	60	$5,7\cdot10^{-4}$	$\bar{1},11$			[246]
CH_3COCH_3	60	$4,3\cdot10^{-5}$	$\bar{3},98$			[254]
	60	$4,1\cdot10^{-4}$	$\bar{2},965$			[255]
	60	$5,0\cdot10^{-5}$	$\bar{2},055$			[246]
$CH_3COC_2H_5$	60	$5,0\cdot10^{-4}$	$\bar{1},055$			
cyclo $-C_5H_8O$	60	$3,3\cdot10^{-4}$	$\bar{2},875$			[255]
cyclo $-C_6H_{10}O$	60	$8,0\cdot10^{-4}$	$\bar{1},255$			

Table 72 (continued)

Compound	t^{o}	C_S	log k	log A	E	Ref.
$CH_3COCH_2COCH_3$	100	$2,0 \cdot 10^{-4}$	1,170			[256]
CH_3CH_2COOH	60	$5,0 \cdot 10^{-6}$	3,05			[246]
Dioxane	60	$2,75 \cdot 10^{-4}$	2,79			[255]
$(CH_3O)_3CH$	100	$0,91 \cdot 10^{-4}$	2,83			
$(CH_3O)_2CHCH_2OCH_3$	100	$0,94 \cdot 10^{-4}$	2,84			
CH_3OCOCH_3	100	$3,3 \cdot 10^{-5}$	2,39			
$C_2H_5OCOCH_3$	100	$3,9 \cdot 10^{-5}$	2,46			
$C_2H_5OCOCH_2C_6H_5$	100	$8,8 \cdot 10^{-4}$	1,815			
$CH_3COCH_2COOC_2H_5$	100	$1,3 \cdot 10^{-4}$	2,98			
$(CH_2COOC_2H_5)_2$	100	$4,6 \cdot 10^{-5}$	2,53			[256]
$(CH_2COOCH_3)_2$	100	$4,2 \cdot 10^{-5}$	2,50			
$C_2H_5OCOCH(C_2H_5)CH_2COOC_2H_5$	100	$7,2 \cdot 10^{-5}$	2,73			
$(C_2H_5CHCOOC_2H_5)_2$	100	$8,8 \cdot 10^{-5}$	2,815			
$C_2H_5OOCCH(C_6H_5)CH_2COOC_2H_5$	100	$3,5 \cdot 10^{-4}$	1,415			
$C_2H_5OCOC(n\text{-}C_4H_9)HCH_2COOC_2H_5$	100	$8,2 \cdot 10^{-5}$	2,785			
$C_2H_5OCOCH(COCH_3)CH_2COOC_2H_5$	100	$1,2 \cdot 10^{-4}$	2,945			
$C_2H_5OCOCH(COCH_3)_2$	100	$3,0 \cdot 10^{-4}$	1,35			

n-$ClC_6H_4CH_2OCH_3$	68	$4 \cdot 10^{-4}$	$\bar{1},07$			
$C_6H_5CH_2OCH_3$	68	$6 \cdot 10^{-4}$	$\bar{1},25$			[257]
p-Br $C_6H_4CH_2OCH_3$	68	$6 \cdot 10^{-4}$	$\bar{1},25$			
p-$CNC_6H_4CH_2OCH_3$	68	$2 \cdot 10^{-3}$	$\bar{1},77$			
(n-$C_3H_7O)_2$	60	$8,4 \cdot 10^{-4}$	$\bar{1},28$			[258]
$[(CH_3)_3CO]_2$	60	$1,3 \cdot 10^{-3}$	$\bar{1},47$			[259]
$(C_6H_5COO)_2$	60	$5,3 \cdot 10^{-2}$	1,08			[260]
	100	0,11	1,91			[261]
	60	$5,5 \cdot 10^{-2}$	1,095			[259]
	40–80	$7,5 \cdot 10^{-2}$	1,375 (70°)	8,745	11,55	[262]
(p-$CH_3OC_6H_4COO)_2$	70	$7,4 \cdot 10^{-2}$	1,37			[262]
(p-$CH_3C_6H_4COO)_2$	50–70	0,19	1,78 (70°)	7,14	8,4	[263]
(p-$CH_3OCOC_6H_4COO)_2$	70	0,208	1,82			
	70	0,187	1,77			
(p-$FC_6H_4COO)_2$	70	0,219	1,84			
(p-$ClC_6H_4COO)_2$	70	0,216	1,835			[262]
(p-Br $C_6H_4COO)_2$	70	0,193	1,785			
(p-$IC_6H_4COO)_2$	70	0,293	1,97			
(m-$FC_6H_4COO)_2$	70	0,246	1,89			
(m-I $C_6H_4COO)_2$	70	0,262	1,92			

Table 72 (continued)

Compound	t°	C_S	log k	log A	E	Ref.
$(m\text{-}ClC_6H_4COO)_2$	70	0,346	2,04			
$(m\text{-}BrC_6H_4COO)_2$	70	0,465	2,17			
$(p\text{-}CNC_6H_4COO)_2$	70	0,804	2,405			
$(m\text{-}NO_2C_6H_4COO)_2$	70	6,2	3,29			
$(p\text{-}NO_2C_6H_4COO)_2$	70	7,4	3,87			[262]
$(o\text{-}CH_3C_6H_4COO)_2$	70	0,175	1,74			
$(o\text{-}FC_6H_4COO)_2$	70	0,40	2,10			
$(o\text{-}ClC_6H_4COO)_2$	70	1,91	2,78			
$(o\text{-}BrC_6H_4COO)_2$	70	2,17	2,84			
$(o\text{-}BrC_6H_4COO)_2$	70	1,0	2,215			[263]
$[CH_3(CH_2)_2COO]_2$	70	$1,8\cdot10^{-2}$	0,755			
$[CH_3(CH_2)_4COO]_2$	70	0,166	1,72			
$[CH_3(CH_2)_6COO]_2$	70	$9,8\cdot10^{-2}$	1,49			
$[CH_3(CH_2)_{10}COO]_2$	70	$2,4\cdot10^{-2}$	0,88			[262]
$[CH_3(CH_2)_{12}COO]_2$	70	0,116	1,565			
$[CH_3(CH_2)_{14}COO]_2$	70	0,142	1,65			
$[CH_3(CH_2)_{16}COO]_2$	70	0,154	1,69			

$(CH_3CH{=}CHCOO)_2$	70	0,146	1,665			[262] (rows 1–4)
$(CH_3CH{=}CH{-}CH{=}CHCOO)_2$	70	1,19	2,575			
$(C_6H_5CH{=}CHCOO)_2$	70	1,10	2,54			
$(C_6H_5CH{=}CH{-}CH{=}CHCOO)_2$	70	5,24	3,22			
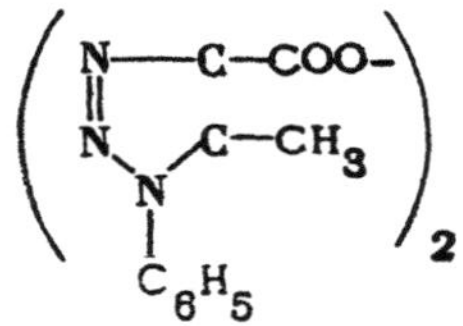	70	0,24	1.88			[262]
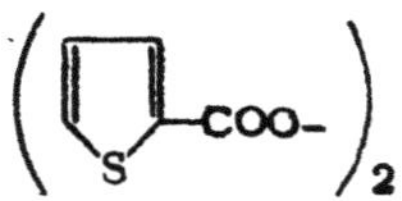	50–70	0,38	2,08 (70°)	10,25	12,8	[263]
CH_2Cl_2	80	$1\cdot10^{-3}$	$\bar{1}$,635			[264]
	60	$1{,}5\cdot10^{-4}$	$\bar{2}$,53			[254]
$CHCl_3$	60	$4\cdot10^{-4}$	$\bar{2}$,955			[257]
	60	$5\cdot10^{-5}$	$\bar{2}$,055			[254]
	60	$5\cdot10^{-5}$	$\bar{2}$,055			[246]
	60	$3{,}4\cdot10^{-4}$	$\bar{2}$,885			[255]
CCl_4	60	$8{,}2\cdot10^{-3}$	0,325			[254]
	80	$1{,}8\cdot10^{-2}$	0,89			[245]
	60	$1{,}1\cdot10^{-2}$	0,395			[246]

Table 72 (continued)

Compound	t°	C_S	log k	log A	E	Ref.
CCl_4	60	$8,7\cdot10^{-3}$	0,295			[255]
$n\text{-}C_4H_9Cl$	60	$4,0\cdot10^{-6}$	$\bar{4},955$			[254]
	60	$4,0\cdot10^{-6}$	$\bar{4},955$			[246]
$ClCH_2(CH_2)_2CH_2Cl$	80	$5,0\cdot10^{-3}$	0,335			[264]
$ClCH{=}CHCl$	80	$1,0\cdot10^{-3}$	$\bar{1},635$			[264]
	60	$4,1\cdot10^{-4}$	$\bar{2},965$			[255]
C_6H_5Cl	100	$5,0\cdot10^{-5}$	$\bar{2},57$			[265]
	60	$1,5\cdot10^{-4}$	$\bar{2},53$			[255]
$p\text{-}ClC_6H_4Cl$	60	$2,6\cdot10^{-4}$	$\bar{2},77$			[255]
$o\text{-}ClC_6H_4Cl$	60	$3,4\cdot10^{-4}$	$\bar{2},885$			[255]
$C_6H_5CH_2Cl$	60	$1,56\cdot10^{-5}$	$\bar{3},545$			
C_6H_5COCl	60	$5\cdot10^{-3}$	0,055			[254]
$C_6H_5CCl_3$	60	$5,8\cdot10^{-3}$	0,12			
$ClCH_2COOH$	68	$2\cdot10^{-2}$	0,655			[257]

$ClCH_2COCl$	60	0,3	1,835			[264]
$Cl_3CCOOC_2H_5$	90	$9\cdot10^{-3}$	0,71			[266]
CH_2Br_2	60	$1,0\cdot10^{-2}$	0,355			[264]
CBr_4	60–90	1,8 (60°)	2,61 (60°)	9,06	9,8	[266]
	60	2,2	2,695			[246]
$n\text{-}C_4H_9Br$	60	$6,0\cdot10^{-6}$	$\bar{3},135$			[254]
C_6H_5Br	60	$1,8\cdot10^{-4}$	$\bar{2},61$			[255]
	155	$3,0\cdot10^{-4}$	$\bar{1},805$			[267]
$Br\,CH_2COOH$	68	$3\cdot10^{-3}$	$\bar{1},95$			[257]
$Br_2CHCOOC_2H_5$	90	0,27	2,185			[266]
$C_6H_5CBr=CBr\,C_6H_5$	60	0,3	1,835			[264]
CH_3COBr	60	0,9	2,31			[264]
CBr_3CHCH_3	90	2,4	3,135			[266]
Br_3CCOOH	90	2,4	3,135			[266]
$CBr_3CH_2CHBr\,C_6H_5$	90	3,65	3,315			[266]
$p\text{-}CH_2Br\,C_6H_4CH_2Br$	60	$1,5\cdot10^{-2}$	0,53			[264]
CH_2I_2	60	$7\cdot10^{-2}$	1,20			[264]
$n\text{-}C_4H_9I$	60	$1,8\cdot10^{-5}$	$\bar{5},61$			[254]
	60	$1,8\cdot10^{-5}$	$\bar{5},61$			[246]

Table 72 (continued)

Compound	t^o	C_S	log k	log A	E	Ref.
CH_2COOH	68	0,8	2,37			[257]
$(CH_3)_3N$	60	$7{,}1 \cdot 10^{-4}$	$\bar{1},205$			[268]
$C_6H_5NH_2$	60	$2{,}0 \cdot 10^{-4}$	$\bar{2},655$			[246]
	50	$2{,}0 \cdot 10^{-3}$	$\bar{1},515$			[269]
$C_6H_5NHCH_3$	50	$1{,}3 \cdot 10^{-3}$	$\bar{1},345$			
$C_6H_5N(CH_3)_2$	50	$1{,}2 \cdot 10^{-3}$	$\bar{1},34$			
$p\text{-}CH_3C_6H_4NH_2$	50	$7{,}8 \cdot 10^{-3}$	0,105			[269]
$p\text{-}CH_3C_6H_4NHCH_3$	50	$1{,}1 \cdot 10^{-3}$	$\bar{1},255$			
$p\text{-}CH_3C_6H_4N(CH_3)_2$	50	$1{,}6 \cdot 10^{-3}$	$\bar{1},42$			
$(CH_3)_2CHCN$	100–130	$2{,}7 \cdot 10^{-4}$ (100°)	$\bar{1},30$ (100°)	5,05	9,8	[253]
H_2S	60	5	3,20			[246]
$n\text{-}C_4H_9SH$	60	22	3,695			[246]
	60	22	3,395			[270]
$(CH_3)_3CSH$	60	3,6	2,915			[271]
	60–100	3,1 (60°)	2,845 (60°)	5,345	3,8	[253]

$CH_3(CH_2)_2CH_2SH$	100	15,4	3,545	
$CH_3(CH_2)_4CH_2SH$	100	15,3	3,54	[261]
$CH_3(CH_2)_5CH_2SH$	100	15,1	3,525	
$CH_3(CH_2)_{10}CH_2SH$	60	19	3,635	[271]
$CH_3(CH_2)_{16}CH_2SH$	100	14,7	4,04	
$CH_3(CH_2)_5CH(CH_3)SH$	100	3,2	3,375	
$C_6H_5CH_2SH$	100	25,5	4,28	[261]
p-$CH_3OC_6H_4CH_2SH$	100	26	4,285	
α-$C_{10}H_7CH_2SH$	100	24,6	4,26	
CH_3OCOCH_2SH	100	27,6	4,31	
$C_2H_5OCH_2CH_2CH_2SH$	60	21	4,19	[271]
$CH_3CH(SH)COOH$	50	6	2,995	[272]
α-$C_{10}H_7CH_2SH$	50	18	3,47	[272]
$C_2H_5OCOCH_2SH$	60	58	4,115	[271]
CS_2	70	10,5	3,52	[273]
$(CH_3S^{\cdot})_2$	60	$9,4\cdot10^{-3}$	0,325	[246]
$(CH_3S)_2$	60	$9,4\cdot10^{-3}$	0,325	[259]
$(C_2H_5S)_2$	100	$4,5\cdot10^{-3}$	0,42	[261]
(n-$C_3H_7S)_2$	60	$2,34\cdot10^{-3}$	$\bar{1}$,725	[259]
$[(CH_3)_2CHS]_2$	60	$6,6\cdot10^{-4}$	$\bar{1}$,175	[259]

Table 72 (continued)

Compound	t°	C_S	log k	log A	E	Ref.
$(n\text{-}C_4H_9S)_2$	60	$2,4 \cdot 10^{-3}$	$\bar{1},735$			[259]
	100	$6,8 \cdot 10^{-3}$	0,70			[261]
$[CH_3CH(C_2H_5)S]_2$	60	$4,4 \cdot 10^{-4}$	$\bar{2},995$			
$[(CH_3)_3CS]_2$	60	$1,4 \cdot 10^{-4}$	$\bar{2},505$			[259]
$[(CH_3)_2CHCH_2S]_2$	60	$2,04 \cdot 10^{-3}$	$\bar{1},665$			
$[CH_3(CH_2)_4CH_2S]_2$	100	$1,04 \cdot 10^{-2}$	0,89			
$[CH_3(CH_2)_6CH_2S]_2$	100	$2,4 \cdot 10^{-2}$	1,25			[261]
$[CH_3(CH_2)_5CH(CH_3)S]_2$	100	$1,04 \cdot 10^{-2}$	0,89			
$(C_6H_5CH_2S)_2$	100	$1,1 \cdot 10^{-2}$	0,91			[261]
	50	$2,0 \cdot 10^{-2}$	0,515			[274]
$(p\text{-}CH_3OC_6H_4CH_2S)_2$	100	$2,1 \cdot 10^{-2}$	1,19			
$(\alpha\text{-}C_{10}H_7CH_2S)_2$	100	$3,3 \cdot 10^{-2}$	1,39			[261]
$(HOOCCH_2S)_2$	100	0,20	2,17			
$(CH_3OCOCH_2S)_2$	100	0,10	1,87			

$(C_6H_5S)_2$	50	1,3	2,33			[274]
$[(CH_3)_2CHCOS]_2$	50	5,3	2,94			[274]
$(C_6H_5COS)_2$	60	$3{,}60\cdot10^{-3}$	$\bar{1}$,91			
$(p\text{-}CH_3OC_6H_4COS)_2$	60	$9{,}60\cdot10^{-3}$	0,335			
$(p\text{-}CH_3C_6H_4COS)_2$	60	$4{,}63\cdot10^{-3}$	0,02			
$(p\text{-}ClC_6H_4COS)_2$	60	$1{,}96\cdot10^{-2}$	0,645			[275]
$(p\text{-}Br\,C_6H_4COS)_2$	60	$7{,}45\cdot10^{-2}$	1,225			
$(p\text{-}CNC_6H_4COS)_2$	60	$3{,}19\cdot10^{-1}$	1,86			
$(p\text{-}NO_2C_6H_4COS)_2$	60	$6{,}65\cdot10^{-1}$	2,18			
$(CH_3)_2NCS-SS-SCN(CH_3)_2$	70	$1{,}4\cdot10^{-2}$	0,645			
$(n\text{-}C_4H_9)_2S$	60	$2{,}2\cdot10^{-3}$	$\bar{1}$,695			[276]
$[(CH_3)_3C]_2S$	60	$2{,}5\cdot10^{-2}$	0,755			
$C_6H_5PH_2$	60	43,9	3,995			
$C_2H_5PH_2$	60	3,6	2,91			[277]
$NCCH_2CH_2PH_2$	60	5,0	3,055			
$(NCCH_2CH_2)_2PH$	60	5,0	3,055			
$[\sim(C_6H_5)CHCH_2\sim]_n$	50	$1{,}9\cdot10^{-4}$	$\bar{2}$,495			[278]
$[\sim(CH_3COO)CHCH_2\sim]_n$	80–100	$6{,}5\cdot10^{-4}$	$\bar{1}$,68 (100°)	5,85	10,5	[279]

TABLE 73. Chain Transfer in Bulk Polymerization of Methyl Methacrylate ($\log k_p = 6.15 - 5.0/\theta$)

Compound	t°	C_S	log k	log A	E	Ref.
C_6H_6	60–80	$4{,}0 \cdot 10^{-6}$ (60°)	$\bar{3}{,}46$ (60°)	6,08	13,1	[280]
	80	$7{,}5 \cdot 10^{-6}$	$\bar{3}{,}925$			[281]
$C_6H_5CH_3$	60–80	$1{,}7 \cdot 10^{-5}$ (60°)	$\bar{2}{,}09$ (60°)	6,51	12,8	[280]
	80	$5{,}25 \cdot 10^{-5}$	$\bar{2}{,}77$			[281]
	60	$2{,}0 \cdot 10^{-5}$	$\bar{2}{,}16$			[282]
$C_6H_5CH_2CH_3$	60–80	$7{,}66 \cdot 10^{-4}$ (60°)	$\bar{2}{,}745$ (60°)	6,175	11,3	[280]
	80	$1{,}35 \cdot 10^{-4}$	$\bar{1}{,}18$			[281]
$C_6H_5CH(CH_3)_2$	80	$1{,}90 \cdot 10^{-4}$	$\bar{1}{,}33$			[281]
	60	$2{,}56 \cdot 10^{-4}$	$\bar{1}{,}27$			[283]
p-$CH_3OC_6H_4CH(CH_3)_2$	60	$3{,}46 \cdot 10^{-4}$	$\bar{1}{,}40$			
p-$(CH_3)_3CC_6H_4CH(CH_3)_2$	60	$2{,}74 \cdot 10^{-4}$	$\bar{1}{,}30$			
p-$(CH_3)_2CHC_6H_4CH(CH_3)_2$	60	$5{,}72 \cdot 10^{-4}$	$\bar{1}{,}62$			[283]
p-$ClC_6H_4CH(CH_3)_2$	60	$3{,}07 \cdot 10^{-4}$	$\bar{1}{,}35$			
p-Br $C_6H_4CH(CH_3)_2$	60	$3{,}71 \cdot 10^{-4}$	$\bar{1}{,}43$			
p-$CNC_6H_4CH(CH_3)_2$	60	$4{,}99 \cdot 10^{-4}$	$\bar{1}{,}56$			

$C_6H_5C(CH_3)_3$	80	$2,6 \cdot 10^{-5}$	$\bar{2},465$			[281]
cyclo-C_6H_{12}	80	$1,0 \cdot 10^{-5}$	$\bar{2},05$			[281]
cyclo-$C_6H_{11}CH_3$	80	$1,95 \cdot 10^{-5}$	$\bar{2},34$			[281]
$CH_2{=}C(CH_3)COOCH_3$	30–80	$6,3 \cdot 10^{-7}$ (60°)	$\bar{4},66$ (60°)	4,09	11,3	[284]
	25–100	$1,0 \cdot 10^{-5}$	$\bar{3},47$ (25°)	3,36	8,0	[285]
$[-CH_2-C(CH_3)COOCH_3]_n$	40–60	$1,75 \cdot 10^{-4}$	$\bar{1},10$ (60°)	5,15	9,2	[279]
$(CH_3)_2CHOH$	60–80	$5,83 \cdot 10^{-5}$	$\bar{2},625$ (60°)	10,805	18,5	[280]
$CH_3(CH_2)_2CH_2OH$	80	$2,50 \cdot 10^{-5}$	$\bar{2},45$			[281]
$(CH_3)_2CHCH_2OH$	80	$2,50 \cdot 10^{-5}$	$\bar{2},45$			[281]
	60–80	$10 \cdot 10^{-5}$	$\bar{3},86$ (60°)	7,46	14,6	[280]
$CH_3CH(OH)C_2H_5$	80	$8,50 \cdot 10^{-5}$	$\bar{2},98$			[281]
$(CH_3)_3COH$	80	$1,0 \cdot 10^{-5}$	$\bar{2},05$			[281]
	60–80	$0,85 \cdot 10^{-5}$	$\bar{3},79$ (60°)	5,55	11,8	[281]
Dioxane	80	$2,22 \cdot 10^{-5}$	$\bar{2},40$			[281]
CH_3COCH_3	60–80	$1,95 \cdot 10^{-5}$	$\bar{2},15$ (60°)	4,07	9,0	[280]
	80	$2,25 \cdot 10^{-5}$	$\bar{2},40$			[281]
$CH_3COC_2H_5$	80	$7,0 \cdot 10^{-5}$	$\bar{2},885$			[281]
$CH_3COCH_2CH(CH_3)_2$	80	$7,0 \cdot 10^{-5}$	$\bar{2},895$			[281]

Table 73 (continued)

Compound	t°	C_S	log k	log A	E	Ref.
$C_2H_5COC_2H_5$	60–80	$8{,}33 \cdot 10^{-5}$	$\bar{2}{,}78$ (60°)	6,12	13,5	[280]
	80	$1{,}78 \cdot 10^{-4}$	$\bar{1}{,}30$			[281]
CH_3COOH	80	$2{,}40 \cdot 10^{-5}$	$\bar{2}.43$			
$(CH_3)_2CHCOOH$	80	$9{,}0 \cdot 10^{-5}$	$\bar{1}{,}005$			[281]
$CH_3COOC_2H_5$	80	$2{,}40 \cdot 10^{-5}$	$\bar{2}{,}43$			
$C_6H_5C(CH_3)_2OOH$	60	0,33				[286]
CH_2Cl_2	60–80	$1{,}0 \cdot 10^{-5}$	$\bar{3}{,}86$ (60°)	7,07	14	[280]
$CHCl_3$	60–80	$4{,}54 \cdot 10^{-5}$	$\bar{2}{,}52$ (60°)	9,44	16,6	[289]
	80	$1{,}40 \cdot 10^{-4}$	$\bar{1}{,}20$			[281]
CCl_4	60–80	$9{,}25 \cdot 10^{-5}$	$\bar{2}{,}83$ (60°)	9,35	16	
$CH_3(CH_2)_2CH_2Cl$	80	$1{,}20 \cdot 10^{-4}$	$\bar{1}{,}13$			
$ClCH_2CH_2Cl$	60–80	$3{,}5 \cdot 10^{-5}$	$\bar{2}{,}405$ (60°)	7,62	14	[280]
$Cl_2CHCHCl_2$	60–80	$1{,}55 \cdot 10^{-5}$	$\bar{2}{,}05$ (60°)	4,50	9,8	
CBr_4	40–60	0,27	2,29 (60°)	7,94	8,6	[287]
$(CH_3)_3N$	60	$8{,}3 \cdot 10^{-4}$	$\bar{1}{,}78$			[268]

$(C_2H_5)_3N$	60	$8,3 \cdot 10^{-4}$	$\bar{1},78$	[288]
$n\text{-}C_4H_9SH$	60	0,67	2,69	[289]
$(CH_3)_2CHSH$	60	0,38	2,44	[290]
$(CH_3)_3CSH$	60	0,18	2,115	
$HSCH_2COOC_2H_5$	60	0,63	2,66	
C_6H_5SH	60	2,7	3,29	
$(XC_6H_4COS)_2$				
H=X	60	$1,0 \cdot 10^{-3}$	$\bar{1},86$	[275]
CH_3O	60	$1,46 \cdot 10^{-3}$	0,025	
CH_3	60	$1,10 \cdot 10^{-3}$	$\bar{1},90$	
Cl	60	$1,00 \cdot 10^{-3}$	$\bar{1},86$	
Br	60	$1,67 \cdot 10^{-3}$	0,08	
CN	60	$2,90 \cdot 10^{-2}$	1,32	
NO_2	60	$6,94 \cdot 10^{-2}$	1,70	
$C_6H_5PH_2$	60	16,1	4,07	[291]
$CNCH_2CH_2PH_2$	60	1,4	3,005	
$(CNCH_2CH_2)_2PH$	60	1,3	2,975	
$C_2H_5PH_2$	60	2,3	3,22	

TABLE 74. Chain Transfer in Bulk Polymerization of Vinyl Acetate ($\log k_p = 6.05 - 4.2/\theta$)

Compound	t°	C_S	log k	Ref.
$CH_2{=}CH{-}OCOCH_3$	0–25	$7{,}7\cdot10^{-5}$	$\bar{5}$,85 (25°) (log A = 3,34 E = 6,1)	[292]
	25	$2{,}3\cdot10^{-4}$	$\bar{1}$,32	[293]
	60	$1{,}75\cdot10^{-4}$	$\bar{1}$,53	[294]
	80	$2{,}0\cdot10^{-4}$	$\bar{1}$,75	[295]
C_6H_6	60	$3{,}0\cdot10^{-4}$	$\bar{1}$,77	[294]
	70	$5{,}3\cdot10^{-4}$	0,095	[296]
$C_6H_5CH_3$	60	$2{,}1\cdot10^{-3}$	0,61	[294, 297]
$C_6H_5C_2H_5$	60	$5{,}5\cdot10^{-3}$	1,03	[294]
$C_6H_5CH(CH_3)_2$	60	$9{,}0\cdot10^{-3}$	1,245	[294]
	70	$1{,}40\cdot10^{-2}$	1,52	[296]
$C_6H_5C(CH_3)_3$	60	$3{,}6\cdot10^{-4}$	$\bar{1}$,85	[294]
cyclo-C_6H_{12}	60	$6{,}6\cdot10^{-4}$	0,11	[294]
	60	$7{,}0\cdot10^{-4}$	0,135	[246]
$CH_3(CH_2)_2CH_2OH$	60	$2{,}0\cdot10^{-3}$	0,59	[294]
$CH_3CH(OH)C_2H_5$	60	$3{,}2\cdot10^{-3}$	0,795	[294]
$(CH_3)_3COH$	60	$5{,}0\cdot10^{-5}$	$\bar{2}$,99	[294]
$CH_3CH_2CH_2CHO$	60	$6{,}5\cdot10^{-2}$	2,10	[246]
CH_3COCH_3	60	$1{,}20\cdot10^{-3}$	0,37	[246]
	60	$1{,}20\cdot10^{-3}$	0,37	[294]
$CH_3COC_2H_5$	60	$7{,}4\cdot10^{-3}$	1,16	[294]
CH_3COOCH_3	60	$1{,}6\cdot10^{-4}$	$\bar{1}$,495	[298]
$CH_3COOC_2H_5$	60	$1{,}3\cdot10^{-4}$	$\bar{1}$,405	[297]
$CH_3CH_2COOCH_3$	60	$2{,}3\cdot10^{-3}$	0,65	[298]
$(CH_3)_2CHCOOCH_3$	60	$5{,}8\cdot10^{-3}$	1,05	[298]

Table 74 (continued)

Compound	t°	C_S	log k	Ref.
$CHCl_3$	60	$1{,}3\cdot10^{-2}$	1,40	[294]
CCl_4	70	0,20	2,67	[296]
$ClCH_2CH_2Cl$	60	$7{,}2\cdot10^{-4}$	0,15	[294]
	70	$1{,}0\cdot10^{-3}$	0,37	[296]
Cl_3CCCl_3	70	0,12	2,45	[296]
$C_6H_5CH_2Cl$	70	$5{,}8\cdot10^{-2}$	2,13	[296]
C_6H_5Cl	60	$8{,}4\cdot10^{-4}$	0,21	[294]
p-ClC_6H_4Cl	70	$1{,}2\cdot10^{-3}$	0,45	[296]
o-$ClC_6H_4CH_3$	70	$9{,}3\cdot10^{-3}$	1,34	[296]
p-$ClC_6H_4CH_3$	70	$1{,}9\cdot10^{-2}$	1,65	[296]
CBr_4	60	>39	$>4{,}88$	[287]
C_6H_5Br	70	$1{,}3\cdot10^{-2}$	1,48	[296]
$(CH_3)_3N$	60	$3{,}7\cdot10^{-2}$	1,86	[268]
$(C_2H_5)_3N$	60	$3{,}7\cdot10^{-2}$	1,86	[288]
	50	$3{,}6\cdot10^{-2}$	1,85	[269]
$C_6H_5NH_2$	50	$2{,}1\cdot10^{-2}$	1,53	[269]
$C_6H_5NHCH_3$	50	$3{,}6\cdot10^{-2}$	1,85	[269]
$C_6H_5N(CH_3)_2$	50	$2{,}6\cdot10^{-2}$	1,625	[269]
p-$CH_3C_6H_4NH_2$	50	$7{,}5\cdot10^{-2}$	2,085	[269]
p-$CH_3C_6H_4NHCH_3$	50	$8{,}3\cdot10^{-2}$	2,13	[269]
p-$CH_3C_6H_4N(CH_3)_2$	50	$3{,}8\cdot10^{-2}$	1,79	[269]
$CH_3(CH_2)_2CH_2SH$	60	48	4,97	[289]

TABLE 75. Chain Transfer in Bulk Polymerization of Various Monomers

Monomer	Compound	t^o	C_S	$\log k_p$	$\log k$	Ref.
$CH_2=CH-COOCH_3$	C_6H_6	80	$3,26 \cdot 10^{-5}$	3,00	$\bar{2},51$	[299]
	$C_6H_5CH_3$	80	$1,78 \cdot 10^{-4}$	3,00	$\bar{1},25$	[299]
	$C_6H_5CH_3$	60	$2,70 \cdot 10^{-4}$	2,74	$\bar{1},17$	[288]
	$C_6H_5C_2H_5$	80	$6,06 \cdot 10^{-4}$	3,00	$\bar{1},78$	[299]
	$C_6H_5CH(CH_3)_2$	80	$7,0 \cdot 10^{-4}$	3,00	$\bar{1},845$	[299]
	cyclo-C_6H_{12}	80	$2,7 \cdot 10^{-6}$	3,00	$\bar{5},43$	[299]
	n-C_4H_9OH	80	$2,75 \cdot 10^{-4}$	3,00	$\bar{1},44$	[299]
	$(CH_3)_2CHCH_2OH$	80	$2,5 \cdot 10^{-4}$	3,00	$\bar{1},40$	[299]
	$CH_3CH(OH)C_2H_5$	80	$1,41 \cdot 10^{-3}$	3,00	0,15	[299]
	$(CH_3)_3COH$	80	$3,89 \cdot 10^{-5}$	3,00	$\bar{2},59$	[299]
	CH_3COCH_3	80	$6,22 \cdot 10^{-5}$	3,00	$\bar{2},795$	[299]
	$CH_3COC_2H_5$	80	$3,24 \cdot 10^{-4}$	3,00	$\bar{1},51$	[299]
	CCl_4	80	$1,32 \cdot 10^{-4}$	3,00	$\bar{1},12$	[299]
	$CHCl_3$	80	$2,14 \cdot 10^{-4}$	3,00	$\bar{1},33$	[299]
	CH_3CCl_3	80	$5,74 \cdot 10^{-5}$	3,00	$\bar{2},76$	[299]

	$CHCl_2CHCl_2$	80	$9{,}32\cdot10^{-5}$	3,00	$\bar{2}{,}97$	[299]
	C_6H_5Cl	80	$9{,}86\cdot10^{-5}$	3,00	$\bar{2}{,}99$	[299]
$CH_2{=}C(CH_3)COOC_2H_5$	C_6H_6	80	$8{,}1\cdot10^{-6}$	2,91	$\bar{3}{,}82$	[299]
	$C_6H_5CH_3$	80	$4{,}36\cdot10^{-5}$	2,91	$\bar{2}{,}55$	
	$C_6H_5C_2H_5$	80	$1{,}43\cdot10^{-4}$	2,91	$\bar{1}{,}065$	
	$C_6H_5CH(CH_3)_2$	80	$2{,}07\cdot10^{-4}$	2,91	$\bar{1}{,}23$	
	$n\text{-}C_7H_{16}$	80	$8{,}65\cdot10^{-5}$	2,91	$\bar{2}{,}85$	
	$cyclo\text{-}C_6H_{12}$	80	$9{,}28\cdot10^{-5}$	2,91	$\bar{2}$,8[illegible]	
	$n\text{-}C_4H_9OH$	80	$4{,}54\cdot10^{-5}$	2,91	$\bar{2}{,}57$	
	$(CH_3)_2CHCH_2OH$	80	$4{,}50\cdot10^{-5}$	2,91	$\bar{2}{,}56$	
	$CH_3CHOHC_2H_5$	80	$1{,}60\cdot10^{-4}$	2,91	$\bar{1}{,}11$	
	$(CH_3)_3COH$	80	$4{,}17\cdot10^{-5}$	2,91	$\bar{2}{,}53$	
	CH_3COCH_3	80	$1{,}02\cdot10^{-5}$	2,91	$\bar{3}{,}92$	
	$CH_3COC_2H_5$	80	$2{,}52\cdot10^{-5}$	2,91	$\bar{2}{,}31$	
	CH_3COOH	80	$9{,}5\cdot10^{-6}$	2,91	$\bar{3}{,}89$	
	$CH_3COOC_2H_5$	80	$9{,}2\cdot10^{-5}$	2,91	$\bar{2}{,}60$	
	CCl_4	80	$9{,}0\cdot10^{-5}$	2,91	$\bar{2}{,}86$	

Table 75 (continued)

Monomer	Compound	t°	C_S	$\log k_p$	$\log k$	Ref.
$CH_2{=}C(CH_3)COOC_2H_5$	$CHCl_3$	80	$7{,}03\cdot10^{-5}$	2,91	$\bar{2},78$	[299]
	CH_3CCl_3	80	$3{,}36\cdot10^{-5}$	2,91	$\bar{2},44$	
	$CHCl_2CHCl_2$	80	$3{,}11\cdot10^{-5}$	2,91	$\bar{2},40$	
	C_6H_5Cl	80	$4{,}26\cdot10^{-5}$	2,91	$\bar{2},54$	
$CH_2{=}CH{-}C{\equiv}N$	$CH_2{=}CHCN$	25	$1{,}05\cdot10^{-5}$	0,65	$\bar{5},87$	[300]
	- " -	40	$1{,}7\cdot10^{-5}$	0,89	$\bar{4},12$	[300]
	- " -	60	$3{,}0\cdot10^{-5}$	1,15	$\bar{4},63$	[301]
	$C_6H_5CH_3$	60	$2{,}6\cdot10^{-4}$	1,15	$\bar{3},565$	[302]
	H_2S	50	1,1	1,02	1,06	[273]
	$(CH_3)_3N$	60	$7{,}9\cdot10^{-2}$	1,15	0,05	[268]
	$(C_2H_5)_3N$	60	0,66	1,15	0,97	
	$(n\text{-}C_3H_7)_3N$	60	1,05	1,15	1,17	
	$(n\text{-}C_4H_9)_3N$	60	0,67	1,15	0,98	
	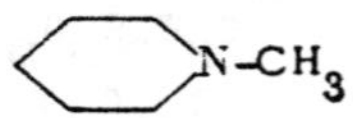	60	0,23	1,15	0,51	
	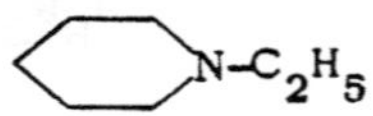	60	0,33	1,15	0,67	

	$C_6H_5N(CH_3)_2$	60	$8{,}7\cdot10^{-2}$	1,15	0,09	[288]
$CH_2{=}C(CH_3)CN$	$C_6H_5CH_3$	60	$1{,}0\cdot10^{-5}$	2,21	$\bar{3},21$	[302]
$CH_2{=}CH{-}CONH_2$	$CH_2{=}CHCONH_2$	25	$1{,}2\cdot10^{-5}$	4,255	$\bar{1},34$	[303]
	H_2O_2	25	$5{,}0\cdot10^{-4}$	4,255	0,96	[303]
$CH_2{=}CH_2$	$CH_2{=}CH_2$	83	$5{,}0\cdot10^{-4}$	2,67	$\bar{1},37$	[304]
	C_6H_6	83	$2{,}0\cdot10^{-3}$	2,67	$\bar{1},97$	[304]
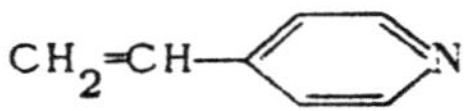	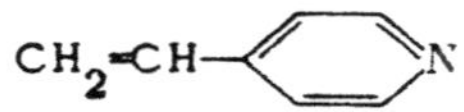	25	$6{,}7\cdot10^{-4}$	1,08	$\bar{3},905$	[305]
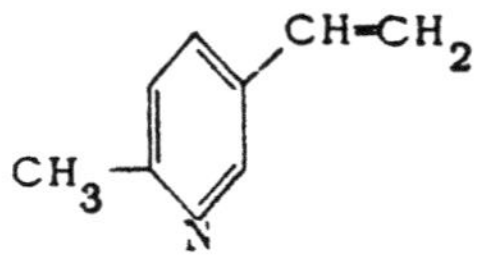	$OCHN(CH_3)_2$	25		1,67	$\bar{3},59$	[306]
$CH_2{=}CH{-}OCOC_6H_5$	$CH_2{=}CH{-}OCOC_6H_5$	60	$6\cdot10^{-4}$			[307]
	- " -	80	$7{,}0\cdot10^{-4}$			[308]
	C_6H_6	60	$7{,}5\cdot10^{-5}$			[307]
	- " -	80	$4{,}0\cdot10^{-5}$			[308]
	CCl_4	80	$7{,}3\cdot10^{-4}$			[308]

TABLE 76. Rate Constants for Reactions of Diphenylpicrylhydrazyl with RH

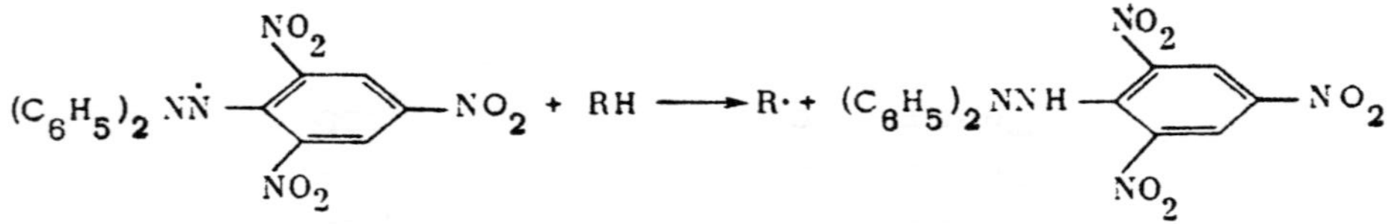

Measured spectrophotometrically by diphenylpicrylhydrazyl consumption

RH	Solvent	t°	log k (30°)	log A	E	Ref.
C_6H_5OH	C_6H_6	30–50	$\bar{2}$,35	5,85	10,4	[309]
	CCl_4	19,6	$\bar{2}$,86			[310]
p-$CH_3C_6H_4OH$	C_6H_6	30–50	$\bar{1}$,575	6,08	9,0	[309]
	CCl_4	19	$\bar{1}$,96			[310]
m-$CH_3C_6H_4OH$	C_6H_6	30–50	$\bar{2}$,83	6,20	10,2	[309]
o-$CH_3C_6H_4OH$	C_6H_6	30–50	$\bar{1}$,255	5,82	9,1	[309]
	CCl_4	19,6	$\bar{1}$,89			[310]
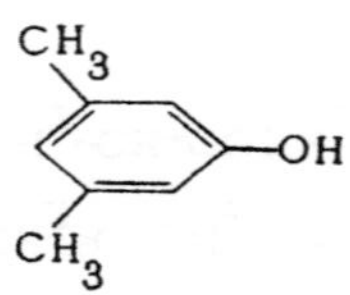	CCl_4	20	$\bar{1}$,88			[310]

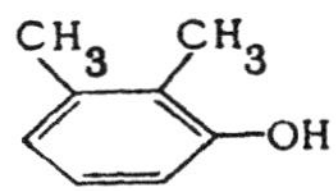	CCl_4	18	0,22			[310]
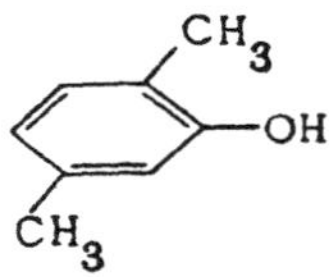	CCl_4	18,6	0,25			[310]
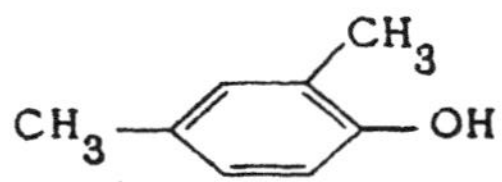	CCl_4	19,4	1,12			[310]
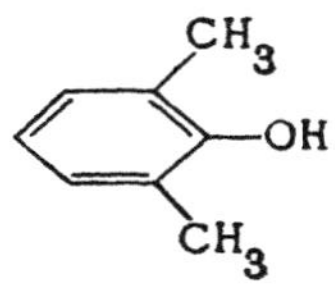	CCl_4	19	0,29			[310]
	C_6H_6	30–50	$\bar{1}$,765	4,18	6,1	[309]
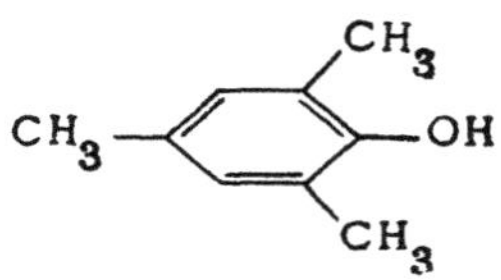	CCl_4	20	1,35			[310]

Table 76 (continued)

RH	Solvent	t°	log k (30°)	log A	E	Ref.
$p-(CH_3)_3CC_6H_4OH$	CCl_4	20	0,14			[310]
	C_6H_6	30–50	$\bar{1}$,53	6,38	9,5	[309]
$p-C_6H_5C_6H_4OH$	C_6H_6	30–50	$\bar{1}$,97	6,25	8,7	
$o-C_6H_5C_6H_4OH$	C_6H_6	30–50	$\bar{2}$,98	5,95	9,7	
$p-ClC_6H_4OH$	- " -	30–50	$\bar{2}$,60	6,18	10,5	
$m-ClC_6H_4OH$	- " -	30–50	$\bar{3}$,70	6,64	12,4	
$o-ClC_6H_4OH$	- " -	30–50	$\bar{2}$,03	5,67	10,6	[309]
$p-FC_6H_4OH$	- " -	30–50	$\bar{2}$,965	5,89	9,6	
$p-NO_2C_6H_4OH$	- " -	30–50	$\bar{3}$,175	7,71	14,6	
$m-NO_2C_6H_4OH$	- " -	30–50	$\bar{3}$,00	7,18	14,1	
$p-CH_3OC_6H_4OH$	C_6H_6	30–50	1,465	4,82	4,8	
	CCl_4	21	2,37			[310]
$m-CH_3OC_6H_4OH$	C_6H_6	30–50	$\bar{1}$,13	7,15	11,1	[309]
$p-CH_3COC_6H_4OH$	- " -	30–50	$\bar{3}$,545	9,15	16,1	[309]
$p-C_6H_5CH_2OC_6H_4OH$	CCl_4	18	2,025			[310]

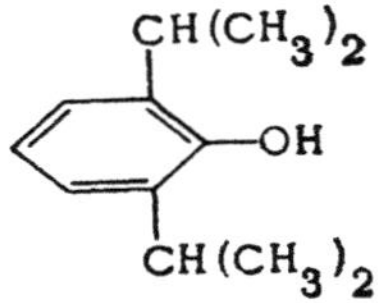	C_6H_6	30–50	$\bar{1},62$	7,25	6,4	[309]
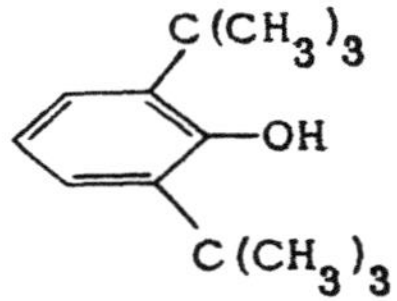	C_6H_6 CCl_4	30–50 18	$\bar{2},255$ 0,94	5,58	6,0	[309] [310]
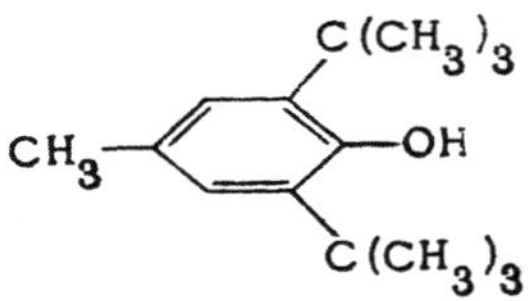	C_6H_6 CCl_4	30–50 18,3	$\bar{1},10$ $\bar{1},72$	3,50	6,1	[309] [310]
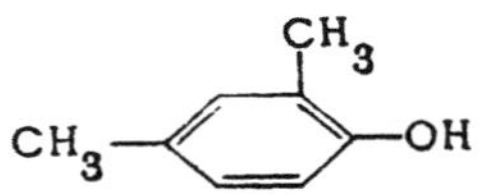	CCl_4	18,5	0,73			[310]

Table 76 (continued)

RH	Solvent	t°	log k (30°)	log A	E	Ref.
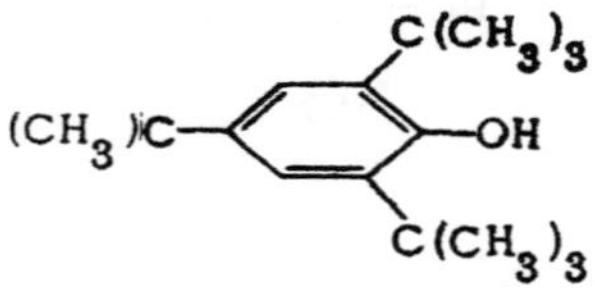	CCl_4	18,2	$\bar{1}$,57			[310]
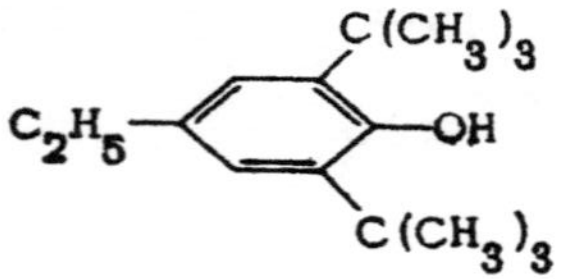	CCl_4	32	$\bar{1}$,725			[310]
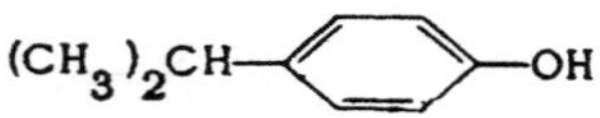	C_6H_6	20–60	$\bar{1}$,95	4,58	6,4	[192]
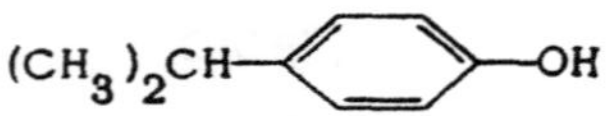	C_6H_6	20–60	$\bar{1}$,43	4,35	6,8	[192]
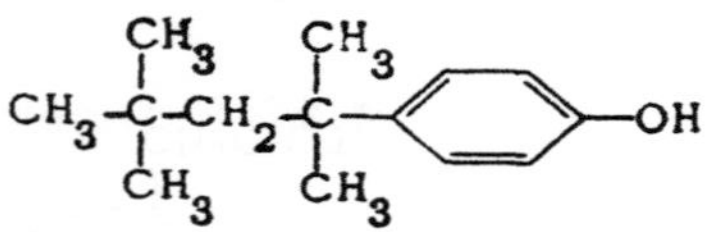	CCl_4	–1,7	0,10			[310]
C_6Cl_5OH	C_6H_6	30–50	$\bar{3}$,785	5,36	10,5	[309]

α-Naphthol	C_6H_6	30–50	1,52	5,04	4,9	[309]
	CCl_4	15,5	2,12			[310]
β-Naphthol	C_6H_6	30–50	0,33	5,08	6,6	[309]
	CCl_4	19	0,80			[310]
$(C_6H_5)_2NH$	C_6H_6	20–77	1,20	7,74	11,8	[311]
$C_6H_5NHCH_3$	- " -	20–77	2,13	8,63	14,5	
α-$C_{10}H_7NHC_6H_5$	- " -	20–77	$\bar{1}$,75	6,62	9,5	
β-$C_{10}H_7NHC_6H_5$	- " -	20–77	0,17	7,62	10,3	
n-$C_4H_9NH_2$	CCl_4	77	$\bar{2}$,18			[310]
$(C_2H_5)_3N$	- " -	40–77	$\bar{3}$,72	2,93	7,2	
$C_6H_5N(CH_3)_2$	- " -	40–77	$\bar{2}$,555	2,00	4,9	
	- " -	77	$\bar{2}$,34			
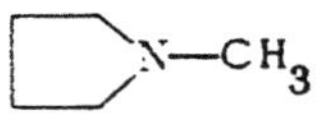	- " -	76	$\bar{2}$,86			
n-C_4H_9SH	C_6H_6	20–60	$\bar{3}$,88	8,45	14,6	[312]
$(CH_3)_2CHCH_2SH$	- " -	20–60	$\bar{3}$,74	8,80	15,3	
$(CH_3)_3CSH$	- " -	20–60	$\bar{4}$,56	7,62	15,3	
n-$C_6H_{13}SH$	- " -	20–60	$\bar{3}$,87	8,79	15,1	
n-$C_8H_{17}SH$	- " -	20–60	$\bar{3}$,69	8,26	14,6	

TABLE 77. Correlation Ratios for Radical Substitution Reactions

Radical	Class of compound	Solvent	t°	Function	ρ	log k_H	Ref.
Cl·	$XC_6H_4CH_3$	$C_6H_5CH_3$	70	σ	−0,76		[77]
	- " -	CCl_4	40	σ	−0,78± ±0,06		[74]
	- " -	- " -	40	$σ^+$	−0,66± ±0,02		[74]
Br·	$XC_6H_4CH_3$ (1)	CCl_4	80	σ	−1,35± ±0,15		[84]
	- " - (2)	- " -	80	σ	−1,36± ±0,16		[84]
	- " - (3)	- " -	80	$σ^+$	−1,28± ±0,12		[84]
	- " - (4)	- " -	80	$σ^+$	−1,21± ±0,10		[84]
	- " - (5)	- " -	80	σ	−1,10± ±0,12		[84]

	- " -	- " -	80	σ	−1,0		[86]
	- " -	- " -	80	σ^+	−1,38		[82]
	- " -	C_6H_6	80	σ^+	−1,39± ±0,03		
	- " -	- " -	19	σ^+	−1,78± ±0,06		[85]
	- " -	$C_6H_5CH_3$	80	σ^+	−1,46		
	$XC_6H_4C_2H_5$	CCl_4	80	σ	−0,69± ±0,03		[87]
Br	$XC_6H_4CH_2CH{=}CH_2$	- " -	69,5	σ^+	−0,76± ±0,03		[88]
	$XC_6H_4CH_2OCH_3$	- " -	80	σ^+	−0,35± ±0,017		[89]
	$XC_6H_4CH(OCH_3)_2$	- " -	80	σ	−0,38± ±0,019		[89]
	$XC_6H_4CH_2OCH_2C_6H_5$	- " -	80	σ	−0,12		[87]
	$XC_6H_4CH(OCH_3)C_6H_5$	- " -	80	σ	0,00		[87]
$HO^\cdot$	H_3CX	H_2O	23	σ	−0,96	8,15	[95]
$(CH_3)_3CO^\cdot$	$XC_6H_4CH_3$	$C_2Cl_3F_3$	45	σ	−0,40		
	- " -	$C_2Cl_3F_3$	45	σ^+	−0,35		[114]
	- " -	C_6H_5Cl	45	σ	−0,36		
	- " -	C_6H_5Cl	45	σ^+	−0,32		

Table 77 (continued)

Radical	Class of compound	Solvent	t^{o}	Function	ρ	$\log k_H$	Ref.
$(CH_3)_3CO$-	$XC_6H_4CH_3$	CH_3CN	45	σ	–0,43		[114]
	- " -	CH_3CN	45	σ^+	–0,39		[114]
	- " -	$C_6H_5CH_3$	110	σ	–0,50		[120]
	- " -	CCl_4	40	σ	–0,68± ±0,3		[119]
	- " -	$\dot{C}Cl_4$	40	σ^+	–0,86± ±0,012		[119]
	- " -	$C_6H_5CH_3$	40	σ	–0,83		[118]
	- " -	C_6H_6	39,6	σ	–0,75± ±0,047		[116]
	- " -	C_6H_6	39,6	σ^+	–0,60± ±0,092		[116]
	cyclo-$C_6H_{11}X$	C_6H_6	135	σ^* (Taft)	–0,43		[123]
	XC_6H_4OH	CCl_4	122	σ^+	-1,19		[313]
	- " -	C_6H_5Cl	122	σ^+	–0,74		[313]
	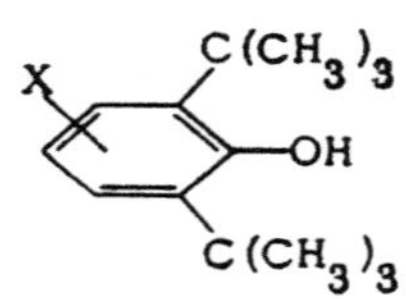	CCl_4	122	σ^+	–0,86		[313]

	$XC_6H_4OCH_3$	C_6H_5Cl	130	σ	−0,39		[115]
	$XC_6H_4SCH_3$	- " -	130	σ	−0,15		
	$XC_6H_4SO_2Cl$	- " -	130	σ	−0,20		
	$XC_6H_4N(CH_3)_2$	- " -	130	σ	−0,42		
$C_6H_5(CH_3)_2OO^{\cdot}$	$XC_6H_4CH(CH_3)_2$	$XC_6H_4CH(CH_3)_2$	30	σ	−0,35± ±0,07	$\bar{1}$,255	[133]
	- " -	- " -	30	$σ^+$	−0,29± ±0,04	$\bar{1}$,255	[133]
	- " -	C_6H_5Cl	60	$σ^+$	−0,41		[74]
	XC_6H_4OH	$C_6H_5CH(CH_3)_2$	62	σ	−3,7		[314]
$C_6H_5CH(OO^{\cdot})CH_2$~	XC_6H_4OH	$C_6H_5CH{=}CH_2$	65	$σ^+$	−1,58	3,465	[166]
	- " -	- " -	65	σ	−1,49		
	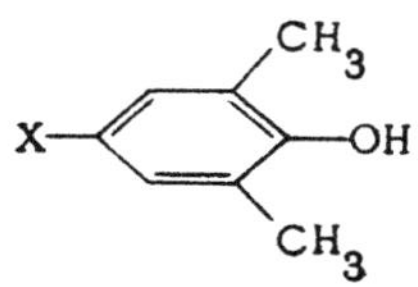	- " -	65	$σ^+$	−1,36	4,58	[168]
	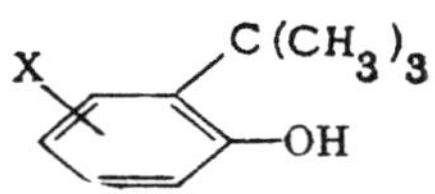	- " -	65	$σ^+$	−1,46	4,22	[168]

Table 77 (continued)

Radical	Class of compound	Solvent	t°	Function	ρ	$\log k_H$	Ref.
$C_6H_5CH(OO^\bullet)CH_2$~	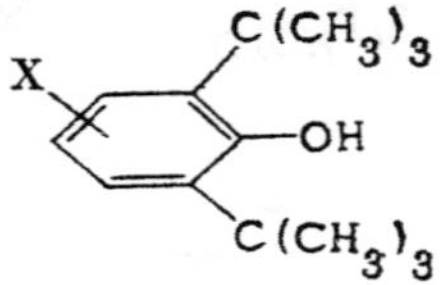	$C_6H_5CH{=}CH_2$	65	σ^+	-1,11	3,70	[168]
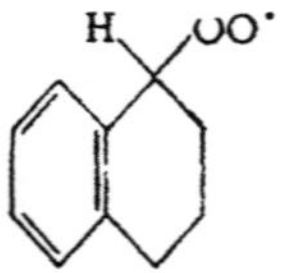	- " -	Tetralin	65	σ^+	-1,36	3,89	[173]
$C_6H_5\dot{C}HCH_2$~	$(XC_6H_4)NH$	o-ClC_6H_4Cl	65	σ	−0,89	4,50	[181]
	$XC_6H\,NHCH_3$	- " -	65	σ^+	-1,6	3,60	[181]
$C_6H_5COOO\cdot$	$C_6H_5OCH_2C_6H_4X$	C_6H_5Cl	60	σ^+	0,00		[74]
	$(XC_6H_4CH_2)_2O$	- " -	60	σ^+	0,00		[74]
$CCl_3\cdot$	$XC_6H_4CH_3$	$C_6H_5CH_3$	50	σ^+	-1,46		[217]
	$XC_6H_4CH_2CH{=}CH_2$	- " -	69,5	σ	−0,63		
	- " -	- " -	69,5	σ^+	−0,39 ± 0,03		[218]

	- " -	- " -	69,5	σ^{o}	−0,58± ±0,03		[218]
C_6H_5	$XC_6H_4CH_3$	CCl_4	60	σ^{+}	0,00		[225]
	- " -	- " -	100	σ^{o}	−0,50		[227]
	- " -	- " -	60	σ	−0,40		[226]
	- " -	- " -	60	σ^{+}	−0,30		[226]
$CH_3C_6H_4\cdot$	$XC_6H_4CH_3$	CCl_4	60	σ	−0,10		[226]
	- " -	- " -	6C	σ^{+}	−0,10		[226]
	- " -	- " -	60	σ	−0,59		[226]
	- " -	- " -	60	σ^{+}	−0,44		[226]
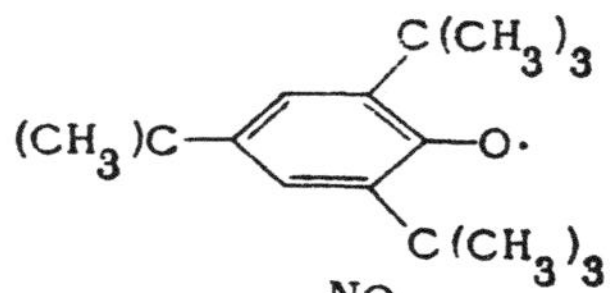	XC_6H_4OH	C_6H_6	24	σ^{+}	−2,72		[189]
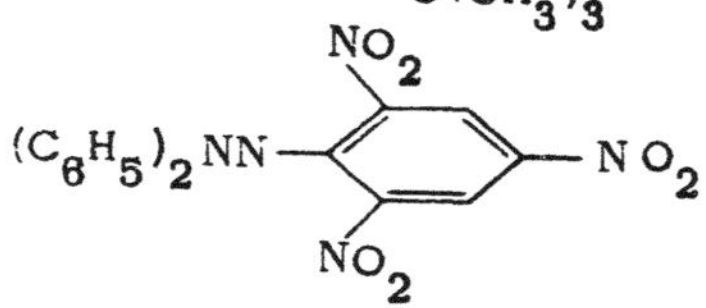	XC_6H_4OH	C_6H_6	30	σ ($\sigma < 0,2$)	−6	2,35	[309]
	- " -	CCl_4	30	σ^{+}	−4,54		[310]
	- " -	CCl_4	20	σ^{+}	−2,77		[310]

CHAPTER VII

ADDITION REACTIONS OF ATOMS AND RADICALS

§1. Addition of Atoms and Radicals to Molecular Oxygen

Hydrogen atoms add to oxygen with a diffusion-rate constant. The addition of oxygen to alkyl radicals is somewhat slower. Under conditions of liquid-phase oxidation, a high rate constant for the reaction of oxygen addition ensures the predominance of peroxy radicals.

§2. Addition at C=C Bond

The C=C double bond is highly reactive toward atoms and free radicals. The addition of atoms to a double bond goes forward with a diffusion-rate constant (see Table 79). The addition of radicals to a double bond is somewhat slower (see Table 80).

When a polymerization is carried out in the presence of two monomers, the resulting copolymer includes monomeric units from both monomers. The copolymer composition depends on the ratio of the rate constants for the chain propagation reactions. In considering the kinetic scheme of copolymerization reactions, it is assumed that the reactivity of the macroradical is determined by the monomer unit completing the macroradical, and that the reactivity is independent of the monomer ratio and of the medium. The scheme of radical copolymerization includes the following reactions of chain propagation:

$$M_1^{\cdot} + M_2 \xrightarrow{k_{11}} M_1M_1^{\cdot}$$

$$M_1^{\cdot} + M_2 \xrightarrow{k_{12}} M_1M_2^{\cdot}$$

$$M_2^{\cdot} + M_1 \xrightarrow{k_{21}} M_2M_1^{\cdot}$$

$$M_2^{\cdot} + M_2 \xrightarrow{k_{22}} M_2M_2^{\cdot}$$

The ratio of monomers incorporated in the polymer is determined by the following expression (copolymerization equation):

$$\frac{d[M_1]}{d[M_2]} = \frac{r_1 m + 1}{r_2 + m}$$

where

$$m = \frac{[M_1]}{[M_2]}; \quad r_1 = \frac{k_{11}}{k_{12}}; \quad r_2 = \frac{k_{22}}{k_{21}}$$

Values of copolymerization constants are listed in Table 84.

Alfrey and Price [390] have proposed a quantitative description of copolymerization constants by means of a Q - e scheme. They proposed that each rate constant of the polymerization is equal to

$$k_{ij} = P_i Q_j e^{-e_i e_j},$$

and e_i and e_j are identical for a macroradical and the corresponding monomer. Then,

$$r_1 = \frac{Q_1}{Q_2} e^{-e_1(e_1-e_2)}$$

and

$$r_2 = \frac{Q_2}{Q_1} e^{-e_2(e_2-e_1)}.$$

Such an approach is approximate, since it does not account for the specific intereffects of reactants in the elementary acts. However, in the majority of cases, such a scheme will be sufficiently accurate in calculating r_1 and r_2 on the basis of Q and e.

Schwan and Price [391] proposed a physical interpretation of the quantities Q and e: Q is related to resonance stabilization of the growing chain due to substituents at the C atom on the end of the chain, $Q = e^{-q/RT}$; and e_i is related to the charge ε included on each of the combined carbon atoms, $e_i = \varepsilon_i/(\ell RT)^{1/2}$, where ℓ is the distance between charges in the activated complex. The values of r_1 and r_2 are related to q and ε by the following equations:

$$\log r_1 = \frac{1}{T}(6.79\,\Delta q - 3.63\,\varepsilon_1 \Delta\varepsilon)$$

$$\log r_2 = \frac{1}{T}(-6.79\,\Delta q + 3.63\,\varepsilon_2 \Delta\varepsilon)$$

The values of q and ε according to Schwan and Price [391] are listed in Table 85.

§3. Addition to Aromatic Compounds

Rate constants are listed in Table 86 for the addition of hydrogen atoms and radicals to aromatic hydrocarbons and their derivatives. Absolute values of k for a hydrogen atom reacting in a medium of $n\text{-}C_6H_{14}$ [70] were calculated from the relative constants on the basis that $k(H^{\bullet} + n\text{-}C_6H_{14}) = 4.9 \cdot 10^6$ liter/mole·sec (23°). Absolute values of k for the addition of a methyl radical were calculated from the ratios $k(CH_3^{\bullet} + \text{aromatic})/k(CH_3^{\bullet} + \text{isooctane})$ and $\log k(CH_3^{\bullet} + \text{isooctane}) = 7.79 - 8.7/\theta$ (see Chapter VI, §3). Absolute values of $k(C_2H_5^{\bullet} + \text{aromatic})$ were calculated from the relative rates and a ratio of reaction constants of methyl and ethyl radicals with isooctane equal to

$$k(CH_3)/k(C_2H_5)\ 10^{1.5/\theta} = 12 \text{ at } 65° \ [392].$$

whence

$$\log k(C_2H_5^{\bullet} + \text{isooctane}) = 7.79 - 10.2/\theta = 1.06 \text{ at } 65°.$$

For the $CF_3^{\bullet}$ radical, absolute rate constants for addition were calculated on the basis of the value $\log k(CF_3^{\bullet} + \text{isooctane}) = 5.05$ at 65° (see Table 66).

Absolute k's for $CCl_3^{\bullet}$ were calculated from the ratios of constants $k(CCl_3^{\bullet} + \text{aromatic})/k(CCl_3^{\bullet} + \text{hexadecene})$ with the assumption that $k(CCl_3^{\bullet} + \text{hexadecene}) = K(CCl_3^{\bullet} + \text{heptene}) = 1.5 \cdot 10^5$ (see Table 81).

In the case of cyanoisopropyl radicals, the ratio $k_{rel} = k/\sqrt{k_1 k_2}$ was measured in the experiment, where k_1 is the rate constant for decomposition of azobisisobutyronitrile into radicals ($k_1 = 4.10^{-6}$ sec^{-1} at 55°), and k_2 relates to the recombination of two cyanoisopropyl radicals, $k_2 = 4.4 \cdot 10^9$ liter/mole•sec (see Table 91). Hence, $k = k_{rel} \cdot 133$.

§4. Addition to Quinones, Carbonyl Compounds, Nitriles, and Nitro Compounds

The addition of alkyl radicals to quinones takes place at an oxygen atom:

The resulting phenoxyl radical is relatively inactive; it is not capable of active chain propagation in reactions of radical polymerization, hence quinones retard the polymerization of unsaturated compounds. Hydroxyl radicals, on the contrary, add at the C=C bond of quinone:

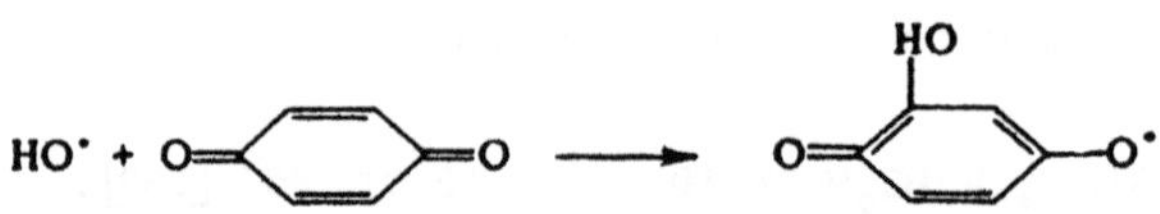

A hydrogen atom will add to the carbonyl group of a ketone or acid in accordance with the reaction

$$H + O = C\!< \longrightarrow HOC\!<$$

TABLE 78. Addition of Atoms and Radicals to Oxygen

Radical (atom)	Solvent	t^{o}	log k	Method	Ref.
H	H_2O	23	10,08	PR	[53]
	- " -	23	10,415	PR	[54]
	- " -	23	10,28	PR	[55]
	- " -	23	10,725	PR	[315]
	- " -	23	10,32	PR	[316]
$CH_3^{\bullet}$	H_2O	25	9,67	PR	[63]
$CH_2{=}C(CH_3)(CH_2)_3C(CH_3){=}CHCH_2$ $CH_2{=}(CH_3)C(CH_2)_3(CH_3)C{=}CHCH^{\bullet}$	RH	25	7,18	NKR	[317]
$CH_3(CH_2)_4CH{=}CH{-}\dot{C}H$ $C_2H_5OOC(CH_2)_7CH{=}\dot{C}H$	RH	25	6,70	NKR	[317]
α-Tetralyl	RH	25	7,83	IIM	[145]
$\sim CH_2\dot{C}HC_6H_5$	RH	50	7,60	MCR	[318]
$\sim CH_2\dot{C}HCN$	H_2O	25	9,28	PR	[319]
$HO\dot{C}_6H_5X$					
X=H	H_2O	25	8,70	PR	[319a]
$HO\dot{C}_6H_5X$ H	- " -	19	8,60	PR	[319b]
X= NO_2	H_2O	25	6,36	PR	[397]
NO_2	- " -	19	6,40	PR	
CN	- " -	19	6,46	PR	
COC_6H_5	- " -	19	6,57	PR	
$COCH_3$	- " -	19	6,58	PR	
SO_2NH_2	- " -	19	6,62	PR	
COOH	- " -	19	6,72	PR	
I	- " -	19	6,84	PR	
Cl	- " -	19	6,90	PR	[319b]
Br	- " -	19	6,98	PR	
OH	- " -	19	6,69	PR	
OCH_3	- " -	19	7,22	PR	
NH_2	- " -	19	7,40	PR	
$N(CH_3)_2$	- " -	19	7,37	PR	
CH_3	- " -	19	8,59	PR	
C_2H_5	- " -	19	8,74	PR	

TABLE 79. Reaction Rate Constants for Addition of Atoms at C = C Bond (MCR method)

Compound	Solvent	t°	log k	Ref.
	Hydrogen atom			
$CH_2=CH_2$	H_2O	25	9,505	[63]
$CH_2=CHCH_3$	-" -	25	9,59	[63]
$CH_2=CHC_2H_5$	H_2O	25	9,71	[63]
$CH_2=C(CH_3)_2$	-" -	25	10,0	[63]
$(CH_3)_2C=C(CH_3)_2$	n-C_6H_{14}	23	8,93	[69]
RCH=CHR	-" -	23	8,715	[69]
$RCH=CH_2$	-" -	23	8,90	[69]
$R_2C=CH_2$	-" -	23	9,05	[69]
$R_2C=CHR$	-" -	23	8,845	[69]
cyclo-$C_6H_{11}CH_2CH=CH_2$	-" -	23	8,88	[69]
cyclo-C_6H_{10}	n-C_6H_4	23	8,69	[69]
	cyclo-C_5H_{10}	23	8,83	[68]
$CH_2=CH-CH=CH_2$	H_2O	25	10,0	[63]
$CH_2=CHCH_2OH$	-" -	23	9,36	[61]
$CH_2=CHCN$	-" -	25	9,26*	[319 a]
	Chlorine atom[2*]			
trans-$CH_3CH=CHCH_3$	cyclo-C_6H_{12}	-9	9,89	[320]
cis-$CH_3CH=CHCH_3$	-" -	-9	9,91	[320]
$CH_2=CHCH_2CH_3$	-" -	9	9,97	[320]
$CH_2=C(CH_3)_2$	-" -	9	9,90	[320]
$CH=CH-CH_2Cl$	-" -	-9	9,67	[320]
cyclo-C_6H_{10}	-" -	25	9,86	[320]
trans-ClCH=CHCl	-" -	25	9,39	[320]
cis-ClCH=CHCl	-" -	25	9,49	[320]
$Cl_2C=CH_2$	-" -	25	9,46	[320]
$Cl_2C=CHCl$	-" -	25	9,25	[320]
$Cl_2C=CCl_2$	-" -	25	8,89	[320]

* PR method

2* Absolute values of constants computed on assumption that chlorine atoms react with cis-$CH_3CH = CHCH_3$ and $CH_2 = CHCH_2CH_3$ with a diffusion rate constant given by log k = 11.80 - 2 4/θ (the same as for iodine atoms in n-hexane; see Table 89).

TABLE 80. Addition of Radicals at C = C Bond

Compound	Solvent	t°	log k	log A	E	Method	Ref.
	Radical $CH_3^{\cdot}$ (85°)						
$CH_2{=}CH_2$	H_2O	23	3,69			PR	[63]
	$(CH_3)_2CHCH_2C(CH_3)_3$	6–95	3,99*	8,70	7,7	MCR	[321]
	- " -	55–85	3,945*	8,905	8,1	MCR	[322]
$CH_2{=}CHCH_3$	H_2O	23	3,725			PR	[63]
	$(CH_3)_2CHCH_2C(CH_3)_3$	6–95	3,81*	8,52	7,7	MCR	[321]
$CH_2{=}CHC_2H_5$	H_2O	23	4,48			PR	[63]
cis-$CH_3CH{=}CHCH_3$	$(CH_3)_2CHCH_2C(CH_3)_3$	54–85	2,99*	7,09	6,7	MCR	[323]
	- " -	65	2,67*			MCR	[200]
trans-$CH_3CH{=}CHCH_3$	$(CH_3)_2CHCH_2C(CH_3)_3$	54–85	3,29*	7,39	6,7	MCR	[323]
	- " -	65	2,99*			MCR	[200]
$CH_2{=}C(CH_3)_2$	$(CH_3)_2CHCH_2C(CH_3)_3$	65	3,705*			MCR	[200]
	- " -	6–95	4,01*	8,54	7,4	MCR	[321]
	H_2O	23	4,59			PR	[63]

Table 80 (continued)

Compound	Solvent	t^{o}	logk	logA	E	Method	Ref.
CH_2=CHC_6H_5	$(CH_3)_2CHCH_2C(CH_3)_3$	6–95	5,29*	8,65	5,5	MCR	[321]
	- " -	65–85	5,285*	8,77	5,7	MCR	[324, 322]
CH_2=$C(CH_3)C_6H_5$	$(CH_3)_2CHCH_2(CH_3)_3$	6–95	5,28*	8,52	5,3	MCR	[321]
	- " -	55–85	5,35*	9,14	6,2	MCR	[324, 322]
cis-C_6H_5CH=CHC_6H_5	$(CH_3)_2CHCH_2C(CH_3)_3$	54–85	3,78*	8,49	7,7	MCR	[323]
trans-C_6H_5CH=CHC_6H_5	- " -	54–85	2,42*	8,27	6,3	MCR	[323, 324]
$(C_6H_5)_2C$=CH_2	- " -	65–85	5,42*	6,765	2,2	MCR	
$(C_6H_5)_2C$=CHC_6H_5	- " -	65–85	4,00	7,12	5,1	MCR	
cis-CH_3OCOCH=$CHCOOCH_3$	- " -	54–85	4,78*	7,39	4,1	MCR	[324]
trans-CH_3OCOCH=$CHCOOCH_3$	- " -	54–85	5,41*	6,39	1,6	MCR	
CH_2=$CHCH_2OCOCH_3$	- " -	65–85	3,225*	5,795	4,2	MCR	[200]
CH_2=$CHCH$=CH_2	H_2O	23	6,10			PR	[63]
	$(CH_3)_2CHCH_2C(CH_3)_3$	6–95	5,65*	9,01	5,5	MCR	[321]
	- " -	55–85	5,69*	9,17	5,7	MCR	[322]
CH_2=$C(CH_3)CH$=CH_2	$(CH_3)_2CHCH_2C(CH_3)_3$	6–95	5,63*	8,87	5,3	MCR	[321]
	- " -	55–85	5,675*	8,61	4,8	MCR	[322]

$CH_3CH{=}CHCH_3CHCH_3$	$(CH_3)_2CHCH_2C(CH_3)_3$	55–85	4,635*	7,995	5,5	MCR	[322]
$(CH_3)_2C{=}CHCH{=}C(CH_3)_2$	- " -	55–85	3,855*	8,505	7,6	MCR	[322]
$CH_2{=}C{=}CH_2$	- " -	55–85	3,63*	7,36	6,1	MCR	[322]
$CH_2{=}C{=}CHCH_3$	- " -	55–85	3,58*	7,74	6,8	MCR	[322]
$F_2C{=}CF_2$	- " -	55–85	4,905*	8,515	5,9	MCR	[325]
$CH_2{=}CH(CH_2)_2CH_3$	- " -	65	3,54*			MCR	[326]
$CH_2{=}CH(CH_2)_4CH_3$	- " -	65	3,56*			MCR	[326]
$CH_2{=}CH(CH_2)_7CH_3$	$(CH_3)_2CHCH_2C(CH_3)_3$	65	3,49*			MCR	[326]
$CH_2{=}CH(CH_2)_{13}CH_3$	- " -	65	3,55*			MCR	[326]
$CH_2{=}CHCH(CH_3)_2$	- " -	65	3,51*			MCR	[326]
	Radical $C_2H_5^{\cdot}$ (60°)						
Cyclohexadiene-1,3	Liquid phase	60	4,60	8,02	5,2	MCR	[327]
C_6H_6	- " -	60	4,175	8,385	6,4	MCR	[327]
Cyclooctatetraene	- " -	60	4,115	7,795	5,6	MCR	[327]
	- " -	60	3,72	8,33	7,0	MCR	[327]
$CH_2{=}CH{-}OCOCH_3$	- " -	60	3,28	7,82	6,9	MCR	[327]

Table 80 (continued)

Compound	Solvent	t°	log k	logA	E	Method	Ref.
	Radical $CCl_3^{\cdot}$ (25°)						
$CH_2{=}CH(CH_2)_4CH_3$	$CH_2{=}CH(CH_2)_4CH_3$	15–40	5,51	9,36	7,0	NKR	[238]
cyclo-C_6H_{10}	cyclo-C_6H_{10}	25	2,89			NKR	[238]
	CCl_3Br	30–50	3,73	5,60	3,4	IIM	[328]
$CH_2{=}CHOCOCH_3$	CCl_3Br	30	3,05			IIM	[328]
	- " -	25	2,90			IIM	[220]
	- " -	25	3,20			NKR	[220]
	$CHCl_3$	25	3,38			IIM	[220]
	$CH_2{=}CHOCOCH_3$	25	3,05			IIM	[329]
	Radical $CF_3^{\cdot}$ 2*						
$CH_2{=}CH_2$	$(CH_3)_2CHCH_2C(CH_3)_3$	65	7,70			MCR	[330]
	$(CH_3)_2CHCH(CH_3)_2$	65	7,81			MCR	[331]
$CH_2{=}CHCH_3$	$(CH_3)_2CHCH_2C(CH_3)_3$	65	7,85			MCR	[330, 322]
	$(CH_3)_2CHCH(CH_3)_2$	65	7,95			MCR	[331]
$CH_2{=}CHC_2H_5$	$(CH_3)_2CHCH_2C(CH_3)_3$	65	7,89			MCR	[330]
	$(CH_3)_2CHCH(CH_3)_2$	65	7,90			MCR	[332]

$CH_2{=}C(CH_3)_2$	$(CH_3)_2CHCH_2C(CH_3)_3$	65	8,285	MCR	[330]
	- " -	65	8,325	MCR	[332]
	$(CH_3)_2CHCH(CH_3)_2$	68	8,40	MCR	[331]
cis-$CH_3CH{=}CHCH_3$	$(CH_3)_2CHCH_2C(CH_3)_3$	65	7,815	MCR	
trans-$CH_3CH{=}CHCH_3$	- " -	65	7,82	MCR	[330]
$CH_2{=}CHCH(CH_3)_2$	- " -	65	7,90	MCR	
$CH_2{=}C(CH_3)CH{=}CH_2$	$(CH_3)_2CHCH(CH_3)_2$	65	8,97	MCR	[331]
$CH_2{=}CHC(CH_3)_3$	$(CH_3)_2CHCH_2C(CH_3)_3$	65	7,87	MCR	[330]
$(CH_3)_2C{=}C(CH_3)_2$	- " -	65	7,78	MCR	[332]
cyclo-C_5H_8	- " -	65	7,595	MCR	[332]
$CH_2{=}CHC_6H_5$	$(CH_3)_2CHCH(CH_3)_2$	65	8,49	MCR	[331]
$CH_2{=}CHF$	$(CH_3)_2CHCH_2C(CH_3)_3$	65	6,915	MCR	
$CH_2{=}CFCH_3$	- " -	65	7,59	MCR	[332]
$CH_2{=}CHCl$	- " -	65	7,485	MCR	
$CF_2{=}CF_2$	- " -	65	6,89	MCR	[330]
$CH_2{=}CHCH{=}CH_2$	- " -	65	8,76	MCR	[332]

Table 80 (continued)

Compound	Solvent	t°	log k	log A	E	Method	Ref.
$CH_2=C=CH_2$	$(CH_3)_2CHCH_2C(CH_3)_3$	65	7,51			MCR	[330]
$CH_3CH=C=CHCH_3$	- " -	65	7,665			MCR	[330]
$HC\equiv CH$	- " -	65	6,89			MCR	[330]
Radical $(CH_3)_3CO\cdot$ 3*							
$CH_2=CHC_2H_5$	RH	40	3,04			MCR	[121]
trans-$CH_3CH=CHCH_3$	RH	40	3,06			MCR	[121]
cis-$CH_3CH=CHCH_3$	RH	40	3,835			MCR	[121]
$CH_2=C(CH_3)C_2H_5$	RH	40	3,09			MCR	[121]
$CH_2=C(CH_3)_2$	RH	40	3,585			MCR	[121]
$CH_2=CHC_6H_5$	RH	40	4,44			MCR	[121]

*Absolute values of constants computed from relative values on assumption that log k($CH_3^\cdot$ + isooctane) = 7.79-8.7/ θ (see Chapter VI, §3).

2*Absolute values of constants computed on basis that log k($CF_3^\cdot$ + isooctane) = 5.05 (at 65°); see Table 66.

3*Absolute values of constants computed from relative values on assumption that log k[$(CH_3)_3CO^\cdot + H\text{-}CH_2R$] = 2.42; see Table 53.

TABLE 81. Addition of $RO_2^{\cdot}$ at C = C Bond in Medium of Corresponding Olefin

RO_2	Olefin	t°	log k (40°)	log A	E	Method	Ref.
$C_6H_5CH(OO^{\cdot})CH_2$~	$C_6H_5CH{=}CH_2$*	13–50	1,845	7,67	8,4	IIM	[170]
p-$CH_3OC_6H_4(OO^{\cdot})CH_2$~	p-$CH_3OC_6H_4CH{=}CH_2$	22–50	2,09	7,115	7,2	IIM	
p-$CH_3C_6H_4CH(OO^{\cdot})CH_2$~	p-$CH_3C_6H_4CH{=}CH_2$	40	1,92			IIM	
p-$ClC_6H_4CH(OO^{\cdot})CH_2$~	p-$ClC_6H_4CH{=}CH_2$	40	2,09			IIM	[172]
m-$ClC_6H_4CH(OO^{\cdot})CH_2$~	m-$ClC_6H_4CH{=}CH_2$	40	2,00			IIM	
p-$CNC_6H_4CH(OO^{\cdot})CH_2$~	p-$CNC_6H_4CH{=}CH_2$	40	1,96			IIM	
$C_6H_5CD(OO^{\cdot})CH_2$~	$C_6H_5CD{=}CH_2$	13–40	1,72	8,36	9,5	IIM	
$C_6H_5C(CH_3)(OO^{\cdot})CH_2$~	$C_6H_5C(CH_3){=}CH_2$*	13–50	1,17	6,82	8,07	IIM	[147]
$CH_3CH_2CH_2CH_2OO$	$C_6H_5CH{=}CH_2$	30	1,145			IIM	[143]
	$C_6H_5C(CH_3){=}CH_2$	30	1,40			IIM	[143]
$C_2H_5CH(CH_3)OO^{\cdot}$	$C_6H_5CH{=}CH_2$	30	0,80			IIM	[157]
	$C_6H_5C(CH_3){=}CH_2$	30	1,08			IIM	[157]
cyclo-$C_6H_{11}OO^{\cdot}$	$C_6H_5CH{=}CH_2$	30	0,83			IIM	[157]
	$C_6H_5(CH_3){=}CH_2$	30	1,15			IIM	[157]
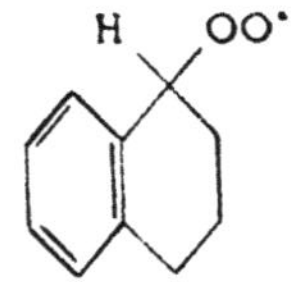	$C_6H_5CH{=}CH_2$	30	0,94			IIM	[157]
	$C_6H_5C(CH_3){=}CH_2$	30	1,20			IIM	[157]

Table 81 (continued)

$RO_2^{\cdot}$	Olefin	t°	log k (40°)	log A	E	Method	Ref.
$(C_6H_5)_2CHOO^{\cdot}$	$C_6H_5CH{=}CH_2$	30	1,20			IIM	[157]
	$C_6H_5C(CH_3){=}CH_2$	30	1,40			IIM	[157]
$C_6H_5CH(OO^{\cdot})CHCH_3$~	$C_6H_5CH{=}CHCH_3$*	30	1,71			IIM	[147]
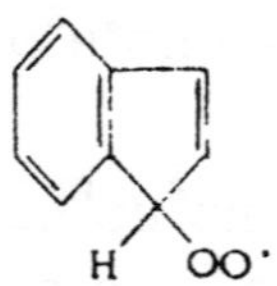	Indene*	30	2,11			IIM	[147]
$C_6H_5COOO^{\cdot}$	$CH_2{=}CH(CH_2)_3CH_3$	20	3,09			MCR	[333]
	$CH_3CH{=}CH(CH_2)_2CH_3$	20	3,13			MCR	[333]
	$CH_2{=}CH(CH_2)_5CH_3$	20	2,32			MCR	[333]
	cyclo-C_5H_8	20	3,45			MCR	[333]
$C_6H_5C(CH_3)_2OO^{\cdot}$	$CH_2{=}CHC_6H_5$	30	0,32			RICR	[133, 143]
	$CH_2{=}C(CH_3)C_6H_5$	30	0,59			RICR	[133]
	$CH(CH_3){=}CHC_6H_5$	30	0,82			RICR	[133]
$(CH_3)_3COO$	$CH_2{=}CHC_6H_5$	30	0,02			IIM	[143]

* In chlorobenzene solution

TABLE 82. Relative Rate Constants for Addition Reactions of Phenyl Radical

$$k_{rel} = \frac{k\,(C_6H_5^{\cdot} + >C=C<)}{k\,(C_6H_5^{\cdot} + CCl_4)}$$ (MCR method [225]; solvent CCl_4; 60°)

Compound	k_{rel}	Compound	k_{rel}	Compound	k_{rel}	Compound	k_{rel}
cis-$CH_3CH{=}CHCH_3$	1,14	$CH_3CH{=}CHCH_2CH_3$	0,80	$CH_2{=}CHCH_3$	2,0	cyclo-C_5H_8	0,7
trans-$CH_3CH{=}CHCH_3$	1,22	$CH_2{=}CH(CH_2)_5CH_3$	2,80	$CH_2{=}C(CH_3)C(CH_3)_3$	5,9	$CH_2{=}CH{-}CH_2C_6H_5$	2,6
$(CH_3)_2C{=}CHCH_3$	1,30	$CH_3CH{=}CH(CH_2)_4CH_3$	1,60	$CH_2{=}C(CH_3)CH(CH_3)_2$	2,5	$(CH_3)_2C{=}CH{-}CH{=}C(CH_3)_2$	2,8
$(CH_3)_2C{=}C(CH_3)_2$	0,30	$CH_2{=}C(CH_3)CH_2CH_3$	5,20	$CH_2{=}CH{-}CH{=}CHCH_3$	24	$CH{\equiv}C{-}CH_3$	1,1
$CH_2{=}CHCH_2CH_3$	2,30	$CH_2{=}CHCH(CH_3)_2$	2,40	$CH_2{=}C(CH_3)C_6H_5$	15	$CH{\equiv}CCH_2CH_3$	1,0
$CH_2{=}CH(CH_2)_2CH_3$	2,60	$CH_3CH{=}CHCH(CH_3)_2$	0,25	$C_6H_5CH{=}CHCH_3$	5,8	$CH_3C{\equiv}CHCH_3$	0,14

TABLE 83. Chain Propagation Rate Constants in Radical Polymerization

Monomer	Solvent	t°	log k	log A	E	Method	Ref.
$CH_2{=}CH_2$	C_6H_6	83	2,67			IIM	[334]
$CH_2{=}CH_2$	Monomer	-20	1,27			IIM	[335]
$C_6H_5CH{=}CH_2$	Monomer	0–25	1,27 (25°)	6,05	6,5	NKR	[249]
	- " -	17–38	1,38 (28°)	5,875	6,99	NKR	[336]
	- " -	15–25	1,29 (25°)	5,92	6,3	NKR	[250]
	- " -	0–60	2,245 (60°)	7,75	7,8	IIM	[251]
		Mean		6,97	7,25		[2]
$C_6H_5CH{=}CH_2$	Emulsion	5–30	1,26	7,61	7,4	EP	[337]

Table 83 (continued)

Monomer	Solvent	t°	log k	logA	E	Method	Ref.
p-$CH_3OC_6H_4CH{=}CH_2$	Monomer	0	0,46			NKR	[338]
$CH_2{=}CHCOOCH_3$	$CH_3COOC_2H_5$	26–50	2,76 (30°)	6,16	4,7	IIM	[339]
	Monomer	28	3,68			NKR	[336]
	- " -	30–60	2,26 (30°)	7,40	7,1	IIM	[340]
		Mean		6,05	4,2		[2]
$CH_2{=}CHCOOC_4H_9$	Monomer	25	2,69			NK	[341]
	- " -	25	3,32			NK	[342]
	- " -	25–35	1,11 (25°)	4,64	2,1	IIM	[343]
	- " -	28	3,58			NKR	[336]
$CH_2{=}C(CH_3)COOCH_3$	Monomer	30–60	2,455 (30°)	6,795	6,0	IIM	[284]
	- " -	17–38	2,30 (28°)	5,22	6,5	NKR	[336]
	- " -	24–50	2,49 (24°)	5,96	4,4		[344]
		Mean		5,82	4,7		[2]
$CH_2{=}C(CH_3)COOCH_3$	H_2O	25	1,875			NKR	[345]
$CH_2{=}C(CH_3)COOC_3H_7$	Monomer	30	2,67			IIM	[346]
$CH_2{=}C(CH_3)COOC_4H_9$	- " -	17–38	2,65 (30°)	5,03	3,29	NKR	[336]
$CH_2{=}C(CH_3)COOC(CH_3)_3$	C_6H_6	30–90	2,61 (30°)	5,79	4,4	IIM	[347]
$CH_2{=}C(CH_3)COOCH_2CH(CH_3)_2$	Monomer	17–38		5,415	3,37	NKR	[336]
	$CH_3COOC_2H_5$	30	2,04			NKR	[348, 349]

$CH_2{=}CHOCOCH_3$	Monomer	25–50	3,28	8,85	7,7	IIM	[350]
$CH_2{=}CHCl$	Tetrahydrofuran	25–55	3,84 (30°)	6,52	8,7	IIM	[351]
$CH_2{=}CHCN$	$HCON(CH_3)_2$	13–40	0,73 (30°)	5,36	6,4	NKR	[306]
	- " -	0–60	2,57 (30°)	5,37	3,88	IIM	[352]
	- " -	60	3,29			IIM	[353]
	Monomer	25	2,10			NKR	[354]
	Emulsion	0–40	4,11 (30°)	8,81	6,5	EP	[355]
$CH_2{=}C(CH_3)CN$	Monomer	60–80	1,45 (30°)	9,78	11,5	IIM	[356]
$CH_2{=}CHCONH_2$	H_2O	25	4,255				[303]
	- " -	10–40	2,92 (30°)	5,60	3,7	IIM	[357]
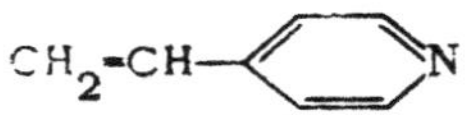	Monomer	25	1,08			NKR	[305]
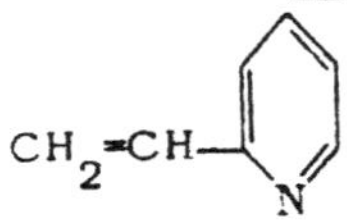	Monomer	25	1,88			NKR	[358]
	- " -	25	2,27			IIM	[358]
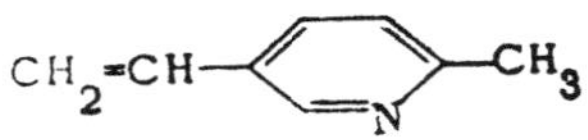	$CHON(CH_3)_2$	25	1,67			NKR	[306]
$CF_2{=}CF_2$	H_2O	30–70	2,38 (30°)	3,33	1,3	NKR	[359]
	- " -	40–60	0,32 (30°)	6,77	4,16	Inhibitor method	[360]
$CH_2{=}CHCH{=}CH_2$	Emulsion	)–30	1,36 (30°)	8,08	9,3	EP	[337]
$CH_2{=}C(CH_3)CH{=}CH_2$	- " -	5–15	1,00 (30°)	8,08	9,8	EP	[337]
$CH_2{=}C(CH_3)C(CH_3){=}CH_2$	- " -	5–60	2,89 (30°)	7,95	7,0	EP	[361]

TABLE 84. Copolymerization Rate Constants

Monomer (1)	Monomer (2)	t^o	k_{11}/k_{12}	k_{22}/k_{21}	Ref.
$C_6H_5CH{=}CH_2$	$CH_2{=}C(CH_3)COOCH_3$	60	0,46	0,48	[362]
	- " -	132	0,60	0,55	[362]
	p-$CH_3C_6H_4CH{=}CH_2$	60	0,83	0,96	[363]
	- " -	60	1,02	1,21	
	o-$CH_3C_6H_4CH{=}CH_2$	60	0,90	0,81	[364]
	m-$CH_3C_6H_4CH{=}CH_2$	60	0,83	0,93	
	$C_6H_5C(CH_3){=}CH_2$	60	0,71	0,14	
	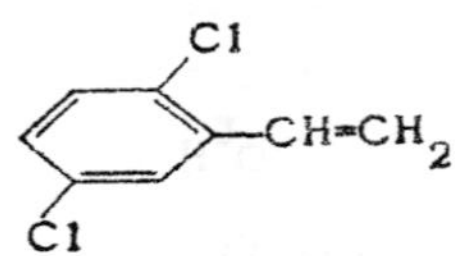	70	0,38	1,77	[365]
	$ClCH{=}CH_2$	60	17	0,02	
	$Cl_2C{=}CH_2$	60	10	0	[302]
	$ClCH{=}CHCl$	30	1,9	0,1	
	$[CH_2OCOC(CH_3){=}CH_2]_2$	60	0,35	0,65	[366]

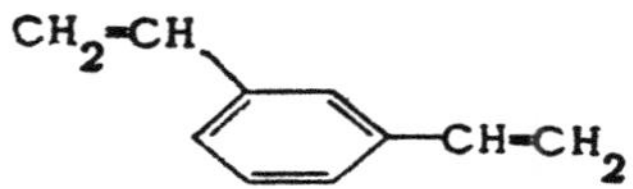	60	0,65	0,60	[366]
$C_6H_5CF{=}CF_2$	50	0,66	0,07	[367]
$C_6H_5CH{=}CHCOOH$	60	1,85	0	[368]
$CH_3COCH{=}CH_2$	60	0,29	0,35	[302]
$CHBr{=}CHOC_2H_5$	50	0,23	0,02	[369]
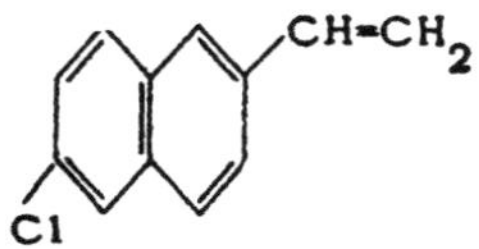	60	0,4	1,5	[370]
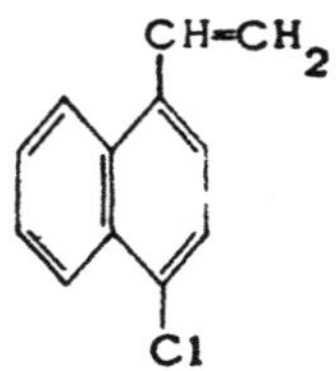	60	0,85	0,8	[370]

Table 84 (continued)

Monomer (1)	Monomer (2)	t°	k_{11}/k_{12}	k_{22}/k_{21}	Ref.
$C_6H_5CH{=}CH_2$	β-$C_{10}H_7CH{=}CH_2$	60	0,5	1,4	[370]
	$CH_2{=}CHCOOCH_2CH_2OC_2H_5$	80	23,5	0	[369]
	$CH_2{=}CHCOOCH_2(CF_2)_2CF_3$	50	0,33	0,07	[371]
	$ClCOCH{=}CHCOCl$	70	0,04	0,00	[372]
	$CH_3SCH{=}CH_2$	60	4,5	0,15	[373]
	$CH_3SO_2CH{=}CH_2$	60	3,3	0,01	[373]
	$CH_3SO(CH{=}CH_2)$	60	4,2	0,01	[374]
	$CH_3OCOCH{=}C(COOCH_3)CH_2COOCH_3$	60	1,10	0,01	[375]
	$CH_2{=}C(CN)_2$	45	0,005	0,001	[376]
	$CH_2{=}CHCH{=}CH_2$	60	0,78	1,4	[302]
	- " -	45	0,65	1,83	[377]
	- " -	5	0,64	1,38	[378]
	- " -	-18	0,38	1,37	[379]
	$CH_2{=}C(CH_3)CH{=}CH_2$	-18	0,48	1,30	[380]
	$CH_2{=}C(CH_3)C(CH_3){=}CH_2$	-18	0,42	0,92	[380]
	$CH_2{=}CFCH{=}CH_2$	50	0,50	1,55	[381]
	- " -	5	0,16	1,61	[381]

$CH_2{=}CHCOOCH_3$	$CH_2{=}CH{-}OCH_2CH_2Br$	80	31,5	0	[369]
	$CH_3SCH{=}CH_2$	60	0,40	0,05	[373]
	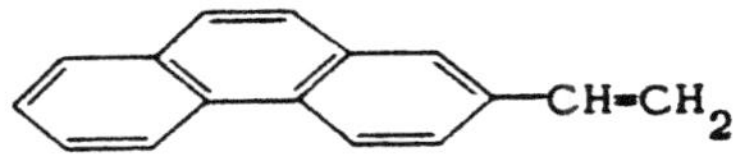	60	0,10	2,0	[370]
	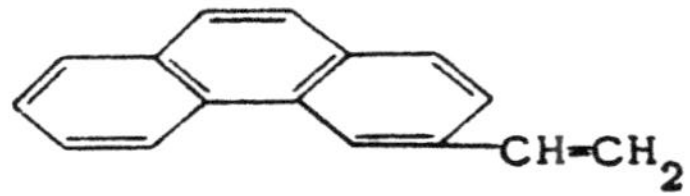	60	0,80	1,75	[370]
	$CH_2{=}CHCOO(CH_2)_{16}CH_3$	70	5,8	0,03	[382]
$CH_2{=}C(CH_3)COOCH_3$	$C_6H_5CH{=}CH_2$	60	0,46	0,52	[383]
	- " -	60	0,49	0,53	[384]
	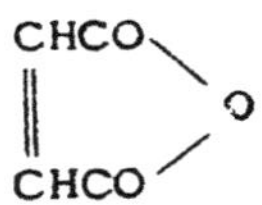	60	3,46	0,03	[385]
	$CH_2{=}C(CH_3)CN$	60	0,67	0,65	[302]
	$CH_2{=}CHCOOCH_3$	60	20	0,03	[302]
	$CH_2{=}CCl_2$	60	2,5	0,24	[302]

Table 84 (continued)

Monomer (1)	Monomer (2)	t°	k_{11}/k_{12}	k_{22}/k_{21}	Ref.
$CH_2=C(CH_3)COOCH_3$	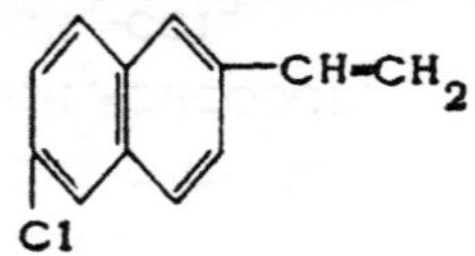	60	0,45	1,6	[370]
	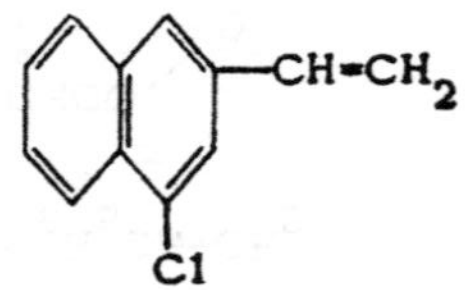	60	0,7	0,7	[370]
	β-$C_{10}H_7CH=CH_2$	60	0,4	1,0	[370]
	$CH_2=CHCOOCH_2(CF_2)_2CF_3$	50	1,4	0,25	[371]
	$CH_3SOCH=CH_2$	60	20	0	[374]
	$CH_2=C(CN)_2$	50	0,046	0,031	[376]
$CH_2=CH-CN$	$CH_2=CCl-CCl=CH_2$	60	0,02	0,1	[302]
	$C_6H_5C\equiv CH$	60	0,26	0,33	[302]
	$C_6H_5CH=CH_2$	60	0,05	0,40	[302]
	$CH_2=CHOCOCH_3$	60	5	0,1	[302]
	$CH_2=CHCl$	60	3,3	0,04	[302]
	$CH_2=CHCH=CHOCOCH_3$	70	0	0,7	[386]

	$CH_2{=}CHCOOCH_2CH_2OC_2H_5$	80	10,5	0,02	[387]
	$CH_2{=}CHOC_2H_5$	80	0,7	0,03	[387]
	$CH_2{=}C(CH_3)CHO$	70	0,06	2,0	[372]
	$CH_2{=}COOC(CH_2)_{16}CH_3$	70	4,3	0,03	[382]
	$CH_3OCOCH{=}C(COOCH_3)CH_2COOCH_3$	60	5,50	0	[375]
	$CH_2{=}CF{-}CH{=}CH_2$	50	0,07	0,59	[381]
$CH_2{=}CH{-}OCOCH_3$	$CH_2{=}CHCH_2OOCCH_3$	60	0,60	0,40	
	$CH_2{=}C(CH_3)CN$	60	0,02	12	[302]
	$CH_2{=}CHOC_2H_5$	60	3,0	0	
	$CH_3COCH{=}CH_2$	70	0,05	7,0	[386]
	$CH_3SO_2CH{=}CH_2$	60	0,28	0,35	[373]
	$CH_3OCOCH{=}CHCOOH$	56	0,522	0,035	[388]
	$CH_2{=}C(CN)_2$	45	0,0054	0,11	[376]
	$CH_2{=}CHOCO(CH_2)_{16}CH_3$	70	0,90	0,73	[388]
	$CH_2{=}CHOCOCF_3$	60	0,60	0,32	[389]
$CH_2{=}C(CN)_2$	$CH_2{=}CHCOOH$	50	0,29	0,26	
	$CH_2{=}CClCH{=}CH_2$	40	0,0017	0,010	[376]
	$ClCH{=}CHCl$	40	30	0,0	
	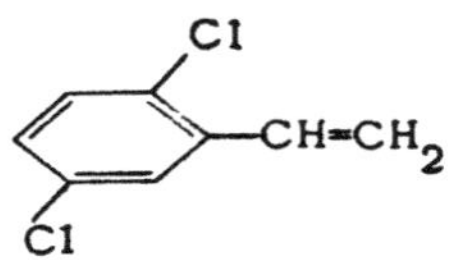	40	0,0092	0,031	[376]

Table 84 (continued)

Monomer (1)	Monomer (2)	t°	k_{11}/k_{12}	k_{22}/k_{21}	Ref.
$CH_2{=}C(CN)_2$	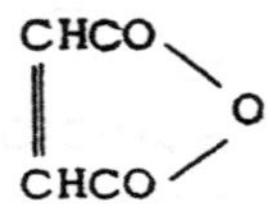	50	45	0	
	$CH_2{=}CClCOOCH_3$	50	0,091	0,41	
	$CH_2{=}CHOCOC_6H_5$	43	0,10	0,02	[376]
	$CH_2{=}CHOCOCH_2Cl$	40	0,13	0,004	
	$CH_2{=}CHCl$	50	0,54	0,017	
	$CH_2{=}CCl_2$	22	0,049	0,012	
$CH_2{=}CF{-}CH{=}CH_2$	$CH_2{=}C(CH_3)CH{=}CH_2$	50	2,05	0,19	[381]
	$CH_2{=}C(CH_3)C_6H_5$	50	1,54	0,64	[381]
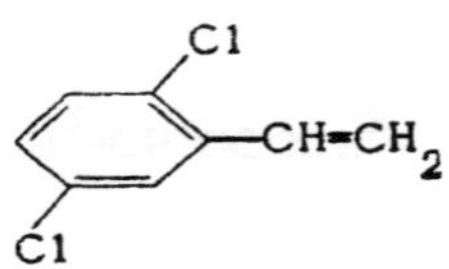	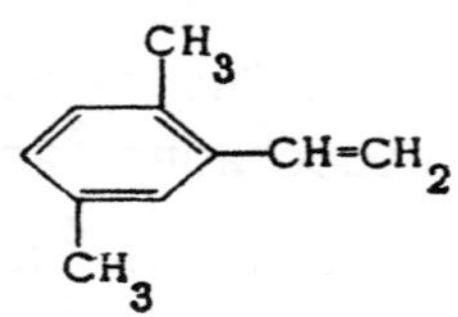	70	1,55	0,27	[386]
$CH_2{=}CHCl$	$CH_3OCOCH{=}C(COOCH_3)CH_2COOCH_3$	60	0,15	0,5	[375]

TABLE 85. Copolymerization Parameters q and ε for Calculating Copolymerization Constants (Note: ε is in elementary charge units; q is value relative to that of phenyl group)

Monomer	ε	q
Double bond in conjugation with aromatic system		
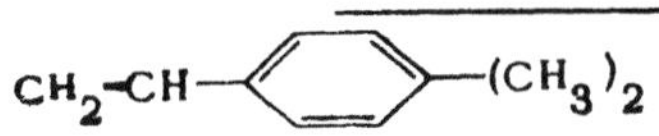	-0,11	1,06
	-0,077	1,06
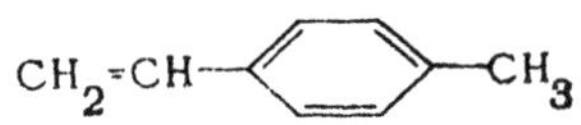	-0,067	1,00
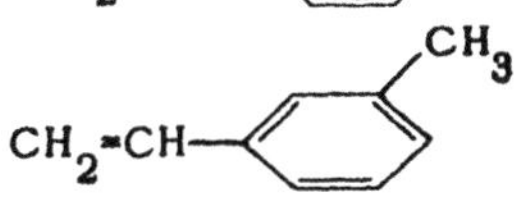	-0,058	0,98
	-0,050	1,00
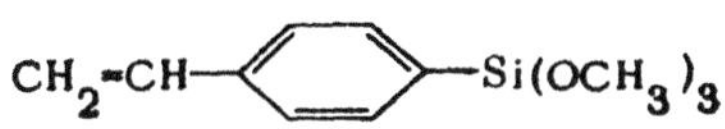	-0,050	1,00
$CH_2{=}CHC_6H_5$	-0,050	1,00
p-I-$C_6H_4CH{=}CH_2$	-0,033	1,00
o-Cl$C_6H_4CH{=}CH_2$	-0,031	1,06
p-Cl$C_6H_4CH{=}CH_2$	-0,029	0,93
p-Br $C_6H_4CH{=}CH_2$	-0,027	1,00
m-Cl$C_6H_4CH{=}CH_2$	-0,013	1,00
2,5-$Cl_2C_6H_3CH{=}CH_2$	-0,008	0,97
p-$CH_3OC_6H_4CH{=}CHCH_3$	-0,13	0,55
$(C_6H_5)_2C{=}CH_2$	-0,13	1,26
$C_6H_5C(CH_3){=}CH_2$	-0,077	0,94
Indene	-0,069	0,74
$C_6H_5CH{=}CHC_6H_5$	-0,056	0,36
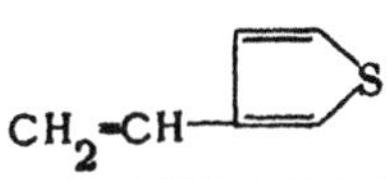	-0,050	1,23
$CH_2{=}C(OCOCH_3)C_6H_5$	-0,042	0,90

Table 85 (continued)

Monomer	ε	q
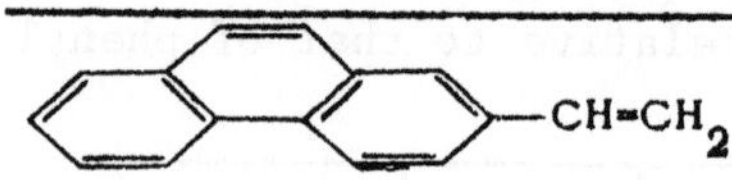	−0,042	1,16
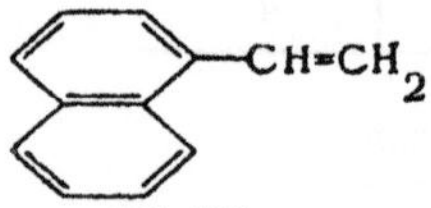	−0,031	1,03
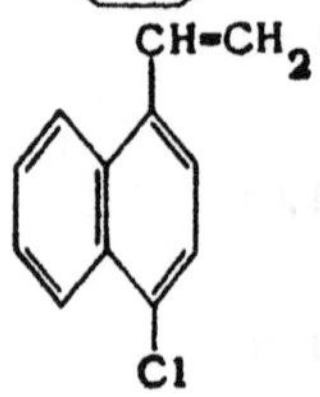	−0,023	0,94
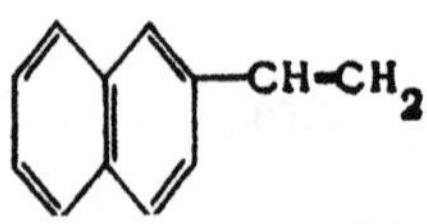	−0,023	0,87
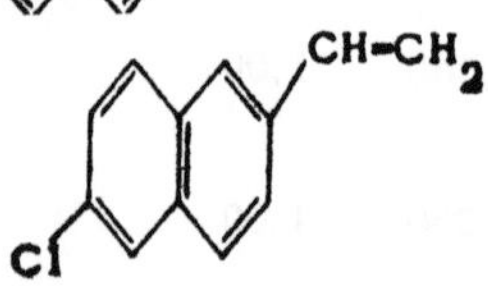	−0,019	0,71
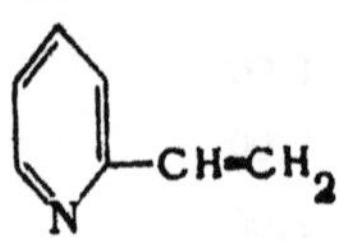	−0,015	1,03
m-Br $C_6H_4CH{=}CH_2$	−0,008	1,00
p-$CNC_6H_4CH{=}CH_2$	0,017	1,10
m-$NO_2C_6H_4CH{=}CH_2$	0,021	1,10
p-$NH_2SO_2C_6H_4CH{=}CH_2$	0,022	1,14
p-$NO_2C_6H_4CH{=}CH_2$	0,027	1,13
$C_6Cl_5CH{=}CH_2$	0,042	0,71
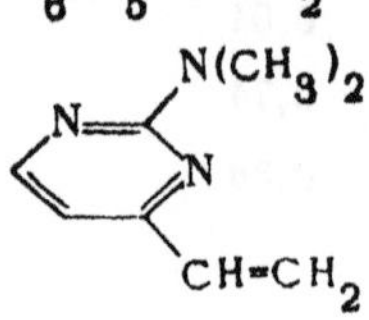	0,003	1,08

Table 85 (continued)

Monomer	ε	q
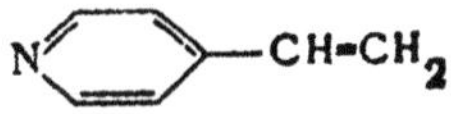	0,019	0,94
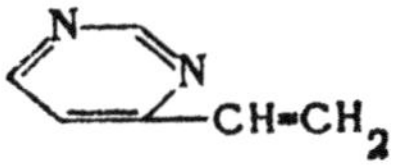	0,03	1,18
Double bond in conjugation with C=O or C=N bond		
Vinylene carbonate	−0,061	0,02
CH_2=C(CH_3)CHO	−0,021	1,16
CH_3COC(CH_3)=CH_2	−0,017	1,03
CH_2=CHCOONa	−0,013	0,87
CH_2=C(CH_3)COOCH_3	0,015	0,87
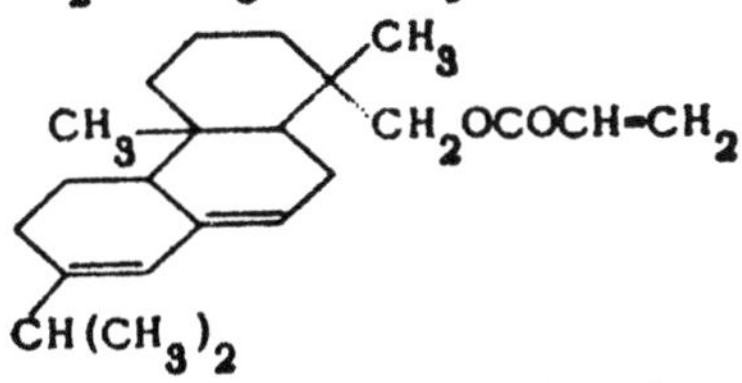	0,033	0,77
CH_2=CHCON(C_2H_5)CH_2(CF_2)$_2CF_3$	0,052	0,97
CH_2=CHCSOCH_3	0,054	1,16
CH_2=CHCOOH	0,058	0,81
CH_2=CHCOOCH_2CH_2Cl	0,060	0,84
CH_2=CHCONH_2	0,063	0,88
CH_3CH=CHCOOH	0,063	0,36
CH_2=C(COOH)CH_2COOH	0,065	0,97
CH_2=CHCOOCH_2(CH_2)$_6CH_3$	0,038	0,77
CH_2=CHCOOCH_3	0,038	0,81
CH_3COCH=CH_2	0,040	0,90
CH_3CH(OCOCH_3)OCOCH=CH_2	0,043	0,83
CH_2=C(CH_3)CONH_2	0,044	0,79
CH_2=CHCHO	0,045	0,74
CH=C(CH_3)COOH	0,046	0,97
CH=C(CH_3)CN	0,048	0,97
trans-CH_3OOCCH=CHCOOCH_3	0,094	0,94
cis-CH_3OOCCH=CHCOOH	0,10	0,16

Table 85 (continued)

Monomer	e	q
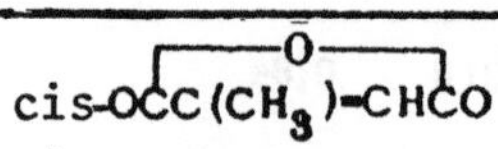	0,11	0,97
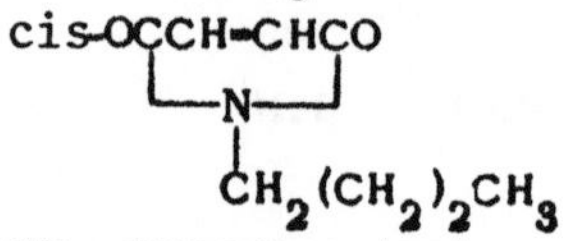	0,101	0,88
$CH_2{=}CHCOOCH_2(CF)_2CF_3$	0,069	0,90
$CH_2{=}CHCN$	0,071	0,87
$CH_2{=}CHCOO(CH_2)_3CH_3$	0,071	0,87
$CH_2{=}CClCOOCH_3$	0,079	1,19
cis-$C_2H_5OOCCH{=}CHCOOC_2H_5$	0,083	0,32
trans-$C_2H_5OOCCH{=}CHCOOC_2H_5$	0,085	0,90
cis-$CH_3OOCCH{=}CHCOOCH_3$	0,088	0,16
cis-$C_2H_5OOCCH{=}CHCOOH$	0,090	1,06
$CH_3OCOCH{=}CH(COOCH_3)CH_2COOCH_3$	0,092	0,77
trans-$C_2H_5OOCCH{=}CHCOOH$	0,094	0,84
trans-$NCCH{=}CHCOOH$	0,112	0,97
cis-$COCH{=}CHCO$	0,15	0,77
$CH_2{=}C(CN)_2$	0,16	1,29
Vinyl group at oxygen		
$CH_2{=}CHOCH_2CH(CH_3)_2$	-0,11	0,22
$CH_2{=}CHOC_2H_5$	-0,10	0,03
$CH_2{=}C(CH_3)OCOCH_3$	-0,046	0,10
$CH_2{=}COCO(CH_2)_{16}CH_3$	-0,046	0,23
$CH_2{=}COCO(CH_2)_{14}CH_3$	-0,046	0,19
$CH_2{=}COCO(CH_2)_2CH_3$	-0,040	0,81
$(C_2H_5)CH{=}COCO(CH_2)_4CH_3$	-0,035	0,74
$CH_2{=}CHOCOCH_2CH_3$	-0,035	0,84
$CH_2{=}CHOCO(CH_2)_8CH_3$	-0,031	0,145
$CH_2{=}CHOCOCH_3$	-0,023	0,23
$CH_2{=}CHOCHO$	-0,023	0,52
$CH_2{=}CHOCOC_6H_5$	-0,013	0,18
$C_2H_5OCH{=}CHCOOC_2H_5$	0,021	0,013
$CH_2{=}CHOCOCF_3$	0,063	0,32

Table 85 (continued)

Monomer	ε	q
Vinyl group at sulfur		
$CH_2{=}CHSCH_3$	–0,09	0,81
$CH_2{=}CHSC_6H_5$	–0,083	0,77
$CH_2{=}CHSCH{=}CH_2$	–0,075	0,88
$CH_2{=}CHSC_6Cl_5$	–0,033	0,65
$BrCH{=}CH{-}SC_2H_5$	0,023	0,02
$CH_2{=}CHSOCH_3$	0,058	0,52
$CH_2{=}CHSO_2CH_3$	0,069	0,39
$CH_2{=}CHSO_2C_6H_5$	0,069	0,39
Olefins and dienes		
$CH_2{=}C(CH_3)_2$	–0,088	0,18
$CH_2{=}C(CH_3)CH{=}CH_2$	–0,075	1,10
$CH_2{=}CH(CH_2)_3CH_3$	–0,083	0,13
$CH_2{=}CClCH{=}CH_2$	–0,031	1,35
$CH_2{=}CHCH{=}CH_2$	–0,050	1,16
$CF_2{=}CFCF{=}CF_2$	–0,033	1,00
$CH_2{=}CClCCl{=}CH_2$	0,0083	1,42
Acetylene derivatives		
$CH{\equiv}C(CH_2)_3CH_3$	–0,071	0,08
$C_6H_5C{\equiv}CC_6H_5$	–0,043	–0,19
$C_6H_5C{\equiv}CH$	–0,04	0,77
Allyl compounds		
$CH_2{=}C(CH_3)CH_2OCOCH_3$	–0,069	0,20
$CH_2{=}CHCH_2OCO(CH_2)_8CH_3$	–0,06	0,32
$CH_2{=}CHCH_2OCOCH_3$	–0,06	0,21
$CH_2{=}C(CH_3)CH_2Cl$	–0,042	0,16
$CH_2{=}CHCH_2Cl$	0,008	0,25
Vinyl group at nitrogen		
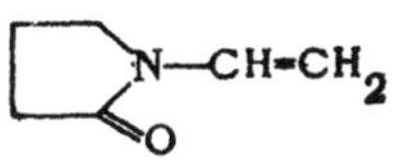	–0,101	0,52
$CH_2{=}CHNHCOOC_2H_5$	–0,075	0,57
$CH_2{=}CHNCO$	–0,031	1,13
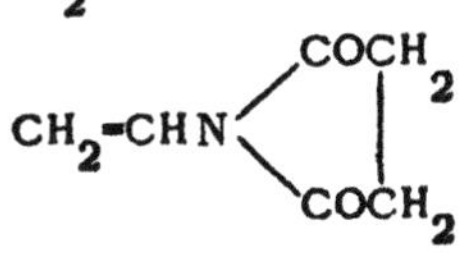	–0,017	0,55

Table 85 (continued)

Monomer	e	q
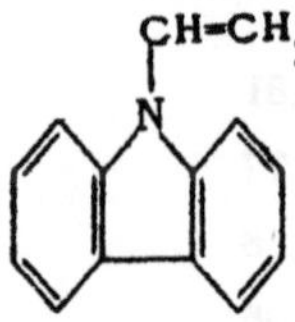	0,069	0,52
Halogen at double bond		
$ClCH=CH_2$	−0,008	0,35
cis-$ClCH=CHCl$	0,067	−0,03
$BrCH=CH_2$	0,006	0,10
$ClCH_2CCl=CH_2$	0,021	0,45
$Cl_2C=CH_2$	0,023	0,68
trans-$ClCH=CHCl$	0,073	0,065
$Cl_2C=CHCl$	0,11	0,10
Organosilicon compounds		
$(CH_3)_3SiCH=CHC_6H_5$	−0,05	1,00
$(CH_3O)_3SiCH=CHC_6H_5$	−0,05	1,00
$CH_2=C(CH_3)COOCH_2Si(CH_3)_2OSi(CH_3)_3$	−0,001	0,94

TABLE 86. Addition of Atoms and Free Radicals to Aromatic Compounds (MCR method)

Compound	Solvent	t°	log k	Ref.
	H˙ atom			
C_6H_6	H_2O	25	7,845	[393]
	- " -	25	9,04*	[394]
	$n\text{-}C_6H_{14}$	23	8,255	[70]
C_6H_5OH	H_2O	25	8,15	[395]
	- " -	25	9,25*	[396]
	$n\text{-}C_6H_{14}$	23	8,62	[70]
$C_6H_5NO_2$	H_2O	25	9,75*	[397]
	$n\text{-}C_6H_{14}$	23	9,10	[70]
$C_6H_5CH_2OH$	H_2O	23	8,81	[61]
	$n\text{-}C_6H_{14}$	23	8,78	[70]

Table 86 (continued)

Compound	Solvent	t°	log k	Ref.
C_6H_5COOH	n-C_6H_{14}	23	9,46	[70]
C_6H_5CN	- " -	23	9,00	[70]
$C_6H_5COCH_3$	n-C_6H_4	23	9,08	
$C_6H_5CH_2OCOCH_3$	- " -	23	9,43	
o-$CH_3C_6H_4CH_3$	- " -	23	8,38	
m-$CH_3C_6H_4CH_3$	- " -	23	8,46	
p-$CH_3C_6H_4CH_3$	- " -	23	8,69	[70]
$C_6H_5(CH_2)_2CH_3$	- " -	23	8,52	
C_6H_5-C_6H_5	- " -	23	9,15	
$C_6H_5CH_2CH_2C_6H_5$	- " -	23	8,64	
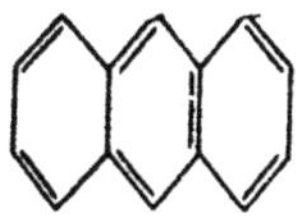	- " -	23	9,49	[70]
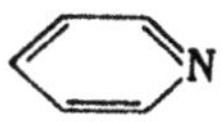	H_2O	23	8,81	[64]
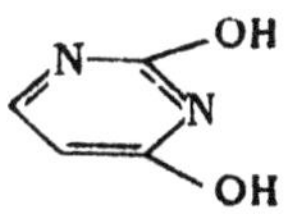	H_2O	23	9,48	[64]
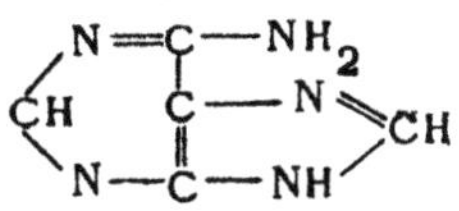	H_2O	23	8,78	[64]
	Radical ·OH			
C_6H_6	H_2O	25	9,63*	[319]
	- " -	25	9,00	[102]
	- " -	25	9,505	[94]
	- " -	25	9,68	[100]
	- " -	25	9,48	[398]
	- " -	25	9,48	[105]
C_6D_6	H_2O	25	9,67*	[319]

Table 86 (continued)

Compound	Solvent	t°	logk	Ref.
C_6H_5OH	H_2O	25	10,15*	[396]
	- " -	25	9,62	[100]
	- " -	25	9,71	[398]
	- " -	25	10,03	[105]
C_6H_5COOH	H_2O	25	9,34	[102]
	- " -	25	9,68*	[90]
	- " -	25	9,59	[100]
	- " -	25	9,32	[398]
	- " -	25	9,52	[105]
$C_6H_5SO_3H$	H_2O	25	9,18	[102]
$C_6H_5NO_2$	H_2O	25	8,97	[102]
	- " -	25	9,175	[100]
	- " -	25	9,30	[398]
	- " -	25	9,67*	[397]
C_6H_5CN	H_2O	25	9,36	[100]
	- " -	25	9,33	[398]
C_6H_5Cl	H_2O	25	9,63	[100]
	- " -	25	9,62	[398]
$C_6H_5NH_2$	H_2O	25	9,69	[100]
	- " -	25	9,725	[398]
$C_6H_5CONH_2$	H_2O	25	9,415	[398]
$C_6H_5SO_2NH_2$	- " -	25	9,22	[398]
$C_6H_5COCH_3$	- " -	25	9,46	[398]
$C_6H_5OCOCH_3$	- " -	25	9,48	[398]
$C_6H_5CH_3$	H_2O	25	9,48	[398]
	- " -	25	9,845	[90]
$C_6H_5OCH_3$	H_2O	25	9,56	[398]
$C_6H_5N(CH_3)_2$	- " -	25	9,725	[398]
p-HOC_6H_4OH	H_2O	25	10,08*	[399]
	- " -	25	10,09	[105]

Table 86 (continued)

Compound	Solvent	t°	log k	Ref.
p-$NH_2C_6H_4COOH$	H_2O	25	10,0	[105]
	H_2O	25	9,255*	[400]
	- " -	25	7,76	[96]
	- " -	25	9,49	[105]
Thymine	H_2O	25	9,48	[401]
Uracil	- " -	25	9,46	[401]
Cytosine	H_2O, pH 2,1	25	9,255	[401]
	H_2O, pH 7,5	25	9,48	[401]
Adenine	H_2O, pH 2,1	25	8,725	[401]
	H_2O, pH 7,5	25	9,48	[401]
Thymidine	H_2O, pH 2,1	25	9,44	[401]
Cytidine	H_2O	25	9,40	[401]
Adenosine	H_2O	25	9,25	[401]
	Radical $CH_3^{\cdot}$			
C_6H_6	$(CH_3)_2CHCH_2C(CH_3)_3$	65	1,50	[402]
	- " -	85	2,05	[403]
C_6H_5F	$(CH_3)_2CHCH_2C(CH_3)_3$	65	1,84	[402]
C_6H_5Cl	- " -	65	2,12	[402]
C_6H_5Br	- " -	65	2,055	[402]
p-ClC_6H_4Cl	- " -	65	2,57	[402]
m-ClC_6H_4Cl	- " -	65	2,59	[402]
$CH_3OC_6H_5$	- " -	65	1,31	[402]
C_6H_5CN	- " -	65	2,59	[402]
$C_6H_5COC_6H_5$	$(CH_3)_2CHCH_2C(CH_3)_3$	65	1,88	[402]
	- " -	85	3,09	[403]
$C_6H_5NO_2$	$(CH_3)_2CHCH_2C(CH_3)_3$	65	6,50	[402]
$C_6H_5OC_6H_5$	- " -	85	2,45	[403]
C_6H_5-C_6H_5	- " -	85	2,75	[403]

Table 86 (continued)

Compound	Solvent	t°	log k	Ref.
$C_6H_5CH{=}CHC_6H_5$	$(CH_3)_2CHCH_2C(CH_3)_3$	85	4,31	[403]
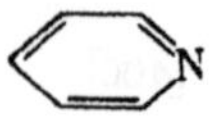	- " -	85	2,53	[403]
$C_{10}H_8$	$(CH_3)_2CHCH_2C(CH_3)_3$	85	3,39	[403]
	- " -	65	3,11	[404]
α-$C_{10}H_7CH_3$	$(CH_3)_2CHCH_2C(CH_3)_3$	65	3,05	[404]
β-$C_{10}H_7CH_3$	- " -	65	3,25	[404]
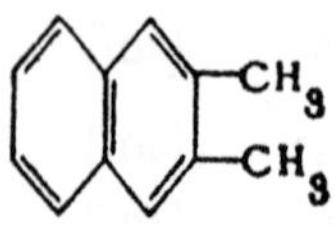	- " -	65	3,16	[404]
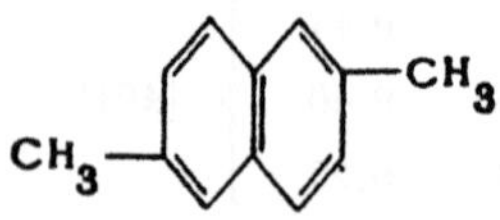	- " -	65	3,28	[404]
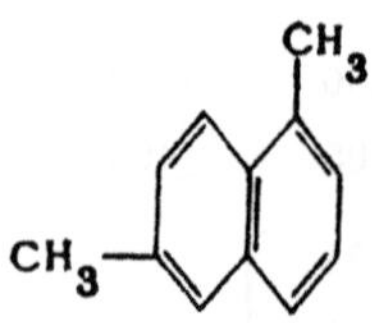	- " -	65	2,905	[404]
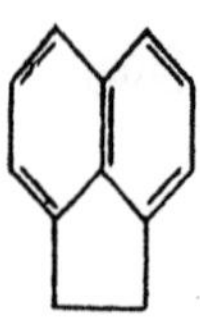	- " -	65	2,81	[404]
α-$C_{10}H_7C_2H_5$	$(CH_3)_2CHCH_2C(CH_3)_3$	65	3,04	[404]
β-$C_{10}H_7C_2H_5$	- " -	65	3,09	[404]
α-$C_{10}H_7F$	- " -	65	3,16	[404]
β-$C_{10}H_7F$	- " -	65	3,26	[404]

Table 86 (continued)

Compound	Solvent	t°	log k	Ref.
α-$C_{10}H_7Cl$	$(CH_3)_2CHCH_2C(CH_3)_3$	65	3,26	[404]
β-$C_{10}H_7Cl$	- " -	65	3,475	
α-$C_{10}H_7Br$	- " -	65	3,04	
β-$C_{10}H_7Br$	- " -	65	2,465	
α-$C_{10}H_7I$	- " -	65	3,405	
β-$C_{10}H_7I$	- " -	65	3,62	
1,4-$Cl_2C_{10}H_6$	- " -	65	3,455	
α-$C_{10}H_7OCH_3$	- " -	65	3,085	
β-$C_{10}H_7OCH_3$	- " -	65	3,015	
α-$C_{10}H_7N(CH_3)_2$	- " -	65	2,685	
α-$C_{10}H_7COCH_3$	- " -	65	3,44	
β-$C_{10}H_7COCH_3$	- " -	65	3,79	
α-$C_{10}H_7CN$	- " -	65	3,63	
β-$C_{10}H_7CN$	- " -	65	3,80	
α-$C_{10}H_7COCH_3$	- " -	65	3,43	
α-$C_{10}H_7COOC_2H_5$	- " -	65	3,485	
β-$C_{10}H_7COOCH_3$	- " -	65	3,745	
α-$C_{10}H_7$-N=C=O	- " -	65	3,37	
Chrysene	- " -	85	3,81	
Pyrene	- " -	85	4,16	
Anthracene	- " -	85	4,96	
Acridine	- " -	85	4,68	
Naphthanthracene	- " -	85	6,02	
Benzanthracene	- " -	85	4,72	
Quinoline	- " -	85	3,51	
Isoquinoline	- " -	85	3,61	
	Radical $C_2H_5^{\cdot}$			
C_6H_6	$(CH_3)_2CHCH_2C(CH_3)_3$	65	1,52	[405]
C_6H_5-C_6H_5	- " -	65	2,10	
trans-$C_6H_5CH{=}CHC_6H_5$	- " -	65	4,22	
$C_{10}H_8$	- " -	65	3,06	
$C_6H_5COC_6H_5$	- " -	65	2,76	

Table 86 (continued)

Compound	Solvent	t°	log k	Ref.
Phenanthrene	$(CH_3)_2CHCH_2C(CH_3)_3$	65	3,155	[405]
Quinoline	- " -	65	3,32	[405]
	Radical $CH_3CH_2CH_2^{\cdot}$			
C_6H_6	$(CH_3)_2CHCH_2C(CH_3)_3$	65	0,30[2*]	[406]
C_6H_5-C_6H_5	- " -	65	0,91[2*]	[406]
$C_6H_5COC_6H_5$	- " -	65	1,525[2*]	[406]
$C_{10}H_8$	- " -	65	1,835[2*]	[406]
Phenanthrene	- " -	65	1,955[2*]	[406]
Quinoline	- " -	65	2,165[2*]	[406]
trans-$C_6H_5CH{=}CHC_6H_5$	- " -	65	2,955[2*]	[406]
	Radical $CF_3^{\cdot}$			
C_6H_6	$(CH_3)_2CHCH_2C(CH_3)_3$	65	5,90	[407]
$C_6H_5CH_3$	- " -	65	6,12	[407]
o-$CH_3C_6H_4CH_3$	- " -	65	6,18	[407]
m-$CH_3C_6H_4CH_3$	- " -	65	6,36	[407]
p-$CH_3C_6H_4CH_3$	- " -	65	6,285	[407]
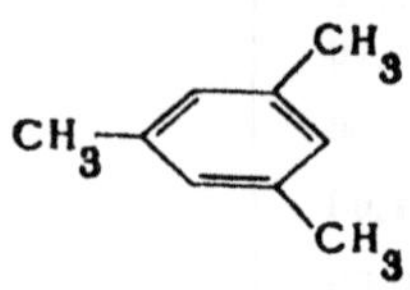	- " -	65	6,665	[407]
$C_6H_5C_2H_5$	- " -	65	6,10	[407]
$C_6H_5CH(CH_3)_2$	- " -	65	6,105	[407]
$C_6H_5C(CH_3)_3$	- " -	65	5,86	[407]
p-$(CH_3)_3CC_6H_4C(CH_3)_3$	- " -	65	5,63	[407]
C_6H_5F	- " -	65	5,58	[407]
C_6H_5Cl	- " -	65	5,71	[407]
C_6H_5Br	- " -	65	5,79	[407]
C_6H_5I	- " -	65	6,12	[407]

Table 86 (continued)

Compound	Solvent	t°	log k	Ref.
$C_6H_5CF_3$	$(CH_3)_2CHCH_2C(CH_3)_3$	65	5,37	[407]
C_6H_5CN	- " -	65	5,53	[407]
$C_6H_5NO_2$	- " -	65	5,555	[407]
$C_6H_5COOC_2H_5$	- " -	65	5,95	[407]
$C_6H_5OCH_3$	- " -	65	5,35	[407]
C_6H_5OH	- " -	65	6,12	[407]
$C_6H_5COCH_3$	- " -	65	5,68	[407]
$C_6H_5NH_2$	- " -	65	5,705	[407]
C_6H_5-C_6H_5	- " -	65	6,25	[408]
$C_{10}H_8$	- " -	65	7,10	[408]
Phenanthrene	- " -	65	7,045	[408]
Pyrene	- " -	65	7,605	[408]
Anthracene	- " -	65	8,255	[408]
	Radical $CCl_3^{\cdot}$			
C_6H_6	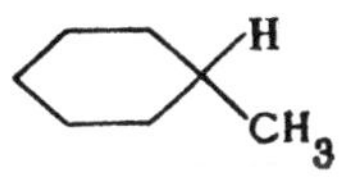	91,5	2,48	[409]
C_6H_5-C_6H_5	- " -	91,5	2,92	[409]
Phenanthrene	- " -	91,5	3,68	[409]
$C_{10}H_8$	- " -	91,5	4,08	[409]
Chrysene	- " -	91,5	4,29	[409]
$C_6H_5CH{=}CHC_6H_5$	- " -	91,5	5,48	[409]
Pyrene	- " -	91,5	5,505	[409]
Anthracene	- " -	91,5	6,84	[409]
Benzanthracene	- " -	91,5	6,98	[409]
1,2,5,6-Dibenzanthracene	- " -	91,5	6,06	[409]
Naphthacene	- " -	91,5	7,505	[409]
	Radical $(CH_3)_2\dot{C}CN$			
Anthracene	cyclo-$C_6H_{11}CH_3$	55	4,53	[410]
9-$CH_3C_{14}H_9$	- " -	55	4,56	[410]
9-$C_2H_5C_{14}H_9$	- " -	55	4,43	[410]

Table 86 (continued)

Compound	Solvent	t°	log k	Ref.
$2\text{-}CH_3C_{14}H_9$	cyclo-$C_6H_{11}CH_3$	55	4,54	[410]
$2,3,6,7\text{-}(CH_3)_4C_{14}H_6$	- " -	55	4,64	[410]
$9\text{-}CH_3OC_{14}H_9$	- " -	55	4,57	[410]
$9\text{-}ClC_{14}H_9$	- " -	55	4,46	[410]
$9\text{-}Br\,C_{14}H_9$	- " -	55	4,38	[410]
$1\text{-}ClC_{14}H_9$	- " -	55	4,17	[410]
$9,10\text{-}(CH_3)_2C_{14}H_8$	- " -	55	3,60	[410]
$9,10\text{-}Cl_2C_{14}H_8$	- " -	55	3,55	[410]
$1,5\text{-}Cl_2C_{14}H_8$	- " -	55	3,76	[410]
$9\text{-}CNC_{14}H_9$	- " -	55	3,86	[410]
$9\text{-}C_6H_5C_{14}H_9$	- " -	55	4,21	[410]
$9\text{-}C_{14}H_9N_2$	- " -	55	4,08	[410]
$1\text{-}N_2C_{14}H_9$	- " -	55	4,65	[410]
Azulene	- " -	55	4,83	[410]
	Radical $C_6H_5C(CH_3)_2OO^{\cdot}$			
Naphthalene	$C_6H_5CH(CH_3)_2$	60	1,79[3*]	[411]
Benzanthracene	- " -	60	0,63[3*]	[411]
	Radical 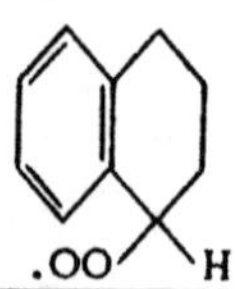			
Tetracene	$C_6H_5CH(CH_3)_2$	60	4,31[3*]	[411]

---

*Measured by PR method.

2* Values cited are for $\log \frac{k(C_3H_7^{\cdot} + \text{aromatic})}{k(C_3H_7^{\cdot} + \text{isooctane})}$.

3* Measured by CINH method.

TABLE 87. Addition of Free Atoms and Radicals to Quinones, Carbonyl Compounds, Acids, and CO (MCR method)

Compound	Solvent	t°	log k	Ref.
	Atom H·			
	H_2O	25	9,49	[58]
	- " -	25	9,08	[412]
CH_3COCH_3	n-C_6H_{14}	23	8,505	[70] *
$CH_3COC_6H_5$	- " -	23	9,08	
cyclo-$C_6H_{10}O$	- " -	23	8,83	
CH_3COOH	- " -	23	8,785	[70] *
CH_3CH_2COOH	- " -	23	8,84	
CF_3COOH	- " -	23	9,20	
	Radical ·OH			
CO	H_2O, pH 0	25	8,83	[412]
	H_2O, pH 1	25	8,92	[412]
CO	H_2O, pH 7	25	8,77	[413]
	H_2O	25	8,65	[97]
	H_2O, pH 7	25	9,08	[399]
	Radical CH_3· 2*			
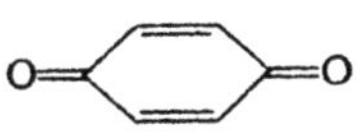	$(CH_3)_2CHCH_2C(CH_3)_3$	65	5,68	[414]
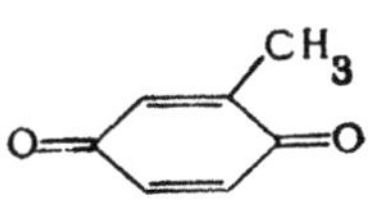	- " -	65	5,52	[414].

Table 87 (continued)

Compound	Solvent	t^o	log k	Ref.
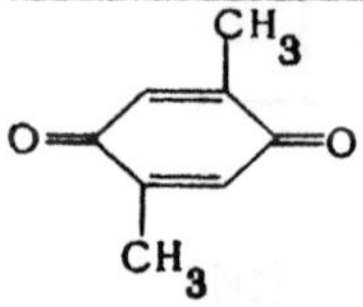	$(CH_3)_2CHCH_2C(CH_3)_3$	65	5,31	[414]
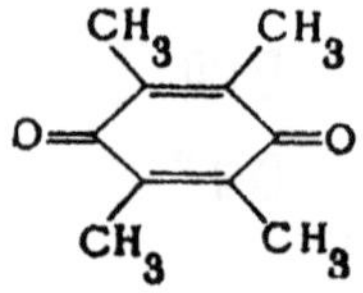	- " -	65	4,40	[414]
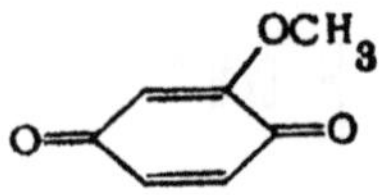	- " -	65	5,405	[414]
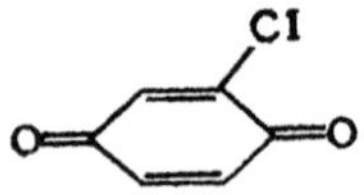	- " -	65	5,93	[414]
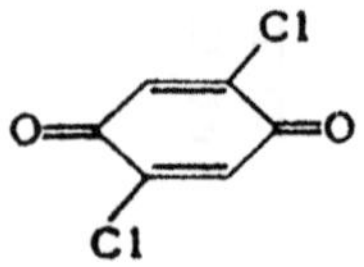	- " -	65	6,09	[414]
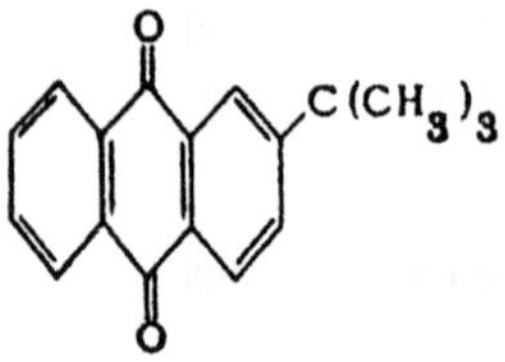	- " -	65	3,455	[414]
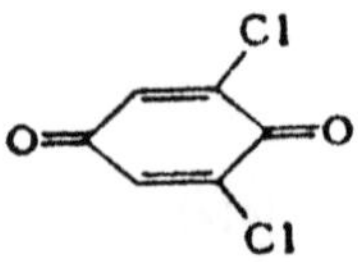	- " -	65	6,085	[414]
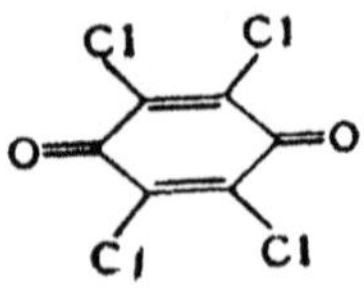	- " -	65	3,98	[414]

Table 87 (continued)

Compound	Solvent	t°	log k	Ref.
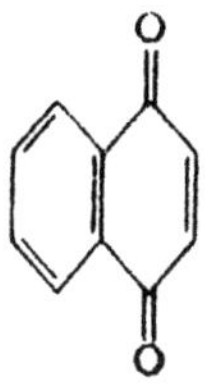	$(CH_3)_2CHCH_2C(CH_3)_3$	65	5,18	[414]
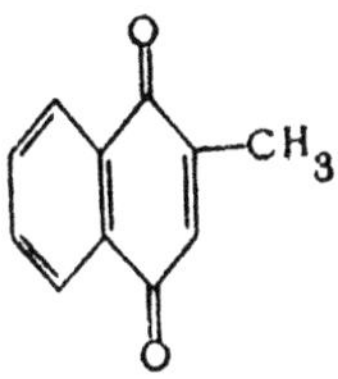	- " -	65	5,03	[414]
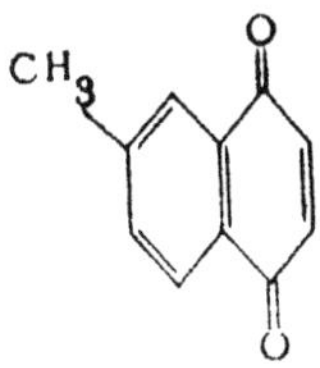	- " -	65	5,11	[414]
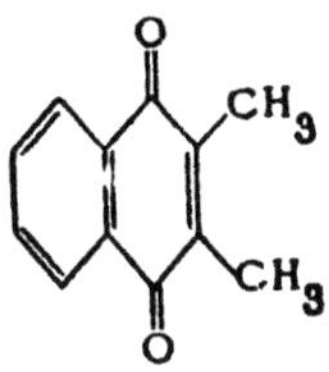	- " -	65	4,24	[414]
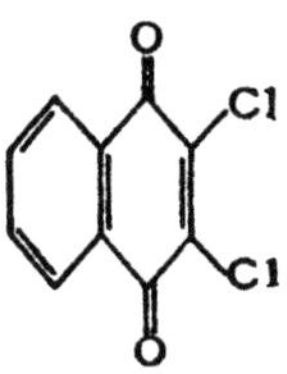	- " -	65	3,455	[414]

Table 87 (continued)

Compound	Solvent	t°	log k	Ref.
o-Naphthoquinone	$(CH_3)_2CHCH_2C(CH_3)_3$	65	5,03	[414]
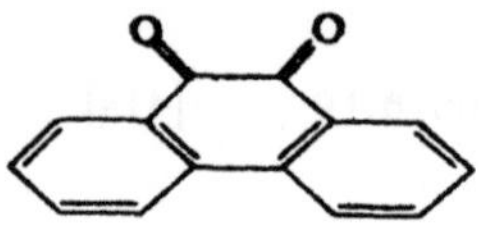	- " -	65	4,345	[414]
Radical $\sim CH_2\dot{C}(CH_3)COOCH_3$				
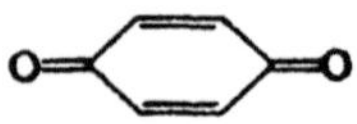	$CH_2{=}C(CH_3)COOCH_3$	44	3,33[3*]	[415]
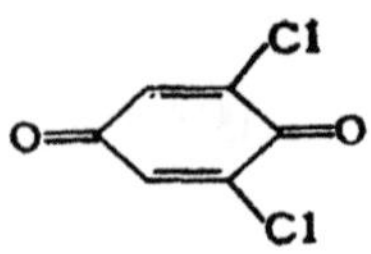	- " -	44	4,075[3*]	[415]
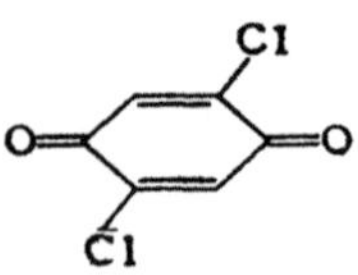	$CH_2{=}C(CH_3)COOCH_3$	44	3,69[3*]	[415]
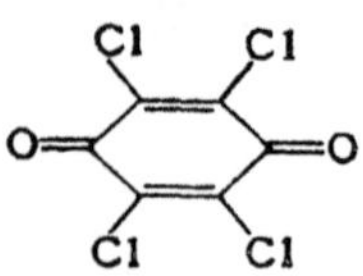	$CH_2{=}C(CH_3)COOCH_3$	44	2,02[3*]	[415]
	$CH_2{=}CHCOOCH_3$	44	3,06[4*]	[416]

Table 87 (continued)

Compound	Solvent	t^{o}	log k	Ref.
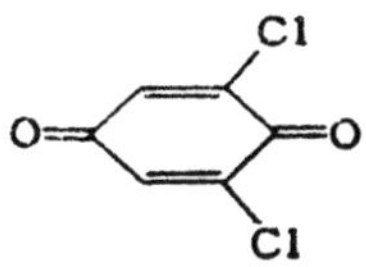	CH_2=$CHCOOCH_3$	44	4,205[4*]	[416]
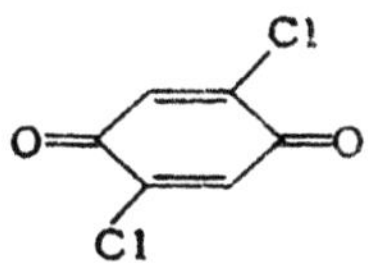	- " -	44	3,99[4*]	[416]
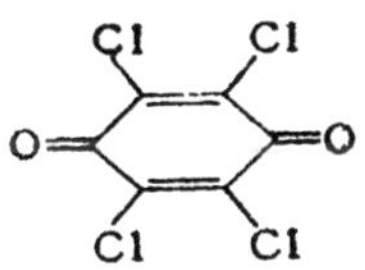	- " -	44	3,29[4*]	[416]

* Absolute constants calculated from relative with value $k(H^{\cdot}+n\text{-}C_6H_{14}) = 4.9\cdot10^6$ (23°).

2* Absolute constants calculated from relative with value $\log k(CH_3^{\cdot} +$ isooctane$) = 2.14$ (65°); see Chapter VI, §3.

3* Measured by CINH method and calculated on assumption that $\log k_p = 2.62$ at 44°.

4* Measured by CINH method and calculated with value $\log k_p = 3.09$ at 44°.

TABLE 88. Addition of Radicals to Nitriles, Nitro Compounds, and Schiff Bases (CINH method)

Compound	log k	Compound	log k
Radical $\sim CH_2CHOCOCH_3$, at 40° [417]; log k_p = 8.85 - 7.7/θ; monomer as solvent.			
CH_3CN	2,615	$C_6H_5CH_2CN$	5,35
C_6H_5CN	4,30	p-CNC_6H_4CN	4,965
$CH_3C_6H_4CN$	4,61	m-CNC_6H_4CN	5,63
Radical $\sim CH_2\dot{C}HCHOCOCH_3$; at 45° [418] ; monomer as solvent.			
$C_6H_5NO_2$	3,84	m-$NO_2C_6H_4NO_2$	5,58
p-$CH_3C_6H_4NO_2$	4,86	p-$NO_2C_6H_4NO_2$	5,99
o-$NO_2C_6H_4NO_2$	5,54		
Radical $\sim CH_2\dot{C}(CH_3)COOCH_3$; at 44° [415] ; log k_p = 2.62; monomer as solvent.			
m-$NO_2C_6H_4NO_2$	0,32	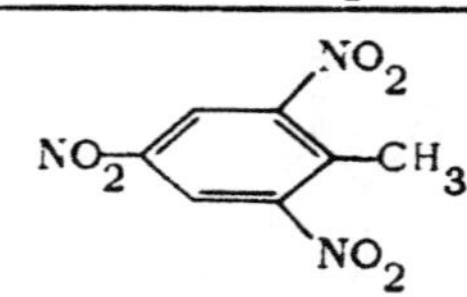	1,32

Radical $C_6H_5CH(CH_3)OO\cdot$ in $C_6H_5C_2H_5$ at 60° [419]

$C_6H_5CH{=}N{-}C_6H_5$	3,54	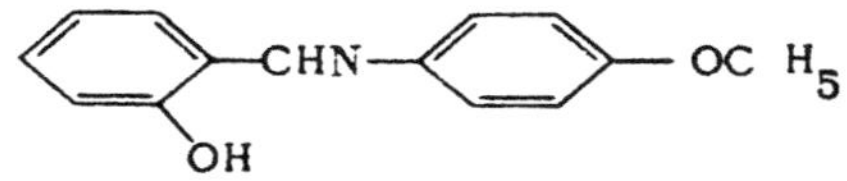	4,415
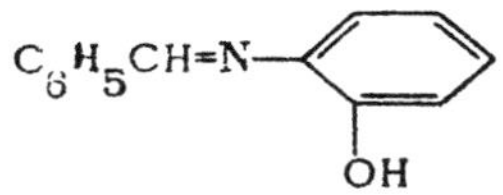	5,53	$C_6H_5C(CH_3){=}N{-}C_6H_5$	4,11
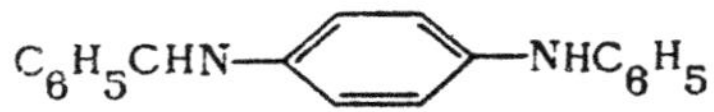	6,54	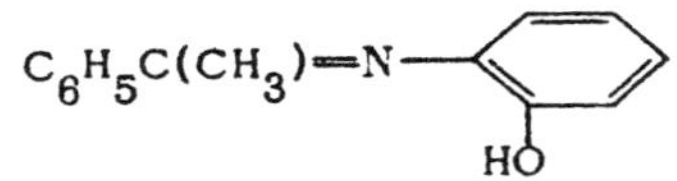	6,18
	6,77	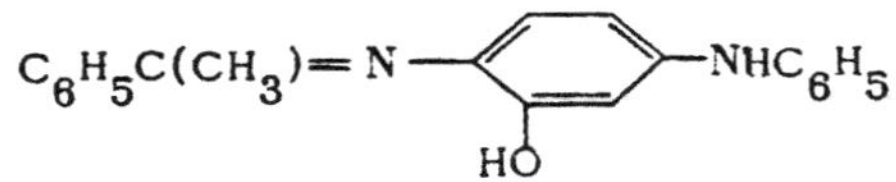	6,08
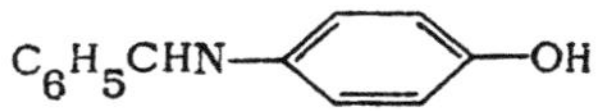	5,28		6,59
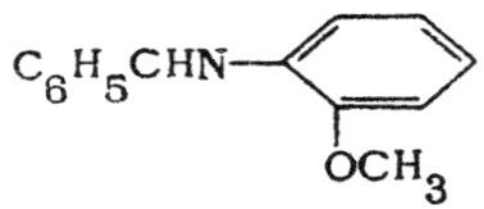	4,67	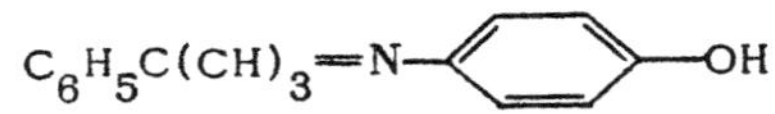	5,28

Table 88 (continued)

Compound	log k	Compound	log k
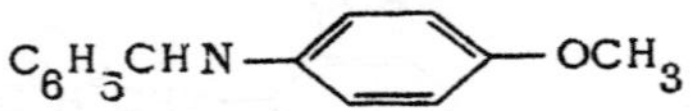	4,53	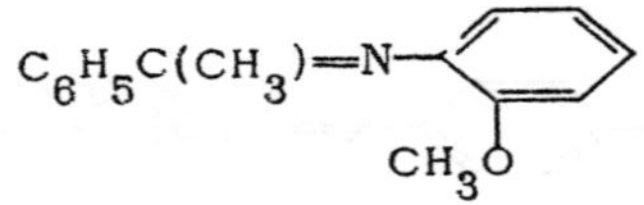	4,04
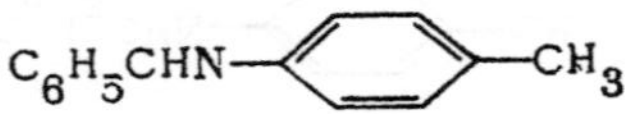	4,255	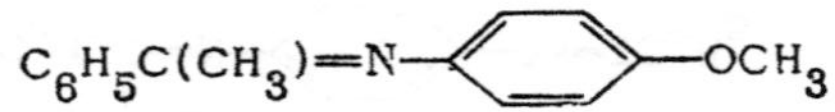	4,45
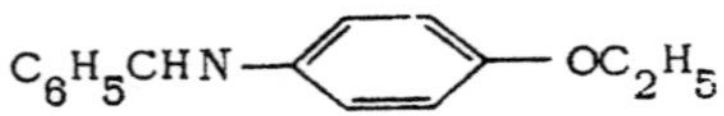	4,60	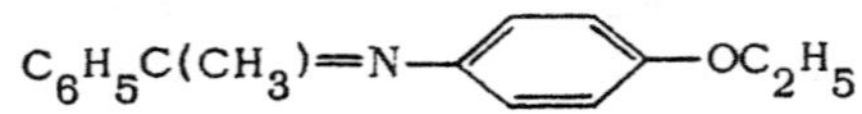	4,54
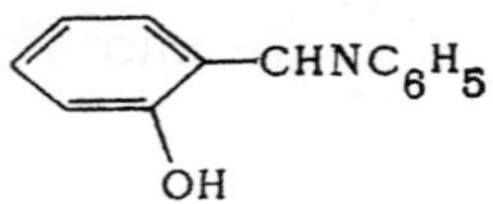	3,46		4,23
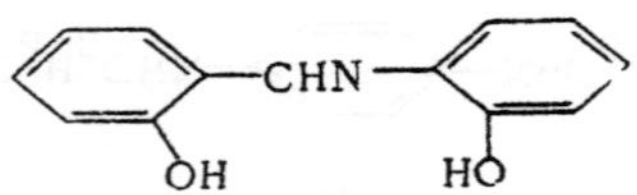	5,77	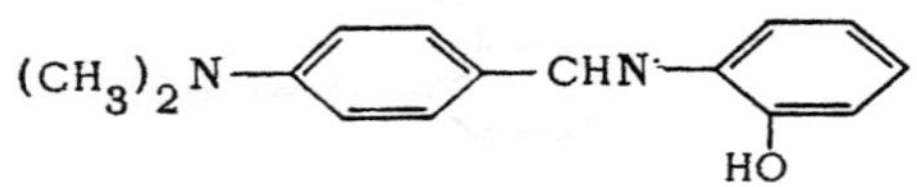	5,20
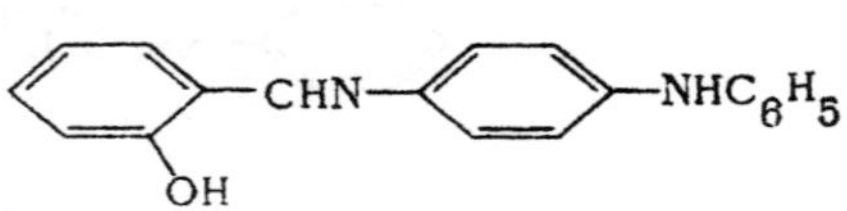	5,91	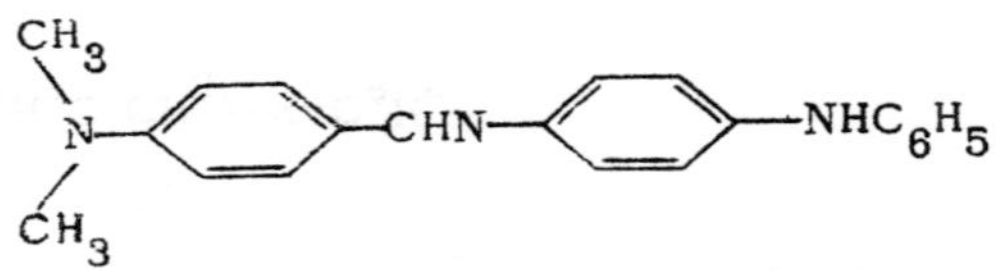	5,58

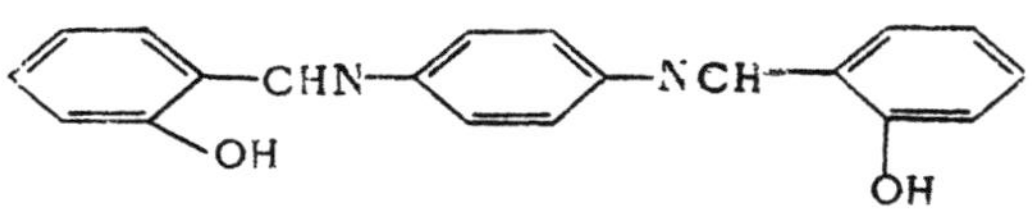	6,43	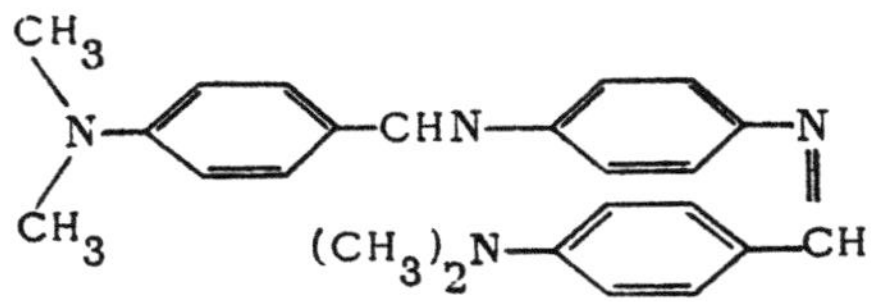	4,49
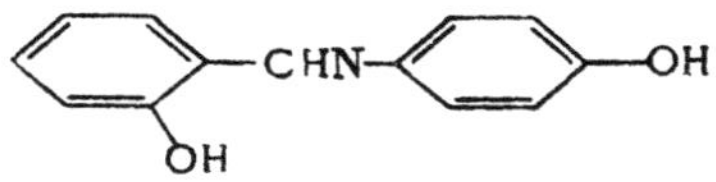	4,81	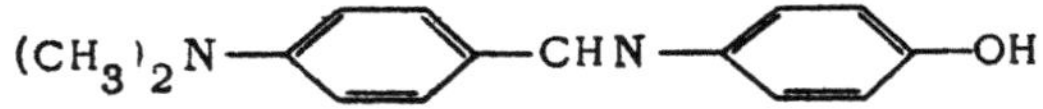	4,49
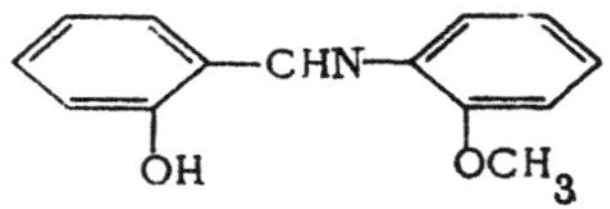	4,15	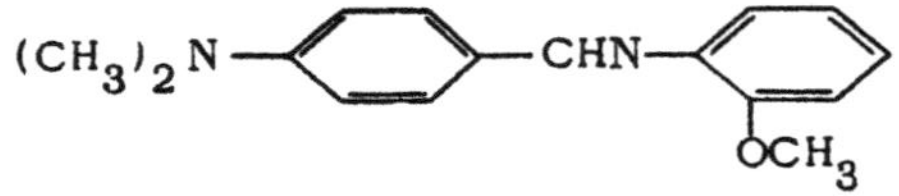	4,34
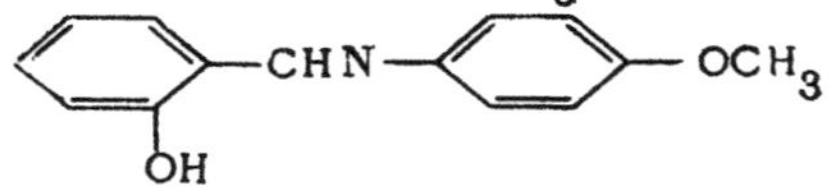	4,34	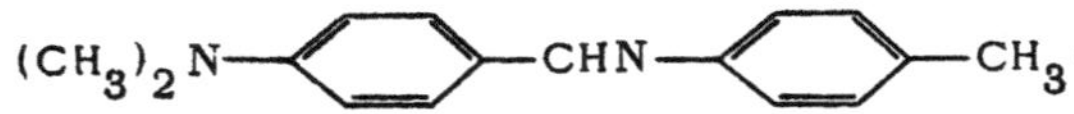	4,40
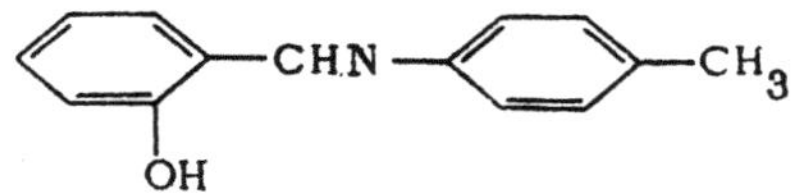	3,54	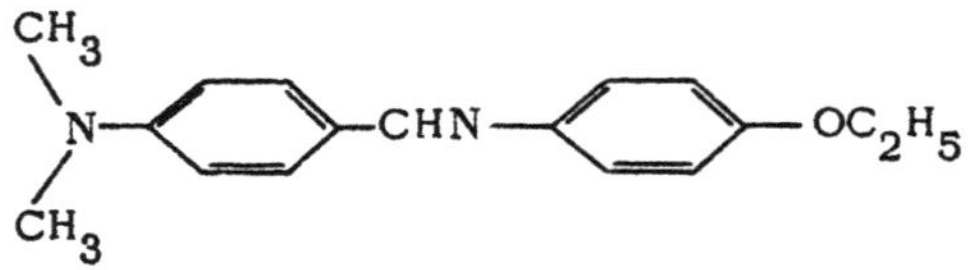	4,62
		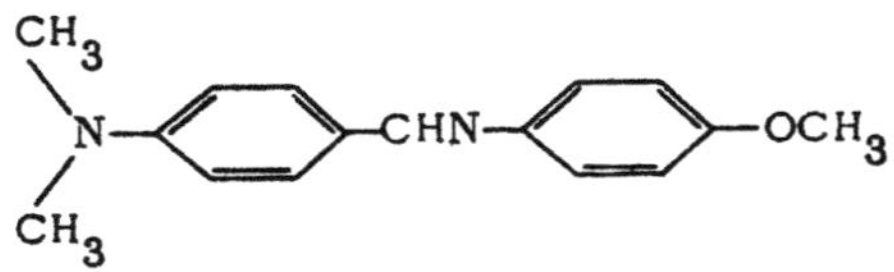	4,42

CHAPTER VIII

RECOMBINATION AND DISPROPORTIONATION OF FREE ATOMS AND RADICALS

In each elementary act of recombination or disproportionation of free radicals, two active centers are destroyed (two chains are broken under chain-reaction conditions). In the stationary reaction regime, $W_i = 2k[X^\cdot]^2$, where k is the rate constant for the reaction of two radicals or atoms $X^\cdot$. Since the quantity 2k is always used in calculations, the quantities listed here in the tables are always these (doubled) values of the reaction rate constants.

§1. Recombination of Atoms

The recombination of atoms in a liquid is limited by their diffusion, and the recombination rate constant is the diffusion rate constant of the atoms. Values of these constants for hydrogen, bromine, and iodine are very similar (Table 89).

§2. Disproportionation and Recombination of Alkyl and RO˙ Radicals

Alkyl radicals react with each other by two routes; they may recombine to form a new C-C bond, and they may disproportionate by a reaction in which an H atom is abstracted by one radical from the other to form an olefin and a paraffin. As an example, we may cite the reaction of a methyl radical with an ethyl:

$$CH_3^\cdot + C_2H_5^\cdot \longrightarrow CH_3CH_2CH_3 + 80.5 \text{ kcal/mole}$$
$$CH_3^\cdot + C_2H_5^\cdot \longrightarrow CH_4 + CH_2{=}CH_2 + 64 \text{ kcal/mole}$$

The heats of these reactions are calculated on the basis of data on the heats of formation of the radicals that are listed in the handbook [428]. The ratio of the rate constants for disproportionation and recombination of alkyl radicals has been measured experimentally: Photochemical decomposition of an azo compound was effected in a solvent, giving the following reactions:

$$R-N=N-CH_2CH_2R' \xrightarrow{h\nu} \overline{R\cdot + \cdot CH_2CH_2R'} + N_2$$

$$\overline{CH_2CH_2R' + R\cdot} \xrightarrow{k_d} RH + CH_2 = CH_2R'$$

$$\overline{R\cdot + \cdot CH_2CH_2R'} \xrightarrow{k_r} RCH_2CH_2R'$$

$$\overline{R\cdot + \cdot CH_2CH_2R'} \xrightarrow{k_D} R\cdot + \cdot CH_2CH_2R'$$

The radical disproportionation and recombination took place in a solvent cage. Since the olefin $CH_2=CHR'$ was obtained only from the disproportionation reaction, and the hydrocarbon RCH_2CH_2R' only from the recombination reaction, the ratio of rate constants was determined from the analytical data on the ratio of olefin to paraffinic hydrocarbon:

$$k_d/k_r = \left[CH_2 = CHR'\right] / \left[RCH_2CH_2R'\right].$$

Upon transition from solution to the solid state, log (k_d/k_r) is no longer temperature-dependent [429].

Hydroxyalkyl radicals in a recombination reaction will form a diol, but in a disproportionation reaction an alcohol and a carbonyl compound forms, for example:

$$2\ CH_3\dot{C}HOH \nearrow CH_3CHOHCHOHCH_3$$
$$2\ CH_3\dot{C}HOH \searrow CH_3CHO + CH_3CH_2OH$$

The ratio of constants k_d/k_r has been computed on the basis of the composition of products obtained by radiolysis of the corresponding alcohols.

Alkyl radicals react with each other with rate constants close to the diffusion constant (10^9 to 10^{10} liter/mole·sec), as can be seen from Table 91, the same as do hydroxyalkyl radicals. It is interesting to note that tertiary alkyl radicals recombine just as rapidly as do primary alkyl radicals. In [445-447], reaction rate constants for $R^{\cdot}$ radicals were determined from data on the oxidation of the corresponding hydrocarbons, i.e., by a rather indirect method, and it is possible that the rate constants so determined are understated.

In the rate constants for macroradicals, a considerable divergence of values is observed. One of the reasons for such deviations is the viscosity effect. Even a comparatively slight increase in viscosity will shift a reaction between macromolecules into the diffusion region. Hence, it is very important to know at what average degree of polymerization the measurement of k is performed. Apparently the closest to the true value will be the maximum (not the mean) values of the rate constants for reactions of macroradicals with each other.

§3. Reactions between Peroxy Radicals

The $HO_2^{\cdot}$ radical has amphoteric properties, and in aqueous solution, depending on the pH, it may exist in the form of $HO_2^{\cdot}$, $O_2^{\overline{\cdot}}$, or $H_2O_2^{\dot{+}}$:

$$H_2O_2^{\dot{+}} \underset{+H^+}{\overset{-H^+}{\rightleftharpoons}} HO_2^{\cdot} \rightleftharpoons H^+ + O_2^{\overline{\cdot}}$$

According to [460], $pH_{HO_2^{\cdot}} = 4.4$, and at $pH \geq 5$ the equilibrium is shifted in the direction of $O_2^{\overline{\cdot}}$; $pK_{H_2O_2^{\dot{+}}} = 1.0$ [461]. As a consequence, the rate constant for HO_2 disproportionation depends on pH. At a pH above 5, $k \approx 1.5 \cdot 10^7$ (the reaction $O_2^{\overline{\cdot}} + O_2^{\overline{\cdot}} \rightarrow O_2^{2-} + O_2$ takes place). With $0.8 < pH < 2$, $k = 2 \cdot 10^6$ (the reaction $HO_2^{\cdot} + HO_2^{\cdot} \rightarrow H_2O_2 + O_2$ takes place). At a pH below 0.5, $k \approx 0.5 \cdot 10^6$ (the reaction $H_2O_2^{\dot{+}} + H_2O_2^{\dot{+}} \rightarrow H_2O_2 + O_2 + 2H^+$ takes place). Table 93 lists the value of pH at which the k's were measured. The higher values of $k(HO_2^{\cdot} + HO_2^{\cdot})$ in organic solvents in comparison with water are explained by the influence of hydrogen bonds of the type H_2O ... HOO and $HOO^{\cdot}$ ... HOH.

The mechanism of the reaction between two tertiary peroxy radicals is quite complex, including several elementary stages [462-469]:

$$2\,RO_2^{\cdot} \longrightarrow ROOOOR \longrightarrow \overline{2\,RO^{\cdot}} + O_2\ \text{(triplet)}$$

$$\overline{2\ RO\cdot} \longrightarrow ROOR$$

$$\overline{2\ RO\cdot} \longrightarrow 2\ RO\cdot$$

For the recombination of two secondary peroxy radicals, the following reaction mechanism has been proposed [470, 471]:

$$2\ RO_2\cdot \longrightarrow ROOOOR \longrightarrow \text{ketone} + \text{alcohol} + O_2\ \text{(singlet)}$$

§4. Reaction of Radicals Having Free Valence on Nitrogen or Tin

See Table 94.

§5. Reactions between Free Radicals of Different Types

See Table 95.

TABLE 89. Rate Constants for Recombination of Atoms in Solution

Atom	Solvent	t°	log (2k)	Method	Ref.
H	H_2O, pH 3	25	10,08	PR	[53]
	H_2O,pH 0,4–3	25	10,415	PR	[54]
	H_2O, pH 0,1– 1,0	25	10,48	PR	[420]
	H_2O, pH 2	25	10,30	PR	[55]
		Mean	10,30 ± 0,13		
Br	CCl_4	25	10,6	FM	[421]
I	CCl_4	15–35	14,40–5,6/θ	IIM	[422]
	– " –	25	10,06	FM	[423]
	– " –	20–50	12,49–3,2/θ	FM	[424]
	– " –	25	10,64	FM	[425]
	– " –	25	10,30	FM	[426]
		Mean at 25°	10,28 ± 0,15		
I	n-C_6H_{14}	15–35	12,20–2,4/θ	IIM	[422]
	– " –	25	10,30	IIM	[426]
	– " –	25	10,64	IIM	[425]
	– " –	25	10,56	IIM	[424]
	$Cl_2C{=}CCl{-}CCl{=}CCl_2$	15–35	12,50–3,3/θ	IIM	[422]
	cyclo-C_6H_{12}	22	10,03	PR	[427]

Table 90. Ratio between Disproportionation and Recombination of Radicals with Free Valence on Carbon (θ = 4.57 T/1000)

Radical	Solvent	t°	log (k_d/k_r)	Ref.
$CH_3\cdot + C_2H_5\cdot$	n-C_7H_{16} + CH_2=CHC_6H_5	-75–+100	$\bar{2}$,416 + 0,40/θ	[430]
	$(CH_3)_2CHCH_2C(CH_3)_3$	0	$\bar{2}$,875	[431]
	$CH_2(OH)CH_2OH$	0	$\bar{2}$,965	[431]
2 C_2H_5	$(CH_3)_2CHCH_2C(CH_3)_3$	65	$\bar{1}$,16	[432]
	- " -	0	$\bar{1}$,25	[429]
	$C_6H_5CH_3$	0	$\bar{1}$,27	[429]
	$(CH_3)_2CHOH$	0	$\bar{1}$,305	[429]
	$CH_2(OH)CH_2OH$	0	$\bar{1}$,415	[429]
	- " -	65	$\bar{1}$,38	[432]
	Gas phase	0	$\bar{1}$,115	[429]
	n-C_7H_{16}	-75 to +99	$\bar{2}$,94 + 0,37/θ	[433]
2 $C_6H_5(CH_3)_2C\cdot$	$C_6H_5CH(CH_3)_2$	20–60	$\bar{2}$,73	[434]
2 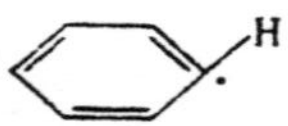	cyclo-C_6H_{12}	24	0,04	[435]
2 $CH_2\dot{C}HOH$	H_2O	23	$\bar{1}$,40	[436]
2 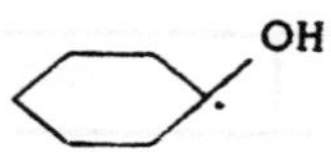	cyclo-$C_6H_{11}OH$	25	0,30	[437]
2 ~$CH_2\dot{C}(CH_3)COOCH_3$	CH_2=$C(CH_3)COOCH_3$	0–60	1,99–2,26/θ	[438]

TABLE 91. Total Rate Constants for Self-Reactions of Alkyl Radicals ($k = k_r + k_d$)

Radical	Solvent	t°	log (2k)	log(2A)	E	Method	Ref.
$CH_3\cdot$	H_2O	25	7,34			JM + EPR	[439]
	cyclo-C_6H_{12}	25	9,95			IIM	[233]
$C_2H_5\cdot$	C_2H_6	-140 to + + 170	9,51 (30°)	10,11	0,83	PR	[440]
n-$C_5H_{11}\cdot$	C_6H_6	25	9,30			IIM	[441]
n-$C_6H_{13}\cdot$	cyclo-C_6H_{12}	25	9,34			IIM	[233]
$C_6H_5CH_2\cdot$	cyclo-C_6H_{12}	23	9,60			PR	[442]
	C_6H_6	25	9,915			IIM	[441]
	Paraffin	23	7,90			FM	[443]
	$C_2H_5OH + \overset{OH}{C}H(CH_2OH)_2$	25	7,80			PR	[444]
$CCl_3\cdot$	$CHCl_3$	24	7,845			IIM	[221]
	- " -	25	8,21			IIM	[220]
	$CH_2{=}CHOCOCH_3$	25	7,725			IIM	[220]
	- " -	25	8,025			NKR	[220]
	- " -	25	8,00			IIM	[329]

Table 91 (continued)

Radical	Solvent	t°	log (2k)	log (2A)	E	Method	Ref.
$CCl_3^{\cdot}$	$CH_2=CHOCOCH_3$	25	8,00			NKR	[238]
$(CH_3)_3C\cdot$	cyclo-C_6H_{12}	25	9,32			IIM	[233]
	n-C_5H_{12}	25	9,73			IIM	[233]
	n-$C_{13}H_{28}$	25	9,255			IIM	[233]
$C_6H_5\dot{C}HCH_3$	$C_6H_5C_2H_5$	60	8,30*			NKR	[445]
$C_6H_5\dot{C}(CH_3)_2$	$C_6H_5CH(CH_3)_2$	60	6,70*			NKR	[445]
$CH_3(CH_2)_4CH=CH$ $CH_2CH=CH-CH$ $CH_2(CH_2)_5\dot{C}OOC_2H_5$	RH	25	7,18			NKR	[446]
CH_3 CH_3 $CH_2=C(CH_2)_3C=CH$ $HC=HC-\dot{C}H-CH_2$ $CH_2(CH_2)_2C=CH_2$ CH_3	RH	25	8,415			NKR	[446]
cyclo-$C_6H_{11}\cdot$	RH	25	9,43			IIM	[233]
	- " -	22	9,40			PR	[427]

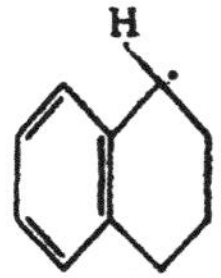	RH	25	6,85			IIM	[447]
$(CH_3)_2\dot{C}CN$	C_6H_5Cl	60	8,30*			NKR	[445]
	C_6H_6	25	9,64			EPR	[448]
	C_6H_6	25	9,64			EPR	[448]
$CH_3\dot{C}HOH$	H_2O	23	9,04			PR	[436]
	- " -	23	9,15			PR	[449]
	- " -	23	9,30			PR	[450]
	cyclo-$C_6H_{11}OH$	25	8,53			PR	[437]
	 + $CH_2{=}C(CH_3)_2$	25	8,32			PR	[437]
$(C_6H_5)_2COH$	$(CH_3)_2CHOH$	25	7,77			FM	[451]
$\sim CH_2 - CH_2\cdot$	C_6H_6	83	9,02			IIM	[334]
	$CH_2{=}CH_2$	-20	8,66			IIM	[335]
$C_6H_5\dot{C}H{-}CH_2\sim$	$CH_2{=}CHC_6H_5$	0–25	6,45 (25°)	8,505	2,8	NKR	[249]
	- " -	28	6,78			NKR	[336]
	- " -	15–25	6,48 (25°)	7,875	1,9	NKR	[250]
	$CH_2{=}CHC_6H_5$	0–60	7,86 (60°)	9,27	2,4	IIM	[251]
		Mean		8,85	1,5		

Table 91 (continued)

Radical	Solvent	t^o	log (2k)	log(2A)	E	Method	Ref.
$p\text{-}CH_3OC_6H_4\dot{C}H\text{-}CH_2\sim$	$CH_2{=}CHC_6H_5OCH_3$	0	6,025			NKR	[338]
$\sim CH_2\text{-}\dot{C}HCOOCH_3$	$CH_3COOC_2H_5$	26–50	6,55			IIM	[339]
	- " -	30–60	6,63 (30°)	10,45	5,3	IIM	[340]
	- " -	28	8,00			NKR	[336]
$\sim CH_3\text{-}\dot{C}HCOOC_4H_9$	$CH_2{=}CHCOOC_4H_9$	25	6,92			NKR	[341]
$\sim CH_2\text{-}\dot{C}HCOOC_4H_9$	$CH_2{=}CHCOOC_4H_9$	25	8,52			NKR	[342]
	- " -	25	4,255			IIM	[343]
	- " -	28	7,53			NKR	[336]
$\sim CH_2\text{-}\dot{C}(CH_3)COOCH_3$	$CH_2{=}C(CH_3)COOCH_3$	30–60	7,39 (30°)	9,41	2,8	IIM	[284]
	- " -	24–50	7,82 (24°)	8,555	1,0	IIM	[344]
	- " -	28	7,41			NKR	[336]
		Mean		7,69	0,4		[2]
$\sim CH_2\text{-}\dot{C}(CH_3)COOC_3H_7$	$CH_2{=}C(CH_3)COOC_3H_7$	30	7,65			IIM	[346]
$\sim CH_2\text{-}\dot{C}(CH_3)COOC_4H_9$	Monomer	28	8,255			NKR	[336]
$\sim CH_2\text{-}\dot{C}(CH_3)COOC(CH_3)_3$	$CH_2{=}C(CH_3)COOC(CH_3)_3$	20–90	7,15 (30°)		1,1	IIM	[347]

$\sim CH_2\dot{C}(CH_3)CH_2CH(CH_3)_2$	$CH_2{=}C(CH_3)CH_2CH(CH_3)_2$	30	7,08			NKR	[348]
	- " -	28	8,00			NKR	[336]
$\sim CH_2\dot{C}HCl$	Tetrahydrofuran	25–55	9,07 (30°)	12,115	4,2	IIM	[351]
$\sim CH_2\dot{C}HCN$	$HCON(CH_3)_2$	13–40	6,26			NKR	[306]
	- " -	0–60	7,94 (60°)	10,11	3,7	IIM	[352]
	- " -	60	8,89			IIM	[353]
	$CH_2{=}CHCN$	25	7,09			NKR	[354]
$\sim CH_2\dot{C}(CH_3)CN$	$CH_2{=}C(CH_3)CN$	60–80	6,23 (30°)	9,855	5,0	IIM	[356]
$\sim CH_2\dot{C}HCONH_2$	H_2O	25	7,16			IIM	[303]
	- " -	10–40	7,21 (30°)	10,11	4,0	IIM	[357]
$\sim CH_2{-}\dot{C}H{-}$ 	$CH_2{=}CH{-}$ 	25	6,48			NKR	[305]
$\sim CH_2{-}\dot{C}H{-}$ 	$CH_2{=}CH{-}$ 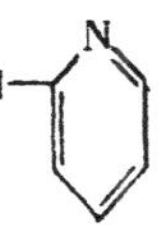	25	7,52			IIM	[358]
	- " -	25	6,95			NKR	[358]
$\sim CH_2CH{-}$  $-CH_3$	$CHON(CH_3)_2$	25	6,545			NKR	[306]
$\sim CF_2{-}\dot{C}F_2$	H_2O	30–70	1,12 (30°)	1,26	0,2	NKR	[359]
$\sim CF_2{-}\dot{C}F_2$	- " -	10–60	1,79 (30°)	4,14	3,25	Inhibitor Method	[360]

Table 91 (continued)

Radical		Solvent	t^o	log(2k)	log(2A)	E	Method	Ref.
HOC_6H_5X	X = H	H_2O	19–80	8,66 (19°)	12,48	3,5	PR	[319b]
	NO_2	- " -	19–80	8,70 (19°)	10,92	6,0	PR	[319b]
	NO_2	- " -	25	8,78			PR	[397]
	CN	H_2O	19–80	8,59 (18°)	11,43	4,7	PR	[319b]
	COC_6H_5	- " -	19–80	8,56 (19°)	11,40	4,7	PR	[319b]
	$COCH_3$	- " -	19–80	8,62 (19°)	11,73	4,3	PR	[319b]
	SO_2NH_2	- " -	19–80	8,57 (19°)	11,05	5,4	PR	[319b]
	COOH	- " -	19–80	8,60 (19°)	11,71	4,3	PR	[319b]
	I	- " -	19–80	8,60 (19°)	11,27	5,0	PR	[319b]
	Cl	- " -	19–80	8,69 (19°)	12,51	3,5	PR	[319b]
	Br	- " -	19–80	8,62 (19°)	12,04	3,9	PR	[319b]
	OH	- " -	19–80	9,04 (19°)	11,04	6,7	PR	[319b]
	OCH_3	- " -	19–80	8,76 (19°)	11,06	5,8	PR	[319b]
	NH_2	- " -	19–80	9,20 (19°)	12,63	3,9	PR	[319b]
	$N(CH_3)_2$	- " -	19–80	8.86 (19°)	11,50	4,6	PR	[319b]
	CH_3	- " -	19–80	8,58 (19°)	13,03	3,0	PR	[319b]
	C_2H_5	- " -	19–80	8,62 (19°)	12,32	3,6	PR	[319b]

*These constants were measured by intensity of chemiluminescence at the moment of feeding oxygen into a hydrocarbon in the presence of an initiator. The values are probably understated, owing to the time lag in saturating the hydrocarbon with oxygen.

TABLE 92. Rate Constants for Self-Reactions of HO˙ and RO˙ Radicals

Radical	Solvent	t°	log (2k)	log(2A)	E	Method	Ref.
HO·	H_2O, pH 7	25	9,9			MCR	[93]
	H_2O, pH 0,4 to 3	25	10,08			PR	[453]
$(CH_3)_3CO^{\cdot}$	CCl_4	10–55	8,32	8,32	0	IIM	[46]
$C_6H_5O\cdot$	H_2O	20	9,03			FM	[454]
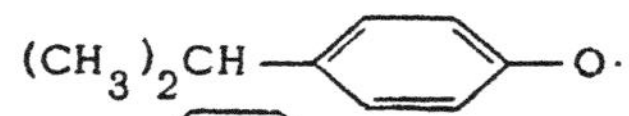	cyclo-C_6H_{12}	25	0,92			Kinetics	[455]
	H_2O, pH 7	25	9,34			PR	[399]
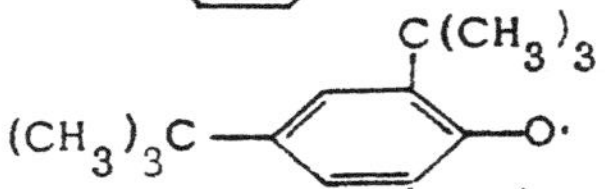	Paraffin	20	7,60			FM	[454]
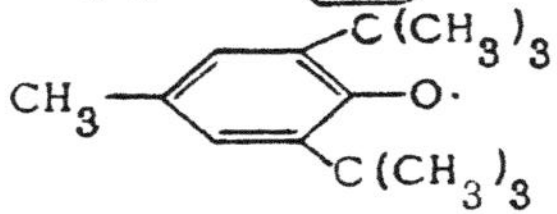	CCl_4	20	8,48			FM	[454]
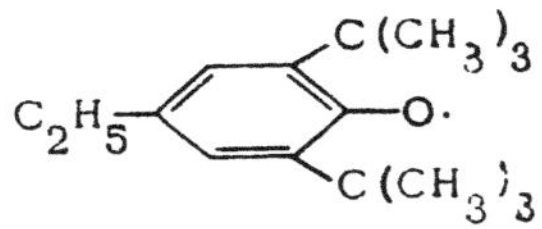	CCl_4	20	7,42			FM	[454]
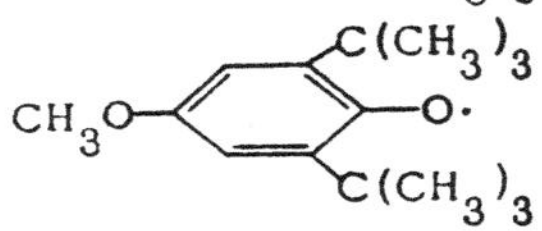	$CHCl_3$	–30	5,70			NMR	[456]

Table 92 (continued)

Radical	Solvent	t°	log (2k)	log(2A)	E	Method	Ref.
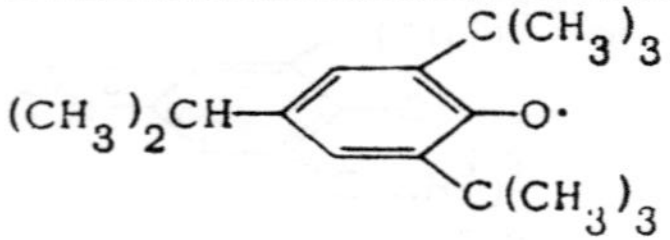	C_6H_6	10–50	0,41 (10°)	5,24	6,23	Kinetics	[457]
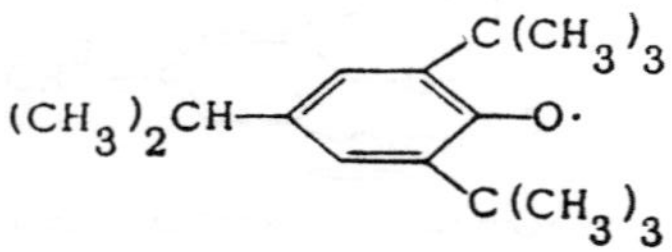	cyclo-C_6H_{12}	10–50	0,75 (10°)	4,91	5,50	Kinetics	[457]
	p-$CH_3OC_6H_4OH$	10–50	0,55 (10°)	6,13	7,20	- " -	[457]
	C_6H_5CN	10–50	0,70 (10°)	7,02	8,16	- " -	[457]
	C_6H_5Cl	10–50	0,50 (10°)	5,88	6,94	- " -	[457]
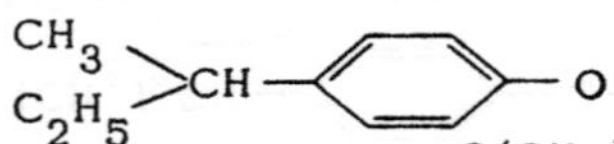	C_6H_6	10–50	$\bar{1}$,78 (10°)	5,22	7,04	Kinetics	[458]
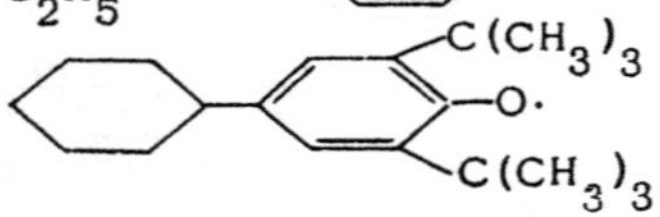	C_6H_6	10–50	0,10 (10°)	6,23	7,93	- " -	[458]
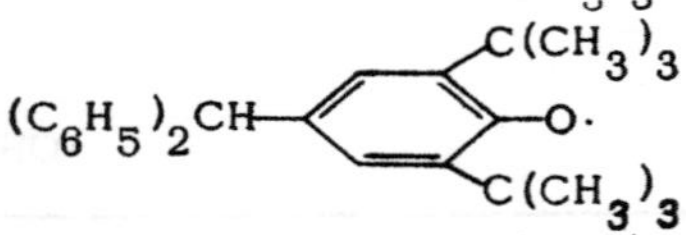	cyclo-C_6H_{12}	25	0,62			- " -	[455]
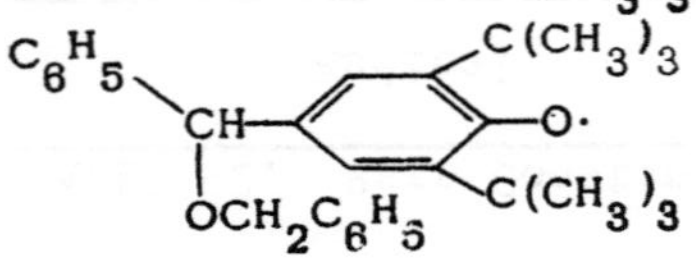	- " -	25	0,145			- " -	[455]

Table 92 (continued)

Radical	Solvent	t°	log(2k)	log(2A)	E	Method	Ref.
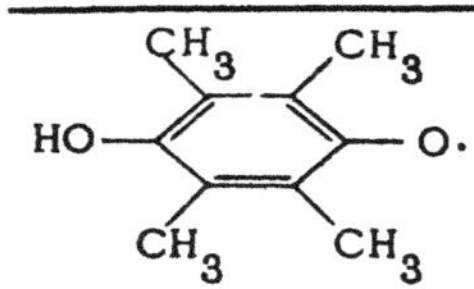	Paraffin	25	9,00			FM	[459]
	$H_2O + C_2H_5OH$	25	9,20			FM	[459]

TABLE 93. Rate Constants for Self-Reactions of Peroxy Radicals (RH is compound for which $RO_2^{\cdot}$ radical is listed)

Radical	Solvent	t°	log (2k)	log (2A)	E	Method	Ref.
$HO_2^{\cdot}$	H_2O, pH 0,5 to 1,55	1–43	6,82 (30°)	11,08	5,9	JM	[472]
	H_2O, pH 2,7	25	6,40			FM	[473]
	H_2O, pH 2	23	6,64			PR	[460]
	H_2O, pH 2	23	6,74			PR	[474]
	H_2O	-10 to +50	6,40 (30°)	9,8	4,7	JM	[475]
	H_2O, pH 2	23	6,30	9,8	4,7	PR	[476]
	CCl_4	30	10,05			IIM	
	C_6H_5Cl	30	9,10			IIM	
	- " -	30	8,85			IIM	[155]
	CH_3CN	30	6,935			IIM	
	n-$C_{10}H_{22}$	30	9,13			IIM	

Table 93 (continued)

Radical	Solvent	t°	log (2k)	log (2A)	E	Method	Ref.
$HO_2\cdot$	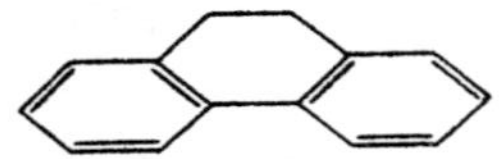	30	7,78			IIM	[139]
	Cyclohexadiene-1,3	30	7,82			IIM	[155]
$CH_3(CH_2)_3OO\cdot$	CH_3OH	7	7,48			JM	[475]
	Tetralin	30	7,60			IIM	[478]
$(CH_3)_2C{=}C(CH_3)CH_2OO\cdot$	RH	30	5,805			IIM	[478]
$C_6H_5CH_2OO\cdot$	RH	30	8,48			IIM	[139]
m-$CH_3C_6H_4CH_2OO\cdot$	RH	30	8,48			IIM	[139]
o-$CH_3C_6H_4CH_2OO\cdot$	RH	30	8,18			IIM	[139]
p-$CH_3C_6H_4CH_2OO\cdot$	RH	30	8,48			IIM	[139]
$CH_3CH(OO\cdot)CH_2CH_3$		30	6,18			IIM	[478]
$(CH_3)_2C(OO\cdot)CH(CH_3)CH(CH_3)_2$	RH	42-78	5,545			IIM	[130]
cyclo-$C_6H_{11}OO\cdot$	Tetralin	30	6,30			IIM	[478]
	- " -	40-70	6,68 (30°)	7,84	1,6	NKR	[131]
	$CH_3COO_2H_5$	40-70	5,37 (30°)	9,64	5,9	NKR	[131]
	CH_3COOH	40-70	5,31 (30°)	8,72	6,1	NKR	[131]

cyclo-$C_6H_{11}OO\cdot$	RH	23	6,505			PR	[479]
$(CH_3)_2C{=}CHCH(OO\cdot)C_2H_5$	RH	40–60	6,48 (30°)	7,415	1,2	NKR	[131]
$CH_2{=}CHCH(OO\cdot)(CH_2)_4CH_3$	RH	25	7,08			PR	[437]
	RH	30	8,41			IIM	[478]
	RH	25–40	5,3			NKR	[132]
$CH_2{=}CHCH(OO\cdot)CH{=}CH_2$	RH	30	9,03			IIM	[480]
$C_2H_5CH{=}CHCH(OO\cdot)C_2H_5$	RH	30	6,81			IIM	[480]
$CH_3(CH_2)_4CH{=}CHCH(OO\cdot)$–$CH{=}CH(CH_2)_7COOC_2H_5$ (printed: $CH_3(CH_2)_4CH{=}CHCH$ / $C_2H_5OOC(CH_2)_7CH{=}CH$)	RH	11	5,70			IIM	[132]
	RH	25	5,48			NKR	[132]
$CH_3(CH_2)_7CH(OO\cdot)CH{=}CH$–$CH_2(CH_2)_6COOCH_3$ (printed: $CH_3(CH_2)_7CHCH{=}CH$ / $CH_3OOC(CH_2)_6CH_2$)	RH	30	6,C2			IIM	[480]

Table 93 (continued)

Radical	Solvent	t°	log (2k)	log (2A)	E	Method	Ref.
OO· | $(CH_3)_2C{=}CHCHCH_2$ $CH_3CH{=}C(CH_3)$	RH	0	5,59			NKR	[132]
		15	5,72			IIM	[132]
		25	5,72			NKR	[132]
$CH_3(CH_2)_4CH{=}CHCHOO\cdot$ $CH_2(CH_2)_6CH{=}CH$ $COOCH_3$	RH	30	6,945			IIM	[480]
$CH_3CH_2CH{=}CHCHOO\cdot$ $CH{=}CHCH_2CH{=}CH$ $CH_2(CH_2)_6COOCH_3$	RH	30	7,555			IIM	[480]
CH_3 CH_3 $CH_2{=}C(CH_2)_3C{=}CH$ $CH_2C{=}CHCH_2CHOO\cdot$ $CH_2CH_2C(CH_3){=}CH_2$	RH	25	5,45			NKR	[132]
		40	5,00			NKR	[132]
$C_6H_5CH(OO\cdot)CH_2\sim$	C_6H_5Cl	40	7,86			IIM	[172]
$C_6H_5CH(OO\cdot)CH_2\sim$	C_6H_5Cl	13–50	7,71 (30°)	9,01	1,8	IIM	[170]
$C_6H_5CD(OO\cdot)CH_2\sim$	C_6H_5Cl	13–50	7,12 (30°)	9,80	3,7	IIM	[172]

p-$CH_3OC_6H_4CH(OO\cdot)CH_2$~	- " -	40	7,64	IIM	[172]
p-$CH_3C_6H_4CH(OO\cdot)CH_2$~	- " -	40	7,75	IIM	
p-$ClC_6H_4CH(OO\cdot)CH_2$~	- " -	40	7,99	IIM	
m-$ClC_6H_4CH(OO\cdot)CH_2$~	- " -	40	8,03	IIM	
p-$CNC_6H_4CH(OO\cdot)CH_2$~	- " -	40	9,45	IIM	
$C_6H_5CH(OO\cdot)CH(CH_3)$~	- " -	30	7,505	IIM	[147]
$C_6H_5CH(OO\cdot)CH{=}CH_2$	- " -	30	8,645	IIM	[138]
$C_6H_5CH(OO\cdot)CH{=}CHCH_3$	RH	30	6,58	IIM	[480]
$(C_6H_5)_2CHOO\cdot$	RH	30	8,20	IIM	[138]
$(C_6H_5)_2CD(OO\cdot)$	RH	30	8,06	IIM	[138]
$C_6H_5CH(OO\cdot)CH_2C_6H_5$	RH	30	7,30	IIM	[139]
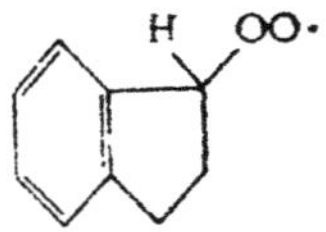	RH	30	6,91	IIM	[139]
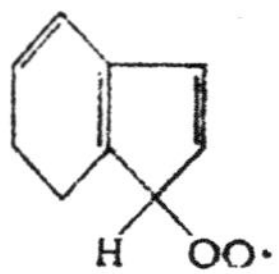	C_6H_5Cl	30	7,70	IIM	[147]

Table 93 (continued)

Radical	Solvent	t°	log(2k)	log(2A)	E	Method	Ref.
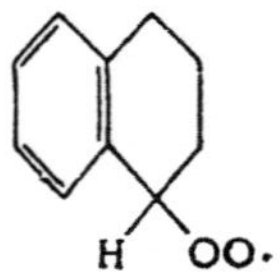	RH	25	7,33			NKR	[145]
	RH	40	6,98			NKR	[136]
	RH	13–50	6,82 (30°)	9,93	4,3	IIM	[138]
	CH_3OH	-10 to + +50	6,67 (30°)	10,0	4,6	JM	[475]
$CH_3(CH_2)_3C{=}C(CH_2CH_2CH_3)CH(OO\cdot)$	RH	30	7,145			IIM	[480]
	CH_3OH	30	7,175			IIM	[475]
	RH	30	6,79			IIM	[480]
	RH	30	9,03			IIM	[480]
	RH	15	5,98			IIM	[132]
	RH	20–50	6,20			NKR	[134]
		56	6,30			IIM	[136]
	RH	15	5,855			IIM	[137]
	RH	30	6,75			IIM	[138]
	RH	30	6,42			NKR	[16]
		Mean	6,25				

	CH_3OH	30	6,45	10,8	6,0	JM	[475]
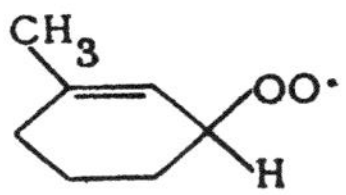	RH	15	5,70			IIM	[132]
	RH	40	5.935			NKR	[134]
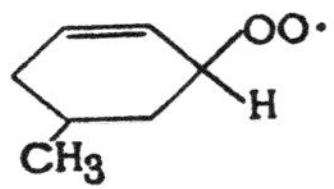	RH	40	6,52			NKR	[135]
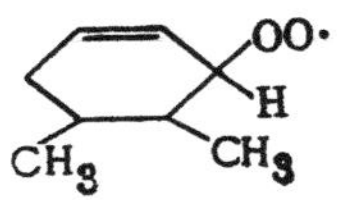	RH	40	6,43			NKR	[135]
$C_6H_5CH(CH_3)OO\cdot$	RH	50–90	7,28			IIM	[140]
	RH	30	7,60			IIM	[138]
	RH	30	7,60			NKR	[445]
$C_6H_5CH(OO\cdot)CH_2CH_2CH_3$	RH	30	7,70			IIM	[139]

Table 93 (continued)

Radical	Solvent	t°	log (2k)	log(2A)	E	Method	Ref.
$(CH_3)_3C\overset{OO\cdot}{C}HC(CH_3)_3$	CH_3OH	-10÷+ +50	3,92 (30°)*	11,0	9,8	JM	[475]
	C_6H_6	-10÷+ +50	4,18 (30°)*	9,2	7,0	JM	[475]
$CH_3CO\overset{OO\cdot}{C}HCH_3$	RH	40-75	6,15 (30°)	7,30	1,6	NKR	[150]
$CH_3CO\overset{OO\cdot}{C}HC_2H_5$	RH	40-90	6,26 (30°)	7,48	1,7	NKR	[151]
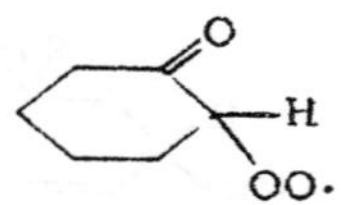	RH	75	6,43			NKR	[149]
$(CH_3)_3COO\cdot$	CH_3OH	-10 to + +50	3,54 (30°)*	10,4	9,2	JM	[475]
	CH_3OH	36-65	2,59 (22°)	10,56	10,2	JM	[466]
	C_6H_6	-10 to + +50	3,11 (30°)*	8,4	4,5	JM	[475]
	$C_6H_5CH(CH_3)_2$	25-65	3,15 (35°)	9,8	7,0	NKR	[142]

	$C_6H_5CH(CH_3)_2$	30	3,11			IIM	[143]
	n-C_6H_{14}	-41	4,115	9,145	7,2	EPR	[481]
	cyclo-C_5H_{10}	-48 to -24	4,06 (41°)	9,99	8,5	EPR	[484]
	cyclo-$C_6H_{11}CH_3$	-48 to -24	4,19 (41°)	10,12	8,5	EPR	[481]
	$C_2H_5CH(CH_3)C_2H_5$	-48 to -24	4,26 (41°)	10,12	8,4	EPR	[481]
	$(CH_3)_3COOH$	-38–0	4,00 (41°)	9,79	8,3	EPR	[484]
$(CH_3)_3CCH_2C(CH_3)_2OO\cdot$	$C_6H_5CH(CH_3)_2$ + + ROOH	30	4,01*			IIM	[143]
$(CH_3)_2CHCH(CH_3)CH(CH_3)_2$	RH	60	5,54			IIM	[130]
$C_6H_5(CH_3)_2COO\cdot$	RH	59	5,255			NKR	[136]
	- " -	80	4,45			IIM	[141]
	- " -	65	4,52			IIM	[141]
	- " -	23	4,64			EPR	[482]
	- " -	40	4,38			NKR	[16]
	- " -	60	4,68			NKR	[445]
	- " -	30	4,175			IIM	[139]
	CH_3OH	-10 to +50	5,09 (30°)*	10,7	7,8	JM	[475]

Table 93 (continued)

Radical	Solvent	t°	log (2k)	log (2A)	E	Method	Ref.
$C_6H_5(CH_3)_2COO\cdot$	C_6H_6	-10 to + +50	4,57 (30°)*	8,8	5,8	JM	[475]
	RH + ROOH	25–65	4,27 (30°)*	8,58	5,96	IIM	[142]
$C_6H_5C(OO\cdot)(CD_3)_2$	RH + ROOH	30	4,08*			IIM	
p-$CH_3OC_6H_4C(OO\cdot)(CH_3)_2$	- " -	30	4,60*			IIM	
m-$CH_3OC_6H_4C(OO\cdot)(CH_3)_2$	- " -	30	4,90*			IIM	[133]
p-$CH_3OCOC_6H_4C(CH_3)_2OO\cdot$	- " -	30	4,48*			IIM	
$(C_6H_5)_2(CH_3)COO\cdot$	RH	30	4,97			IIM	[139]
	RH + ROOH	30	4,81*			IIM	[143]
	RH	50–88	5,98 (30°)	11,04	7,0	EPR	[144]
	C_6H_6	-10 to + +50	6,34 (30°)	8,1	2,5	JM	[475]
$C_6H_5(CH_3)(C_2H_5)COO\cdot$	CH_3OH	-10 to + +50	5,46 (30°)*	9,4	5,5	JM	[475]
	RH + ROOH*	30–56	4,43 (30°)*	8,48	5,6	IIM	[143]
$C_6H_5C(CH_3)(CH_2CH_2CH_3)OO\cdot$	RH	30	5,64			IIM	[139]

	RH	30	5,20			IIM	[139]
$C_6H_5C(OO\cdot)(CH_3)CH_2$~	C_6H_5Cl	13–50	5,82 (30°)	8,50	3,7	IIM	[147]
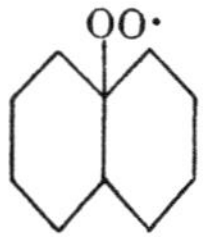	RH	60	5,18			NKR	[415]
$(CH_3)_2C(OO\cdot)CN$	C_6H_5Cl	80	7,20			NKR	[483]
$CH_3CO(CH_3)_2COO\cdot$	RH	40–80	6,32 (30°)	7,48	1,6	NKR	[151]
n-$C_9H_{19}COOO\cdot$	n-$C_{10}H_{22}$	5	7,53			IIM	[172]
	RH	5	6,875			IIM	[172]
$C_6H_5COOO\cdot$	RH	5	8,32			IIM	[153]
$CH_3CH(OH)OO\cdot$	RH	23	8,08			PR	[484]
	RH	23	7,255			PR	[437]
	RH	50–75	5,37 (30°)	7,10	2,4	NKR	[148]

*In these studies, k was measured in the presence of hydroperoxide ROOH, which ensured the predominance of a single sort of peroxy radical, namely ROO•. Other radicals appearing in the system would be rapidly converted to ROO• by the reaction $R'\cdot + HOOR \rightarrow R'H + ROO\cdot$.

TABLE 94. Recombinations of Radicals Having Free Valence on Nitrogen or Tin (at 25°)

Radical	Solvent	log(2k)	Method	Ref.
$C_6H_5\dot{N}H$	H_2O	9,18	FM	[485]
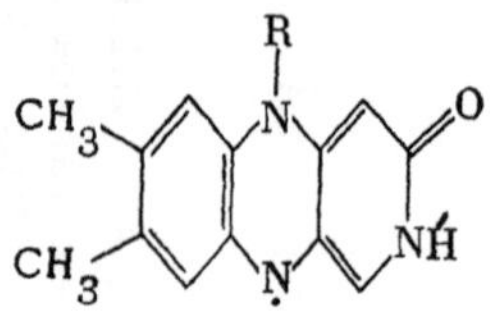	H_2O	8,48	FM	[486]
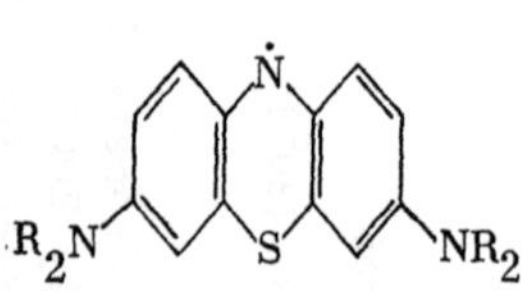	H_2O, acid	9,38	FM	[487]
Semiquinone of methylene blue	- " -	9,18	FM	[488]
$C_6H_5\dot{N}O_2H$	H_2O	8,78	PR	[488a]
$(C_6H_5)_3Sn\cdot$	- " - C_6H_{12}	9,45	IIM	
$(CH_3)_3Sn\cdot$	- " -	9,49	IIM	
$(n\text{-}C_4H_9)_3Sn\cdot$	- " -	9,15	IIM	[233]
$(n\text{-}C_4H_9)_2ClSn\cdot$	- " -	9,56	IIM	
$(n\text{-}C_4H_9)_2HSn\cdot$	- " -	9,20	IIM	

TABLE 95. Rate Constants for Reactions of Different Types of Radicals with Each Other

Reaction	Solvent	t°	log k	Method	Ref.
$H\cdot + HO\cdot$	H_2O	25	10,08	PR	[53]
	H_2O, pH 0,4 to 3	25	10,505	PR	[54]
	H_2O, 0,8N H_2SO_4	25	9,85	PR	[489]
	H_2O, 0,4N H_2SO_4	25	10,65	PR	[315]
$H\cdot + HO_2\cdot$	H_2O, acid	25	10,30	PR	[490]
$HO\cdot + HO_2\cdot$	H_2O, pH 0,4 to 3	25	10,175	PR	[55]
	H_2O	25	10,04	PR	[490]
	H_2O, pH 0,5 to 6,7	25	9,85	PR	[491]
$C_6H_5CH_2\cdot + CH_3CHOH$	C_2H_5OH + + $CH_2(OH)CH(CH_2OH)OH$	25	8,30	PR	[444]
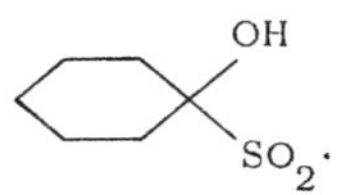 + $HSO_2\cdot$	cyclo-$C_6H_{11}OH$	25	8,45	PR	[437]

Table 95 (continued)

Reaction	Solvent	t°	log k	Method	Ref.
$(C_6H_5)_3C\cdot + (C_6H_5)_3COO\cdot$		30	8,175	IIM	[143]
$R\cdot + RO_2\cdot$ in ethyl linoleate	RH	25	7,70	NKR	[132]
$R\cdot + RO_2\cdot$ in digeranyl	RH	25	7,00	NKR	[132]
a-$C_{10}H_{11}OO\cdot + C_6H_5C(OO\cdot)(CH_3)_2$	Tetralin + $C_6H_5CH(CH_3)_2$	90	6,78	MCR	[492]
$C_6H_5COOO\cdot$ + p-$C_9H_{19}COOO\cdot$	C_6H_5CHO + p-$C_9H_{19}CHO$	5	7,93	MCR	[493]
$C_6H_5C(OO\cdot)(CH_3)_2 + C_6H_5CH(OO\cdot)CH_3$	$C_6H_5CH(CH_3)_2 + C_6H_5C_2H_5$	60	5,87	MCR	[158]
	- " -	80	5.945	MCR	[158]
p-$CH_3OC_6H_4O\cdot$ + 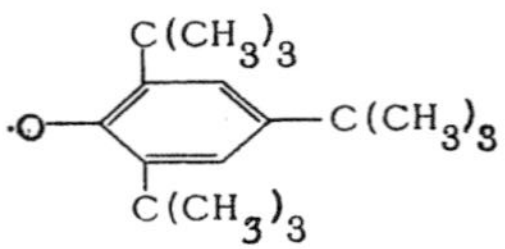	C_6H_5Cl	24	7,725	TsINR	[186]
p-$CH_3OC_6H_4O\cdot$ + 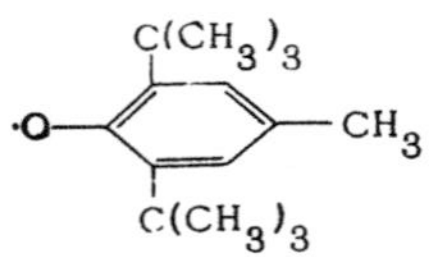	C_6H_5Cl	24	8,00	TsINR	[186]
$CH_3\cdot$ + 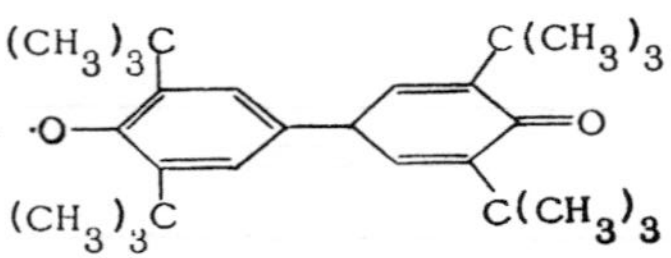	$(CH_3)_3CCH_2CH(CH_3)_2$	25	7,30	MCR	[494]

CHAPTER IX

EFFECT OF SOLVENT ON FREE RADICAL REACTIONS

§1. Solvent Viscosity

Solvent viscosity has a substantial influence on the elementary reactions for which the rates are diffusion-limited. Such reactions include recombination of atoms and reactions between free radicals that have rate constants on the order of 10^9 or 10^{10} liter/mole•sec. A decrease in the diffusion coefficient of iodine atoms with increasing viscosity was demonstrated in [422]. Often used to calculate k is the Stokes-Einstein formula $D = kT/3\pi\eta\sigma$, which is substituted into the formula $k = 4 \cdot 10^{-3} \pi\sigma\, D\, N$ to give $k = 8\, RT/3000\, \eta$. For atoms, this formula is, strictly speaking, inapplicable, as is evident from Table 96: The product $k\eta/T$ changes from solvent to solvent instead of remaining constant.

Macroradicals diffuse far more slowly than do ordinary radicals; hence, even in relatively nonviscous media, diffusion may prove to be the limiting process in the reaction between two macroradicals. Since the solution viscosity increases as polymerization progresses, a timewise decrease will be observed in the rate constant for chain breaking [495]. As an example, we may cite data on the photopolymerization of methyl methacrylate at 32° [342]:

% polymer	0	15	25	35
k, liter/mole•sec · 10^{-4}	570	140	35	8.8

§2. Internal Pressure of Liquid

Since the combination of two radicals into one molecule is accompanied by a decrease in free volume, one should expect an increase in the rate constant for a reaction between two free radicals when the internal pressure of the liquid increases. Unfortunately, no systematic measurements in this direction have been performed. A very evident effect of internal pressure of the liquid upon a reaction between two radicals was established in the example of competing disproportionation and recombination of ethyl radicals [432], where a linear relation was shown between $\log(k_d/k_r)$ and $P_S^{1/2}$, where P_S is the internal pressure of the solvent:

$$\log \frac{k_d}{k_r} = -1.08 + 4.76 \cdot 10^{-3} P_S^{1/2} \quad (65°).$$

This is explained on the basis that the volume of the activated complex is somewhat less in the case of disproportionation than in the case of recombination. For ethyl radicals;

$$V_r^{\neq} - V_d^{\neq} = 2.4 \text{ cm}^3/\text{mole at } 0° \; [429].$$

§3. Nonspecific Solvation

A polar particle (molecule or radical) in a polar solvent is solvated, i.e., it is surrounded by a "jacket" of solvent molecules. Nonspecific solvation is caused by electrostatic interaction between molecules. The rate constant of a reaction depends on the solvation of the original particles and the activated complex; the degree of particle solvation is governed in particular by orientative interaction and depends on the dipole moment of the particle and the polarity of the solvent. If the activated complex is more polar than the original particle, the reaction will be faster in a polar solvent. The influence of polarity of the medium on the rate constant of a reaction A + B → activated complex can be accounted for quantitatively by means of the Kirkwood formula [496]:

$$\ln k = \ln k_0 - \frac{1}{kT} \frac{\varepsilon - 1}{2\varepsilon + 1} \left(\frac{\mu_A^2}{r_A^3} + \frac{\mu_B^2}{r_B^3} - \frac{\mu^2}{r_{\neq}^3} \right).$$

Such a treatment of the effect of solvent on the rate of radical reactions was effected for the first time for the reactions $RO_2^{\bullet}$ + RH and $RO_2^{\bullet}$ + $RO_2^{\bullet}$ in the work of Zaikov and Maizus [497-500].

A linear relation between log $k(RO_2^{\bullet} + RH)/k^{1/2}(RO_2^{\bullet} + RO_2^{\bullet})$ and $(\varepsilon-1)/(2\varepsilon + 1)$ was observed in the oxidation of methyl ethyl ketone [500], styrene [501], cyclohexane [502], cumene [502], and tetralin [503]. In [504], in the oxidation of methyl ethyl ketone, $k(RO_2^{\bullet} + RH)$ and $k(RO_2^{\bullet} + RO_2^{\bullet})$ were measured by chemiluminescence in mixtures of methyl ethyl ketone and benzene, and the following relations were obtained (at 60°):

$$\log k(RO_2\cdot + RH) = -1.025 + 1.3\,\frac{\varepsilon - 1}{2\varepsilon + 1}\,.$$

$$\log k(RO_2\cdot + RO_2\cdot) = 3.36 + 6.0\,\frac{\varepsilon - 1}{2\varepsilon + 1}\,.$$

Along with solvent polarity, the polarizability of the solvent molecules plays a substantial role. Readily polarized molecules will solvate polar particles to a greater degree because of the induced dipole moment. Hence, in the oxidation of styrene, for easily polarized aromatic solvents the ratio

$$\log \frac{k(RO_2\cdot + RH)}{k^{1/2}(RO_2\cdot + RO_2\cdot)} : \frac{\varepsilon - 1}{2\varepsilon + 1}$$

is substantially larger than for nonaromatic solvents [501].

For the disproportionation reaction of 2,6-di-tert-butyl-4-isopropylphenoxy radicals [487], the activation energy of the reaction is observed to increase with increasing polarity of the medium. This comes about because desolvation of one of the phenoxyl radicals is essential to the formation of an activated complex, since the activated complex has the following form:

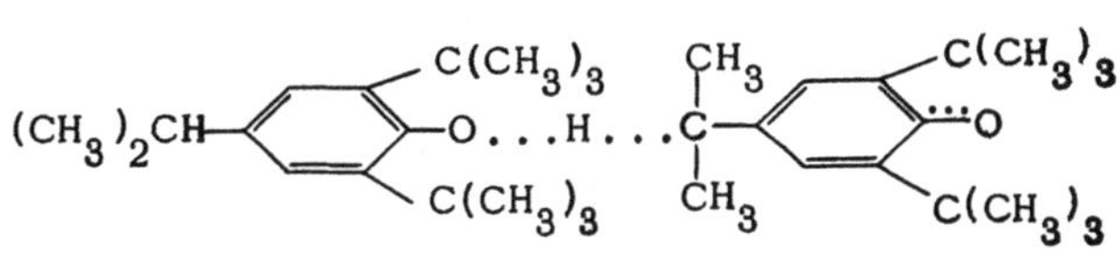

The formation of an activated complex amy be regarded as the joining of two dipoles "head to tail." The energy of such a compound is expressed as function of the dielectric constant by the formula [457]:

$$E = -\frac{28.9\,\mu^2\,(\varepsilon - 1)}{\varepsilon r^2}\,.$$

In the experiment, a linear relation is observed between the activation energy of the reaction of phenoxyl radical disproportionation and the parameter $(\varepsilon - 1)/\varepsilon$.

In a quantitative interpretation of experimental data on the effect of solvent on k, it must be kept in view that the formulas cited here are approximate. They account for only electrostatic interaction between the medium and the reacting particles, but do not account for van der Waal's forces or any formation of a complex with participation of the solvent. The molecule (or radical) is regarded as a rigid dipole, and the solvent is regarded as an isotropic medium with a uniform dielectric constant, whereas it really consists of molecules that possess a dipole moment and are capable of being polarized.

§4. Hydrogen Bond between Molecules

The association of molecules through hydrogen-bond formation will change the physical properties of a liquid very markedly. The formation of a hydrogen bond of the type X-H...B is reflected also in the reactivity of molecules, or more precisely, the reactivity of the hydrogen-bonded H atom in the molecule. A radical substitution reaction at an X-H bond is strongly retarded by hydrogen bonding. As an example, we may cite the reaction [207]:

$$CH_3^{\cdot} + (CH_3)_3CO-T \xrightarrow{k_{O-T}} CH_3T + (CH_3)_3CO^{\cdot}$$

The relative rate constant of this reaction k_{O-T}/k_{C-H} (where k_{C-H} is the rate constant for $CH_3^{\cdot}$ with a secondary C-H bond in heptane) increases as the alcohol is diluted with heptane, increasing from $1.02 \cdot 10^{-2}$ with [ROH] = $2.275 \cdot 10^4$ mole/liter up to $14.3 \cdot 10^{-2}$ with [ROH] = $8.6 \cdot 10^{-4}$ mole/liter (at 90°). Calculations show that the O-T ... O bond is essentially unreactive with methyl radicals, and only the free O-T bonds react. From the kinetic data, the association constant of the molecules can be computed. If the association proceeds according to the scheme:

$$ROH + (ROH)_n \rightleftharpoons (ROH)_{n+1}\,,$$

then the association constant is given by $K = (1-\sqrt{\alpha})/\alpha(ROH)$, where α is the fraction of the molecules that are not hydrogen bonded. From the kinetic data, α is found as the ratio $\alpha = k/k_\infty$, where k_∞ is the value of k at infinite dilution. In Table 97, reactions are listed for which an effect of hydrogen bonding has been observed, along with the values of k calculated from the kinetic data.

§5. Radical Hydrogen Bond

In the formation of a hydrogen bond X-H ... B, the polar X-H bond takes a part. It may be expected that radicals with a free valence on a heteroatom will also form hydrogen bonds and that this will be reflected in their activities [506]. A hypothesis regarding an intraradical hydrogen bond of the type

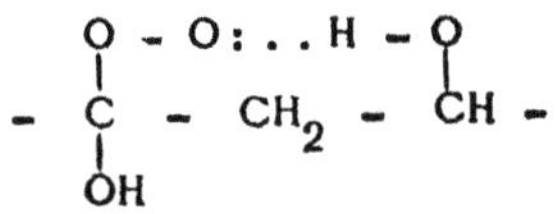

was advanced in [507] on the basis of results from the autoxidation of 2,4-dimethylpentane and pentanediol-2,4. Evidence of hydrogen bonding between a nitroxy radical and solvents with O-H groups was obtained by an EPR method in the work of Buchachenko [508], where a change was observed in a_N (splitting between components of the nitrogen triplet in the EPR spectrum) when the solvent was changed from CCl_4 to H_2O. On the basis of the relation of a_N to H_2O concentration in acetone mixtures, the equilibrium constant was measured for the reaction

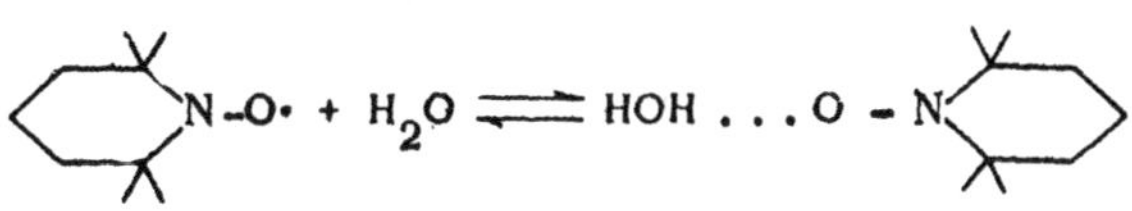

which proved to be $\log K = \bar{3}.08 + 2.4/\theta$.

The role of the radical hydrogen bond in reactions of peroxy radicals was studied in detail in the work of Zaikov, Maizus, and Émanuél',[131, 509-512] in the examples of the oxidation of methyl ethyl ketone, cyclohexane, and 2-methylpentene-2. In the case of methyl ethyl ketone, the formation of a hydrogen bond $RO_2^{\bullet}$ ... HOH gives a tenfold reduction in the rate constant of the $RO_2^{\bullet}$ + RH and $RO_2^{\bullet}$ + $RO_2^{\bullet}$ reactions. For cyclohexylperoxy radicals, the difference in reactivity between free and hydrogen-bonded radicals is not great. The results are anomalous for 2-methylpentene-2; the hydrogen-bonded radicals proved to be more active than the free radicals. Apparently this is explained by the existence of an intramolecular π-bond in the peroxy radical of methylpentene:

Alcohol, forming a hydrogen bond with the peroxy radical, destroys the π-complex, and the hydrogen-bonded peroxy radicals are more active than the peroxy radicals with intraradical π-bond.

§6. Formation of π-Complexes

Kinetic evidence for the formation of a complex between a chlorine atom and benzene was obtained by Russell [513] in studying the photochlorination of 2,3-dimethylbutane. It was established that the ratio of rate constants for the reactions of chlorine with tertiary and primary C-H bonds of 2,3-dimethylbutane (k_t/k_p) is increased substantially when the reaction is conducted in an aromatic solvent. For example, $k_t/k_p = 3.5$ in CCl_4 and 14.6 in C_6H_6. The higher values of k_t/k_p in the presence of aromatic molecules are explained by the fact that in a complex with an aromatic molecule the chlorine atom is less active and reacts more selectively than does a free chlorine atom. The complex of chlorine with benzene may be either a π-complex or a σ-complex. Assumption of a σ-complex is contradictory to the experimental data, since the reaction products do not contain any chlorobenzene or dichlorocyclohexadiene, which should have been formed from the σ-complex by the reactions

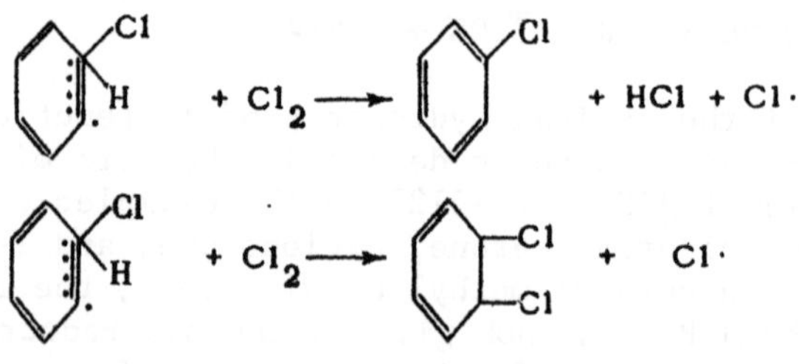

Hence, the complex of $Cl^{\bullet}$ with C_6H_6 is apparently a π-complex. The dependence of k_t/k_p on $[C_6H_6]$ has the following form:

$$k_t/k_p = (k_t/k_p)_0 + a[C_6H_6] + b[C_6H_6]^2.$$

which is explained by the formation of complexes with one and two molecules of benzene:

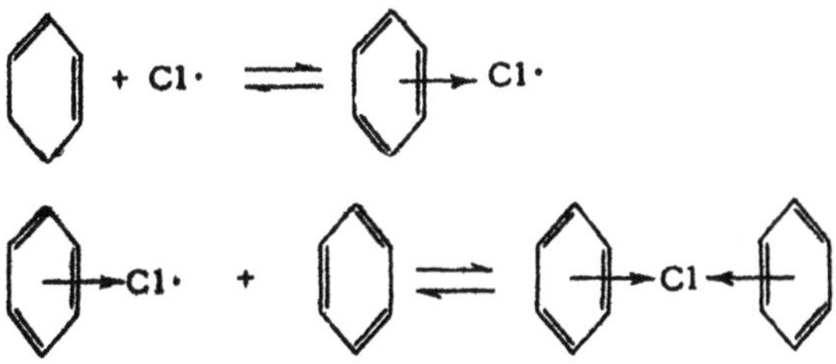

Increases of k_t/k_p in the presence of alcohols, ethers, and sulfur compounds can be explained on the basis of complex formation:

$$Cl\cdot + OR_2 \rightleftharpoons Cl\cdot \rightarrow OR_2$$

$$Cl\cdot + S{=}C{=}S \rightleftharpoons Cl \longrightarrow S{=}C{=}S$$

The high selectivity of Cl• in the presence of iodobenzene may be explained by the equilibrium

$$Cl\cdot + IC_6H_5 \rightleftharpoons Cl - I - C_6H_5$$

An analogous phenomenon is observed in the chlorination of ethylbenzene and cumene [514]; dilution of the alkylaromatic hydrocarbon with solvent decreases the ratio k_α/k_β as a consequence of increasing the concentration of free chlorine atoms, which react less selectively than does the π-complex of chlorine.

$$Cl\cdot + C_6H_5CHRCH_3 \xrightarrow{k_\alpha} HCl + C_6H_5CRCH_3$$

$$Cl\cdot + C_6H_5CHRCH_3 \xrightarrow{k_\beta} HCl + C_6H_5CHRCH_2\cdot$$

The formation of a π-complex of an iodine atom with hexamethylbenzene in carbon tetrachloride was demonstrated spectroscopically and kinetically [515]; the equilibrium constant at room temperature was found to be 2.7 liter/mole.

In [516], a detailed investigation was made of the kinetic manifestation of π-complexes of nitroxy radicals in reaction with 2,6-di-tert-butylphenol. It was established that the rate constants of the reaction

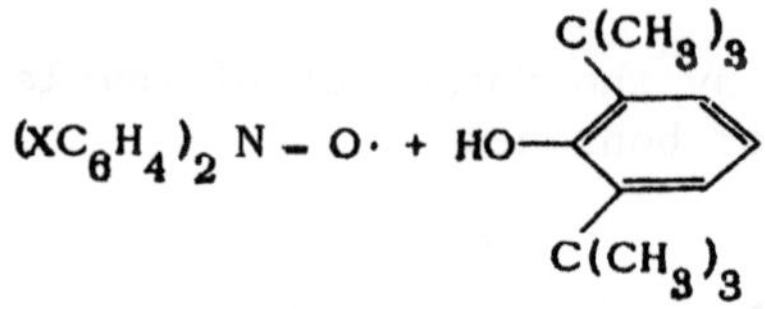

in benzene and ethylbenzene were substantially lower than in n-heptane. From the kinetic data, the equilibrium constant was determined for the formation of a π-complex between $(p\text{-}CH_3OC_6H_4)_2N\text{-}O^{\bullet}$ and C_6H_6: $\log K = \bar{5}.39 + 6.4/\theta$.

TABLE 96. Diffusion Coefficients of Atomic Iodine at 25° [422]

Solvent	$10^5 D$, cm^2/sec	$k\eta/T10^{-5}$
$n\text{-}C_6H_{14}$	4,0	1,29
CCl_4	2,5	2,22
C_4Cl_6	1,9	6,15

TABLE 97. Manifestation of Molecular Hydrogen Bonds in Reactions of Radical Substitution

Reaction	Solvent	t°	Measured Value	log K (liter/mole)	Ref.
$CH_3\cdot + T\text{–}OC(CH_3)_3$	$(CH_3)_3COH + n\text{-}C_7H_{16}$	55–90	k_{O-T}/k_{C-H}	$\bar{4},10 + 1,7/\theta$	[207]
$CH_3\cdot + T\text{–}OCH_3$	$CH_3OH + n\text{-}C_7H_{16}$	80	k_{O-T}/k_{C-H}	0,11	[206]
$CH_3\cdot + T\text{–}OCH(CH_3)_2$	$(CH_3)_2CHOH + n\text{-}C_7H_{16}$	80	k_{O-T}/k_{C-H}	$\bar{1},78$	
$CH_3\cdot + T\text{–}OOC(CH_3)_3$	$(CH_3)_3COOH + n\text{-}C_7H_{16}$	60–90	k_{O-T}/k_{C-H}	$\bar{4},54 + 8,14/\theta$	[208]
$CH_3\cdot + H\text{–}OC_6H_5$	$C_6H_5OH + n\text{-}C_7H_{16}$	60–90	k_{O-H}/k_{C-H}	$\bar{2},23 + 4,4/\theta$	[211]
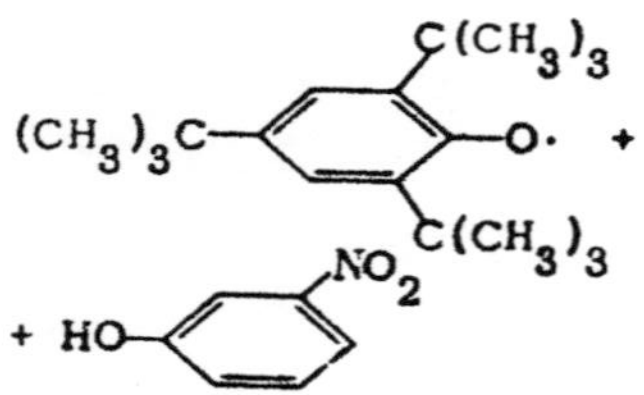	$CH_3COCH_3 + n\text{-}C_7H_{16}$	22	k/K	$\log(k/K) = \bar{3},23\ (sec^{-1})$	[185]
	$CH_3OH + C_7H_{16}$	22	k/K	$\log(k/K) = \bar{4},725\ (sec^{-1})$	[185]

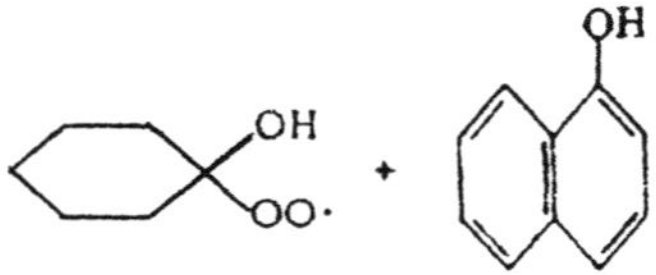	$C_6H_{11}OH + C_6H_{12}$	75	$\frac{k(RO_2\cdot + InH)}{k^{1/2}(RO_2\cdot + RO_2\cdot)}$	1,115	[175]
	$C_6H_{11}OH + C_6H_5Cl$	75	- " -	1,34	[175]
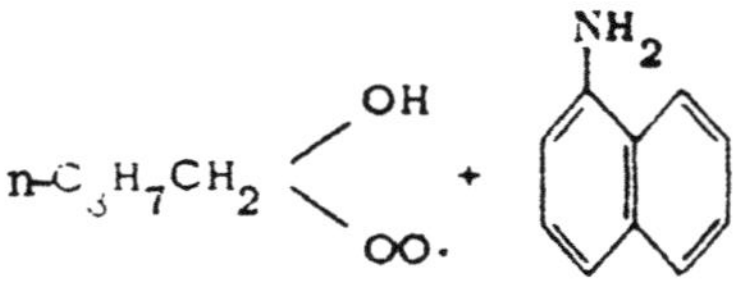	$n\text{-}C_4H_9OH + n\text{-}C_7H_{16}$	74	$\frac{k(RO_2\cdot + InH)}{k^{1/2}(RO_2\cdot\ RO_2\cdot)}$	0,40	[175]
$CH_3COCH(CH_3)OO\cdot + InH$	$CH_3COC_2H_5 + H_2O$	60	$k(RO_2\cdot + InH)$		
InH = Hydroquinone		60		1,93	[505]
α-Naphthol		60		2,65	[505]
2,4,6-Trimethylphenol		60		1,78	[505]
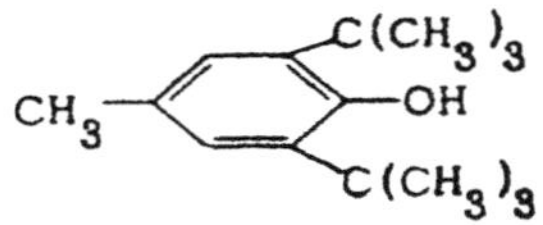		60		1,86	[505]

Table 97 (continued)

Reaction	Solvent	t°	Measured value	log k (liter/mole)	Ref.
$CH_3COCH(CH_3)OO\cdot$ + InH	$CH_3COC_2H_5 + H_2O$	60	$k(RO_2\cdot + InH)$		
InH= 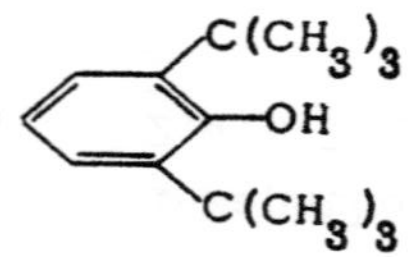		60		1,67	[505]
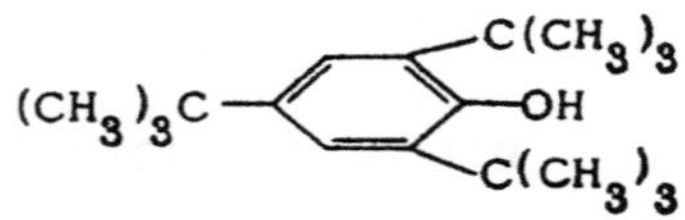		60		1,91	[505]
Azolactone 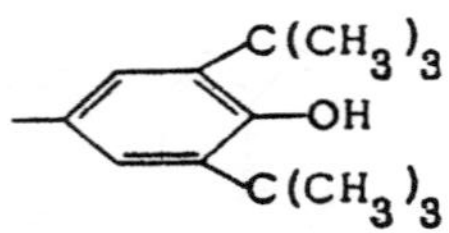		60		1,88	[505]
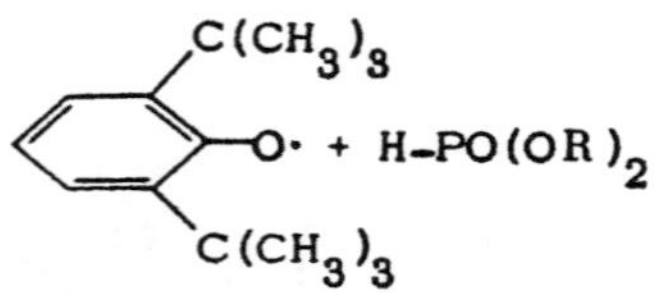	C_6H_6	20	k	0,65	[184]

TABLE 98. Reactions of Hydrogen-Bonded $RO_2^{\cdot}$

Reaction	k (60°)	log A	E or Q
	RH = methyl ethyl ketone		
$RO_2\cdot + RH$	$3{,}9\cdot10^{-1}$	5,10	8,4 ± 0,5
$RO_2\cdot \ldots H_2O + RH$	$2{,}0\cdot10^{-2}$	9,11	16,5 ± 1
$RO_2\cdot + RO_2\cdot$	$1{,}8\cdot10^{6}$	7,30	1,6 ± 0,8
$RO_2\cdot \ldots H_2O + RO_2\cdot \ldots H_2O$	$7{,}0\cdot10^{4}$	7,86	4,6 ± 1,0
$RO_2\cdot + H_2O \rightleftharpoons RO_2\cdot \ldots H_2O$	1,1 liter/mole	$\bar{4}{,}88$	4,8 ± 0,4
$RO_2\cdot \ldots HOCH_3 + RH$	$6{,}7\cdot10^{-2}$		
$RO_2\cdot \ldots HOCH_3 + RO_2\cdot$	$8\cdot10^{5}$		
$2RO_2\cdot \ldots HOCH_3$	$7\cdot10^{5}$		
$RO_2\cdot + HOCH_3 \rightleftharpoons RO_2\cdot \ldots HOCH_3$	0,70 liter/mole		
$RO_2\cdot \ldots HOC(CH_3)_3 + RH$	$9{,}0\cdot10^{-2}$		
$RO_2\cdot \ldots HOC(CH_3)_3 + RO_2\cdot$	$8\cdot10^{5}$		
$2RO_2\cdot \ldots HOC(CH_3)_3$	$7\cdot10^{5}$		
$RO_2\cdot + HOC(CH_3)_3 \rightleftharpoons RO_2\cdot \ldots HOC(CH_3)_3$	0,84 liter/mole		

Table 98 (continued)

Reaction	k(60°)	log A	E or Q
$RO_2\cdot \ldots H_2O$ + 	$4{,}0\cdot 10^4$		
+ $-C_{10}H_7OH$	$3{,}4\cdot 10^4$		
+ 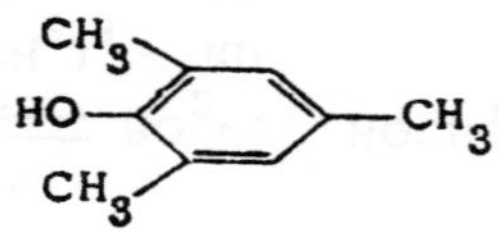	$2{,}2\cdot 10^4$		
+ 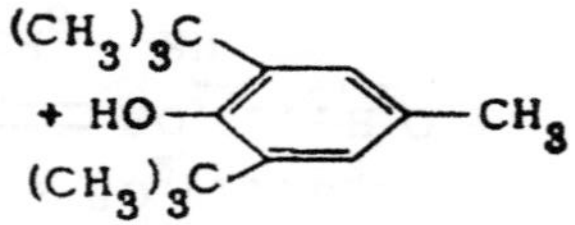	$0{,}30\cdot 10^4$		
+ 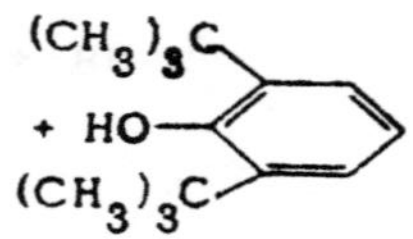	$0{,}15\cdot 10^4$		

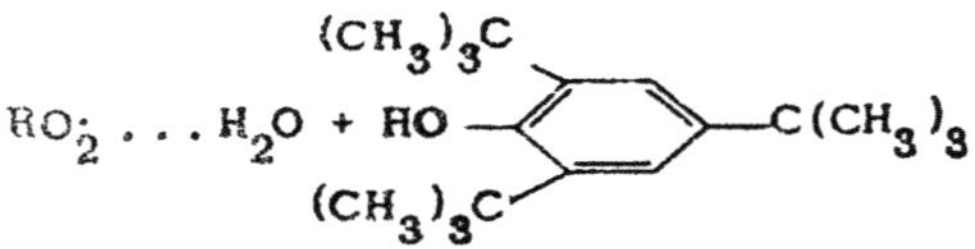	$0{,}61 \cdot 10^4$		
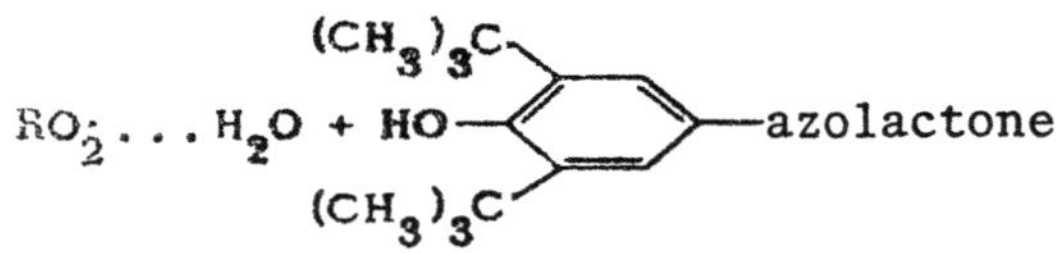	$0{,}12 \cdot 10^4$		
	RH - cyclohexane		
$RO_2\cdot + RH$	0,53		
$RO_2\cdot \ldots HOC(CH_3)_3 + RH$	0,22		
$RO_2\cdot \ldots HOC(CH_3)_3 \rightleftharpoons RO_2\cdot \ldots HOC(CH_3)_3$	0,9 liter/mole	$\bar{3}{,}62$	3,5
$RO_2\cdot + RO_2\cdot$	$5{,}0 \cdot 10^6$	7,724	$1{,}6 \pm 0{,}6$
$RO_2\cdot \ldots HOC(CH_3)_3 + RO_2\cdot$	$4{,}0 \cdot 10^6$	8,518	$3{,}0 \pm 0{,}8$
$2RO_2\cdot \ldots HOC(CH_3)_3$	$1{,}6 \cdot 10^6$	9,740	$5{,}4 \pm 0{,}8$
	RH - 2-methylpentene-2		
$RO_2\cdot + RH$	1,77 (40°)		
$RO_2\cdot \ldots HOC(CH_3)_3 + RH$	13,3 (40°)		
$RO_2\cdot + HOC(CH_3)_3 \rightleftharpoons RO_2\cdot \ldots HOC(CH_3)_3$	0,76 liter/mole	$\bar{3}{,}22$	4,22
$RO_2\cdot + RO_2\cdot$	$4{,}3 \cdot 10^6$		
$RO_2\cdot \ldots HOC(CH_3)_3 + RO_2\cdot$	$9{,}0 \cdot 10^6$		
$2RO_2\cdot \ldots HOC(CH_3)_3$	$13{,}2 \cdot 10^6$		

LITERATURE CITED (PART II)

1. N. M. Émanuél',E. T. Denisov, and Z. K. Maizus, Chain Reactions of Liquid-Phase Hydrocarbon Oxidation, Moscow, "Nauka" (1965). [English translation: Liquid-Phase Oxidation of Hydrocarbons, New York, Plenum Press (1967).]

2. Kh. S. Bagdasar'yan, Theory of Radical Polymerization, Moscow, "Nauka" (1966).

3. B. Khuvink and A. Staverman, Chemistry and Technology of Polymers, Vol. 1, Moscow (1965).

4. M. Levy and M. Szwarc, J. Am. Chem. Soc., 77:1949 (1955).

5. T. J. Hardwick, J. Phys. Chem., 65:101 (1961).

6. Kh. S. Bagdasar'yan and R. I. Milyutinskaya, Zh. Fiz. Khim., 34:234 (1960).

7. I. V. Berezin, Doctoral Dissertation, Moscow (1962).

8. V. L. Antonovskii and I. V. Berezin, Dokl. Akad. Nauk SSSR, 127:124 (1959).

9. Ya. S. Lebedev, V. F. Tsepalov, and V. Ya. Shlyapintokh, Dokl. Akad. Nauk SSSR, 139:1409 (1961); Kinetika i Kataliz, 5:64 (1964).

10. V. F. Shuvalov, Ya. S. Lebedev, V. F. Tsepalov, and V. Ya. Shlyapintokh, Zh. Fiz. Khim., 38:1287 (1964).

11. M. Bersohn and J. R. Thomas, J. Am. Chem. Soc., 86:959 (1964).

12. A. L. Buchachenko, Stable Radicals, Moscow, Izd. Akad. Nauk SSSR (1963).

13. D. Ingram, Free Radicals as Studied by Electron Spin Resonance [Russian translation] Moscow, IL (1961). [Original English-language edition: New York, Academic Press.]

14. L. A. Blyumenfel'd, V. V. Voevodskii, and A. G. Semenov, Application of Electron Paramagnetic Resonance in Chemistry, Novosibirsk, Izd. Siber. Otd. Akad. Nauk SSSR (1962).

15. E. F. Caldin, Fast Reactions in Solution [Russian translation], Moscow, "Mir" (1966). [Original English-language edition: New York, Wiley.]

16. H. Berger, A. M. W. Blaauw, M. M. Al, and P. Smael, Advances in Chemistry Series, 75:346 (1968).

17. V. Ya. Shlyapintokh, O. N. Karpukhin, L. M. Postnikov, I. V. Zakharov, A. A. Vichutinskii, and V. F. Tsepalov, Chemiluminescence Methods of Investigating Slow Chemical Processes, Moscow, "Nauka" (1966).

18. C. M. Burrell, T. G. Majury, and H. W. Melville, Proc. Roy. Soc. (London), A205:309 (1951).

19. N. Grassie and H. W. Melville, Proc. Roy. Soc. (London), A207:285 (1951).

20. W. I. Bengough and H. W. Melville, Proc. Roy. Soc. (London), A225:330 (1954).

21. J. C. Robb and B. Shahin, J. Inst. Petrol., 44:283 (1958).

22. W. I. Bengough, Nature, 180:1120 (1957).

23. C. H. Bamford and M. J. S. Dewas, Proc. Roy. Soc. (London), A192:309, 329 (1948).

24. L. Bateman and G. Gee, Trans. Faraday Soc., 47:155 (1951).

25. A. A. Vichutinskii, Dokl. Akad. Nauk SSSR, 157:150 (1964).

26. G. M. Burnett and H. W. Melville, Technique of Organic Chemistry, 8:133 (1953).

27. G. Porter, Z. Elektrochem., 64:59 (1960).

28. G. Porter, Investigation of Rates and Mechanisms of Reactions, Vol. 2, New York, Interscience (1963), chap. 19.

29. G. Porter, Radiation Res. Suppl., 1:479, 489 (1959).

30. R. L. McCarthy and A. MacLachlan, Trans. Faraday Soc., 56:1187 (1960).

31. A. K. Pikaev, Pulse Radiolysis of Water and Aqueous Solutions, Moscow, "Nauka" (1965).

32. Pulse Radiolysis (M. Ebert, ed.), New York, Academic Press (1965).

33. B. H. J. Bielsky and E. Saito, J. Phys. Chem., 66:2266 (1962).

34. J. R. Thomas, J. Am. Chem. Soc., 87:3935 (1965).

35. W. V. Smith and R. H. Ewart, J. Chem. Phys., 16:592 (1948).

36. C. S. Johnston, Advances in Magnetic Resonance, Vol. 1, New York, Academic Press (1965).

37. D. J. Williams and R. Kreilick, J. Am. Chem. Soc., 89:3408 (1967).

38. R. M. Noyes, J. Am. Chem. Soc., 81:566 (1959).

39. R. H. Freydlina, Advances in Free Radical Chemistry, 1:211 (1965).

40. J. E. Bennett, Nature, 186:385 (1960).

41. C. Ruchardt, Chem. Ber., 94:2599 (1961).

42. C. Ruchardt, Chem. Ber., 94:2609 (1961).

43. C. Ruchardt and S. Eichler, Chem. Ber., 95:1921 (1962).

44. C. Ruchardt and H. Trautwein, Chem. Ber., 96:160 (1963).

45. C. Ruchardt, Tetrahedron Letters, 1962, p. 961.

46. D. J. Carlsson and K. U. Ingold, J. Am. Chem. Soc., 89:4891 (1967).

47. C. Walling and P. Wagner, J. Am. Chem. Soc., 85:2333 (1963).

48. J. K. Kochi, J. Am. Chem. Soc., 84:1193 (1962).

49. J. D. Bacha and J. K. Kochi, J. Org. Chem., 30:3272 (1965).

50. D. G. Hoare and W. A. Waters, J. Chem. Soc., 1964, p. 2552.

51. H. C. Suton, in: Pulse Radiolysis (M. Ebert, ed.), New York, Academic Press (1965), p. 61.

52. C. J. Hochanadel, Radiation Res., 17:286 (1962).

53. J. K. Thomas, J. Phys. Chem., 67:2593 (1963).

54. H. Fricke and J. K. Thomas, Radiation Res. Suppl., 4:35 (1964).

55. J. P. Sweet and J. K. Thomas, J. Phys. Chem., 68:1363 (1964).

56. G. Czapski and J. Jortner, Nature, 188:50 (1960).

57. F. S. Dainton and S. A. Sills, Proc. Chem. Soc., 1962, p. 223.

58. J. H. Baxendale and D. H. Simithies, Z. Physik. Chem. (Frankfurt), 7:242 (1956).

59. J. Rabani, J. Phys. Chem., 66:361 (1962).

60. J. Rabani and G. Stein, J. Chem. Phys., 37:1865 (1962).

61. A. Appleby, G. Scholes, and M. Simic, J. Am. Chem. Soc., 85:3891 (1963).

62. M. Anbar and D. Meyerstein, J. Phys. Chem., 68:3184 (1964).

63. J. K. Thomas, J. Phys. Chem., 71:1919 (1967).

64. G. Scholes and M. Simic, J. Phys. Chem., 68:1731, 1738 (1964).

65. S. Nehari and J. Rabani, J. Phys. Chem., 67:1609, (1963).

66. J. Jortner and J. Rabani, J. Phys. Chem., 66:2078, 2081 (1962).

67. P. E. M. Allen, H. W. Melville, and J. C. Robb, Proc. Roy. Soc. (London), A218:311 (1953).

68. T. J. Hardwick, J. Phys. Chem., 65:101 (1961).

69. T. J. Hardwick, J. Phys. Chem., 66:291 (1962).

70. T. J. Hardwick, J. Phys. Chem., 66:117 (1962).

71. T. J. Hardwick, J. Phys. Chem., 66:2246 (1959).

72. J. H. Knox and R. L. Nelson, Trans. Faraday Soc., 55:937 (1959).

73. H. O. Pritchard, J. B. Pyke, and A. E. Trotman-Dickenson, J. Am. Chem., 77:2629 (1955).

74. G. A. Russell and R. C. Wiliamson, J. Am. Chem. Soc., 86:2357 (1964).

75. G. A. Russell, A. Ito, and D. G. Hendry, J. Am. Chem. Soc., 85:2976 (1963).

76. G. A. Russell, J. Am. Chem. Soc., 80:4997 (1958).

77. C. Walling and B. Miller, J. Am. Chem. Soc., 79:4181 (1957).

78. G. A. Russell and H. C. Brown, J. Am. Chem. Soc., 77:4578 (1955).

79. H. R. Anderson, H. A. Scheraga, and E. R. Van Artsdalen, J. Chem. Phys., 21:1258 (1953).

80. R. L. Huang and K. H. Lee, J. Chem. Soc., C, 1966, p. 932.

81. G. A. Russell and K. M. Desmond, J. Am. Chem. Soc., 85:3139 (1963).

82. C. Walling, A. L.Rieger, and D. D. Tanner, J. Am. Chem. Soc., 85:3129 (1963).

83. G. A. Russell and C. De Boer, J. Am. Chem. Soc., 85:3136 (1963).

84. C. Walling and A. L. Rieger, J. Am.Chem. Soc., 85:3134 (1963).

85. R. E. Pearson and J. C. Martin, J. Am. Chem. Soc., 85:354, 3142 (1963).

86. E. C. Kooymen, R. Van Helden, and A. F. Bickel, Koninkl. Ned. Akad. Wetenschap., Proc., B56:75 (1953).

87. R. L. Huang and K. H. Lee, J. Chem. Soc., C, 1966, p. 935.

88. M. M. Martin and G. J. Gleicher, J. Org. Chem., 28:3266 (1963).

89. R. L. Huang and K. H. Lee, J. Chem. Soc., 1964, p. 5963.

90. M. Anbar and P. Neta, Radioact. Isotopes, 16:227 (1965).

91. C. J. Hochanadel, J. Phys. Chem., 56:587 (1952).

92. D. Bunn, F. S. Dainton, G. A. Salmon, and T. J. Hardwick, Trans. Faraday Soc., 55:1760 (1959).

93. H. A. Schwarz, J. Phys. Chem., 66:255 (1962).

94. J. K. Thomas, Trans. Faraday Soc., 61:702 (1965).

95. M. Anbar, D. Meyerstein, and P. Neta, J. Chem. Soc., B, 1966, p. 742.

96. J. H. Merz and W. A. Waters, J. Chem. Soc. Suppl. Ser., 1949, p. 15.

97. E. J. Hart, J. K. Thomas, and S. Gordon, Radiation Res. Suppl., 4:74 (1964).

98. G. Scholes and R. L. Willson, Trans. Faraday Soc., 63:2983 (1967).

99. G. E. Adams, J. W. Boag, and B. D. Michael, Trans. Faraday Soc., 61:1417 (1965).

100. R. W. Matthews, and D. F. Sangster, J. Phys. Chem., 69:1938 (1965).

101. I. Kraljic and C. N. Trumbore, J. Am. Chem. Soc., 87:2547 (1965).

102. J. H. Merz and W. A. Waters, J. Chem. Soc., 1949, p. 2427.

103. J. Rabani and G. Stein, Trans. Faraday Soc., 58:2150 (1962).

104. G. E. Adams and J. W. Boag, Proc. Chem. Soc., 1964, p. 112.

105. G. E. Adams, J. W. Boag, J. Currant, and B. D. Michael, in: Pulse Radiolysis (M. Ebert, ed.), New York, Academic Press (1965), p. 131.

106. E. J. Hart, J. Am. Chem. Soc., 73:68 (1951).

107. E. J. Hart, J. Am. Chem. Soc., 74:4174 (1952).

108. J. L. Weeks and M. S. Matheson, J. Am. Chem. Soc., 78:1273 (1956).

109. T. J. Sworski, Radiation Res., 6:645 (1957).

110. G. Scholes, P. A. Shaw, R. L. Wilson, and M. Ebert, in: Pulse Radiolysis (M. Ebert, ed.) New York, Academic Press (1965), p. 151.

111. J. Rabani and G. Stein, Trans. Faraday Soc., 58:2150 (1962).

112. M. Anbar, R. A. Munoz, and P. Roha, J. Phys. Chem., 67:2708 (1963).

113. J. H. T. Brook, Trans. Faraday Soc., 53:327 (1957).

114. H. Sakurai and A. Hosomi, J. Am. Chem. Soc., 89:458 (1967).

115. K. Uneyama, H. Namba, and S. Oae, Bull. Chem. Soc., Japan, 41:1928 (1968).

116. R. D. Gilliom and B. F. Ward, J. Am. Chem. Soc., 87:3944 (1965).

117. S. L. Sosin, V. V. Korshak, and D. G. Val'kovskii, Dokl. Akad. Nauk SSSR, 155:376 (1964).

118. C. Walling and B. B. Jacknow, J. Am. Chem. Soc., 82:6113 (1960).

119. R. R. Kennedy and K. U. Ingold, Can. J. Chem., 44:2381 (1966).

120. K. M. Johnston and G. H. Williams, J. Chem. Soc., 1960, p. 1446.

121. C. Walling and W. Thaler, J. Am. Chem. Soc., 83:3877 (1961).

122. A. L. Williams, E. A. Oberright, and J. W. Brooks, J. Am. Chem. Soc., 78:1190 (1956).

123. E. L. Patmore and R. J. Gritter, J. Org. Chem., 27:4196 (1962).

124. T. J. Wallace and R. J. Gritter, Tetrahedron, 19:657 (1963).

125. K. Schweitlick and U. Pietzsch, J. Prakt. Chem., 22:125 (1963).

126. K. Schweitlick and R. Spitz, J. Prakt. Chem., 30:218 (1965).

127. K. U. Ingold, Can. J. Chem., 41:2816 (1963).

128. E. Niki, Y. Kamiya, and N. Ohta, Bull. Chem. Soc., Japan, 41:1466 (1968).

129. J. Scanlan and D. K. Thomas, J. Polymer Sci., A1:1015 (1963).

130. A. L. Buchachenko, K. Ya. Kaganskaya, M. B. Neiman, and A. A. Petrov, Kinetika i Kataliz, 2:44 (1961).

131. G. E. Zaikov, Doctoral Dissertation, Moscow (1968).

132. L. Bateman and G. Gee, Trans. Faraday Soc., 47:155 (1951).

133. J. A. Howard, K. U. Ingold, and M. Symonds, Can. J. Chem., 46:1017 (1968).

134. J. C. Robb and M. Shahin, J. Inst. Petrol. 44:283 (1958).

135. J. C. Robb and M. Shahin, Trans. Faraday Soc., 55:1753 (1959).

136. J. A. Howard and J. C. Robb, Trans. Faraday Soc., 59:1590 (1963).

137. L. Bateman and G. Gee, Proc. Roy. Soc., (London), A195:391 (1948).

138. J. A. Howard and K. U. Ingold, Can. J. Chem., 44:1119 (1966).

139. J. A. Howard and K. U. Ingold, Can. J. Chem., 45:793 (1967).

140. V. F. Tsepalov and V. Ya. Shlyapintokh, Kinetika i Kataliz, 3:870 (1962).

141. H. W. Melville and S. Richards, J. Chem. Soc., 1954, p. 944.

142. D. G. Hendry, J. Am. Chem. Soc., 89:5433 (1967).

143. J. A. Howard and K. U. Ingold, Can. J. Chem., 46:2655 (1968).

144. A. F. Guk, V. F. Tsepalov, V. F. Shuvalov, and V. Ya. Shlyapintokh, Izv. Akad. Nauk SSSR, Ser. Khim., 1968, p. 2250.

145. C. H. Bamford and M. J. S. Dewar, Proc. Roy. Soc., (London), A198:252 (1949).

146. B. S. Middleton and K. U. Ingold, Can. J. Chem., 45:191 (1967).

147. J. A. Howard and K. U. Ingold, Can. J. Chem., 44:1113 (1966).

148. A. L. Aleksandrov and E. T. Denisov, Izv. Akad. Nauk SSSR, Ser. Khim., 1966, p. 1737.

149. A. L. Aleksandrov and E. T. Denisov, Kinetika i Kataliz, 10:912 (1969).

150. G. E. Zaikov, A. A. Vichutinskii, and Z. K. Maizus, Kinetika i Kataliz, 8:675 (1967).

151. G. E. Zaikov, Kinetika i Kataliz, 9:1166 (1968).

152. H. R. Cooper and H. W. Melville, J. Chem. Soc., 1951, p. 1984.

153. T. A. Ingles and H. W. Melville, Proc. Roy. Soc., (London), A218:175 (1953).

154. J. A. Howard, W. J. Schwalm, and K. U. Ingold, Advances in Chemistry Series, 75:6 (1968).

155. J. A. Howard and K. U. Ingold, Can. J. Chem., 45:785 (1967).

156. F. S. Dainton and J. Rowbottom, Trans. Faraday Soc., 49:1160 (1953).

157. J. A. Howard and K. U. Ingold, Can. J. Chem., 46:2661 (1968).

158. V. F. Tsepalov, V. Ya. Shlyapintokh, and Chzhou Péi-khuan, Zh. Fiz. Khim., 38:52 (1964).

159. G. A. Russell, J. Am. Chem. Soc., 77:4583 (1955).

160. R. Haitt, C. W. Gould, and F. R. Mayo, J. Org. Chem., 29:3461 (1964).

161. T. A. Ingles and H. W. Melville, Proc. Roy. Soc., (London), A218:163 (1953).

162. J. R. Thomas and C. A. Tolman, J. Am. Chem. Soc., 84:2079 (1962).

163. V. E. Agabekov, E. T. Denisov, and N. I. Mitskevich, Izv. Akad. Nauk SSSR, Ser. Khim., 1968, p. 2254.

164. O. N. Karpuhkin, V. Ya. Shlyapintokh, and N. V. Zolotova, Izv. Akad. Nauk SSSR, Otd. Khim. Nauk, 1963, p. 1722.

165. G. V. Karpukhina, Z. K. Maizus, and L. I. Matienko, Neftekhimiya, 6:603 (1966).

166. J. A. Howard and K. U. Ingold, Can. J. Chem., 41:1744 (1963).

167. D. S. Davies, H. L. Goldsmith, A. K. Gupta, and G. R. Lester, J. Chem. Soc., 1956, p. 4926.

168. J. A. Howard and K. U. Ingold, Can. J. Chem., 41:2800 (1963).

169. L. M. Andronov, G. E. Zaikov, and Z. K. Maizus, Teor. Eksperim. Khim., 3:620 (1967).

170. L. M. Andronov and G. E. Zaikov, Izv. Akad. Nauk SSSR, Ser. Khim., 1968, p. 2261.

171. J. A. Howard and K. U. Ingold, Can. J. Chem., 43:2729 (1965).

172. J. A. Howard and K. U. Ingold, Can. J. Chem., 40:1851 (1962).

173. J. A. Howard and K. U. Ingold, Can. J. Chem., 43:2737 (1965).

174. J. A. Howard and K. U. Ingold, Can. J. Chem., 43:2724 (1965).

175. J. L. Bolland and P. ten Have, Trans. Faraday Soc., 43:201 (1947).

176. E. T. Denisov, A. L. Aleksandrov, and V. P. Shcheredin, Izv. Akad. Nauk SSSR, Ser. Khim., 1964, p. 1583.

177. E. T. Denisov, Kinetika i Kataliz, 4:508 (1963).

178. E. T. Denisov and A. L. Aleksandrov, Zh. Fiz. Khim., 38:491 (1964).

179. G. V. Karpukhina, Candidate's Dissertation, Moscow (1968).

180. A. L. Buchachenko, K. Ya. Kaganskaya, and M. B. Neiman, Kinetika i Kataliz, 2:161 (1961).

181. I. T. Brownlie and K. U. Ingold, Can. J. Chem., 44:861 (1966).

182. I. T. Brownlie and K. U. Ingold, Can. J. Chem., 45:2419 (1967).

183. I. T. Brownlie and K. U. Ingold, Can. J. Chem., 45:2427 (1967).

184. M. B. Neiman, Yu. G. Mamedova, P. Blenke, and A. L. Buchachenko, Dokl. Akad. Nauk SSSR, 144:392 (1962).

185. A. L. Buchachenko, E. I. Sdobnov, S. R. Rafikov, and M. B. Neiman, Izv. Akad. Nauk SSSR, Ser. Khim., 1963, p. 1118.

186. A. L. Buchachenko, M. B. Neiman, O. P. Sukhanova, and Yu. G. Mamedova, Zh. Fiz. Khim., 37:221 (1963).

187. L. R. Mahoney and M. A. Da Rooge, J. Am. Chem. Soc., 89:5619 (1967).

188. M. A. Da Rooge and L. R. Mahoney, J. Org. Chem., 32:1 (1967).

189. J. C. McGowan and T. Powell, J. Chem. Soc., 1960, p. 238.

190. A. M. Vasserman and A. L. Buchachenko, Izv. Akad. Nauk SSSR, Ser. Khim., 1967, p. 1947.

191. V. A. Bidzilya, V. D. Pokhodenko, and A. I. Brodskii, Dokl. Akad. Nauk, SSSR, 166:1099 (1966).

192. P. B. Ayscough and K. E. Russel, Can. J. Chem., 45:3019 (1967).

193. R. W. Kreilick and S. J. Weissman, J. Am. Chem. Soc., 84:306 (1962).

194. R. W. Kreilick and S. J. Weissman, J. Am. Chem. Soc., 88:2645 (1966).

195. Yu. G. Mamedova, A. L. Buchachenko, and M. B. Neiman, Izv. Akad. Nauk SSSR, Ser. Khim., 1965, p. 911.

196. K. H. Gieb and P. Harteck, Z. Phys. Chem. Bodensteinfest band (1931), p. 849.

197. F. W. Evans and M. Szwarc, Trans. Faraday Soc., 57:1905 (1961).

198. I. V. Berezin and O. Dobish, Dokl. Akad. Nauk SSSR, 142:105 (1962).

199. O. Dobish, Candidate's Dissertation, Moscow (1962).

200. V. A. Antonovskii and I. V. Berezin, Dokl. Akad. Nauk SSSR, 134:860 (1960).

201. R. J. Fox, F. W. Evans, and M. Szwarc, Trans. Faraday Soc., 57:1915 (1961).

202. R. P. Buckley, F. Leavitt, and M. Szwarc, J. Am. Chem. Soc., 78:5557 (1956).

203. J. Gresser, A. Rajbenbach, and M. Szwarc, J. Am. Chem. Soc., 83:3005 (1961).

204. J. A. Meyer, V. Stannett, and M. Szwarc, J. Am. Chem. Soc., 83:25 (1961).

205. A. E. Eachus, J. A. Meyer, J. Pearson, and M. Szwarc, J. Am. Chem. Soc., 90:3646 (1968).

206. E. Kalatzis and G. H. Williams, J. Chem. Soc., B, 1966, p. 1112.

207. J. Gresser, J. H. Binks, and M. Szwarc, J. Am. Chem. Soc., 81:5004 (1959).

208. V. Keler, N. F. Kazanskaya, and I. V. Berezin, Vestn. Mosk. Gos. Univ., No. 2, p. 29 (1966).

209. I. V. Berezin, K. Vatsek, and N. F. Kazanskaya, Dokl. Akad. Nauk SSSR, 144:139 (1962).

210. I. V. Berezin, N. F. Kazanskaya, and N. N. Ugarova, Zh. Fiz. Khim., 40:766 (1966).

211. I. V. Berezin, V. L. Ivanov, N. F. Kazanskaya, and N. N. Ugarova, Zh. Fiz. Khim., 39:3011 (1965).

212. O. Dobis, I. Nemes, and R. Kerepes, Acta. Chim. Acad. Sci. Hungary, 55:215 (1968).

213. L. N. Shishkina and I. V. Berezin, Vetsn. Mosk. Gos. Univ., No. 2, p. 13 (1966).

214. V. Keler, Candidate's Dissertation, Moscow (1965).

215. W. Köhler, N. F. Kazanskaya, L. G. Nagler, and I. V. Berezin, Ber. Buns. Phys. Chem., 71:736 (1967).

216. L. N. Shishkina and I. V. Berezin, Zh. Fiz. Khim., 39:2547 (1965).

217. D. J. Carlsson and K. U. Ingold, J. Am. Chem. Soc., 90:7047 (1968).

218. R. H. Schuler and R. R. Kuntz, J. Phys. Chem., 67:1004 (1953) [sic].

219. E. C. Kooyman, Discussions Faraday Soc., 10:163 (1951).

220. E. S. Huyser, J. Am. Chem. Soc., 82:394 (1960).

221. M. M. Martin and G. J. Gleicher, J. Am. Chem. Soc., 86:233 (1964).

222. M. M. Martin and G. J. Gleicher, J. Am. Chem. Soc., 86:238 (1964).

223. W. I. Bengough and R. A. M. Thomson, Trans. Faraday Soc., 57:1928 (1961).

224. D. J. Carlsson, J. A. Howard, and K. U. Ingold, J. Am. Chem. Soc., 88:4726 (1966).

225. S. W. Charles and E. Whittle, Trans. Faraday Soc., 56:794 (1960).

226. A. P. Stefani, L. Herk, and M. Szwarc, J. Am. Chem. Soc., 83:4732 (1961).

227. I. M. Whittemore, A. P. Stefani, and M. Szwarc, J. Am. Chem. Soc., 84:3799 (1962).

228. R. F. Bridger and G. A. Russell, J. Am. Chem. Soc., 85:3754 (1963).

229. W. A. Pryor, J. T. Echols, and K. Smith, J. Am. Chem. Soc., 88:1189 (1966).

230. É. A. Trosman and Kh. S. Bagdasar'yan, Zh. Fiz. Khim., 38:2698 (1964).

231. M. V. Bazilevskii and N. I. Piskun, Zh. Fiz. Khim., 39:951 (1965).

232. M. V. Bazilevskii and N. I. Piskun, Zh. Fiz. Khim., 39:762 (1965).

233. W. A. Pryor and P. K. Platt, J. Am. Chem. Soc., 85:1496 (1963).

234. W. A. Pryor and H. Guard, J. Am. Chem. Soc., 86:1150 (1964).

235. J. K. Kochi and R. V. Subramanian, J. Am. Chem. Soc., 87:4855 (1965).

236. D. J. Carlsson and K. U. Ingold, J. Am. Chem. Soc., 90:7047 (1968).

237. É. A. Trosman and M. V. Bazilevskii, Kinetika i Kataliz, 9:684 (1968).

238. De Los, F. De Tar, and D. V. Wells, J. Am. Chem. Soc., 82:5839 (1960).

239. Y. Schaafsma, A. F. Bickel, and E. C. Kooyman, Rec. Trav. Chim., 76:180 (1957).

240. H. W. Melville, J. C. Robb, and R. C. Tutton, Discussions Faraday Soc., 10:154 (1951).

241. Kh. S. Bagdasar'yan and A. F. Revzin, Kinetika i Kataliz, 4:844 (1963).

242. E. M. Kosower and I. Schwager, J. Am. Chem. Soc., 86:4493 (1964).

243. E. M. Kosower and M. Mohammad, J. Am. Chem. Soc., 90:3271 (1968).

244. E. S. Huyser and D. C. Neckers, J. Am. Chem. Soc., 85:3641 (1963).

245. R. A. Gregg and F. R. Mayo, Discussions Faraday Soc., 2:328 (1947).

246. G. Henrici-Olive and S. Olive, Fortsch. Hochpolymer-Forsch., 2:496 (1961).

247. T. Yamamoto, Bull. Chem. Soc., Japan, 40:642 (1967).

248. S. Loshaek, E. Broderick, and P. Bernstein, J. Polymer Sci., 39:223 (1959).

249. C. H. Bamford and M. J. S. Dewar, Proc. Roy. Soc. (London), A192:308 (1948).

250. G. M. Burnett, Trans. Faraday Soc., 46:772 (1950).

251. M. S. Matheson, E. E. Auer, E. B. Bevilacqua, and E. J. Hart, J. Am. Chem. Soc., 73:1700 (1951).

252. H. W. Melville and L. Valentine, Trans. Faraday Soc., 46:210 (1950).

253. M. Morton, J. A. Cala, and I. Piirma, J. Am. Chem. Soc., 78:5394 (1956).

254. R. A. Gregg and F. R. Mayo, J. Am. Chem. Soc., 75:3530 (1953).

255. S. L. Kapar, J. Polymer Sci., 11:399 (1953).

256. J. I. G. Cadogam, D. H. Hey, and J. T. Sharp, J. Chem. Soc., B, 1966, p. 933.

257. R. Hiddema, Proefschrift Rijksuniveristät Groningen, (1953).

258. W. A. Pryor and E. P. Pultinas, J. Am. Chem. Soc., 85:133 (1963).

259. W. A. Pryor and T. L. Pickering, J. Am. Chem. Soc., 84:2705 (1962).

260. F. R. Mayo, R. A. Gregg, and M. S. Matheson, J. Am. Chem. Soc., 73:1691 (1951).

261. V. A. Dinaburg and A. A. Vansheidt, Zh. Obshch. Khim., 24:840 (1954).

262. W. Cooper, J. Chem. Soc., 1952, p. 2408.

263. J. W. Breitenbach and A. Schindler, Montash. Chem., 83:724 (1952).

264. J. A. Gannon, E. M. Fettes, and A. V. Tobolsky, J. Am. Chem. Soc., 74:1854 (1952).

265. F. R. Mayo, J. Am. Chem. Soc., 65:2324 (1943).

266. J. W. Breitenbach, O. F. Olaj, and A. Schindler, Montash. Chem., 91:205 (1960).

267. F. R. Mayo, J. Am. Chem. Soc., 75:6133 (1953).

268. C. H. Bamford and E. F. T. White, Trans. Faraday Soc., 52:716 (1965).

269. F. Hrabak and L. Jirešova, Collection Czech. Chem. Commun., 26:1283 (1961).

270. C. Walling, J. Am. Chem. Soc., 70:2561 (1948).

271. R. A. Gregg, D. M. Alderman, and F. R. Mayo, J. Am. Chem. Soc., 70:3740 (1948).

272. R. M. Pearson, A. J. Constanza, and H. A. Weinstein, J. Polymer Sci., 17:221 (1955).

273. A. Staverman, Chemistry and Technology of Polymers, Vol. 1, Moscow-Leningrad, "Khimiya" (1965) p. 212.

274. A. J. Constanza, R. M. Pierson, C. S. Marvel, and C. King, J. Polymer Sci., 17:319 (1955).

275. K. Tsuda and T. Otsu, Bull. Chem. Soc., Japan, 39:2206 (1966).

276. T. E. Ferington and A. V. Tobolsky, J. Am. Chem. Soc., 77:4510 (1955).

277. J. Pellon, J. Polymer Sci., 43:537 (1950).

278. G. Henrici-Olive, S. Olive, and G. V. Schulz, Z. Physik. Chem. (Frankfurt), 20:176 (1959).

279. M. Morton and I. Piirma, J. Polymer Sci., A1:3043 (1963). (1963).

280. R. N. Chadha, J. S. Shukla, and G. S. Misra, Trans. Faraday Soc., 53:240 (1957).

281. S. Basu, J. N. Sen, and S. R. Palit, Proc. Roy. Soc. (London), A202:485 (1950).

282. S. Basu, J. N. Sen, and S. R. Palit, Proc. Roy. Soc. (London), A214:247 (1952).

283. T. Yamamoto and T. Otsu, J. Polymer Sci., A1:6, 49 (1968).

284. M. S. Matheson, E. E. Auer, E. B. Bevilacqua, and E. J. Hart, J. Am. Chem. Soc., 71:497 (1949).

285. T. E. Ferington and A. V. Tobolsky, J. Colloid Sci., 10:536 (1955).

286. N. G. Saha, U. S. Nandi, and S. R. Palit, J. Chem. Soc., 1956, p. 427.

287. N. Fuhrman and R. B. Mesrobian, J. Am. Chem. Soc., 76:3281 (1954).

288. C. H. Bamford and E. F. T. White, Trans. Faraday Soc., 52:716 (1956).

289. C. Walling, J. Am. Chem. Soc., 70:2561 (1948).

290. J. L. O'Brien and F. Cornick, J. Am. Chem. Soc., 77:4757 (1955).

291. J. Pellon, J. Polymer Sci., 43:537 (1960).

292. G. Dixon-Lewis, Proc. Roy. Soc. (London), A198:510 (1949).

293. H. Kwart, H. S. Broadbent, and P. D. Bartlett, J. Am. Chem. Soc., 72:1060 (1950).

294. S. R. Palit and S. K. Das, Proc. Roy. Soc. (London), A226:82 (1954).

295. M. Matsumato and M. Maeda, J. Polymer Sci., 17:438 (1955).

296. A. A. Vansheidt and D. Khardi, Acta Chim. Acad. Sci. Hungary, 20:261, 381 (1959).

297. R. N. Chadba and G. S. Misra, Trans. Faraday Soc., 54:1227 (1958).

298. M. Matsumoto, J. Ukida, G. Takayama, T. Eguchi, and K. Mukumoto, Makromol. Chem., 32:13 (1959).

299. S. R. Chatterjee, S. N. Khanna, and S. R. Palit, J. Indian Chem. Soc., 41:622 (1964).

300. P. F. Onyon, J. Polymer Sci., 22:19 (1956).

301. A. Prevost-Bernas and J. Sobban-Danon, J. Chim. Phys., 53:418 (1956).

302. C. H. Bamford, W. Barb, A. Jenkins, and P. Onyon, Kinetics of Vinyl Polymerization by Radical Mechanisms [Russian translation]. Moscow, IL (1961). [Original English-language edition: New York, Academic Press.]

303. F. S. Dainton and M. Tordoff, Trans. Faraday Soc., 53:499 (1957).

304. Z. Laita and Z. Machachek, J. Polymer Sci., 38:247, 459 (1959).

305. P. F. Onyon, Trans. Faraday Soc., 51:400 (1955).

306. A. F. Revzin and Kh. S. Bagdasar'yan, Zh. Fiz. Khim., 38:1020 (1964).

307. A. Vrancken and G. Smets, Makromol. Chem., 30:197 (1959).

308. S. Banarjee and M. S. Muthana, J. Polymer Sci., 37:469 (1959).

309. J. S. Hogg, D. H. Lohmann, and K. E. Russell, Can. J. Chem., 39:1588 (1961).

310. J. C. McGowan, T. Powell, and R. Raw, J. Chem. Soc., 1959, p. 3103.

311. J. E. Hazell and K. E. Russell, Can. J. Chem., 36:1729 (1958).

312. K. E. Russell, J. Phys. Chem., 58:437 (1954).

313. K. U. Ingold, Can. J. Chem., 41:2816 (1963).

314. G. S. Hammond, C. E. Boozer, C. E. Hamilton, and J. N. Sen, J. Am. Chem. Soc., 77:3238 (1955).

315. A. K. Pikaev, P. Ya. Glazunov, and V. I. Spitsyn, Dokl. Akad. Nauk SSSR, 150:1077 (1963).

316. S. Gordon, E. J. Hart, and J. K. Thomas, J. Phys. Chem., 68:1262 (1964).

317. L. Bateman, J. L. Bolland, and G. Gee, Trans. Faraday Soc., 47:274 (1951).

318. A. A. Miller and F. R. Mayo, J. Am. Chem. Soc., 78:1017 (1956).

319. K. W. Chambers, E. Collinson, F. S. Dainton, W. A. Seddon, and F. Wilkinson, Trans. Faraday Soc., 63:1966 (1967); Chem. Commun., 1968, p. 1250.

319a. L. M. Dorfman, I. A. Taub, and R. E. Bohler, J. Chem. Phys., 36:549, 3051 (1962).

319b. B. Cercek, J. Phys. Chem., 72:3832 (1968).

320. M. L. Poutsma, J. Am. Chem. Soc., 87:2172 (1965).

321. M. Feld and M. Szwarc, J. Am. Chem. Soc., 82:3791 (1960).

322. A. Rajbenbach and M. Szwarc, Proc. Roy. Soc. (London), A251:394 (1959).

323. A. R. Bader, R. P. Buckley, F. Leavitt, and M. Szwarc, J. Am. Chem. Soc., 79:5621 (1957).

324. F. Leavitt, M. Levy, M. Szwarc, and V. Stannett, J. Am. Chem. Soc., 77:5493 (1955).

325. R. P. Buckley and M. Szwarc, J. Am. Chem. Soc., 78:5696 (1956).

326. R. P. Buckley and M. Szwarc, Proc. Roy. Soc. (London), A240:396 (1957).

327. E. S. Ferdinandi, W. P. Garby, and D. G. L. James, Can. J. Chem., 42:2568 (1964).

328. H. W. Melville, J. C. Robb, and T. C. Tutton, Discussions Faraday Soc., 10:154 (1951).

329. H. W. Melville, J. C. Robb, and R. C. Tutton, Discussions Faraday Soc., 14:154 (1951).

330. A. P. Stefani, L. Herk, and M. Szwarc, J. Am. Chem. Soc., 83:4732 (1961).

331. G. E. Owen, J. M. Pearson, and M. Szwarc, Trans. Faraday Soc., 60:564 (1964).

332. P. S. Dixon and M. Szwarc, Trans. Faraday Soc., 59:112 (1963).

333. T. Ikawa, H. Tomizawa, and T. Yanagihara, Can. J. Chem., 45:1900 (1967).

334. Z. Laita, J. Polymer Sci., 38:247 (1959).

335. W. Rabel and K. Ueberreiter, Ber. Buns. Phys. Chem., 67:514, 710 (1963).

336. K. M. Gibov and G. P. Gladyshev, Izv. Akad. Nauk KazSSR, Ser. Khim., 1967, 53, 57; Vysokomolekul. Soedin., 9B:331 (1967).

337. M. Morton, P. Salatiello, and H. Landfield, J. Polymer Sci., 8:279 (1952).

338. D. W. E. Axford, Proc. Roy. Soc. (London), A197:374 (1949).

339. Z. A. Sinitsyna and Kh. S. Bagdasar'yan, Zh. Fiz. Khim., 32:1319 (1958).

340. M. S. Matheson, E. E. Auer, E. B. Bevilaqua, and E. J. Hart, J. Am. Chem. Soc., 73:5395 (1951).

341. W. I. Bengough and H. W. Melville, Proc. Roy. Soc. (London), A225:330 (1954).

342. W. I. Bengough and H. W. Melville, Proc. Roy. Soc. (London), A249:445 (1959).

343. H. W. Melville and A. F. Bickel, Trans. Faraday Soc., 45:1049 (1949).

344. M. H. Mackay and H. W. Melville, Trans. Faraday Soc., 45:323 (1949).

345. B. Atkinson and G. R. Cotten, Trans. Faraday Soc., 54:877 (1958).

346. G. M. Burnett, P. Evans, and H. W. Melville, Trans. Faraday Soc., 49:1096, 1105 (1953).

347. D. H. Grant and N. Grassie, Trans. Faraday Soc., 55:1042 (1959).

348. A. M. North and G. E. Reed, J. Polymer Sci., A1:1311 (1963).

349. A. F. Revzin, Usp. Khim., 53:173 (1966).

350. M. S. Matheson, E. E. Auer, E. B. Bevilacqua, and E. J. Hart, J. Am. Chem. Soc., 71:2610 (1949).

351. G. M. Burnett and W. W. Wright, Proc. Roy. Soc., (London), A221:28, 37, 41 (1954).

352. N. Colebourne, E. Collinson, D. J. Currie, and F. S. Dainton, Trans. Faraday Soc., 59:1357 (1963).

353. G. H. Bamford, A. D. Jenkins, and R. Johnston, Proc. Roy. Soc., (London), A241:364 (1957).

354. W. I. Bengough, J. Polymer Sci., 28:475 (1958).

355. W. M. Thomas and R. L. Webb, J. Polymer Sci., 25:124 (1957).

356. N. Grassie and E. Vance, Trans. Faraday Soc., 52:727 (1957).

357. F. S. Dainton and W. D. Sisley, Trans. Faraday Soc., 59:1369 (1963).

358. W. I. Bengough and W. Henerson, Trans. Faraday Soc., 61:141 (1965).

359. É. F. Nosov, Kinetika i Kataliz, 8:680 (1967).

360. A. N. Plyusnin and N. M. Chirkov, Teor. Eksperim. Khim., 2:777 (1966).

361. M. Morton and W. E. Gibbs, J. Polymer Sci., A1:2679 (1963).

362. R. H. Wiley and E. E. Sale, J. Polymer Sci., 42:479 (1960).

363. R. H. Wiley and B. Davis, J. Polymer Sci., 46:423 (1960).

364. L. A. Wall, D. W. Brown, and V. E. Hart, J. Polymer Sci., 15:157 (1955).

365. R. H. Wiley and B. Davis, J. Polymer Sci., 62:132 (1962).

366. R. H. Wiley and E. E. Sale, J. Polymer Sci., 42:491 (1960).

367. D. I. Livingston, P. M. Kamath, and R. S. Corley, J. Polymer Sci., 20:485 (1956).

368. C. A. Barson, J. Polymer Sci., 62:128 (1962).

369. C. C. Price and T. C. Schwan, J. Polymer Sci., 16:577 (1955).

370. C. C. Price, B. D. Halpern, and S. T. Yoong, J. Polymer Sci., 11:575 (1953).

371. C. L. Sandberg and F. A. Bovey, J. Polymer Sci., 15:533 (1955).

372. H. C. Haas and M. S. Simon, J. Polymer Sci., 9:309 (1952).

373. C. C. Price and H. Morita, J. Am. Chem. Soc., 75:4747 (1953).

374. C. C. Price and R. D. Gilbert, J. Am. Chem. Soc., 74:2073 (1952).

375. C. S. Marvel, J. W. Johnson, J. Economy, and G. P. Scott, J. Polymer Sci., 20:437 (1956).

376. H. Gilbert, F. F. Miller, S. J. Averill, E. J. Carlson, V. L. Folt, H. J. Heller, F. D. Stewart, R. F. Schmidt, and H. Z. Trumbull, J. Am. Chem. Soc., 78:1669 (1956).

377. J. M. Mitchell and H. L. Williams, Can. J. Res., 27:35 (1949).

378. R. D. Gilbert and H. L. Williams, J. Am. Chem. Soc., 74:4114 (1952).

379. R. J. Orr and H. L. Williams, Can. J. Chem., 29:270 (1951).

380. R. J. Orr and H. L. Williams, Can. J. Chem., 30:108 (1952).

381. R. J. Orr and H. L. Williams, Can. J. Chem., 33:1328 (1955).

382. L. P. Witnauer, N. Watkins, and W. S. Port, J. Polymer Sci., 20:213 (1956).

383. J. K. Allen and J. C. Bevington, Trans. Faraday Soc., 56:1762 (1960).

384. R. H. Wiley and B. Davis, J. Polymer Sci., 62:132 (1962).

385. D. C. Blackley and H. W. Melville, Makromol. Chem., 18:16 (1956).

386. H. C. Haas and M. S. Simon, J. Polymer Sci., 9:309 (1952).

387. C. C. Price and T. C. Schwan, J. Polymer Sci., 16:577 (1955).

388. S. N. Ushakov, S. P. Mitsengendler, and V. M. Poliyatskaya, "Chemistry and Physical Chemistry of High Molecular Weight Compounds," Proceedings of Conference on High Molecular Weight Compounds (1952), p. 19.

389. H. C. Haas, E. S. Emerson, and N. W. Schuler, J. Polymer Sci., 22:291 (1956).

390. T. Alfrey and C. C. Price, J. Polymer Sci., 2:101 (1947).

391. T. C. Schwan and C. C. Price, J. Polymer Sci., 40:457 (1959).

392. M. Szwarc, J. Phys. Chem., 61:40 (1957).

393. P. V. Phung and M. Burton, Radiation Res., 7:199 (1957).

394. M. C. Sauer and B. Ward, J. Phys. Chem., 71:3971 (1967).

395. K. C. Kurien, P. V. Phung, and M. Burton, Radiation Res., 11:283 (1959).

396. E. J. Land and M. Ebert, Trans. Faraday Soc., 63:1181 (1967).

397. K. D. Asmus, B. Cerček, M. Ebert, A. Henglein, and A. Wigger, Trans. Faraday Soc., 63:2435 (1967).

398. M. Anbar, D. Meyerstein, and P. Neta, J. Phys. Chem., 70:2660 (1966).

399. G. E. Adams and B. D. Michael, Trans. Faraday Soc., 63:1171 (1967).

400. B. Cerček and M. Ebert, Trans. Faraday Soc., 63:1687 (1967).

401. G. Scholes, P. Shaw, R. L. Wilson, and M. Ebert, in: Pulse Radiolysis (M. Ebert, ed.), New York, Academic Press (1965), p. 151.

402. W. J. Heilman, A. Rembaum, and M. Szwarc, J. Chem. Soc., 1957, p. 1127.

403. M. Levy and M. Szwarc, J. Am. Chem. Soc., 77:1949 (1955).

404. J. Gresser, J. H. Binks, and M. Szwarc, J. Am. Chem. Soc., 81:5004 (1959).

405. J. Smid and M. Szwarc, J. Am. Chem. Soc., 78:3322 (1956).

406. J. Smid and M. Szwarc, J. Am. Chem. Soc., 79:1534 (1957).

407. I. M. Whittemore, A. P. Stefani, and M. Szwarc, J. Am. Chem. Soc., 84:3799 (1962).

408. A. P. Stefani and M. Szwarc, J. Am. Chem. Soc., 84:3661 (1962).

409. E. C. Kooyman and E. Farenhorst, Trans. Faraday Soc., 49:58 (1953).

410. E. Farenhorst and E. C. Kooyman, Rec. Trav. Chim., 81:816 (1962).

411. L. R. Mahoney, J. Am. Chem. Soc., 86:444 (1964).

412. F. S. Dainton and T. J. Hardwick, Trans. Faraday Soc., 53:333 (1957).

413. G. Buxton and W. K. Wilmarth, J. Phys. Chem., 67:2835 (1963).

414. A. Rembaum and M. Szwarc, J. Am. Chem. Soc., 77:4468 (1955).

415. J. L. Kice, J. Am. Chem. Soc., 76:6274 (1954); J. Polymer Sci., 19:123 (1956).

416. C. H. Bamford and E. F. T. White, Trans. Faraday Soc., 52:716 (1956).

417. T. Ya. Smirnova, S. R. Rafikov, and G. P. Gladyshev, Vysokomolekul. Soedin., 10B:341 (1968).

418. P. D. Bartlett and H. Kwart, J. Am. Chem. Soc., 72:1051 (1950).

419. L. M. Postnikov, N. G. Zarubin, E. V. Vichutinskaya, V. Ya. Shlyapintokh, N. V. Martynov, and Ya. É. Bryuske, Izv. Akad. Nauk SSSR, Ser. Khim., 1968, p. 260.

420. H. A. Schwarz, J. Phys. Chem. 67:2827 (1963).

421. R. L. Strong, J. Am. Chem. Soc., 87:3563 (1965).

422. H. Rosman and R. M. Noyes, J. Am. Chem. Soc., 80:2410 (1958).

423. R. L. Strong and J. E. Willard, J. Am. Chem. Soc., 79:2098 (1957).

424. S. Aditya and J. E. Willard, J. Am. Chem. Soc., 79:2680 (1957).

425. R. Marshall and N. Davidson, J. Chem. Phys., 21:2086 (1953).

426. J. Zimmerman and R. M. Noyes, J. Chem. Phys., 18:658 (1950).

427. M. Ebert, J. P. Keene, E. J. Land, and A. J. Swallow, Proc. Roy. Soc. (London), A287:1 (1965).

428. Handbook: Rupture Energies of Chemical Bonds. Ionization Potential and Electron Affinity. Moscow, Izd. Akad. Nauk SSSR (1962).

429. P. S. Dixon, A. P. Stefani, and M. Szwarc, J. Am. Chem. Soc., 85:2551 (1963).

430. S. Fujita, S. Kodama, and O. Toyama, Bull. Chem. Soc. Japan, 39:1323 (1966).

431. P. S. Dixon, A. P. Stefani, and M. Szwarc, J. Am. Chem. Soc., 85:3344 (1963).

432. A. P. Stefani, J. Am. Chem. Soc., 90:1694 (1968).

433. S. Kodama, S. Fujita, J. Takeishi, and O. Toyama, Bull. Chem. Soc. Japan, 39:1009 (1966).

434. S. F. Nelsen and P. D. Bartlett, J. Am. Chem. Soc., 88:137 (1966).

435. W. A. Cramer, J. Phys. Chem., 71:1171 (1967).

436. I. A. Taub and L. M. Dorfman, J. Am. Chem. Soc., 84:4053 (1962).

437. R. L. McCarthy and A. MacLachlan, Trans. Faraday Soc., 57:1107 (1961).

438. J. C. Bevington, H. W. Melville, and R. P. Taylor, J. Polymer Sci., 12:449 (1954); 14:463 (1954).

439. D. Mickwich and J. Turkerich, J. Phys. Chem., 72:3703 (1968).

440. R. W. Fessenden, J. Phys. Chem., 68:1508 (1964).

441. R. D. Burkhart, J. Am. Chem. Soc., 90:273 (1968).

442. R. J. Hagemann and H. A. Schwarc, J. Phys. Chem., 71:2694 (1967).

443. G. Porter and M. W. Windsor, Nature, 180:187 (1957).

444. R. L. McCarthy and A. MacLachlan, Trans. Faraday Soc., 56:1187 (1960).

445. A. A. Vichutinskii, Candidate's Dissertation, Moscow (1965).

446. L. Bateman, J. L. Bolland, and G. Gee, Trans. Faraday Soc., 47:274 (1951).

447. C. H. Bamford and M. J. S. Dewar, Proc. Roy. Soc., (London), A198:252 (1949).

448. S. Weiner and G. S. Hammond, J. Am. Chem. Soc., 90:1659 (1968).

449. L. M. Dorfman and I. A. Taub, J. Am. Chem. Soc., 85;2370 (1963).

450. W. A. Seddon and A. O. Allen, J. Phys. Chem., 71:1914 (1967).

451. A. Beckett and G. Porter, Trans. Faraday Soc., 59:2038 (1963).

452. L. M. Strigun, L. S. Vartanyan, and N. M. Émanuél', Usp. Khim., 37:969 (1968).

453. J. K. Thomas, J. Phys. Chem. 67:2593 (1963).

454. E. J. Land and G. Porter, Trans. Faraday Soc., 59:2016 (1963).

455. A. Hubele, H. Suhr, and U. Heilmann, Chem. Ber., 95:639 (1962).

456. D. J. Williams and R. Kreilick, J. Am. Chem. Soc., 89:3408 (1967).

457. C. D. Cook and B. E. Norcross, J. Am. Chem. Soc., 81:1176 (1959).

458. C. D. Cook and N. D. Gilmour, J. Org. Chem., 25:1429 (1960).

459. N. K. Bridge and G. Porter, Proc. Roy. Soc. (London), A244:276 (1958).

460. G. Czapski and B. H. Bielski, J. Phys. Chem., 67:2180 (1963).

461. B. H. J. Bielski and A. O. Allen, Proceedings Society Tihany Symposium Radiation Chemistry, Budapest, (1967), p. 81.

462. H. S. Blanchard, J. Am. Chem. Soc., 81:4548 (1959).

463. T. G. Traylor and P. D. Bartlett, Tetrahedron Letters, No. 24, p. 30 (1960).

464. P. D. Bartlett and T. G. Traylor, J. Am. Chem. Soc., 85:2407 (1963).

465. R. Hiatt and T. G. Traylor, J. Am. Chem. Soc., 87:3766 (1965).

466. J. R. Thomas, J. Am. Chem. Soc., 87:3935 (1965).

467. T. G. Traylor and C. A. Russell, J. Am. Chem. Soc., 87:3698 (1965).

468. N. A. Milas and S. M. Djokic, Chem. Ind. (London), 1962, p. 405.

469. P. D. Bartlett and G. Guaraldi, J. Am. Chem. Soc., 89:4799 (1967).

470. G. A. Russell, J. Am. Chem. Soc., 79:3871 (1957).

471. J. A. Howard and K. U. Ingold, Can. J. Chem., 43:2737 (1965); J. Am. Chem. Soc., 90:1056 (1968).

472. B. H. J. Bielski and E. Saito, J. Phys. Chem., 66:2266 (1962).

473. J. H. Baxendale, Radiation Res., 17:312 (1962).

474. G. Czapski and L. M. Dorfman, J. Phys. Chem., 68:1169 (1964).

475. J. R. Thomas and K. U. Ingold, Advances in Chemistry Series, 75:258 (1968).

476-477. D. J. Currie and F. S. Dainton, Trans. Faraday Soc., 61:1156 (1965).

478. J. A. Howard and K. U. Ingold, J. Am. Chem. Soc., 90:1058 (1968).

479. R. L. McCarthy and A. MacLachlan, J. Chem. Phys., 35:1625 (1961).

480. J. R. Thomas and K. U. Ingold, International Symposium on Oxidation, Vol. 1, San Francisco, American Chemical Society (1967), p. 517.

481. W. J. Maguire and R. C. Pink, Trans. Faraday Soc., 63:1097 (1967).

482. J. J. Zwolenik, J. Phys. Chem., 71:2464 (1967).

483. V.M. Gol'dberg and L. K. Obukhova, Dokl. Akad. Nauk SSSR, 165:860 (1965).

484. A. MacLachlan, J. Am. Chem. Soc., 87:960 (1965).

485. E. J. Land and G. Porter, Trans. Faraday Soc., 59:2027 (1963).

486. B. Hölmström, Bull Soc. Chim. Belges, 71:869 (1962).

487. C. G. Hatchard and C. A. Parker, Trans. Faraday Soc., 57:1093 (1961).

488. C. A. Parker, J. Phys. Chem., 63:26 (1959).

488a. K. D. Asmus, A. Wigger, and A. Henglein, Ber. Bunsen. Phys. Chem., 70:862 (1966).

489. J. K. Thomas, Trans. Faraday Soc., 61:702 (1965).

490. H. A. Schwarz, Radiation Res. Suppl., 4:89 (1964).

491. K. Schested and O. L. Rasmussen, J. Phys. Chem., 72:626 (1968).

492. G. A. Russell, J. Am. Chem. Soc., 77:4583 (1955).

493. T. A. Ingles and H. W. Melville, Proc. Roy. Soc. (London), A218:163 (1953).

494. R. H. Schuler, J. Phys. Chem., 68:3873 (1964).

495. G. P. Gladyshev and K. M. Gibov, Polymerization with High Conversion and Methods of Studying this Process. Alma-Ata, "Nauka" KazSSR (1968).

496. S. Glasstone, K. Laidler, and H. Eyring, Theory of Rate Processes [Russian translation]. Moscow, GIIL (1948). [Original English-language edition: New York, McGraw-Hill (1941).]

497. G. E. Zaikov and Z. K. Maizus, Izv. Akad. Nauk SSSR, Ser. Khim., 1962, p. 1175.

498. G. E. Zaikov, Neftekhimiya, 3:381 (1963).

499. G. E. Zaikov and Z. K. Maizus, Dokl. Akad. Nauk SSSR, 150:116 (1963).

500. G. E. Zaikov, Z. K. Maizus, and N. M. Émanuél', Kinetika i Kataliz, 7:401 (1966).

501. J. A. Howard and K. U. Ingold, Can. J. Chem., 42:1044 (1964).

502. D. G. Hendry and G. A. Russell, J. Am. Chem. Soc., 86:2368 (1964).

503. Y. Kamiya, Bull. Chem. Soc. Japan, 38:2156 (1965).

504. G. E. Zaikov, A. A. Vichutinskii, Z. K. Maizus, and N. M. Émanuél', Dokl. Akad. Nauk SSSR, 168:1096 (1966).

505. L. M. Andronov, G. E. Zaikov, Z. K. Maizus, and N. M. Émanuél', Zh. Fiz. Khim., 41:2002 (1967).

506. E. T. Denisov, Izv. Akad. Nauk.SSSR, Otd. Khim. Nauk, 1960, p. 53.

507. F. F. Rust and E. A. Youngman, J. Org. Chem., 27:3778 (1962).

508. A. L. Buchachenko, Dokl. Akad. Nauk SSSR, 158:932 (1964).

509. G. E. Zaikov, Z. K. Maizus, and N. M. Émanuél', Dokl. Akad. Nauk SSSR, 173:859 (1967).

510. G. E. Zaikov, L. M. Andronov, Z. K. Maizus, and N. M. Émanuél, Dokl. Akad. Nauk SSSR, 174:127 (1967).

511. L. M. Andronov, G. E. Zaikov, and Z. K. Maizus, Zh. Fiz. Khim., 41:1122 (1967).

512. G. E. Zaikov, Z. K. Maizus, and N. M. Émanuél', Neftekhimiya, 8:217 (1968); Izv. Akad. Nauk SSSR, Ser. Khim., 1968, p. 2265.

513. G. A. Russell, J. Am. Chem. Soc., 79:2977 (1957); 80:4987, 4997, 5002 (1958).

514. G. A. Russell, A. Ito, and D. G. Hendry, J. Am. Chem. Soc., 85:2976 (1963).

515. R. L. Strong and J. Perano, J. Am. Chem. Soc., 83:2843 (1961).

516. A. L. Buchachenko, O. P. Sukhanova, L. A. Kalashnikova, and M. B. Neiman, Kinetika i Kataliz, 6:601 (1965).

PART THREE

IONIC HOMOLYTIC REACTIONS

CHAPTER X

OXIDATION-REDUCTION REACTIONS OF IONS WITH MOLECULES

§1. Oxidation of Organic Compounds by Variable-Valence Metal Ions

Polar organic compounds (alcohols, ketones, hydroxy acids, acids) are oxidized by variable-valence ions (Co^{3+}, Ce^{4+}, Mn^{3+}, VO_2^+) through a one-electron mechanism with the formation of free radicals. In water, water-alcohol, and acetic acid solutions, the oxidation normally includes two stages. First, the organic compound enters into the internal coordination sphere of the ion-oxidant by way of an equilibrium reaction. Then, in this complex, the oxidation-reduction reaction takes place. The scheme of such a reaction is represented in simplified form as follows:

$$RH + Me^{n+1} \underset{}{\overset{K}{\rightleftharpoons}} HR\ldots Me^{n+1} \xrightarrow{k_1} R\cdot + Me^{n+} + H^+$$

If the equilibrium is shifted to the left, then

$$W = k[RH][Me^{n+1}],$$

where $K = k_1K$. If a considerable number of the metal ions are bound in a complex, then

$$W = \frac{k_1K[Me^{n+1}][RH]}{1 + K[RH]} \text{ and } \frac{[Me^{n+1}]}{W} = \frac{1}{k_1} + \frac{1}{k_1K[RH]}$$

As a rule, the rate of the oxidation reaction depends on the hydrogen-ion concentration. In the case of alcohol oxidation, this is governed by the presence of two or several forms of the oxidant ions: Co^{3+} and $CoOH^{2+}$, Ce^{4+}, and $CeOH^{3+}$, etc., related to each other by the equilibrium

$$Co^{3+}_{aq} \rightleftharpoons CoOH^{2+} + H^{+}$$

The oxidation rate is given by

$$W = k[RH][Co^{3+}] + k''[RH][CoOH^{2+}] =$$
$$= k'[RH][Co^{3+}] + \frac{K_{H+}k''[RH][Co^{3+}]}{[H^{+}]} \approx$$
$$\approx \frac{K_{H^+}k''[RH][Co^{3+}]}{[H^{+}]}$$

with the condition that $k' \ll (K_H k'')/[H^+]$. Organic acids themselves dissociate into ions; in solution, both the molecules and the anions of these acids are oxidized, so that the reaction rate depends on pH. When mineral acids with anions that form stable complexes with metal ions are introduced into the solution, this complicates the kinetic relationships still more, since in this case the solution contains a collection of different complexes of the metal, all having different reactivities. The reaction rate in this case depends on both pH and the anion concentration.

The oxidation of ketones and aldehydes very often is independent of ion-oxidant concentration and is limited by the rate of enolization. Aromatic and unsaturated hydrocarbons are oxidized by metal ions in polar solvents by a reaction with electron transfer: $RH + Me^{n+1} \rightarrow R\cdot + Me^{n} + H^{+}$. For oxygen-containing compounds, replacement of the hydrogen atoms in the vicinity of the oxygen group by deuterium will slow down the oxidation rate somewhat (isotope effect). This indicates that the α-C-H bond participates in the elementary oxidizing act in the oxidation of alcohols, formic acid, and other compounds. For carbonyl compounds, the isotope effect is explained by the enolization as the limiting stage in the oxidation process.

The rate constants for reactions of ions with molecules have usually been measured by kinetic methods based on the ion-oxidant consumption. The substance to be oxidized is usually taken in excess, so that the ion-oxidant is consumed in accordance with a first-order reaction:

$$-\frac{d[Me^{n+1}]}{dt} = k'[Me^{n+1}]$$

where k is the apparent rate constant, which depends on pH, [RH], the ion strength μ, and the concentrations of ions and of compounds forming complexes with the oxidant.

Oxidation by Pentavalent Vanadium. In aqueous solutions, ions of pentavalent vanadium exist in the form VO_2^+ and $V(OH)_3^{2+}$, the ratio between these two forms depending on pH because of the equilibrium $VO_2^+ + H_3O^+ \rightleftharpoons V(OH)_3^{2+}$. The oxidation of an alcohol (or aldehyde or acid) is preceded by complex formation:

$$RH + VO_2^+ \overset{K_1}{\rightleftharpoons} \text{complex} \longrightarrow V(IV) + \text{products}$$

$$RH + V(OH)_3^{2+} \overset{K_2}{\rightleftharpoons} \text{complex} \longrightarrow V(IV) + \text{products}$$

The reaction rate constant measured in an experiment, k = W/[V(V)] RH , is an over-all constant that includes K_1 and k_1 if VO_2^+ is the oxidant, K_{H+}, K_2, and k_2 if $V(OH)_3^{2+}$ is the oxidant, or all five constants if the oxidant consists of both VO_2^+ and $V(OH)_3^{2+}$. In the oxidation of malonic acid and formaldehyde, it was established that $W \sim [V^V] + a[V^V]^2$), and it was concluded that these compounds are oxidized in both mononuclear and binuclear complexes:

$$RH + V(OH)_3^{2+} \overset{K_2}{\rightleftharpoons} X_1 \xrightarrow{k_2} \text{products}$$

$$X_1 + V(OH)_3^{2+} \overset{K_3}{\rightleftharpoons} X_2 \xrightarrow{k_3} \text{products}$$

In Table 99, experimental values are listed for k = W/[V^V]RH and their theoretical treatment within the framework of the scheme set forth above. In the presence of acids with anions that form stable complexes with vanadium, the mechanism of the oxidation is more complicated. For example, in the oxidation of alcohols in the presence of H_2SO_4, it was established [1] that $W \sim [H_2SO_4]^2$ and $W \sim ([NaHSO_4] + [H_2SO_4])^2$, and it was concluded that the alcohols are oxidized by the complex $V(OH)_2(HSO_4)^{2+}$ or $V(OH)(H_2O)(HSO_4)(SO_4)^+$.

The following abbreviations are used to denote the ligands in complexes:

En: $NH_2-CH_2-CH_2-NH_2$

Phen: 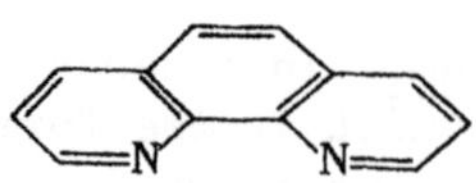

Dipy: 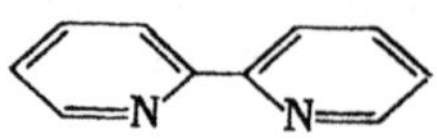

Tripy: 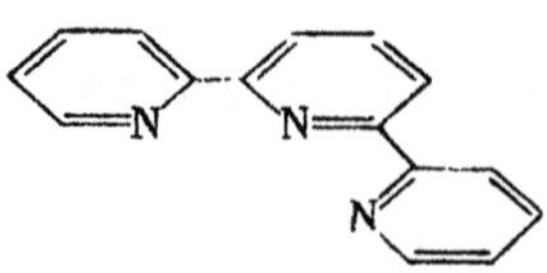

EDTA : 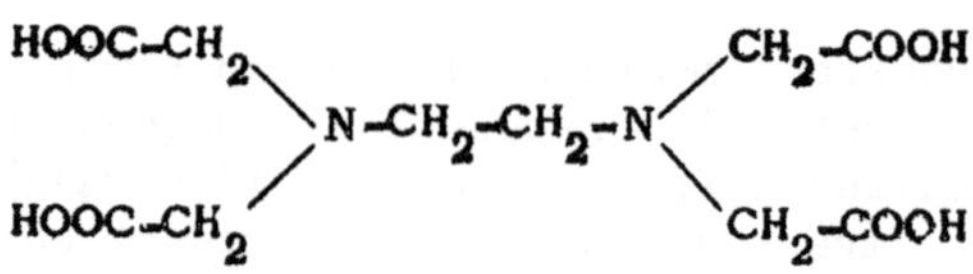

PDTA : 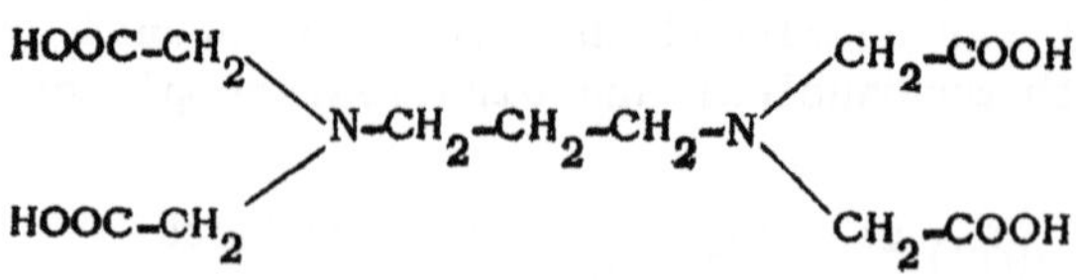

EDTAOH : 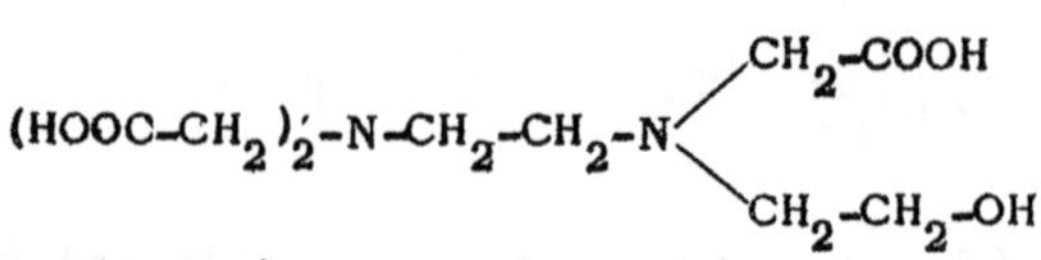

TEA: $N(CH_2-CH_2-OH)_3$

Tetren : $NH_2-(CH_2)_2-NH-(CH_2)_2-NH-(CH_2)_2-NH-(CH_2)_2-NH_2$

Cyt C : Cytochrome C

Hb : Hemoglobin

NTA: Nitriletriacetate

Ac Ac: $CH_3C(O^-) = CHC(=O)CH_3$

Oxin : 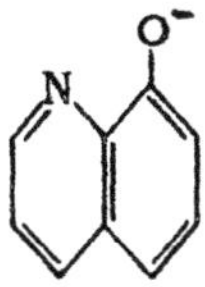

α-$C_{10}H_{11}OOH$: α-tetralyl hydroperoxide

$C_{10}H_8$: Naphthalene

$C_{12}H_{10}$: Biphenyl

$C_{14}H_{10}$: Anthracene

$C_{18}H_{12}$: Naphthacene

$C_{18}H_{14}$: 

Oxidation by Trivalent Manganese. The trivalent manganese ion is a strong oxidizer. Oxidation reactions involving sulfate and pyrophosphate complexes of trivalent manganese have been studied in aqueous solution. Carbonyl compounds as a rule are oxidized in the form of enols, and the oxidation rate is often limited by the enolization rate. Acids and hydroxy acids first form a complex with trivalent manganese, within which the oxidation-reduction reaction is accomplished. The most widely accepted scheme for the oxidation is

$$RH + Mn(III) \underset{}{\overset{K}{\rightleftharpoons}} \text{complex} \xrightarrow{k_1} \text{products } Mn(II)$$

Therefore, the experimental bimolecular rate constants listed in Table 100 are close to k_1K.

According to [20, 21], the oxidation of aromatic compounds proceeds through the preliminary formation of a complex with charge transfer according to the equilibrium

$$RH + Mn(OOCCH_3)_3 \overset{K}{\rightleftharpoons} RH^+ + CH_3COO^- + Mn(OOCCH_3)_2$$

$$RH^+ + CH_3COO^- \longrightarrow R\cdot + CH_3COOH$$

Corresponding to this mechanism, the reaction rate is given by W ∿ [RH][Mn(III)]/[Mn(II)]. In Table 100, values are listed for k = W/[Mn(III)][RH].

Oxidation by Trivalent Iron. The oxidation of ascorbic acid by complexes of trivalent iron was studied in [38]. The anion of the acid is oxidized, and the reaction rate constant $k = k_1K/[H^+]$, where K is the dissociation constant of the acid. The experiments were run at temperatures of 0.4 to 25°, μ = 0.1 M, and pH 1.8-3.45. Values of the constants are as follows:

Complexing agent	log k (25°)	log A	E
$HOOCCH_2N(OH)CH_2CH_2N(CH_2COOH)_2$	4,33	8,44	5,6
$(HOOCCH_2)_2NCH_2CH_2N(CH_2COOH)_2$	3.60	7.36	5.1
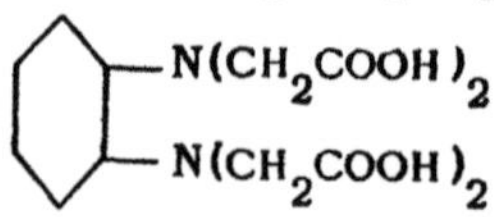	3.11	6.71	4.9
$[(HOOCCH_2)_2NCH_2CH_2]_2NCH_2COOH$	2.94	6.44	4.7

Oxidation by Trivalent Cobalt. The trivalent cobalt ion is a strong oxidizer, oxidizing even water. Reactions of oxidation by Co(III) ions are studied in strongly acidic solutions. The oxidation rate depends on pH as a result of the equilibrium: $Co_{aq}^{3+} \rightleftharpoons CoOH^{2+} + H^+$. The scheme for oxidation by Co^{3+} ions is the same as that for trivalent manganese. As the hydrogen-ion concentration increases, the reaction is slower, since the value of k_1K for $CoOH^{2+}$ is larger than that for Co^{3+}. In the presence of sulfuric acid, the reaction rate depends on both the hydrogen-ion concentration and on $[HSO_4^-]$.

Oxidation by Copper Ions. The oxidation of carbonyl compounds (aldehydes, ketones, sugars) by copper complexes is as a rule limited by the enolization rate and does not depend on copper ion concentration. One of the exceptions is the oxidation of α-hydroxyacetophenone by copper ions in the presence of pyridine [60]. The oxidation rate is given by $W = W_{en} + k[Cu^{2+}][C_6H_5COCH_2OH]$, where W_{en} is the enolization rate. The following oxidation mechanism is postulated:

$$C_6H_5COCH_2OH + Cu^{2+} \overset{K}{\rightleftharpoons} \text{complex} \xrightarrow{k_1} Cu^{+} + \text{radical}$$

and $k = k_1K$. At 25° in 50% pyridine with [pyridine H^+] = 0.02 M and μ = 0.062 M, $k = 2.97 \cdot 10^{-3}$ liter/mole·sec, log A = 11.77, and E = 19.5 kcal/mole (measured over the range 15-35°).

Oxidation by Tetravalent Cerium. The Ce^{4+} ion is a strong oxidizer. The rate of oxidation by cerium ions depends on pH as a consequence of the equilibrium:

$$Ce_{aq}^{4+} \rightleftharpoons CeOH^{3+} + H^{+}$$

along with the difference in activity between Ce^{4+} and $CeOH^{3+}$. The oxidation of oxygen-containing compounds proceeds through the preliminary formation of a complex.

Oxidation by Iridium Ions. The oxidation of phenols by tetravalent iridium ions was studied in [92]. The reaction is bimolecular. The pH dependence of the reaction rate is contingent on dissociation of the phenol and the rapid reaction between ArO^- and $IrCl_6^{2-}$. Trivalent iridium ions retard the reaction; this is related to their reducing action:

$$H^{+} + ArO\cdot + IrCl_6^{3-} \longrightarrow ArOH + IrCl_6^{2-}$$

For the reaction $C_6H_5OH + IrCl_6^{2-}$, log k = $\bar{1}$.875 (20.8°), log A = 7.425, E = 0.8 kcal/mole (11-50°, [$HClO_4$] = 1 M). For the reaction 2,6-dimethylphenol + $IrCl_6^{2-}$, log k = 1.60 (20.4°), log A = 9.42, E = 10.5 kcal/mole (3.6-27.2°, [$HClO_4$] = 1 M).

§2. Reaction of Variable-Valence Metal Ions with Oxygen, Peroxides, and Quinones

Variable-valence ions in the reduced form can be oxidized by such oxidants as molecular oxygen, ozone, peroxides, and quinones. The oxidation rate depends on pH, and also on concentration of complexing anions. The rate constants have been measured by conventional kinetic methods.

<u>Oxidation by Molecular Oxygen.</u> The vast majority of the studies in this field relate to the oxidation of iron ions by molecular oxygen. The oxidation of Fe^{2+} is first-order with respect to oxygen and, depending on conditions, first or second-order with respect to Fe^{2+}. The limiting stage includes transfer of an electron from Fe^{2+} to oxygen. The postulated reaction mechanism is as follows:

$$Fe^{2+}_{aq} + O_2 \rightleftharpoons FeO_2^{2+}$$

$$FeO_2^{2+} \longrightarrow Fe^{3} + O_2^{-} \quad \text{(limiting stage)}$$

$$O_2^{-} + Fe^{2+} \longrightarrow O_2^{2-} + Fe^{3+}$$

$$O_2^{2-} + 2H^{+} \rightleftharpoons H_2O_2$$

$$H_2O_2 + Fe^{2+} \longrightarrow Fe^{3+} + HO^{-} + HO\cdot$$

$$HO\cdot + Fe^{2+} \longrightarrow HO^{-} + Fe^{3+}$$

The second order with respect to Fe^{2+} is explained by the formation of a binuclear complex and its more rapid oxidation:

$$FeO_2^{2+} + Fe^{2+} \rightleftharpoons Fe^{2+}O_2Fe^{2+}$$

$$Fe^{2+}O_2Fe^{2+} \longrightarrow Fe^{3+} + FeO_2^{+} \quad \text{(limiting stage)}$$

$$FeO_2^{+} \longrightarrow Fe^{3+} + O_2^{2-}$$

With increasing hydrogen-ion concentration, the rate of Fe^{2+} oxidation decreases; this is related to hydrolysis of Fe^{2+} [34]:

$$Fe^{2+}_{aq} \rightleftharpoons FeOH^{+} + H^{+}$$

and the more rapid oxidation of $FeOH^{+}$. However, in concentrated hydrochloric acid, an increase in oxidation rate is observed as the acid concentration is increased. This is explained by the formation and rapid oxidation of $Fe^{2+}\cdot HCl$ complexes [93]:

$$Fe^{2+}_{aq} + HCl \rightleftharpoons Fe^{2+}ClH$$

$$Fe^{2+}ClH + O_2 \longrightarrow FeCl^{2+} + HO_2\cdot$$

Oxidation by Ozone. According to [106] , the rate constant for the reaction $Co^{2+} + O_3$ is equal to $5.5\cdot10^{-5}$ (0°), $\log k = 3.44 - 9.6/\Theta$.

Reactions with Peroxides. Peroxides oxidize variable-valence ions through a one-electron mechanism--for example,

$$H_2O_2 + Fe^{2+} \longrightarrow HO\cdot + HO^- + Fe^{3+}$$

Complex formation precedes the oxidation:

$$Fe^{2+}_{aq} + H_2O_2 \rightleftharpoons Fe\cdot H_2O_2^{2+} \longrightarrow Fe^{3+} + HO + HO^-$$

In the majority of cases, such a complex has a 1 : 1 composition; but in certain cases the reaction rate is second-order with respect to metal ion, and it is possible that the conversion takes place in a binuclear complex. By far the greatest number of rate constants for peroxide reactions with metal ions have been measured by conventional kinetic methods. In some cases, jet methods have been used.

TABLE 99. Oxidation by Pentavalent Vanadium Ions

Compound	Conditions	t°	Expression for k	log k (50°)	log A	E	Ref.
	Oxidation in aqueous solutions						
$n\text{-}C_3H_7OH$	$[HClO_4] = 1\ M$; $\mu = 3{,}1\ M$	50		$\bar{5},37$			[2]
	$[HClO_4] = 2\ M$	30–60		$\bar{7},92$	8,29	21,2	[1]
	$[H_2SO_4] = 2\ M$	30–60		$\bar{5},30$	10,95	23,1	[1]
$(CH_3)_2CHOH$	$[HClO_4] = 1\ M$; $\mu = 3{,}1\ M$	50		$\bar{6},88$			[2]
$n\text{-}C_4H_9OH$	- " -	50		$\bar{5},36$			[2]
cyclo-$C_6H_{11}OH$	$[HClO_4] = 1\ M$; $\mu = 3{,}1\ M$	50		$\bar{5},38$			[2]
	$[HClO_4] = 4{,}77\ M$	25–60		$\bar{6},85$	10,10	22,5	[3]
$CH_2{=}CHCH_2OH$	$[HClO_4] = 1\ M$; $\mu = 3{,}1\ M$	50		$\bar{4},85$			[2]
$CH_3CH{=}CHCH_2OH$	- " -	50		$\bar{4},86$			[2]
$C_6H_5CH_2OH$	60% CH_3COOH	25–45		$\bar{4},70$	5,65	13,22	[4]
	80% CH_3COOH	25–45		$\bar{3},35$	5,26	11,68	[4]

$HOCH_2CH_2CH_2OH$	$[HClO_4] = 2\ M$	30–60	$\bar{5},00$	13,23	26,9	[1]
	$[H_2SO_4] = 2\ M$	30–60	$\bar{5},58$	11,11	22,9	
$HOCH_2(CH_2)_2CH_2OH$	$[HClO_4] = 2\ M$	30–60	$\bar{5},04$	11,31	24,0	
	$[H_2SO_4] = 2\ M$	30–60	$\bar{5},50$	10,08	21,6	
$CH_3CH(OH)CH_2OH$	$[H_2SO_4] = 1\ M$	60	$\bar{4},59$			[5]
$CH_3CH(OH)CH(OH)CH_3$	- " -	60	$\bar{4},58$			
$(CH_3)_2C(OH)CH(OH)CH_3$	- " -	60	$\bar{3},41$			
cis 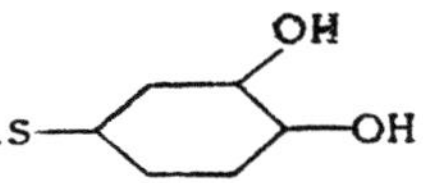	$[H_2SO_4] = 1\ M$	60	$\bar{4},68$			
trans 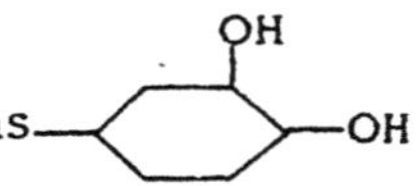	- " -	60	$\bar{4},15$			
trans 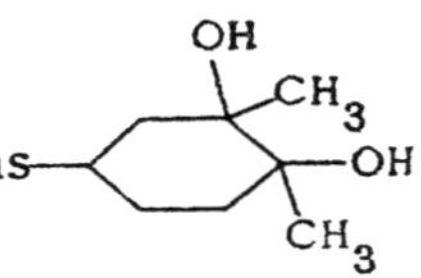	- " -	60	$\bar{4},76$			
cis 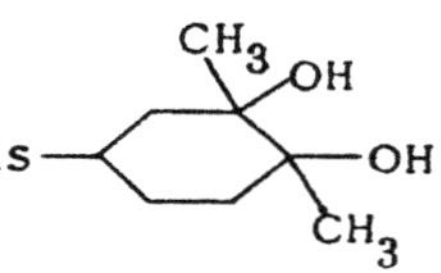	- " -	60	$\bar{3},04$			[5]

Table 99 (continued)

Compound	Conditions	t°	Expression for k	log k (50°)	log A	E	Ref.
$(CH_3)_2C(OH)C(CH_3)_2OH$	$[HClO_4]$ = =0,5–5 M; μ = 5 M	25		$k = k_1 + k_2 h_o$; $\log k_1 = \bar{1},52$; $\log k_2 = \bar{4},31$			[6]
$CH_3OC(CH_3)_2C(CH_3)_2OH$	$[HClO_4]$ = = 1 – 5 M	50		$k = k_1 [H^+]$; $\log k_1 = \bar{3},21$			[6]
CH_2O	$[HClO_4]$ = = 0,7÷5,4 M; μ =5M	50	$k_1 [H^+] + k_2 [H^+]^2 [V]$	$\log k_1 = \bar{5},365$; $\log k_2 = \bar{4},05$			[3]
CH_3CH_2CHO	$[HClO_4]$ = 1 M; μ = 3,1 M	25	$k_1 K/(1 + K [RH])$	$\log k_1 = \bar{4},59$; $\log K = 0,16$			[3] [7]
$CH_3(CH_2)_2CHO$	$[HClO_4]$ = 1 M, μ =3,1 M	25	$k_1 K(1 + K [RH])$	$\log k_1 = \bar{4},52$; $\log K = 0,13$			[7]
$(CH_3)_2CHCHO$	$[HClO_4]$ = 1 M, = 3,1 M	25		$\bar{6},98$			

$cyclo\text{-}C_5H_8O$	$[H_2SO_4]$ = 2,6 M	40		$\bar{3}$,915			[8]
$cyclo\text{-}C_6H_{10}O$	$[H_2SO_4]$ = 0,55 M	35		$\bar{3}$,31			[8]
	$[HClO_4]$ = 1÷5,5 M; μ = 6,0M	35	$k_1[H^+]$	log k_1 = 3,34			[8]
$CH_3COCH(OH)CH_3$	$[HClO_4]$ = 1 M; μ = 3,1 M	25		$\bar{2}$,48			[9]
$(CH_3)_2C(OH)COCH_3$	– " –	25		$\bar{3}$,82			[9]
HCOOH	$[H_2SO_4]$ = 0,46 M	33–58		$\bar{6}$,26	9,72	22,8	[10]
$HOOCCH_2COOH$	$[H_2SO_4]$ = 1,81 M	56,5	k_1 + + $k_2[V]$	log k_1 = $\bar{5}$,04; log k_2 = $\bar{4}$,76			[10]
$HOCH_2COOH$	$[H_2SO_4]$ = 1,72 M	25–60		$\bar{5}$,58(50°)	8,92	19,7	[10]
	$[HClO_4]$ = 2,13 M; μ = 2,5 M	38	k_1 + + k_2h_o	log k_1 = $\bar{4}$,14; log k_2 = $\bar{5}$,43	9,01	16,7	[11]
	$[H_2SO_4]$ = 1,47 M	26,6		$\bar{3}$,865			[12]
$CH_3CH(OH)COOH$	$[H_2SO_4]$ = 1,47 M	26,6		$\bar{3}$,715			[12]
	$[HClO_4]$ = = 0.5 to 4 M; μ = 3 M	25	k_1 + + k_2h_o	log k_1 = 4,81; log k_2 = $\bar{5}$,69			[6]

Table 99 (continued)

Compound	Conditions	t°	Expression for k	log k (50°)	log A	E	Ref.
$CH_3CH(OH)COOH$	$[HClO_4]$ = 0,67 M	30–40		$\bar{5}$,97	7,77	16,5	[13]
$(CH_3)_2C(OH)COOH$	$[HClO_4]$ = 2 M	30–50		$\bar{2}$,12	13,02	22,0	[14]
	$[H_2SO_4]$ = 2 M	20–45		$\bar{3}$,97	13,57	23,0	[14]
	$[HClO_4]$ = 2 to 5 M; μ = 5 M	18–35	$k_1[H^+]$	$\log k_1$ = $\bar{3}$,87	12,65	21,8	[15]
	$[HClO_4]$ = 1 M; μ = 3,1 M	25		$\bar{4}$,81			[12]
$C_6H_5CH(OH)COOH$	$[HClO_4]$ = 0,5–2,6 M; μ = 5,0 M	25	$k_1 + k_2h$	$\log k_1$ = $\bar{2}$,51; $\log k_2$ = $\bar{4}$,61			[6]
	$[H_2SO_4]$ = 1,47 M	26,6		$\bar{1}$,21			[12]
$HOOCCH_2CH(OH)COOH$	$[HClO_4]$ = 0,5–4,6 M; μ = 5,0 M	25	$k_1 + k_2h_o$	$\log k_1$ = $\bar{4}$,94; $\log k_2$ = $\bar{5}$,75			[6]
$HOOCCH_2CH(SH)COOH$	pH 2,4–4,4	15	$k_1 + k_2[RH]$	$\log k_1$ = 1,70; $\log k_2$ = 2,60			[16]

COC(OH)=C(OH)CHCHOH, $HOCH_2$ (ring closed through O)	pH 1,75–2,85	25–43	$k[H^+]$	2,25 (25°)	7,84	7,6	[17]
$CH(OH)(CH_2COOH)_2$	$[HClO_4]$ = 1 M	25		$\bar{2},26$	13,78	21,2	[18]
	H_2SO_4 = 1 M	25		$\bar{2},23$	12,73	19,2	[18]
	Oxidation in nonaqueous solutions						
$C_6H_5CH_3$	70% CH_3COOH +	60		$\bar{3},76$			[19]
$m\text{-}NO_2C_6H_4CH_3$	+ 30% RH +	60		$\bar{5},11$			[19]
$p\text{-}NO_2C_6H_4CH_3$	+ 4 M H_2SO_4	60		$\bar{6},95$			[19]

Note: μ is ion strength; h_o is acidity function; and RH is the substance being oxidized.

TABLE 100. Oxidation by Trivalent Manganese Ions

Compound	Conditions	t°	Expression for k	log k	log A	E	Ref.
	Oxidation in aqueous solutions						
cyclo-$C_6H_{11}OH$	$[Mn(II)]$ = 0,167 M; $[H_2SO_4]$ = 1,97 M;	24–54		$\bar{4},30$ (25°)	17,50	28,9	[22]
CH_2O	$[Mn(II)]$ = 0,20 M; $[H_2SO_4]$ = 2,025 M	25–57		$\bar{4},34$ (25°)	14,81	25,2	[22]

Table 100 (continued)

Compounds	Conditions	t°	Expression for k	log k	log A	E	Ref.
Glucose	$[H_2P_2O_7^{2-}]$ = 0,112 M; $[H^+]$ = 4,5 M	20–35	k = $k_1K/(1 + K[RH])$	$\log k_1 = \bar{3},46$ (30°); log K = 1,79 (30°)	14,33	23,3	[23]
HCOOH	$[H_2SO_4]$ = 4,5 M	40		$\bar{6},89$			[24]
COOH–COOH	$[Mn(II)]$ = 0,01 M; μ = 2 M	18–36	$MnC_2O_4^+ \xrightarrow{k_1}$	$\bar{1},29$ (25°)	14,30	18,3	[22, 25]
			$Mn(C_2O_4)_2^- \xrightarrow{k_2}$	$\bar{4},88$ (25°)			[25]
			$Mn(C_2O_4)_3^{3-} \xrightarrow{k_3}$	$\bar{4},53$ (25°)	14,48	22,2	[25]
$HOOCCH_2COOH$	$[H_2SO_4]$ = 2,95 M; $[Mn(II)]$ = 0,37 M	25	$k = k_1K/(1 + K[RH])$	$\log k_1 = 3,29$; log K = 1,285			[26]
	$[RH]$ = 0,0113 M; $[Mn(II)]$ = 0,094 M	25	$k[RH] = k_1 + k_2[H^+]$	$\log k_1 = 1,70$; $\log k_2 = 1,25$			[26]
	$[RH]$ = 0,017 M; $[H_2SO_4]$ = 2,95 M	25	$1/k[RH] = k_1 + k_2[Mn(II)]$	$\log k_1 = 1,48$; $\log k_2 = 2,86$			[26]
$HOCH_2COOH$	$[H_2SO_4]$ = 2,47 M	24,4		$\bar{3},755$			[12]
$CH_3CH(OH)COOH$	- " -	24,4		$\bar{2},98$			[12]

$C_6H_5CH(OH)COOH$	$[H_2SO_4]$ = 3,38 M; $[Mn(II)]$ = 0,43 M	23,2		$\bar{2},24$			[12]
$(CH_3)_2C(OH)COOH$	$[H_2SO_4]$ = 3,35 M; $[Mn(II)]$ = 0,338 M	21–36		$\bar{1},70$ (24,4°)		20,0	[12]
$CH_3COCOOH$	$[H_2P_2O_7^{2-}]$ = 0,435 M; $[Mn(III)]$ = $1,79 \cdot 10^{-2}$ M,	10	$1/k = k_1 + k_2[RH]$	$\log k_1 = \bar{1},49$; $\log k_2 = \bar{1},52$			[27]
$CH_3COCOOH$	$[Mn(II)]$ = $1,08 \cdot 10^{-2}$ M; $[RH]$ = 3,92 M; pH 1,52; $[Mn(II)]$ = 0,437 M	10	$1/k = k_1 + k_2[H_2P_2O_7^{2-}]$	$\log k_1 = \bar{0},33$; $\log k_2 = 1,325$			[27]
1-$HOOCCH(OH)CH_2COOH$	$[H_2P_2O_7^{2-}]$ = 0,29 M; $[Mn(II)]$ = $4,9 \cdot 10^{-3}$; pH 1,9	35	$k = k_1K/(1 + K[RH])$	$\log k_1 = \bar{3},09$			[28]
+$HOOC[CH(OH)]_2COOH$	- " -	35	- " -	$\log k_1 = \bar{3},28$			[28]
meso-$HOOC[CH(OH)]_2COOH$	- " -	35	- " -	$\log k_1 = \bar{3},25$			[28]
	- " -	26–40	- " -	$\log k_1 = \bar{4},57$ (26°)	11,34	20,16	[29]

Table 100 (continued)

Compound	Conditions	t°	Expression for k	log k (50°)	log A	E	Ref
			Oxidation in nonaqueous solutions				
p-$CH_3OC_6H_4CH_3$	CH_3COOH	100	$k = k_1$/Mn (II)	$\bar{5}$,72			[20]
	- " -	70	- " -	$\bar{6}$,45			[20]
p-$CH_3OC_6H_4CH_2OH$	CH_3COOH	70	$k = k_1$/Mn (II)	$\bar{6}$,43			[20]
p-$CH_3OC_6H_4OCH_3$·	- " -	70	- " -	$\bar{7}$,75			[20]
p-$CH_3OC_6H_4OCOCH_3$	- " -	70	- " -	$\bar{8}$,99			[20]
α - naphthyl-OCH_3	- " -	100	- " -	$\bar{4}$,20			[21]
β - naphthyl-OCH_3	- " -	100	- " -	$\bar{4}$,08			[21]
$(n\text{-}C_4H_9)_2CO$	CH_3COOH + + n-$C_{10}H_{22}$	140		$\bar{2}$,81			[30]
$CH_3CH(OH)COCH_3$	C_6H_5Cl, Mn $(OOCCH_3)_3$	120		$\bar{1}$,70			[30]
$CH_3(CH_2)_4COO^-$	n-$C_{10}H_{22}$ + + $CH_3(CH_2)_4COOH$	120		0,30			[31]
$CH_3CH(OH)COOH$	CH_3COOH	25–50		$\bar{3}$,09 (25°)	7,36	14,0	[32]
$C_6H_5CH(OH)COOH$	- " -	30		$\bar{1}$,53			[32]

TABLE 101. Oxidation by Trivalent Iron Ions

Compound	Conditions	t°	Expression for k	log k	log A	E	Ref.
	Oxidation by Fe^{3+} ions in aqueous solutions						
$CH_3CH(OH)COCH_3$	$[HClO_4]$ = I M	25–50	$k = k_1 + k_2K/[H^+]$		13,56	24,7	[33]
		40		$\log k_1 = \bar{4},35$; $\log k_2 = \bar{3},85$; $\log K = \bar{3},545$			
C_6H_5OH	$[HClO_4]$ = 0,05–0,3 M	200	$k = k_1/[H^+]$	$\bar{4},545$			[34]
p-HOC_6H_4OH	pH 0	25		$\bar{1},69$	18,50	25,6	[35]
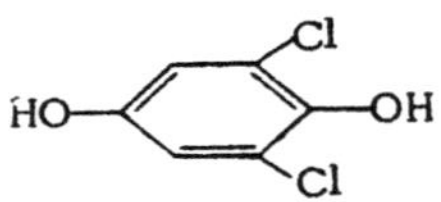	- " -	15–35		0,04 (25°)	17,34	23,5	[35]
							[35]
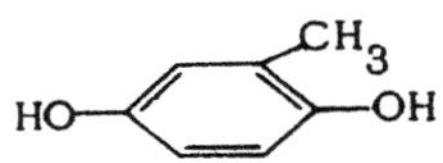	- " -	25–35		0,32 (25°)	16,96	22,6	[35]

Table 101 (continued)

Compound	Conditions	t°	Expression for k	log k	log A	E	Ref.
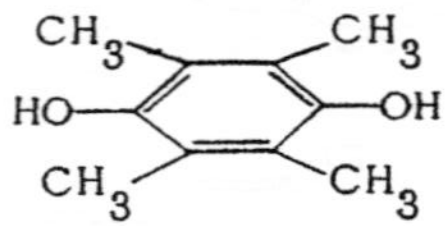	pH 0	15–35		0,18 (25°)	17,09	23,0	[35]
H_2NNH_2	$[H^+]$ = 1 M; μ = = 3,5 M	25		$\bar{2}$,78			[35a]
		35		$\bar{1}$,54			[35a]
	Oxidation by $Fe(Phen)_3$ in aqueous solutions						
cyclo-$C_6H_{10}O$	$[H_2SO_4]$ = 1,5 M	5–40		$\bar{1}$,43 (25°)	8,72	12,6	[36]
CH_3CHO	- " -	25		$\bar{5}$,91			[36]
cyclo-$C_6H_{11}OH$	- " -	25		$\bar{4}$,82			[36]
C_6H_5OH	- " -	25		$\bar{1}$,65°			[36]
$CH_3CH(OH)COCH_3$	- " -	25		$\bar{3}$,08			[36]
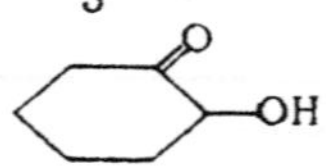	$[H_2SO_4]$ = 1,5 M	25		$\bar{2}$,83			[36]
$CH_3COCOCH_3$	- " -	25		$\bar{3}$,32			[36]

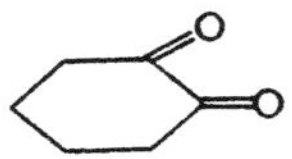	$[H_2SO_4] = 1,5$ M	25	0,98	[36]
$CH_3C(OH)(CH_3)C(OH)(CH_3)_2$	- " -	25	$\bar{4},08$	
HOOCCOOH	- " -	25	$\bar{3},08$	
$HOOCCH_2COOH$	- " -	25	$\bar{4},41$	
$CH_3COCOOH$	- " -	25	$\bar{3},82$	
$C_6H_5CH(OH)COOH$	- " -	25	$\bar{4},00$	
$CH_3COCH_2COOC_2H_5$	- " -	25	$\bar{4},83$	
Oxidation by $Fe[OOC(CH_2)_8CH_3]_3$ in nonaqueous solutions				
$C_6H_5CH_2SH$	$CH_3C_6H_4CH_3$	35	$\bar{1},875$	[37]
$CH_3(CH_2)_8CH_2SH$	$CH_3C_6H_4CH_3$	55	$\bar{1},38$	
	CH_3COCH_3	35	$\bar{2},77$	

TABLE 102. Oxidation by Trivalent Cobalt Ions

Compound	Conditions	t°	Expression for k or W	log k	log A	E	Ref.
			Oxidation in aqueous solutions				
H_2O	H_2SO_4	25	$k_1[Co^{3+}] +$	$\log k_1 = \bar{2},25$;	22,9	33,6	[39]
			$+ k_2[Co^{3+}]^2$	$\log k_2 = 0,75$	26,5	35,1	
	$HClO_4$	25	$W = k_2[Co^{3+}]^2$	$\log k_2 = 0,25$	29,3	39,6	[39]
	$HClO_4$	25	$W = k[Co^{3+}]^{3/2}[H^+]^{-2}$	$\bar{3},41$			[40]
$CH_2=C(CH_3)CH_2CH_3$	$[H_2SO_4] = 0,09$ M	25		0,34	20,20	27,1	[41]
$CH_2=CH(CH_2)_2CH_3$	- " -	25		$\bar{1},955$			[41]
$CH_2=CH(C_2H_5)_2$	- " -	25		$\bar{1},875$			[41]
$CH_2=CH(CH_2)_3CH_3$	- " -	25		$\bar{1},505$	20,75	29,0	[41]
$CH_2=CH(CH_2)_4CH_3$	- " -	25		$\bar{1},58$			[41]
$CH_2=CH(CH_2)_5CH_3$	- " -	25		$\bar{1},58$	20,49	28,5	[41]
$CH_2=CHC_6H_5$	- " -	25		0,87			[41]
$CH_2=C(CH_3)CH=CH_2$	- " -	25		0,79			[41]
$C_6H_5CH_3$	50% CH_3CN; $[HClO_4] = 1$ M	0,3–31		$\bar{3},89$ (25°)	18,50	28,0	[42]
	84% CH_3CN; $[HClO_4] = 1$ M	0,3–31		$\bar{2},58$ (25°)	15,25	22,7	[42]

cyclo-C_6H_{12}	50% CH_3CN; $[HClO_4] = 1$ M; $\mu = 1{,}05$ M; $[RH] = 2{,}8 \cdot 10^{-2}$	20		$\bar{3}{,}99$	[42]
$C_6H_5C_2H_5$	70% CH_3CN; $[HClO_4] = 0{,}77$ M; $\mu = 0{,}9$ M	15		$\bar{3}{,}36$	[42]
$(C_6H_5)_2CH_2$	- " -	15		$\bar{3}{,}77$	[42]
p-$NO_2C_6H_4CH_3$	- " -	15	$W = k[RH]^{1/2}[Co^{3+}]$	$\bar{4}{,}415$	[42]
p-$(CH_3)_3CC_6H_4CH_3$	50% CH_3CN; $[HClO_4] = 1$ M; $\mu = 1{,}04$ M	20		$\bar{2}{,}58$	[42]
$(C_6H_5CH_2)_2$	60% CH_3CN; $[HClO_4] = 1$ M; $\mu = 1{,}02$ M	15		$\bar{3}{,}30$	[42]
$(C_6H_5)_2$	50% CH_3CN; $[HClO_4] = 1$ M; $\mu = 1{,}05$ M	20		$\bar{2}{,}96$	[42]
Naphthalene	- " -	20,2		1,48	[42]

Table 102 (continued)

Compound	Conditions	t^{o}	Expression for k or W	log k	log A	E	Ref.
CH_3OH	$[HClO_4]$ = 1,57 M	15		$\bar{1}$,07			
C_2H_5OH	- " -	15		$\bar{1}$,455			
n-C_3H_7OH	- " -	15		$\bar{1}$,46			
$(CH_3)_2CHOH$	- " -	15		$\bar{1}$,57	22,02	29,6	
$(CH_3)_2CHCH_2OH$	- " -	15		$\bar{1}$,53			[43]
$(C_2H_5)_2CHOH$	- " -	15		$\bar{1}$,025	21,625	29,8	
$C_2H_5CH(CH_3)OH$	- " -	15		$\bar{2}$,80			
n-$C_3H_7CH(CH_3)OH$	- " -	15		$\bar{2}$,80			
$(CH_3)_2CHCD(CH_3)OH$	$[HClO_4]$ = 1,57 M	15		$\bar{1}$,40			
$(C_2H_5)CDOH$	- " -	15		$\bar{1}$,02			
cyclo-$C_6H_{11}OH$	$[HClO_4]$ = 1,57 M;	15		$\bar{2}$,695	22,60	31,5	[43]
	μ = 3,4 M	10	$k = k_1/[H^+] + k_2/[H^+]^2$	log k_1 = $\bar{2}$,73; log k_2 = $\bar{2}$,00			[44]
$(CH_3)_3COH$	$[HClO_4]$ = 1,57 M;	15		$\bar{2}$,43			[43]
	μ = 3,25	10	$k = k_1/[H^+] + k_2/[H^+]^2$	log k_1 = $\bar{3}$,26; log k_2 = $\bar{5}$,845			[44]

	$[HClO_4] = 1{,}57$ M	15		$\bar{3}$,335	21,68	32,1	[45]
	$HClO_4$; $\mu = 1{,}5$ M	10	$k = k_1/[H^+]$	$\bar{3}$,09			[46]
	H_2SO_4	9	$k = k_1 + k_2/[H^+]$	$\log k_1 = \bar{4}$,66;	$\bar{1}$,48	6,2	[46]
				$\log k_2 = \bar{3}$,27	18,17	26,9	
$(C_2H_5)(CH_3)_2COH$	$[HClO_4] = 2{,}3$ M	10	$k_1/[H^+]$	$\bar{2}$,55			
	H_2SO_4	9	$k_1 + k_2/[H^+]$	$\log k_1 = \bar{3}$,49;	21,24	30,6	[46]
				$\log k_2 = \bar{3}$,12	29,93	42,3	[46]
$(CH_3)_2(C_2H_5)COH$	$[HClO_4] = 1{,}57$ M	15		$\bar{2}$,70			[45]
$(CH_3)_2CHC(CH_3)_2OH$	$[HClO_4] = 1{,}57$ M	15		$\bar{2}$,875			[45]
$(CH_3)_3C(CH_3)_2OH$	$[HClO_4] = 1{,}57$ M	15		$\bar{2}$,05			[45]
n-$C_3H_7(CH_3)(C_2H_5)COH$	$[HClO_4] = 1{,}57$ M	15		$\bar{2}$,81			[45]
n-$C_5H_{11}(CH_3)(C_2H_5)COH$	– " –	15		$\bar{2}$,92			[45]
$CH_3OCOC(CH_3)(C_2H_5)OH$	– " –	15		$\bar{3}$,935			[45]
$CH_3OCH_2C(C_2H_5)OH$	– " –	15		$\bar{1}$,20			[45]
$(C_2H_5)_2CO$	$\mu = 3{,}4$	10	$k = k_1 + k_2/[H^+]$	$\log k_1 = \bar{4}$,48; $\log k_2 = \bar{3}$,62			[47]

Table 102 (continued)

Compound	Conditions	t°	Expression for k or W	log k	log A	E	Ref.
p-$NO_2C_6H_4CHO$	45% CH_3CN; μ = 1,75	5–27	$k = k_1 / [H^+]$	log k_1 = = 0,87 (25°)	16,07	20,7	[48]
m-$NO_2C_6H_4CHO$	- " -	5–27	- " -	log k_1 = = 0,67 (25°)	16,29	21,3	[48]
CH_2O	$[HClO_4]$ = =1 - 6M μ = 5,98 M	18–29	$1/k = 1/k_1 + K/k_1 [H+]$	log k_1 = = 1,30 (22°);	17,73	22,0	[49]
				log K = $\bar{1}$,845 (22°)	1,34	Q = -2,0	[49]
$(C_6H_5CH_2)_2O$	60% CH_3CN, μ = 1 M; $[HClO_4]$ = 1 M	15,3		$\bar{2}$,38			[50]
$C_6H_5CH_2OCH_3$	- " -	0–25		$\bar{3}$,97	18,77	27,4	[50]
cyclo-C_6H_{11}-$OCH_2C_6H_5$	- " -	15		$\bar{2}$,48			[50]
$C_6H_5OC_6H_5$	- " -	15		$\bar{2}$,96			[50]

HCOOH	$[H_2SO_4]$ = 1–5N; μ = 4 M	0–30	$k = k_1 + k_2/[H^+]$	$\log k_1 = \bar{2},11$ (26°)	17,83	26,9	[51]
				$\log k_2 = 2,38$ (26°)	18,28	21,7	
	μ = 5,4 M; $[HClO_4]$ = 1 M			$\bar{1},04$			[3]
CH_3CH_2COOR	$[HClO_4]$ = 1 M; $[NaClO_4]$ = 2 M	10–25	$1/k = [RH]/k_1 + [H^+]/k_1K$	$\log k_1K = \bar{2},59$ (25°)	17,88	26,7	[52]
$(CH_3)_2CHCOOH$	– " –	10–25	– " –	$\log k_1K = \bar{3},415$ (25°)	17,05	23,6	[52]
$(CH_3)_3CCOOH$	– " –	10–25	– " –	$\log k_1K = \bar{3},64$ (25°)	14,46	18,7	[52]
$C_6H_5CH_2COOH$	– " –	10–25	– " –	$\log k_1K = \bar{3},94$	16,70	21,4	[52]
$C_6H_5C(CH_3)_2CH_2COOH$	30% CH_3CN; $[HClO_4]$ = 3,14 M	15		$\bar{1},59$			[52]

Table 102 (continued)

Compound	Conditions	t°	Expression for k or W	log k	log A	E	Ref.
$C_6H_5(CH_2)_n COOH$	50% CH_3CN		$1/k = [RH]/k_1 + [H^+]/k_1K$				
n = 3		20		$\log k_1 = \bar{2},505$; $\log K = 0,67$			[53]
n = 4		20	- " -	$\log k_1 = \bar{2},28$; $\log K = 1,23$			[53]
n = 5		20	- " -	$\log k_1 = \bar{2},32$; $\log K = 1,15$			[53]
$CH_3COCOOH$	$[HNO_3] = 1$ M; $\mu = 2,1$ M	5		$\bar{1},473$	18,37	24,0	[54]
	$[H_2SO_4] = 1$ M; $\mu = 2,1$ M	5		$\bar{2},89$	21,54	28,8	[54]
$HOOCCH_2CH(OH)COOH$	$[HClO_4] = 0,25$ M; $\mu = 0,25$ M	0–12	$CoRH^{3+} \longrightarrow$	6,41 (25°)	16,64	15,3	[55]
			$CoRH(OH)^{2+} \longrightarrow$	7,44 (25°)	25,84	25,1	[55]
$CH_3CH{=}CHCOOH$	$[HClO_4] = 1$ M	10–25	$k = k_1/[H^+]$	$\log k_1 = \bar{2},76$ (25°)	16,06	23,6	[53a]

$C_6H_5CH{=}CHCOOH$	- " -	10–25	- " -	$\log k_1 = 0{,}78$ (25°)	20,23	26,5	[53a]
C_6H_6	$HClO_4$; $\mu = 2{,}0$	15,44	$k = k_1 + k_2/[H^+]$	$\log k_1 = \bar{3},30$ (25°);	11,255	19,0	[56]
				$\log k_2 = \bar{3},64$ (25°)	18,94	29,0	[56]
Oxidation by Co(III) acetylacetonate in nonaqueous solutions							
$(CH_3)_2CHOH$	$n\text{-}C_7H_{14}$; $[RH] < 2{,}4$ M	112–130		$\bar{4},23$ (121°)	12,20	28,0	[57]
	C_6H_6	114–133		$\bar{5},83$ (121°)	13,43	31,0	[57]
$n\text{-}C_4H_9OH$	C_6H_5Cl	89–118		$\bar{5},43$ (118°)	17,175	39,4	[58]
cyclo-$C_6H_{11}OH$	- " -	88–118		4,66 (118°)	13,63	31,0	[58]
C_6H_5OH	$[RH] < 0{,}1$ M; C_6H_5Cl	71–120	$1/k = [RH]/K + 1/k_1K$	$\log k_1 = \bar{2},07$ (120°);	16,28	33,0	[59]
				$\log K = 0{,}11$ (120°)	$\bar{3},61$	Q = 4,5	

TABLE 103. Oxidation by Cerium Ions in Aqueous Solutions

Substance	Conditions	t°	Expression for k	log k	log A	E	Ref.
$CH_3C_6H_5$	50% CH_3COOH	40		$\bar{3},92$			
$m\text{-}NO_2C_6H_4CH_3$	$[HClO_4] = 1$ M	40		$\bar{3},77$			[61]
$p\text{-}NO_2C_6H_4CH_3$	- " -	40		$\bar{4},55$			
$m\text{-}ClC_6H_4CH_3$	- " -	40		$\bar{3},275$			
$o\text{-}ClC_6H_4CH_3$	- " -	40		$\bar{3},42$			
$p\text{-}ClC_6H_4CH_3$	- " -	40		$\bar{3},98$			
CH_3OH	$[HClO_4] = 1$ M	7–27	$\frac{k_1K}{1 + K[CH_3OH]}$ $\frac{1}{K} = \frac{1}{K_o} + \frac{K_{H^+}}{K_o[H^+]}$	$\log k_1 = \bar{4},64\ (20°)$ $\log K = 0,37\ (20°)$ $\log K_o = 0,825\ (20°)$ $\log K_{H^+} = 0,55\ (20°)$	6,08	12,2	[62]
	$[H_2SO_4] = 1$ M	40–70	$k = \frac{k_1}{[H_2SO_4]}$	$\log k_1 = \bar{5},46\ (40°)$	10,74	21,85	[63]

C_2H_5OH	$[HClO_4] = 3{,}2M$	20	$\frac{k_1K}{1 + K[RH]}$	$\log k_1 = \bar{3},825$; $\log K = 0{,}63$			[64]
	$[H_2SO_4] = 1\ M$	40–70	$k = \frac{k_1}{[H_2SO_4]}$	$\log k_1 = \bar{5},97$ (40°)	9,10	18,8	[63]
$n\text{-}C_3H_7OH$	$[H_2SO_4] = 1\ M$	40–70	$k = \frac{k_1}{[H_2SO_4]}$	$\log k_1 = 4{,}00$ (40°)	6,32	13,86	[63]
$(CH_3)_2CHCH$	$[HNO_3] = 1\ M$; $[RH] = 0{,}5\ M$	15–34		$\log k_{15°} = \bar{5},845$; $\log k_{34°} = \bar{3},30$			[65]
$C_2H_5CH(OH)CH_3$	$[HNO_3] = 1\ M$; $[RH] = 0{,}5\ M$	15–34		$\log k_{15°} = \bar{4},09$; $\log k_{34°} = \bar{3},38$			[65]
$n\text{-}C_4H_9OH$	$[HClO_4] = 0{,}5$–$2\ M$; $\mu = 2{,}1\ M$	10–15	$\frac{1}{k} = \frac{[RH]}{k_1} + \frac{1}{k_1K_1} + \frac{K_2}{k_1K_1[H^+]}$		17,685 ($\Delta S_1 = 11{,}6$; $\Delta S_2 = 56{,}8$)	26,76 ($\Delta H_1 = 1{,}54$; $\Delta H_2 = 15{,}9$)	[66]
$C_6H_5CH_2OH$	$[HClO_4] = 1\ M$; $\mu = 2{,}1\ M$	10–20		0,00 (40°)	13,215	18,92	[66]

Table 103 (continued)

Substance	Conditions	t°	Expression for k	log k	log A	E	Ref.
cyclo-C_5H_9OH	$[H_2SO_4]$ = 0,25 M; $[HClO_4]$ = 0,75 M	10–30		$\bar{3}$,61 (40°)	16,83	27,5	[67]
cyclo-$C_6H_{11}OH$	$[H_2SO_4]$ = 1,48 M; μ = 0,24	24–59		$\bar{4}$,55 (44,5°)	11,85	22,2	[3]
	$[HClO_4]$ = 1 M	15	$\frac{k_1K}{1 + K[RH]}$	log k_1 = $\bar{3}$,34; log K = 0,46			[67]
	- " -	50	- " -	log k_1 = $\bar{4}$,14; log K = 1,11			[68]
trans- 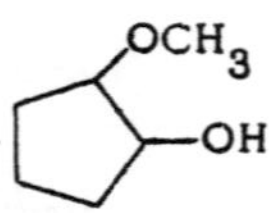	$[H_2SO_4]$ = 0,25 M; $[HClO_4]$ = 0,75 M	10–30		$\bar{2}$,59 (15°)	16,04	23	[67]

trans - 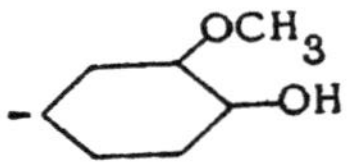	$[HClO_4]$ = 1 M	15	$\frac{k_1 K}{1 + K[RH]}$	$\log k_1 = \bar{2},74$; $\log K = 0,32$			[67]
	$[H_2SO_4]$ = 0,25 M; $[HClO_4]$ = 0,75 M	10–30		$\bar{4},775$ (30°)	18,925	30,7	[67]
cis - 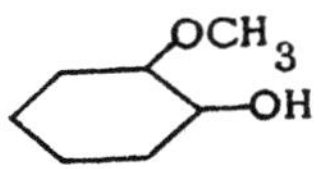	$[H_2SO_4]$ = 0,25 M; $[HClO_4]$ = 0,75 M	10–30		$\bar{3},36$ (30°)	18,22	28,9	[67]
$CH_2(OH)CH_2OH$	H_2SO_4	35	$k_0 + k_1[SO_4^{2-}] + k_2[SO_4^{2-}]^2 + k_3[SO_4^{2-}]^3$	$\log k_2 = 7,40$; $\log k_3 = 7,70$			[69]
$CH_3CH(OH)CH(OH)CH_3$	$[HNO_3]$ = 0,5 M	0	$\frac{k_1 K}{1 + K[RH]}$	$\log k_1 = \bar{3},485$; $\log K = 1,38$			[70]

Table 103 (continued)

Substance	Conditions	t°	Expression for k	log k	log A	E	Ref.
$CH_2(OH)CH(OH)CH_2OH$	$[HSO_4^-]$ = 0,2 M; $[HClO_4]$ = 0,2 M	20		$\bar{5},825$			[71]
	$[HClO_4]$ = 0,5 M; $[NaClO_4]$ = 0,1 M	20	$\frac{k_1K}{1 + K[RH]}$	$\log k_1 = \bar{2},14$; $\log K = 1,30$			[71]
$(CH_3)_2C(OH)C(OH)(CH_3)_2$	pH 0,07 $[SO_4^{2-}]$ = 0,29 M;	20–30		$\bar{2},58$ (25°)	15,09	22,5	[72]
cis- 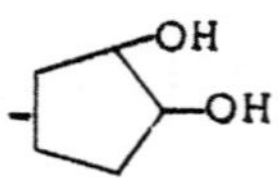	$[H_2SO_4]$ = 0,25 M	10–30		$\bar{1},665$ (30°)	15,76	23,3	[67]
	$[H_2SO_4]$ = 0,25 to 0,75 M;	15	$\frac{W}{[RH]}$ =	$\log k_1 = 2,93$; $\log k_2 = 0,66$;			[67]

	$[HClO_4] =$ $= 0{,}75$ M		$= k_1[CeSO_4^{2+}] + k_2[Ce(SO_4)_2]^+ + k_3[Ce(SO_4)_3^{2-}]$	$\log k_3 = \bar{1}{,}08$			
trans- 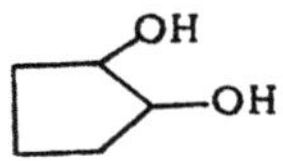	$[H_2SO_4] = 0{,}25$; $[HClO_4] = 0{,}75$ M	10–30		$\bar{1}{,}87$ (30°)	16,62	23,2	[67]
	$[H_2SO_4] = 0{,}25$ to $0{,}75$ M;	15	$\frac{W}{[RH]} = k_1[CeSO_4^{2+}] + k_2[Ce(SO_4)_2]^+ + k_3[Ce(SO_4)_3^{2-}]$	$\log k_1 = 3{,}175$; $\log k_2 = 0{,}73$; $\log k_3 = \bar{1}{,}38$			[67]
cis- 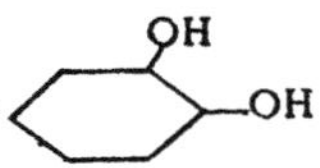	$[HClO_4] = 1$ M	15	$\frac{k_1 K}{1 + K[RH]}$	$\log k_1 = \bar{2}{,}15$; $\log K = 1{,}46$			
	$[H_2SO_4] = 0{,}25$ M; $[HClO_4] = 0{,}75$ M	10–30		$\bar{3}{,}25$			[67]

Table 103 (continued)

Substance	Conditions	t°	Expression for k	log k	log A	E	Ref.
trans - 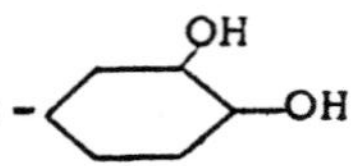	$[HClO_4]$ = 1 M	15	$\frac{k_1 K}{(1 + K[RH])}$	log k_1 = $\bar{3}$,78; log K = 1,255			[67]
	$[H_2SO_4]$ = = 0,25 M; $[HClO_4]$ = = 0,75 M	10–30		$\bar{4}$,725			[67]
$H_2C(OH)_2$	$HClO_4$			1,23 (22°)	24,99	32	[73]
	$[HClO_4]$ = 1 M	10–20		$\bar{1}$,43 (20°)	10,86	15,3	[74]
CH_3CHO	$[HClO_4]$ = 1 M	13–25		$\bar{2}$,65 (20°)	5,86	9,65	[74]
CCl_3CHO (hydrate)	- " -	40–50		$\bar{3}$,04 (45°)	15,49	24,72	[74]
CH_3COCH_3	$HClO_4$	25	$CH_3C(OH)$= =CH_2+$CeOH^{3+}$	2,08			[75]

Arabinose ($C_5H_{10}O_5$)	$[H_2SO_4]$ = 0,25 M	35	$\frac{k_1K}{(1+K[RH])}$	log k_1 = $\bar{3}$,745; log K = 0,95			[76]
HOOCCOOH	$[H_2SO_4]$ = 1 M	19–35		1,99 (35°)	13,71	16,5	[77]
	$[H_2SO_4]$ = 2 M	19–35		1,32 (35°)	13,04	16,5	[77]
	$[H_2SO_4]$ = 0,78 M			$\bar{1}$,36 (35°)	8,46	12,8	[78]
	$[H_2SO_4]$ = 1,96			$\bar{2}$,82 (35°)	11,32	17,6	[78]
$HOOCCH_2COOH$	$[HClO_4]$ = 0,5 M	9–30		2,54 (25°)	9,29	9,2	[79]
	$[H_2SO_4]$ = 0,25 M	9–30		$\bar{2}$,86 (25°)	10,66	16,1	[79]
	$[H_2SO_4]$ = 2 M	20–30	$\frac{k_1K}{(1+K[RH])}$	log k_1 = $\bar{1}$,80 (25°) log K = 0,88 (25°)	8,61	12,0	[80]
$HOCH_2COOH$	$[H_2SO_4]$ = 1 M	25–50		$\bar{3}$,08 (35°)	4,26	10,1	[81]
	$[H_2SO_4]$ = 1,66 M	26,6		$\bar{4}$,00			[12]
	$[H^+]$ = 0,53 M; $[SO_4^{2-}]$ = $4\cdot10^{-2}$	30–40	$\frac{k_1[H^+]}{3,94+[H^+]}$	$\bar{1}$,36 (35°)	15,56	22,8	[82]

Table 103 (continued)

Substance	Conditions	t°	Expression for k	log k	log A	E	Ref.
$CH_3CH(OH)COOH$	$[H^+] = 0{,}53$ M; $[SO_4^{2-}] = 4 \cdot 10^{-2}$	26,6		$\bar{4}{,}99$			[12]
$CH_3CH(OH)COOH$	$[H_2SO_4] = 0{,}5$ M	25		$\bar{1}{,}155$			[83]
	$[HClO_4] = 0{,}98$	5–25	$k = k_1 [H^+]$	$\bar{2}{,}49$ (25°)	8,61	13,8	[84]
$CH_3CH(OH)COOH$	$[HClO_4] = 0{,}7$ M	25	$\frac{k_1 K}{1 + K[RH]}$	$\log k_1 = \bar{4}{,}87$; $\log K = 1{,}44$			[85]
$C_2H_5CH(OH)COOH$	$[H_2SO_4] = 1$ M	25–45		$\bar{3}{,}51$ (25°)	14,23	22,8	[86]
$C_6H_5CH(OH)COOH$	$[H_2SO_4] = 1$ M	0–50		$\bar{1}{,}35$ (25°)	13,22	18,9	[86]
	$[SO_4^{2-}] = 0{,}025$ M; pH 1,36	25		$\bar{1}{,}52$	4,22	6,4	[87]
	$[HClO_4] = 0{,}9$ M	5		$\bar{2}{,}44$			[87]
	$[H_2SO_4] = 1{,}66$ M	26,6		$\bar{2}{,}835$			[12]
	$[H_2SO_4] = 1{,}0$ M	20		0,53			[83]

$(C_6H_5)_2C(OH)COOH$	$[H_2SO_4] = 1$ M	20–35	0,07 (25°)	14,97	20,3	[88]
$HOOCCH(OH)CH_2COOH$	$[H_2SO_4] = 0,5$ M	25	$\bar{1},89$			[83]
$HOOCCH(OH)CH(OH)COOH$	$[H_2SO_4] = 0,25$ M	25	$\bar{1},17$ (25°)	17,32	24,75	[89]
$HOOCCH(OH)CH(OH)COOH$	$Ce(SO_4)_2$; pH 0,86	5–25	$\bar{1},66$ (25°)	7,74	11,0	[90]
	$[H_2SO_4] =$ = 0,49 M	11–23	$\bar{1},43$ (23°)	15,13	21,2	[90]
CH_2COOH $HOCHCOOH$ CH_2COOH	$[H_2SO_4] =$ 1 to 5		$\bar{1},88$		19,8	[91]
$CH_3COCOOH$	$Ce(SO_4)_2$; pH 1,26	0–15	$\bar{2},69$ (15°)	6,91	11,0	[90]

TABLE 104. Oxidation of Variable-Valence Metal Ions by Oxygen

Equation for rate	Conditions	t°	log k	log A	E	Ref.
	Fe^{2+} ion					
$k[Fe^{2+}][H_2PO_4^-]^2 p_{O_2}$ (liter²/mole²·atm·sec)	pH 2–3	20–30	$\bar{3}$,10 (30°)	11,55	20	[94]
$k[Fe^{2+}]^2 p_{O_2}$ (liter/mole·atm·sec)	$[FeSO_4]$ = 0,5 M	60	$\bar{5}$,79			[95]
	$[FeSO_4]$ = 1 M	20–60	$\bar{5}$,685 (60°)	6,38	16,3	
	$[FeSO_4]$ = $[H_2SO_4]$ = 1 M	20–60	$\bar{5}$,395 (60°)	5,14	14,8	
$k[Fe^{2+}]^2 p_{O_2}$ (liter/mole·atm·sec)	$[FeCl_2]$ = 0,5 M	60	$\bar{6}$,94			[95]
	$[FeCl_2]$ = 1 M	20–60	$\bar{5}$,13 (60°)	4,01	13,5	
	$[FeCl_2]$ = 1 M; $[HCl]$ = 2 M	20–60	$\bar{5}$,71 (60°)	5,39	14,7	
$(k_1 + k_2[H^+]^{-1})[Fe^{2+}][O_2]$; k_1 (liter/mole·sec); k_2 (sec⁻¹)	$Fe(NH_4)_2(SO_4)_2$; pH 0,76–1,34	200	log k_1 = $\bar{3}$,63; log k_2 = $\bar{5}$,954			[34]

$k[Fe^{2+}][O_2]$ $k = k'([HCl] - 5{,}5)$	$[HCl] = 8$ M $[HCl] > 6$ M	0–35	$\bar{1}{,}86$ (30°)	10,41	14,6	[93]
$k[Fe^{2+}]^2 p_{O_2}$ (liter/mole·atm·sec)	$[HClO_4] = 0{,}51$ M	25-40	$\bar{7}{,}975$ (35°)	6,36	17,4	[96]
$k[Fe^{2+}][H_2P_2O_7^{2-}][O_2]$ (liter²/mole²·sec)	$HClO_4$; pH 2–3	30	0,71			[97]
$k_1[Fe^{2+}]p_{O_2} + k_2[Fe^{2+}]^2 p_{O_2}$ k_1 (atm⁻¹·sec⁻¹) k_2 (liter/mole·atm·sec)	$[H_2SO_4] = 1N$	140–180	$\log k_1 = \bar{5}{,}285$ (159°); $\log k_2 = \bar{3}{,}205$ (159°)	2,08 5,47	13,4 16,3	[98]
$k_2 = \frac{(k_3 + k_4K[SO_4^{2-}])}{(1 + K[SO_4^{2-}])}$	$\mu = 1$ M	30,5	$\log k_3 = \bar{6}{,}145$; $\log k_4 = \bar{5}{,}49$; $\log K = 0{,}04$			[98]

Table 104 (continued)

Equation for rate	Conditions	$t°$	log k	log A	E	Ref.
$k[Fe^{2+}]^2 p_{O_2}$ (liter/mole·atm·sec)	$[H_2SO_4] = 0{,}5\ M$	30	$\bar{6},58$			[99]
$k[Fe^{2+}]^2 p_{O_2}$ (liter/mole·atm·sec)	CH_3OH	30	$\bar{3},08$			[100]
	Ti^{3+} ion					
$k[Ti^{3+}] p_{O_2}$ ($atm^{-1} \cdot sec^{-1}$)	$[HCl] = 1\ M$	20–50	$\bar{5},80$ (30°)	8,08	17	[101]
	V^{3+} ion					
$k[V^{3+}] p_{O_2} / [H^+]$	$[HClO_4] = 1{,}6$ to $8{,}3\ 10^{-2}\ M$	25	$\bar{7},16$	7,91	20,1	[102]
	Co(ℓ-histidine)$_2$					
$k[Co(II)][O_2]$ (liter/mole·sec)	pH 8–11; $\mu = 0$ to $1\ M$	25	3,545	7,215	5,0	[103]
	Co(dℓ-histidine)$_2$					
$k[Co(II)][O_2]$	pH 8–11; $\mu = 0$ to $1\ M$	25	3,415	7,085	5,0	[103]

$k[Co(II)][O_2]$	H_2O	$[Co(glycylglycine)_2(OH)_2]^{2-}$	25	3,00			[104]
$k[Cu^+][O_2]$	HCl	Cu^+	25	3,00			[105]
$k[CuCl_2^-][O_2]$	1 M HCl + KCl	$CuCl_2^-$	25	2,4			[105]
$k[Cu(I)][NH_3][O_2]$	$\mu = 1$	$Cu(NH_3)_2^+$	25	4,21			[105a]
$k[Cu(I)][imidazole][O_2]$	$\mu = 1$	$Cu(imidazole)_2$	25	3,82		~1,6	[105a]
$k[Cu(Dipy)_2^+][O_2]$	pH 5,0	$Cu(Dipy)_2^+$	25	3,81			[105b]
$k[U(IV)][O_2]/[H^+]$	0,5 M $HClO_4$	U(IV)	20–60	$\bar{2}$,40 (30°)	14,6	22,	[105c]

TABLE 105. Reactions of Ions with Hydrogen Peroxide in Aqueous Solutions

Ion	Conditions	t°	log k	log A	E	Method	Ref.
Ti^{3+}	H_2O	25	2,70			JM	[107]
$Cr_2O_7^{2-}$	H_2O	18	3,28			JM	[107]
MnO_4^-	0,1N H_2SO_4	18	3,48			JM	[107]
Mn^{3+}	μ = 2,23 to 4,90	25	4,86			JM	[107a]
$MnOH^{2+}$	- " -	25	4,50			JM	[107a]
Fe^{2+}	H_2O	12–35	1,90 (25°)	9,25	10,1	RUCR	[108]
	1,35 M $HClO_4$	0–25	1,75 (25°)	8,65	9,4	CSM	[109]
	0,8N H_2SO_4	18	1,82			CSM	[110]
	- " -	25	1,81			CSM	[110]
	1N H_2SO_4	20	1,79			CSM	[111]
	0,5M H_2SO_4	25	1,78			CSM	[112]
	0,5 - 0,8N H_2SO_4	15–41	1,71 (20°)	8,02	8,46	CSM	[113]
	0,01–1N $HClO_4$	0–40	1,67 (20°)	8,725	9,45	CSM	[114]
	0,8N H_2SO_4	0–40	1,70 (20°)	8,98	9,75	CSM	[114]
	$(HClO_4) = 4\cdot10^{-3}$ - 1M	0–45	1,70 (25°)	7,145	7,3	CSM	[115]

FeF_2	$\mu = 1$ M	0–45	2,21 (25°)	12,40	14,0	CSM	
$FeCl^+$	$\mu = 1$ M	10–45	1,95 (25°)	8,08	8,4	CSM	
$FeBr^+$	$\mu = 1$ M	0–25	1,90 (25°)	8,95	9,6	CSM	[115]
$FePF_6^+$	$\mu = 1$ M	0–25	2,12	10,20	11,0	CSM	
$Fe(Dipy)^{2+}$	pH 6	17	2,36			CSM	[116]
Cu^+	H_2O; pH 2,3	25	3,36	9,36	10,0	IIM	[116a]
Cu (I)	$\mu = 1$; Cl^-	25	1,69			POL	[116b]
$Cu(Dipy)_2$	pH 5,0	25	2,93			CSM	[117]
Ce^{4+}	$Ce(SO_4)_2$	25	5,935			JM	[107]
	0,8N H_2SO_4	25	6,0			JM	[118]
U(IV)	2M $HClO_4$	2,4–34,2	$\bar{1}$,83 (25°)	12,5	17,3	CSM	[119]
NpO_2^+	$\mu = 3$	5–25	0,95 (25°)	10,1	12,4	KEPF	[120]
AmO_2^+	0,1M $HClO_4$	25–35	$\bar{3}$,61 (25°)	7,4	13,3	CSM	[121]

TABLE 106. Reactions of Metal Ions and Complexes with Peroxides

Peroxide	Conditions	t°	Expression for k	log k	log A	E	Ref.
		Fe^{2+} ion					
$(CH_3)_3COOH$		25		1,11	8,31	9,8	[122]
$C_6H_5C(CH_3)_2OOH$	H_2O			1,22 (25°)	8,55	9,07	[123]
	D_2O			1,00 (25°)	8,97	10,84	[123]
	H_2O	0–25		1,20 (25°)	10,03	12,0	[124]
	H_2O; O_2	0–25		1,43 (25°)	9,59	11,1	[125]
	H_2O	25		$\bar{1}$,44	7,60	11,1	[126]
	$Fe(EDTA)^{2-}$	0		1,38	9,71	10,4	[127]
	$Fe(P_2O_7)^{2-}$	0		1,77	8,58	8,5	[127]
$p-NO_2C_6H_4C(CH_3)_2OOH$	H_2O; pH 4,2	0–25		1,52 (25°)	10,90	13,1	[128]
$p-(CH_3)_2CHC_6H_4C(CH_3)_2OOH$	- " -	11–26		1,66 (25°)	9,60	10,8	[129]
$p-(CH_3)_3CC_6H_4C(CH_3)_2OOH$	- " -	0–15		2,00 (25°)	9,255	9,9	[129]
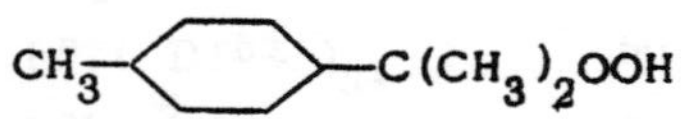	H_2O; pH 4,2	0–16		1,72 (25°)	9,80	11,0	[128]

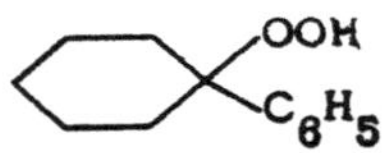	- " -	0–25		1,17 (25°)	9,38	10,6	[128]
$(C_6H_5COO)_2$	C_2H_5OH	15–32		1,09 (30°)	11,34	14,2	[130]
	- " -	10–30		0,755 (30°)	9,36	12,0	[131]
	- " -	17–27		0,265 (17°)	9,00	11,58	[132]
	CH_3OH	25–40		$\bar{1}$,36 (30°)	13,52	19,8	[131]
	n-C_3H_7OH	5–15		0,54 (15°)	11,94	15,0	
	n-C_4H_9OH	5–15		0,48 (15°)	9,875	12,4	
$(p\text{-}CH_3OC_6H_4COO)_2$	C_2H_5OH	17–27		0,46 (17°)	7,75	9,66	[131]
$(p\text{-}CH_3C_6H_4COO)_2$	- " -	17–27		0,44 (17°)	7,75	9,70	
$(p\text{-}ClC_6H_4COO)_2$	- " -	17–27		0,20 (17°)	11,35	14,77	
$(p\text{-}NO_2C_6H_4COO)_2$	- " -	17–27		0,165 (17°)	10,77	14,09	
		Co^{2+} ion					
$(CH_3)_3COOH$	$CH_3COOH:H_2O$ = 1 : 1; $Co(EDTA)^{2-}$	35	$k_1 + k_2[Co(II)]$	$\log k_1 = \bar{5},86$; $\log k_2 = \bar{3},785$		19,8 Co(II)= $=7\cdot10^{-3}$	[133]
	CH_3COOH	60	- " -	$\log k_1 = \bar{2},34$; $\log k_2 = \bar{1},79$			[134]

Table 106 (continued)

Peroxide	Conditions	t°	Expression for k	log k	log A	E	Ref.
$(CH_3)_3COOH$	C_6H_5Cl; $Co[OOCCH(C_2H_5)(CH_2)_3CH_3]_2$	20		log k_1 = 1,02; log k_2 = 5,28			[134]
	CH_3COOH	40,4		log k_1 = $\bar{3}$,785; log k_2 = $\bar{1}$,46			[135]
	C_6H_5Cl; $Co[OOC(CH_2)_6CH_3]_2$	20		$\bar{1}$,43	8,38	12	[126]
	n-C_5H_{12}; $Co[OOC(CH_2)_6CH_3]_2$	38	$k_1[Co(II)]$	0,06			[126]
$C_6H_5C(CH_3)_2OOH$	C_6H_5Cl; $Co[OOC(CH_2)_6CH_3]_2$	45		$\bar{1}$,32	7,58	12	[126]
	n-C_9H_{20}; cobalt palmitate	101		$\bar{2}$,505	3,86	9,4	[136]
	n-C_9H_{20}; cobalt phthalocyanine	50-120		1,61 (98,2)	7,21	9,5	[137]

sec-$C_{10}H_{21}OOH$	n-$C_{10}H_{22}$; $Co[OOC(CH_2)_{16}CH_3]_2$	80–100	$\frac{k_1K[Co(II)]}{(1+K[ROOH])}$	$\log k_1$ = 3,00 (100°)	17,30	24,5	[138]
				$\log K$ = 1,00 (100°)	$\bar{5}$,78	Q=10,0	
α-$C_{10}H_{11}OOH$	CH_3COOH + RH	50	$k_1[Co(II)]$	1,60			[139]
	$CH_3COOH + C_6H_6$	50	$\frac{k_1[Co(II)]}{[CH_3COOH]^2}$	1,14			[140]
-$C_{10}H_{11}OOH$ [sic]	C_6H_5Cl; $Co(OOCCH_3)_2$	50		0,79			[141]
	RH; $Co(OOCCH_3)_2$	50		0,46			[141]
	$CH_3C_6H_4CH_3$; $Co(OOC_8H_{15})_2$	50	$k_1[Co(II)]$	0,88	12,67	17,4	[142]
	RH: cobalt naphthenate	49,7	$k_1[Co(II)]^{1,68}$	1,775			[143]
	RH: cobalt dibromostearate	34	$k_1[Co(II)]^{1,30}$	1,225			[143]
	RH: cobalt oleate	34	$k[Co(II)]^{0,8}$	0,16			[143]
		45		0,33			[143]

Table 106 (continued)

Peroxide	Conditions	t°	Expression for k	log k	log A	E	Ref.
		Co^{3+} ion					
$(CH_3)_2C{=}CHCH_2OOH$	H_2O; H_2SO_4	20	$k_1/[H^+]$	0,45			[144]
$(CH_3)_3COOH$	H_2O; H_2SO_4	20	$k_1/[H^+]$	0,15			[144]
	CH_3COOH	20		$\bar{2}$,13			[144]
		Cu^{2+} ion					
$(CH_3)_3COOH$	C_6H_5Cl; $Cu(OOCC_7H_{15})_2$	60	$k_1/[Cu(II)]^{1/2} \times [ROOH]^{1/2}$	$\bar{1}$,585			[145]
$C_6H_5CH(CH_3)OOH$	RH; $Cu[OOC(CH_2)_{16}CH_3]_2$	70–90		0,13 (80°)	9,925	15,8	[146]
sec-$C_{10}H_{21}OOH$	RH; $Cu[OOC(CH_2)_{16}CH_3]_2$	50–90	$\frac{k_1K[ROOH]}{1+K[ROOH]^2}$	log k_1 = $\bar{1}$,21 (80°);	8,81	15,5	[147]
				log K = = 4,09 (80°)	9,04	8,0	
α-$C_{10}H_{11}OOH$	RH: copper oleate	57,4		$\bar{2}$,00			[148]

$C_6H_5C(CH_3)_2OOH$	Copper phthalocyanine / $C_6H_5CH(CH_3)_2$	94-127	$k_1[ROOH]$	$\bar{4}$,60 (94,4°)	12,01	25,9	[137]
α-$C_{10}H_{11}OOH$	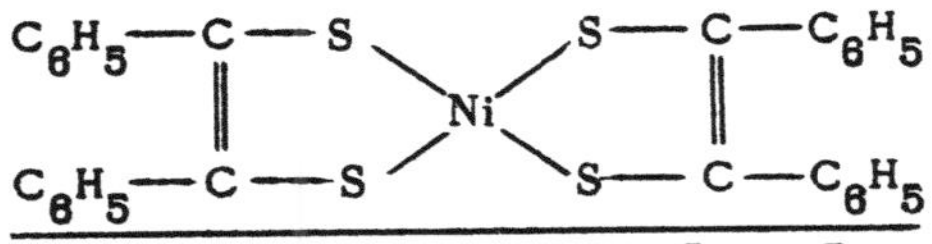 / $C_6H_5COOC_2H_5$	50-80	$k_1[ROOH]$	$\bar{1}$,655 (50°)	9,415	14,4	[148]
sec-$C_{10}H_{21}OOH$	$Mn[OOC(CH_2)_{16}CH_3]_2$ / RH	80-104		0,61 (90°)	11,45	18,0	[149]
α-$C_{10}H_{11}OOH$	$Cr[OOC(CH_2)_7CH{=}CH(CH_2)_7CH_3]_2$ / RH	34	$k_1[Cr(III)]^{0,45}$	$\bar{4}$,62			[143]

TABLE 107. Oxidation of Fe^{2+} Ions by Quinones

Reaction	Conditions	t°	log k	log A	E	Ref.
Fe^{2+} + 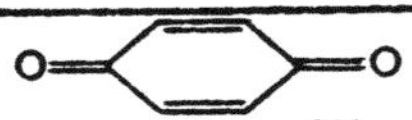	H_2O; pH 0	25	$\bar{3}$,04	4,65	7,9	[35]
Fe^{2+} + 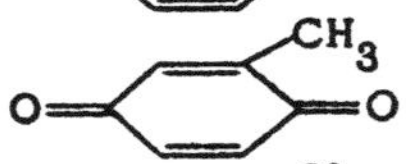	- " -	25–35	$\bar{4}$,45 (25°)	2,845	7,8	[35]
Fe^{2+} + 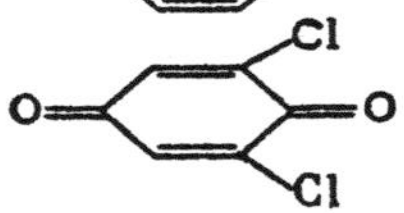	H_2O; pH 0	15–35	$\bar{2}$,63 (25°)	2,125	8,2	[35]

CHAPTER XI

REACTIONS OF ATOMS AND RADICALS WITH IONS

Atoms and active free radicals react very rapidly with metal ions and organic ions. Rate constants are measured for these reactions by the same methods applied to the reactions of free radicals with molecules and with each other (see Chapter IV). The errors in measuring the absolute values of the rate constants for reactions of radicals with ions amount to 20-40%.

§1. Reactions of Hydrogen Atom

Quantitative data on the reactions of the hydrogen atom with ions have been obtained mainly in radiochemical studies. In some instances, the hydrogen atoms have been obtained by passing an electrical discharge through gaseous hydrogen, then rapidly bubbling hydrogen through the solution being investigated. Hydrogen atoms have a dual reactive capacity: They oxidize reducer-ions and reduce oxidizer-ions. For example,

$$Fe^{3+} + H\cdot \longrightarrow Fe^{2+} + H^+$$

$$Fe^{2+}_{aq} + H\cdot \longrightarrow FeOH^{2+} + H_2$$

With organic ions, the hydrogen atom enters into the same reactions as with molecules, i.e., substitution and addition.

§2. Reactions of Free Radicals

The reaction of a free radical with a variable-valence metal cation consists of electron transfer from the ion to the radical, or from the radical to the ion--for example:

$$Fe^{2+} + HO\cdot \longrightarrow Fe^{3+} + OH^-$$

$$Cu^{2+} + HO\cdot_2 \longrightarrow Cu^+ + O_2 + H^+$$

The reaction of a radical with an inorganic anion is analogous:

$$HO\cdot + CO_3^{2-} \longrightarrow HO^- + CO_3^{-}\cdot$$

$$HO\cdot + Br^- \longrightarrow HO^- + Br\cdot$$

Reactions of radicals with organic ions are analogous to reactions with molecules:

$$HO\cdot + CH_3COO^- \longrightarrow HOH + \cdot CH_2COO^-$$

$$HO\cdot + C_6H_5COO^- \longrightarrow HO\dot{C}_6H_5COO^-$$

(analogous to the reaction: $HO\cdot + C_6H_6 \longrightarrow HO\dot{C}_6H_6$)

$$HO\cdot + CH_3NH_3^+ \longrightarrow H_2O + \cdot CH_2NH_3^+$$

The situation is more complicated with metal complexes having organic ligands. With such complexes, a free radical may enter into the following three reactions: It may take away or donate an electron to the central metal atom; it may abstract an H· atom or an electron of the ligand with a subsequent, very rapid oxidation-reduction reaction within the complex; or it may react with the ligand the same as with a molecule.

TABLE 107a. Reactions of Hydrogen Atoms with Ions at Room Temperature (20-25°)

Ion	Conditions	log k	Method	Ref.
H^+	H_2O; pH < 7	$\bar{1}$,7–0,7	MCR, H·	[150]
	H_2O; pH 3,5–11	3,415	FM	[151]
$Cr(NH_3)_5Cl^{2+}$	H_2O; pH 5	6,255	MCR $(CH_3)_2CDOH$	[152]
$Cr(H_2O)_4Cl_2^+$	- " -	7,30	- " -	[152]
Mn^{2+}	H_2O; pH 6	7,40	MCR, $(CH_3)_2CDOH$	[152]
Fe^{2+}	H_2O; pH 0,4 (H_2SO_4)	7,20	MCR, O_2	[152a]
	H_2O; pH 2,1 (H_2SO_4)	7,115	- " -	[152a]
	- " -	7,13	MCR, D_2	[153]
	H_2O; pH 0,1–1	7,30	PR	[154]
Fe^{3+}	H_2O; pH 0,1	7,90	MCR, O_2	[155]
	H_2O; pH 0,4	6,08	- " -	[152a]
	H_2O; pH 1,57	7,34	- " -	[152a]
	H_2O; pH 2,1	7,98	- " -	[152a]
	- " -	7,98	MCR, D_2	[153]
	H_2O; 0,4 M (H_2SO_4)	6,20	MCR Fe^{2+}	[156]
	H_2O; pH 2 (H_2SO_4)	7,955	PR	[157]
$FeOH^{2+}$	H_2O; pH 0,7–2,7	8,88	MCR CH_3OH	[158]

Table 107a (continued)

Ion	Conditions	log k	Method	Ref.
$FeCl^{2+}$	H_2O; pH 0,1	9,655	MCR, O_2	[154]
$FeCl_2^{1+}$	- " -	9,955	- " -	[154]
$Fe(CN)_6^{3-}$	H_2O; pH 1–3	9,60	MCR, $HCOO^-$	[159]
$Fe(CN)_6^{3-}$	H_2O; pH 7	9,60	MCR, $(CH_3)_2CHOH$	[160]
	- " -	9,48	MCR, C_2H_5OH	[160a]
	H_2O; pH 1,1	9,68	- " -	[160a]
	H_2O; pH < 7	9,56	MCR, $(CH_3)_2CHOH$	[160 b]
$Co(NH_3)_6^{3+}$	H_2O; pH 4–6	6,27	MCR, H·	[161]
	- " -	6,20	MCR, CD_3OH	[162]
$Co(NH_3)_5H_2O^{3+}$	H_2O; pH 4–6	5,65	MCR, H·	[163]
$Co(NH_3)_5OH^{2+}$	H_2O	7,49	MCR, H·	[161]
	- " -	7,40	MCR, $(CH_3)_2CDOH$	[162]
$Co(NH_3)_5F^{2+}$	H_2O; $HClO_4$ 0,1 M	⩽ 6,30	MCR, C_2H_5OH	[163a]
	H_2O; pH 4–6	6,04	MCR, H·	[161]
	- " -	6,08	MCR, CD_3OH	[162]
$Co(NH_3)_5Cl^{2+}$	H_2O; pH 4–6	7,86	MCR, H·	[161]

Table 107a (continued)

Ion	Conditions	log k	Method	Ref.
$Co(NH_3)_5Br^{2+}$	H_2O; pH 4–6	8,66	MCR, H·	[161]
	H_2O; pH 1	10,15	MCR, C_2H_5OH	[163a]
	H_2O; pH 4–6	9,00	MCR, $(CH_3)_2CDOH$	[162]
$Co(NH_3)_5I^{2+}$	H_2O; pH 4–6	9,52	MCR, $(CH_3)_2CDOH$	[161]
	– " –	9,48		[162]
	H_2O; pH 1	10,52	MCR, C_2H_5OH	[163a]
$Co(NH_3)_5CN^{2+}$	H_2O; pH 4–6	7,57	MCR, $(CH_3)_2CDOH$	[162]
	H_2O; pH 1	7,785	MCR, C_2H_5OH	[163a]
$Co(NH_3)_5NCS^{2+}$	H_2O; pH 1	9,80	MCR, C_2H_5OH	[163a]
$Co(NH_3)_5N_3^{2+}$	H_2O; pH 4–6	8,59	MCR, $(CH_3)_2CDOH$	[162]
	H_2O; pH 1	9,04	MCR, C_2H_5OH	[163a]
$Co(NH_3)_5NO_2^{2+}$	H_2O; pH 1	8,20	MCR, C_2H_5OH	[163a]
$Co(NH_3)_3(NO_2)_3$	– " –	8,98	– " –	[163a]
$Co(NH_3)_5PO_4$	– " –	$\leqslant$ 5,83	– " –	[163a]
$Co(NH_3)_5OCOCH_3^{2+}$	– " –	$\leqslant$ 6,11	– " –	[163a]
$Co(NH_3)_5OCOCF_3^{2+}$	H_2O; pH 1	$\leqslant$ 5,43	– " –	[163a]
$Co(NH_3)_5$fumarate-H^{2+}	– " –	9,785	– " –	[163a]
$Co(CN)_6^{3-}$	H_2O; pH 1	7,30	– " –	[163a]
	H_2O; pH 5	8,955	MCR, $(CH_3)_2CDOH$	[162]

Table 107a (continued)

Ion	Conditions	log k	Method	Ref.
$Co(En)_2Cl_2^+$	H_2O; pH 8	8,84	MCR, $(CH_3)_2CDOH$	[152]
$Co(En)_2CO_3^+$	- " -	< 8,30	- " -	[152]
Ni^{2+}	H_2O; pH 7	< 5,0	MCR, C_2H_5OH	[160a]
Cu^{2+}	H_2O	7,77	PR	[152]
	H_2O; pH 7	8,70	MCR, $(CH_3)_2CDOH$	[152]
	H_2O; pH 1	7,76	MCR, HCO_2H	[163]
	- " -	8,18	MCR, CH_3COCH_3	[164]
	H_2O; pH 7	8,78	MCR, $(CH_3)_2CHOH$	[160]
	H_2O; pH 0,7-2,7	7,78	MCR, CH_3OH	[158]
Zn^{2+}	H_2O; pH 7	< 5,0	MCR, C_2H_5OH	[160a]
As^{3+}	H_2O; pH 2,7	9,00	MCR, H_2O_2	[165]
Ag^+	H_2O; pH 7	9,60	MCR, C_2H_5OH	[160a]
	- " -	10,04	PR	[166]
Cd^{2+}	H_2O; pH 7	< 5,0	MCR, C_2H_5OH	[160a]
Sn^{2+}	H_2O; pH 1	10,90	MCR, Fe^{3+}	[167]
Sn^{4+}	- " -	6,40	- " -	[167]
SnF_3^-	H_2O; pH 6	9,30	MCR, $(CH_3)_2CDOH$	[152]
SnF_6^{2-}	- " -	6,956	- " -	[152]

Table 107a (continued)

Ion	Conditions	log k	Method	Ref.
Hg^{2+}	H_2O; pH 7	10,40	MCR, C_2H_5OH	[160a]
Tl^{+}	H_2O; pH 1,5	8,00	MCR, $(CH_3)_2CDOH$	[152]
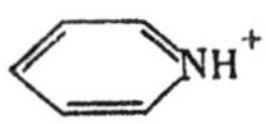	H_2O	8,23	PR	[168]
OH^-	H_2O; pH 11-13	7,36	MCR, CH_3COCH_3	[164]
OH^-	H_2O	7,30	PR	[162]
	H_2O; pH 7,5-12,5	7,08	MCR, $ClCH_2CO_2^-$	[169]
	H_2O; pH 11,5	~7,30	PR	[170]
CN^-	H_2O pH 7	9,415	MCR, $(CH_3)_2CHOH$	[152]
CNS^-	- " -	8,23	- " -	[152]
HCO_3^-	- " -	~ 4,48	MCR, CH_3OH	[164]
NO_2^-	H_2O; pH 6	8,78	MCR, $(CH_3)_2CHOH$	[152]
	H_2O; pH 7	8,78	MCR, CH_3OH	[171]
	- " -	8,12	MCR, $(CH_3)_2CHOH$	[160]
	- " -	9,00	MCR, CH_3OH	[160a]
	H_2O; pH 8	9,38	MCR, C_2H_5OH	[163a]
	H_2O; pH 6	8,955	MCR, $(CH_3)CDOH$	[152]
NO_3^-	H_2O; pH 11-13	6,845	MCR, CH_3COO^-	[164]
	H_2O; pH 7	7,38	MCR, $(CH_3)_2CHOH$	[160]

Table 107a (continued)

Ion	Conditions	log k	Method	Ref.
NO_3^-	H_2O; pH 1,1	7,08	PR	[161]
	H_2O; pH 10–13	6,58	- " -	[161]
	H_2O; pH 7	6,845	MCR, C_2H_5OH	[160a]
	H_2O; pH < 7	7,36	MCR, $(CH_3)_2CHOH$	[160b]
F^-	H_2O	4,175	PR	[172]
CrO_4^{2-}	H_2O; pH 7	10,0	MCR, C_2H_5OH	[160a]
$Cr_2O_7^-$	- " -	10,20	- " -	[160a]
MnO_4^-	H_2O; pH 3	10,45	PR	[205]
	- " -	10,39	- " -	[211]
	H_2O	10,27	- " -	[166]
	- " -	10,28	- " -	[173]
BrO_3^-	H_2O; pH 7	7,30	MCR $(CH_3)_2CHOH$	[152]
IO_3^-	- " -	7,77	- " -	[152]
I_2^-	H_2O;pH 3,5–11	7,255	FM	[150]
$HCOO^-$	H_2O; pH 1–3	8,40	MCR, $Fe(CN)_6^{3-}$	[174]
	H_2O; pH 7	8,34	- " -	[172]
	H_2O; pH 11–13	8,40	MCR, OH^-	[164]
	H_2O; pH 7	8,18	MCR, DCO_2^-	[160]
	- " -	7,34	MCR, C_2H_5OH	[160]
	- " -	7,545	MCR, HCO_2^-	[175]
CH_3COO^-	H_2O; pH 7	5,43	MCR, NO_2^-	[171]

Table 107a (continued)

Ion	Conditions	log k	Method	Ref.
CH_3COO^-	H_2O; pH 9	5,43	MCR, $NOC_6H_4N(CH_3)_2$	[175]
	H_2O;pH 11-13	5,48	MCR, OH^-	[164]
	H_2O; pH 7	5,43	MCR, $DCOO^-$	[160]
CD_3COO^-	H_2O; pH 7	4,08	MCR, $CH_3CO_2^-$	[175a]
$H + ClCH_2CO_2^- \rightarrow H_2$	- " -	6,415	MCR, H·	[169]
$H + ClCH_2CO_2^- \rightarrow HCl$	- " -	5,415	- " -	[169]
$ClCH_2CO_2^-$	H_2O; pH 7	6,08	MCR, DCO_2^-	[160]
$C_2H_5COO^-$	H_2O; pH 9	7,04	MCR, $NOC_6H_4N(CH_3)_2$	
$C_3H_7COO^-$	- " -	7,08	- " -	[175]
$(CH_3)_2CHCOO^-$	- " -	7,45	- " -	[175]
$CH_3(CH_2)_3COO^-$	- " -	7,20	- " -	[175]
$(CH_3)_2CHCH_2COO^-$	- " -	7,34	- " -	[175]
$C_2H_5CH(CH_3)COO^-$	- " -	7,45	- " -	[175]
$(CH_3)_3CCOO^-$	- " -	6,23	- " -	[175]

TABLE 108. Reactions of Radicals with Ions

Ion	Conditions	t°	log k	log A	E	Method	Ref.
	Radical ·OH						
Cr^{3+}	H_2O; pH < 7	25	8,49			MCR, H_2	[160a]
Mn^{2+}	H_2O; pH 7	25	>8,15			PR	[176]
Fe^{2+}	H_2O; pH 1	25	8,48			MCR, H_2	[176a]
	- " -	25	8,505			- " -	[177]
	H_2O; pH 1,6	23	8,505			- " -	[152]
	H_2O; pH 2,1	23	8,40			- " -	[152]
	H_2O; pH 2	23	8,415			PR	[178]
	H_2O; pH 0,4	23	8,43			- " -	[179]
	H_2O; pH 0	18	8,52			MCR	[180]
$Fe(CN)_6^{4-}$	H_2O; pH 2,5-10,5	23	9,32			MCR, HCO_2^-	[181]
	H_2O; pH 7	23	10,23			PR	[182]
	H_2O; pH 10,7	23	9,91			MCR, $C_6H_5COO^-$	[183]

	H_2O; pH 9	23	9,875	MCR, $NOC_6H_4N(CH_3)_2$	[184]
	H_2O; pH 3–10	23	10,03	PR	[184a]
Cu^{2+}	H_2O; pH 7	23	8,545	PR	[185]
Sn^{2+}	H_2O; pH 1	23	9,30	MCR, Fe^{2+}	[167]
Tl^{+}	H_2O; pH 7	23	9,88	PR	[185a]
	H_2O; pH 1	23	9,95	MCR, Ce^{3+}	[186]
Ce^{3+}	H_2O	23	8,34	MCR	[187]
	H_2O	23	8,54	MCR, H_2	[152]
OH^-	H_2O	23	8,555	FM	[188]
NO_2^-	H_2O; pH 7	23	9,60	MCR	[152]
	– " –	23	9,40	MCR, H_2O_2	[189]
	– " –	23	9,555	MCR, CO_3^{2-}	[189a]
	– " –	23	9,40	PR	[190]
	– " –	23	9,38	MCR, NO	[190a]
	H_2O; pH 9	23	9,69	MCR, $NOC_6H_4N(CH_3)_2$	[184]
NO_3^-	H_2O; pH 9	23	<5,7	MCR, $NOC_6H_4N(CH_3)_2$	[152]
N_3^-	– " –	23	9,81	– " –	[184]

Table 108 (continued)

Ion	Conditions	t°	log k	log A	E	Method	Ref.
CN^-	H_2O; pH 9	23	9,43			MCR, $NOC_6H_4N(CH_3)_2$	
PO_3^{3-}	H_2O; pH 10,7	23	9,28			MCR, CO_3^{2-}	[189]
$H_2PO_2^-$	H_2O; pH 10,7	23	8,97			- " -	[189]
	H_2O; pH 7	23	$<7,0$			MCR, I^-	[182]
SO_3^{2-}	H_2O; pH 7	23	9,08			MCR, CO_3^{2-}	[190]
HSO_3^-	- " -	23	9,32			MCR, HSO_3^-	[190]
H_2SO_4	H_2O; pH 1	23	5,60			MCR	[191]
HSO_4^-	H_2O; pH 1	23	7,52			MCR, HCO_2H	[167]
	H_2O	23	5,30			MCR, H_2	[192]
	H_2O	23	5,70			- " -	[192a]
	H_2O	23	7,30			FM	[192]
CNS^-	H_2O; pH 7	23	9,82			PR	[189]
	- " -	23	9,115			- " -	[190]
	H_2O; pH 13,5	23	9,00			- " -	[193]
	H_2O; pH 2	23	9,76			MCR, thymine	[194]

Cl^-	H_2O; pH 0	23	9,60	MCR, thymine	[195]
	H_2O; pH 1	23	8,60	- " -	[195]
	H_2O; pH 2	23	7,60	- " -	[195]
	pH 4	23	6,60	MCR, thymine	[195]
	H_2O; pH 3	23	7,20	MCR	[195a]
	H_2O; pH 9	23	<5,85	MCR, $NOC_6H_4N(CH_3)_2$	[184]
	H_2O; pH 0,4	23	9,60	MCR, ·OH	[196]
	- " -	23	9,60	PR	[197]
	pH 3	23	7,30	- " -	[197]
AsO_2^-	H_2O; pH 10,7	23	9,66	MCR, CO_3^{2-}	[189a]
	H_2O; pH 8	23	9,66	MCR, $NOC_6H_4N(CH_3)_2$	[184]
SeO_3^{2-}	H_2O; pH 7	23	9,43	PR	[193]
Br^-	H_2O; pH 0,5	23	10,20	MCR, ·OH	[196]
	H_2O; pH 0–2	23	10,555	MCR, H_2	[197a]
	H_2O; pH 7	23	8,78	MCR, C_2H_5OH	[198]
	H_2O; pH 2	23	9,70	PR	[199]
	H_2O; pH 3	23	9,34	MCR	[195a]

Table 108 (continued)

Ion	Conditions	t°	log k	log A	E	Method	Ref.
Br^-	H_2O; pH 7	23	9,00			PR	[199]
	H_2O; pH 5–9	23	9,08			- " -	[200]
	H_2O; pH 7	23	9,20			MCR	[190a]
	- " -	23	9,13			MCR, H_2O_2	[201]
	- " -	23	8,115			PR	[190]
	H_2O; pH 10,5	23	8,89			MCR, $C_6H_5CO_2^-$	[183]
	H_2O; pH 9	23	8,81			MCR, $NOC_6H_4N(CH_3)_2$	[202]
	H_2O; pH 1,3	23	9,765			MCR, C_2H_5OH	[203]
BrO^-	H_2O; pH 13	23	9,68			MCR, CO_3^{2-}	[204]
Cu^{2+}	H_2O; pH 7	23	8,54			PR	[205]
I^-	H_2O; pH 7	23	9,20			FM	[188]
	H_2O; pH 7	23	9,845			PR	[204]
	H_2O; pH 3	23	9,52			MCR	[195a]
	H_2O; pH 7	23	10,01			PR	[182]
	H_2O; pH 10,5	23	9,96			MCR, $C_6H_5CO_2^-$	[183]
	H_2O; pH 9	23	9,935			MCR, $NOC_6H_4N(CH_3)_2$	[184]

CO_3^{2-}	H_2O; pH 7	23	7,90	PR	[190]
	H_2O; pH 10,7	23	8,30	MCR, CNS^-	[189]
	H_2O; pH < 11,6	23	8,62	PR	[206]
	H_2O; pH 13	23	7,64		[152]
	H_2O; pH 11	23	8,48	PR	[182]
	H_2O; pH 10,5	23	8,48	MCR, $C_6H_5CO_2^-$	[183]
HCO_3^-	H_2O; pH 7	23	7,00	PR	[207]
	H_2O; pH < 11,6	23	7,18	- " -	[206]
HCO_2^-	H_2O; pH 9	23	9,38	MCR, $NOC_6H_4N(CH_3)_2$	[184]
	H_2O; pH 10,7	23	9,38	MCR, $C_6H_5CO_2^-$	[183]
	H_2O; pH 10	23	9,38	PR	[182]
	H_2O; pH 4	23	8,30	MCR, $(NH_2CH_2)_2$	[162]
	H_2O; pH 2,5–10,5	23	9,00	MCR, $Fe(CN)_6^{4-}$	[181]
$CH_3CO_2^-$	H_2O; pH 9	23	7,63	MCR, $NOC_6H_4N(CH_3)_2$	[175]
	- " -	23	7,725	- " -	[184]
	H_2O; pH 10,7	23	8,63	MCR, $C_6H_5CO_2^-$	[183]

Table 108 (continued)

Ion	Conditions	t°	log k	log A	E	Method	Ref.
$C_2O_4^{2-}$	H_2O; pH 9	23	7,70			MCR, $NOC_6H_4N(CH_3)_2$	[184]
	H_2O	23	6,78			MCR, C_2H_5OH	[203]
$C_2O_4H^-$	- " -	23	7,545			MCR, C_2H_5OH	[203]
$C_2O_4H_2$	- " -	23	<7,0			- " -	[203]
$C_2H_5COO^-$	H_2O; pH 9	23	8,67			MCR, $NOC_6H_4N(CH_3)_2$	[175]
$CH_3(CH_2)_2COO^-$	- " -	23	9,04			- " -	[175]
$(CH_3)_2CHCOO^-$	- " -	23	9,23			- " -	[175]
$CH_3(CH_2)_3COO^-$	- " -	23	9,23			- " -	[175]
$(CH_3)_2CHCH_2COO^-$	- " -	23	9,145			- " -	[175]
$CH_3CH_2CH(CH_3)COO^-$	- " -	23	9,115			- " -	[175]
$(CH_3)_3CCOO^-$	- " -	23	8,935			- " -	[175]
$CH_3(CH_2)_4COO^-$	- " -	23	9,36			- " -	[175]
$(CH_3)_3CCH_2COO^-$	- " -	23	9,00			- " -	[175]
cyclo-$C_4H_7COO^-$	- " -	23	9,255			- " -	[175]
cyclo-$C_5H_9COO^-$	- " -	23	9,38			- " -	[175]

cyclo-$C_6H_{11}COO^-$	- " -	23	9,505	- " -	[175]
$^-OOCC(OH)_2C(OH)_2COO^-$	- " -	23	8,885	- " -	[175]
$C_6H_5COO^-$	H_2O; pH 10,7	23	9,59	MCR, $C_6H_5COO^{14-}$	[183]
	H_2O; pH 9	23	9,40	MCR, $NOC_6H_4N(CH_3)$	[208]
p-$NO_2C_6H_4COO^-$	H_2O; pH 9	23	9,07	MCR, $NOC_6H_4N(CH_3)$	[208]
p-Br $C_6H_4COO^-$	- " -	23	9,27	- " -	[208]
p-$^-OOCC_6H_4COO^-$	- " -	23	9,285	- " -	[208]
p-Cl $C_6H_4COO^-$	- " -	23	9,285	- " -	[208]
p-F $C_6H_4COO^-$	- " -	23	9,32	- " -	[208]
p-$CH_3C_6H_4COO^-$	- " -	23	9,42	- " -	[208]
p-$CH_3OC_6H_4COO^-$	- " -	23	9,47	- " -	[208]
p-$HOC_6H_4COO^-$	- " -	23	9,51	- " -	[208]
p-$NH_2C_6H_4COO^-$	- " -	23	9,675	- " -	[208]
o-$HOC_6H_4COO^-$	- " -	23	9,75	- " -	[184]
	H_2O; pH 10,7	23	9,60	MCR, $C_6H_5C^{14}OO^-$	[183]

Table 108 (continued)

Ion	Conditions	t°	log k	log A	E	Method	Ref.
$C_6H_5O^-$	H_2O; pH 10,7	23	9,81			MCR, $C_6H_5C^{14}OO^-$	[183]
$C_6H_5CH_2COO^-$	H_2O; pH 9	23	9,415			MCR, $NOC_6H_4N(CH_3)_2$	[208]
$C_6H_5SO_3^-$	- " -	23	9,245			- " -	[208]
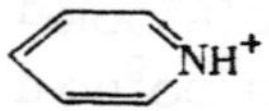	H_2O; pH 1-7	23	7,30			PR	[168]
$CH_3NH_3^+$	H_2O; pH 5	23	7,04			MCR, $NOC_6H_4N(CH_3)_2$	[175]
$H_3^+NCH_2COO^-$	- " -	23	7,43			- " -	[175]
$NCCH_2COO^-$	H_2O; pH 9	23	7,96			- " -	[175]
FCH_2COO^-	- " -	23	7,255			- " -	[175]
$BrCH_2COO^-$	- " -	23	7,415			- " -	[175]
$ClCH_2COO^-$	- " -	23	7,52			- " -	[175]
$^-OOCCH_2COO^-$	- " -	23	7,52			- " -	[175]
$CH_3OCH_2COO^-$	- " -	23	8,56			- " -	[175]
$HOCH_2COO^-$	- " -	23	8,62			- " -	[175]
$NH_2CH_2COO^-$	H_2O; pH 12	23	9,08			- " -	[175]

	Radical $HO_2^{\cdot}$						
Fe^{3+}	H_2O; pH 2,05	25	5,52			MCR	[109]
Fe^{2+}	H_2O; 0,5 M H_2SO_4	23	5,86			PR	[209]
Cu^{2+}	H_2O; pH 2,3	23	7,18			PR	[210]
	- " -	25	8,04	9,7	3,0	IIM	[116a]
Cu^{+}	H_2O	23	9,63	11,3	3,0	- " -	[116a]
MnO_4^{-}	H_2O; pH 3	23	6,90			PR	[211]
Br_2	H_2O	23	9,175			PR	[212]
Br_2^{-}	H_2O	23	9,58			- " -	[212]
	H_2O	23	9,20			- " -	[213]
Br_3^{-}	- " -	23	8,00			PR	[213]
CNS^{-}	H_2O; pH 1	23	9,20			PR	[183]
	Radical $C_6H_5CH(CH_3)OO^{\cdot}$						
$[MnEDTA]^{2-}$	CH_3COOH + + $C_6H_5CH_2CH_3$	60	6,00			CINH	[214]
$[CoEDTA]^{2-}$	- " -	60	5,58			- " -	[214]
$Co(Oxin)_2$	- " -	60	7,955			- " -	[214]

Table 108 (continued)

Ion	Conditions	t°	log *k*	log A	E	Method	Ref.
Cu (Oxin)$_2$	CH_3COOH +	60	5,90			CINH	[214]
Mn (Oxin)$_2$	+ $C_6H_5CH_2CH_3$	60	5,62			- " -	
Ni (Oxin)$_2$	- " -	60	6,35			- " -	
Complexes of Mn^{2+} with azo compounds							
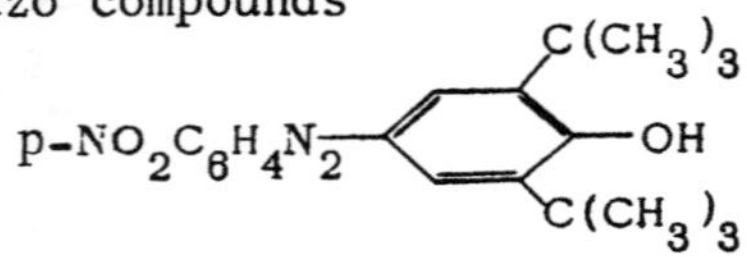	CH_3COOH + + $C_6H_5CH_2CH_3$	60	5,575			CINH	[214]
	- " -	60	6,52			- " -	[214]
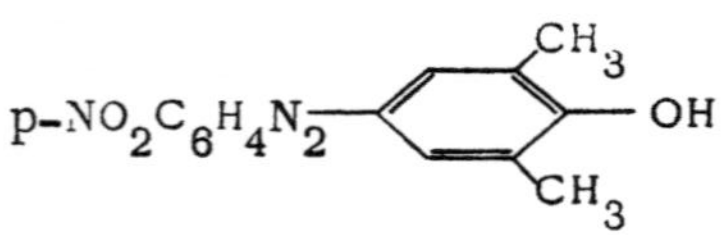	- " -	60	7,29			- " -	[214]
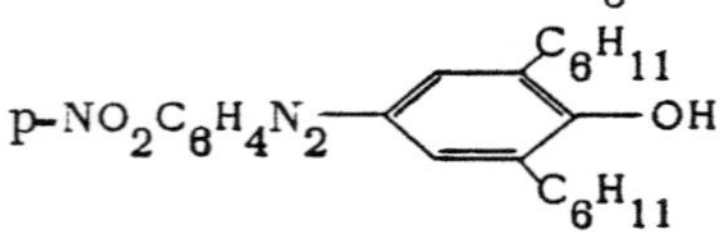	- " -	60	7,48			- " -	[214]

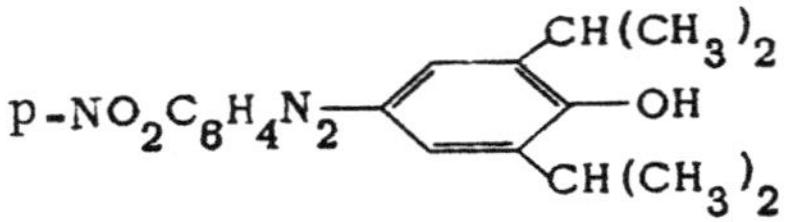	- " -	60	5,96			- " -	[214]
Ferrocene	- " -	60	5,38			- " -	[215]
Ethylferrocene	- " -	60	5,255			- " -	[215]
Phenylferrocene	- " -	60	4,92			- " -	[215]
Acetylferrocene	- " -	60	4,52			- " -	[215]
Cyanoferrocene	- " -	60	3,60			- " -	[215]
Ferrocenecarboxylic acid	- " -	60	4,08			- " -	[215]
	Radical $(CH_3)_2(CN)COO\cdot$						
$Mn[OOC(CH_2)_{16}CH_3]_2$	C_6H_5Cl	60–80	8,45	8,45	4,8	NKR	[216]
$Co(AcAc)_2$	- " -	74	4,36			Kinetics	[217]
	Radical $\sim CH_2C(OO\cdot)HC_6H_5$						
$Co(AcAc)_2$	C_6H_5Cl	55–85	4,69 (80°)	9,46	7,8	Kinetics	[217]
	Radical 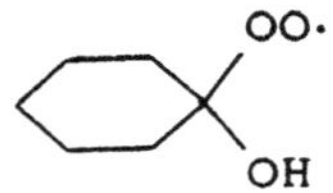						
Cu^{2+}	cyclo-$C_6H_{11}OH$ + + 10% H_2O	75	6,505			CINH	[218]

Table 108 (continued)

Ion	Conditions	t°	log k	log A	E	Method	Ref.
	Radical $(CH_3)_2\dot{C}H$						
$Cu(NCCH_3)_4^{2+}$	$CH_3CN:CH_3COOH$ = = 1:1,5	25,5	6,70			MCR	[219]
	Radical $CH_3CH_2CH_2\dot{C}H_2$						
$Cu(NCCH_3)_4^{2+}$	$CH_3CN:CH_3CO_2H$= = 1:1,5	25,5	6,49			MCR	[219]
$Cu(Dipy)^{2+}$	- " -	25,5	7,23			MCR	[219]
	Radical $(CH_3)_3C\dot{C}H_2$						
$Cu(NCCH_3)_4^{2+}$	$CH_3CN:CH_3CO_2H$= = 1:1,5	25,5	5,65			MCR	[219]
$Cu(Dipy)^{2+}$	- " -	25,5	4,40			MCR	[219]
	Radical $C_6H_5CH_2\dot{C}H_2$						
$Cu(NCCH_3)_4^{2+}$	$CH_3CN:CH_3CO_2H$= = 1:1,5	25,5	6,20			MCR	[219]
$Cu(Dipy)^{2+}$	- " -	25,5	7,145			MCR	[219]

Radical $\sim CH_2\dot{C}HCONH_2$

Ti^{3+}	H_2O; 0,8 N H_2SO_4	25	2,76	12,0	11,1	CINH	[220]
	D_2O; 0,8 N H_2SO_4	25	2,81			- " -	[220]
VO_2^+	H_2O; 0,08 N H_2SO_4	25	3,04			CINH	
V^{2+}	- " -	25	5,04			- " -	[220]
Cr^{2+}	- " -	25	5,45				
Fe^{3+}	H_2O; pH 1	25	3,30			CINH	[221]
	H_2O; pH	25	3,30	2,01	2,35	- " -	[222]
	D_2O; pH 0	25	3,28	1,84	2,44	- " -	[222]
$FeOH^{2+}$	H_2O	25	4,04			CINH	[221]
	H_2O	25	4,325			- " -	[220]
$FeCl^{2+}$	H_2O	25	4,91			CINH	
$FeCl_2^+$	- " -	25	4,23			- " -	[220]
$FeCl_3$	- " -	25	6,00			- " -	
$FeBr^{2+}$	- " -	25	6,22			- " -	

Table 108 (continued)

Ion	Conditions	t°	log k	log A	E	Method	Ref.
FeN_3^{2+}	H_2O	25	6,19			CINH	
$FeCNS^{2+}$	- " -	25	7,13			- " -	
$Fe(Dipy)_3^{3+}$	- " -	25	4,91			- " -	
$Fe(o\text{-}Phen)_3^{3+}$	- " -	25	5,49			- " -	[220]
$Fe\text{-}(5\text{-}CH_3\text{-}o\text{-}Phen)_3^{3+}$	- " -	25	5,415			- " -	
$Fe\text{-}(5\text{-}C_6H_5\text{-}o\text{-}Phen)_3^{3+}$	- " -	25	5,71			- " -	
$Fe\text{-}(5\text{-}Cl\text{-}o\text{-}Phen)_3^{3+}$	- " -	25	5,36			- " -	
$Fe(CN)_6^{3-}$	- " -	25	5,93			- " -	
Cu^{2+}	H_2O; 1 N $HClO_4$	25	3,07	7,04	5,4	CINH	[222]
	D_2O; 1 N $HClO_4$	25	3,145	7,00	5,3	- " -	[222]
Mo^{3+}	D_2O; 1 N $HClO_4$	25	3,845			CINH	[220]
$CeOH^{3+}$	H_2O; pH 0–1	25	3,52			CINH	[222]
Eu^{2+}	H_2O; 0,8 N H_2SO_4	25	4,925			- " -	[220]

Hg^{2+}	H_2O; 0,1 N $HClO_4$	25	0,02	4,62	6,2	- " -	[222]
Tl^{3+}	- " -	25	$\bar{1}$,53	1,32	2,5	- " -	[222]
	Radical p-$CH_3OC_6H_4CH_2\dot{C}H_2$						
$Cu(NCCH_3)_4^{2+}$	$CH_3CN:CH_3CO_2H$ = 1:1,5	25,5	4,32			MCR	[219]
$Cu(Dipy)^{2+}$	- " -	25,5	6,48			MCR	[219]
	Radical $\sim CH_2\dot{C}(CH_3)CONH_2$						
Fe^{3+}	H_2O	25	1,725	7,4	7,7	CINH	[223]
$FeOH^{2+}$	- " -	25	3,24	11,3	11	- " -	[223]
$[Fe\text{-}NH_2COC(CH_3){=}CH_2]^{3+}$	- " -	25	3,785	7,7	5,5	- " -	[223]
	Radical $\sim CH_2\dot{C}HCN$						
$Fe(CN)_6^{3-}$	H_2O	25	6,83			PR	[224]
$FeCl_3$	$CHON(CH_3)_2$	25	3,81			CINH	[225]
	- " -	25	3,91			- " -	[226]

Table 108 (continued)

Ion	Conditions	t°	logk	log A	E	Method	Ref.
	Radical $\sim CH_2\text{-}\dot{C}(CH_3)CN$						
$FeCl_3$	$CHON(CH_3)_2$	25	2,79			CINH	[225]
	Radical $CH_2\dot{C}HC_6H_5$						
$FeCl_3$	$CHON(CH_3)_2$	25	4,73			CINH	[225]
	Radical $\sim CH_2\dot{C}(CH_3)COOCH_3$						
$CuCl_2$	$CHON(CH_3)_2$	60	5,89			CINH	[227]
	Radical 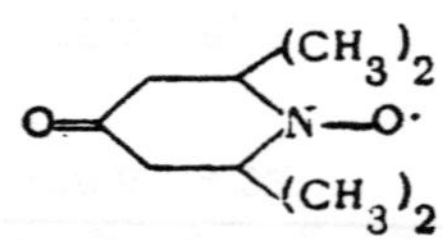						
Fe^{2+}	H_2O; 0,1 N HCl; 0,4 M NaCl	25	5,77	10,91	7,0	Kinetics	[228]

	H_2O; 0,05 HCl; 0,45 *M* NaCl	25	5,22	10,51	7,2	- " -	[228]
	Radical 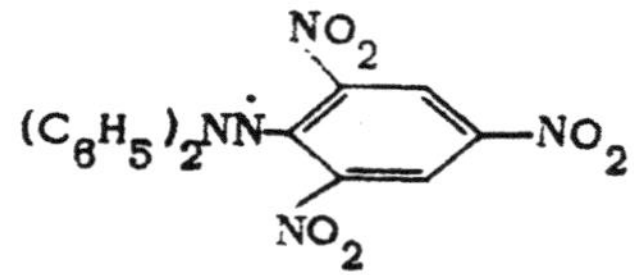						
Fe^{2+}	$H_2O : C_2H_5OH =$ = 1 : 1	20–50	1,38 (20°)	11,44	13,5	Kinetics	[229]
	$H_2O : C_2H_5OH =$ = 1 : 4	12–30	2,03 (20°)	8,08	8,1	- " -	[230]
	CH_3OH	22,5	2,825			- " -	[231]
$FeCl_2$	CH_3OH	22,5	5,565			Kinetics	[231]
Fe^{2+}	C_2H_5OH	22,5	1,46			- " -	
$FeCl_2$	C_2H_5OH	22,5	2,875			- " -	
Fe^{2+}	$n\text{-}C_3H_7OH$	22,5	1,505			- " -	
$FeCl_2$	$n\text{-}C_3H_7OH$	22,5	3,85			- " -	
Fe^{2+}	$(CH_3)_2CHOH$	22,5	2,00			- " -	
$FeCl_2$	$(CH_3)_2CHOH$	22,5	3,70			- " -	

Table 108 (continued)

Ion	Conditions	t°	log k	log A	E	Method	Ref.
Fe^{2+}	n-C_4H_9OH	22,5	1.505			Kinetics	[231]
$FeCl_2$	n-C_4H_9OH	22,5	3,59			- " -	
Fe^{2+}	n-$C_5H_{11}OH$	22,5	1,20			- " -	
$FeCl_2$	n-$C_5H_{11}OH$	22,5	3,54			- " -	
Fe^{2+}	n-$C_7H_{15}OH$	22,5	1,84			- " -	
$FeCl_2$	n-$C_7H_{15}OH$	22,5	3,475			- " -	
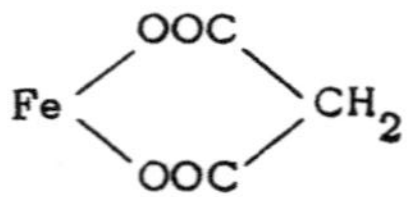	n-C_4H_9OH; 0,5 M $NaClO_4$	20	3,33			Kinetics	[231]
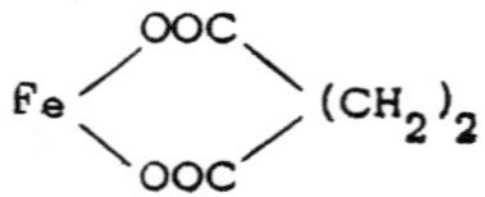	- " -	20	2,185			Kinetics	[231]
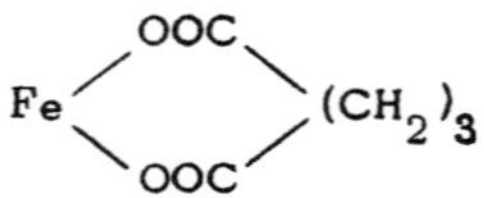	- " -	20	2,04			Kinetics	[231]

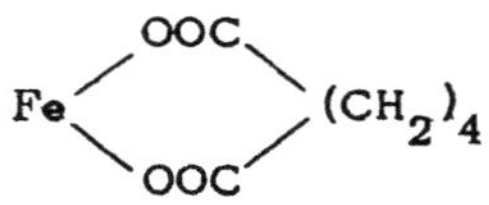	- " -	20	1,895	Kinetics	[231]
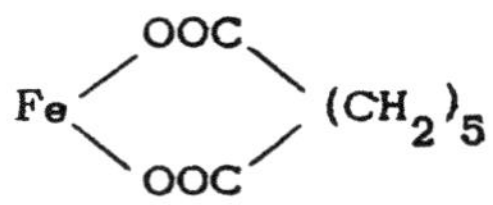	- " -	20	1,875	Kinetics	[231]
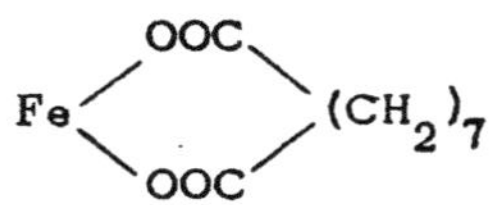	- " -	20	1,855	Kinetics	[231]
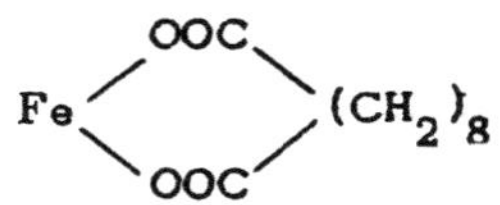	- " -	20	2,00	Kinetics	[231]
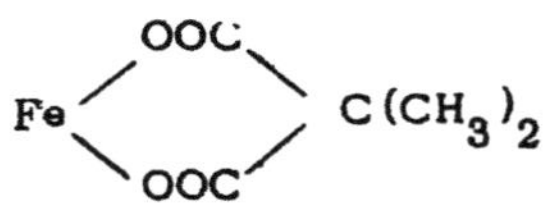	- " -	20	2,095	Kinetics	[231]

CHAPTER XII

REACTIONS OF ION-RADICALS AND SOLVATED ELECTRONS

§1. Reactions of Ion-Radicals

Ion-radicals are formed by the action of alkali metals on aromatic compounds in aprotonic solvents. The ion-radicals exist in solution and react in the form of free particles and in the form of ion pairs, for example, $C_{10}H_8^-Na^+$. An aromatic ion-radical will rapidly exchange an electron with an aromatic molecule. The rate constants for exchange between an ion-radical and a molecule of identical structure are in the range of 10^6 to 10^8 liter/mole·sec. The electron exchange rate is measured by EPR on the basis of absorption band broadening in the EPR spectrum (see Chapter IV, §2). Ion-radicals can bring about the anionic polymerization of unsaturated compounds, acting as initiator-ions and entering into a reaction of radical substitution, for example

$$C_{10}H_8^- + ROH \longrightarrow C_{10}H_9\cdot + RO^-$$

Rate constants for reactions of ion-radicals with molecules are listed in Table 109.

§2. Solvated Electron

A hydrated electron $\bar{e}_{aq}$ appears as a labile intermediate particle when aqueous (or alcoholic) solutions are irradiated with electrons, X-rays, or ultraviolet light. The rate constants for reactions of a solvated electron are measured by PR or MCR methods. The consumption of $\bar{e}_{aq}$ is followed by high-speed spectrophotometry (the absorption maximum of $\bar{e}_{aq}$ lies in the 700-mμ region). The rate constants for the reactions of $\bar{e}_{aq}$ with the majority of ions and organic compounds are on the order of 10^{10} and are diffusion-limited. The diffusion coefficient of $\bar{e}_{aq}$, according to [253], is $4.7\cdot10^{-5}$ $cm^2\cdot sec$, and its radius r = 2.5 to 3Å [254]. The reaction of $\bar{e}_{aq}$ with water is comparatively slow. The rate constant for the reaction $\bar{e}_{aq} + H_2O \rightarrow H^{\cdot} + OH^-$ is 16 sec^{-1} (23°), and log k = 6.09 - 6.7/θ, measured within the temperature range of 5-81° [255] (Tables 111, 112).

TABLE 109. Reactions of Ion-Radicals with Molecules

Reaction	Conditions	t^{o}	log k	log A	E	Ref.
$O^{-} + H_2 \longrightarrow OH^{-} + H$	H_2O; pH 13,1	25	7,90			[232]
$O^{-} + O_2 \longrightarrow O_3^{-}$	H_2O; pH 13	25	9,415			[233]
$C_6H_6^{\overline{\cdot}} + C_6H_6$ (exchange of $\bar{e}$)	$\overbrace{CH_2(CH_2)_2CH_2}^{O}$ + + $(CH_3OCH_2)_2$	18	7,885	9,96	2,8	[234]
$C_{10}H_8^{\overline{\cdot}} + C_{10}H_8$ (exchange of $\bar{e}$)	$(CH_3OCH_2)_2$	23	9,20	12,85	5,00	[235]
	- " -	-23 to + + 21	6,99 (−23°)	23,00	18,5	[236]
	- " -	-23 to + + 21	6,45 (−23°)	17,53	12,8	[236]
$C_{10}H_8^{\overline{\cdot}}.Na + C_{10}H_8$ (exchange of $\bar{e}$)	$\overbrace{CH_2(CH_2)_2CH_2}^{O}$	23	7,26			[237]
	- " -	-23 to + + 21	6,64 (21°)	19,60	17,6	[236]
	- " -	23	<7,48			[235]

Table 109 (continued)

Reaction	Conditions	t°	log k	log A	E	Ref.
$C_{10}H_8^{-\cdot}Na^+ + C_{10}H_8$ (exchange of $\bar{e}$)	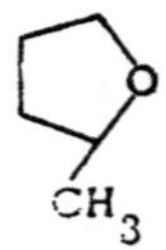	-23 to + 21	6,65 (21°)	15,79	12,4	[236]
$C_{10}H_8^{-\cdot}K^+ + C_{10}H_8$ (exchange of $\bar{e}$)	$(CH_3OCH_2)_2$	23	8,00			[235]
	- " -	23	7,88			[237]
	$\overset{O}{\overline{CH_2(CH_2)_2CH_2}}$	23	7,78			[235]
	- " -	23	7,76			[237]
$C_{12}H_{10}^{-\cdot} + CH_3OH \longrightarrow C_{12}H_{11}\cdot + CH_3O^-$	CH_3OH	25	4,84			[238]
$C_{12}H_{10}^{-\cdot} + C_2H_5OH \longrightarrow C_{12}H_{11}\cdot + C_2H_5O^-$	C_2H_5OH	25	4,415			[238]
$C_{12}H_{10}^{-\cdot} + n\text{-}C_3H_7OH \longrightarrow C_{12}H_{11}\cdot + n\text{-}C_3H_7O^-$	$n\text{-}C_3H_7OH$	25	4,505			[238]
$C_{12}H_{10}^{-\cdot} + (CH_3)_2CHOH \longrightarrow C_{12}H_{11}\cdot + (CH_3)_2CHO^-$	$(CH_3)_2CHOH$	25	3,74			[238]

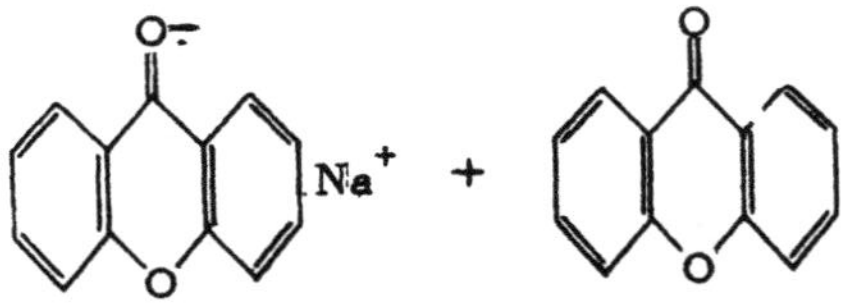 (exchange of $\bar{e}$)	$(CH_3OCH_2)_2$	37	8,66	12,21	5,1	[239]
	$\overline{CH_2(CH_2)_2CH_2O}$	25	8,65	11,77	4,3	
	$\underline{OCH_2(CH_2)_2CH}(CH_3)$	0	8,635			
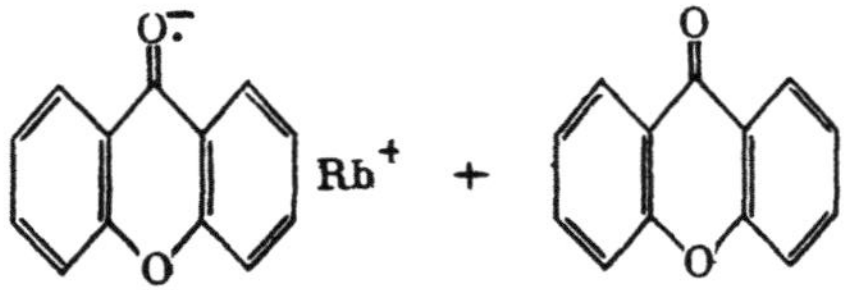	$(CH_3OCH_2)_2$	25	8,40			[239]
$(C_6H_5)_2CO^{\overline{\cdot}}\ Rb^+ + (C_6H_5)_2CO$ (exchange of $\bar{e}$)	$(CH_3OCH_2)_2$	25	8,19			
$(C_6H_5)_2CO^{\overline{\cdot}}\ K^+ + (C_6H_5)_2CO$	$\overline{CH_2(CH_2)_2CH_2O}$	25	8,40			
$(C_6H_5)_2CO^{\overline{\cdot}}\ Na + (C_6H_5)_2CO$	$CHON(CH_3)_2$	25	8,04	12,62	6,3	[239]
	$\overline{CH_2(CH_2)_2CH_2O}$	12	8,06	12,62	6,0	[239]
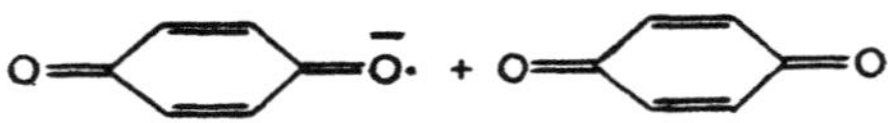 (exchange of $\bar{e}$)	$CHON(CH_3)_2$	25	8,58			[240]

Table 109 (continued)

Reaction	Conditions	t°	log k	log A	E	Ref.
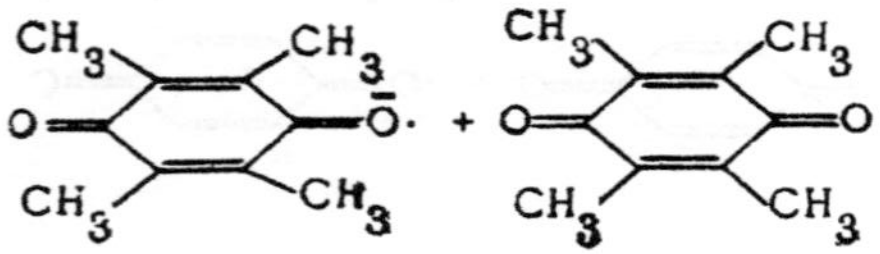 (exchange of ē)	$CHON(CH_3)_2$	25	7,79			[240]
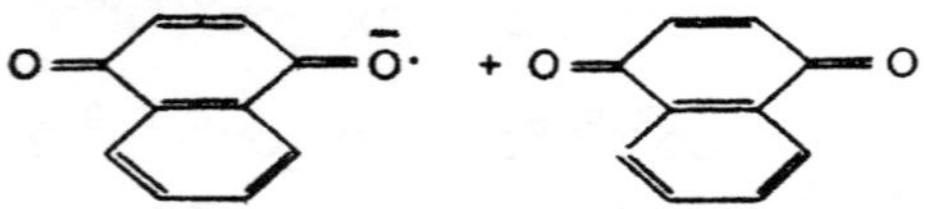 (exchange of ē)	$CHON(CH_3)_2$	25	8,625			[241]
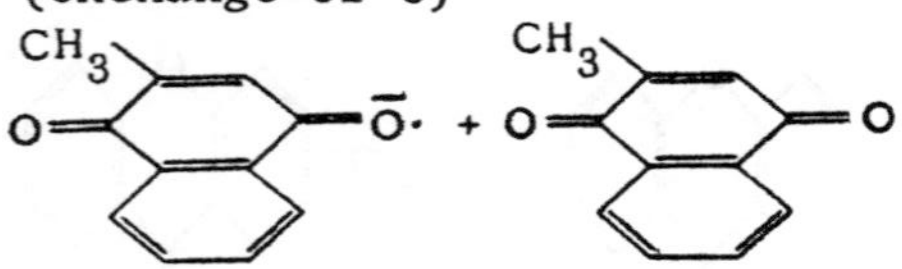 (exchange of ē)	$CHON(CH_3)_2$	25	8,60			[241]
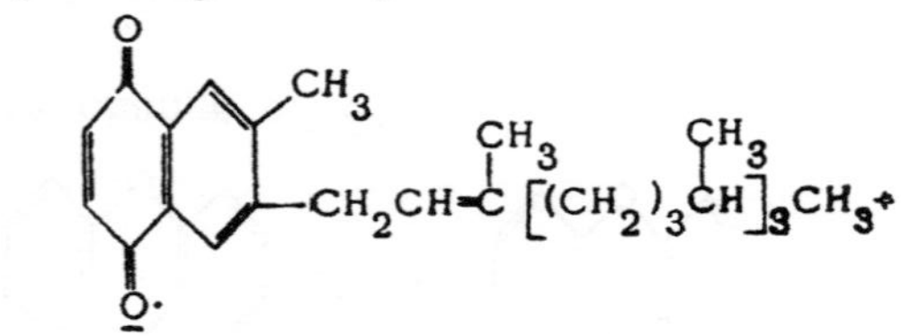	$CHON(CH_3)_2$	25	8,115			[241]

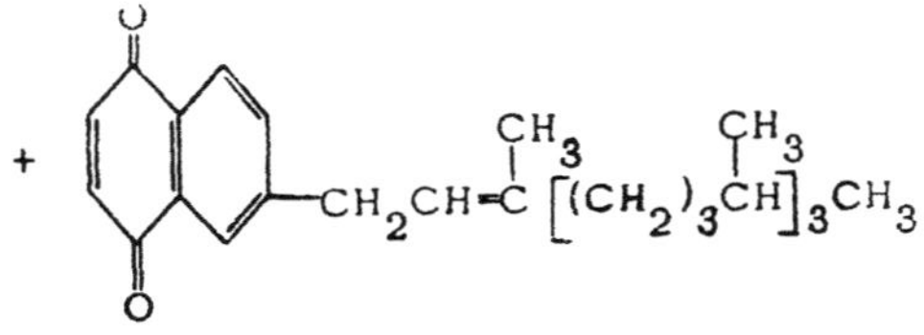

(exchange of ē)

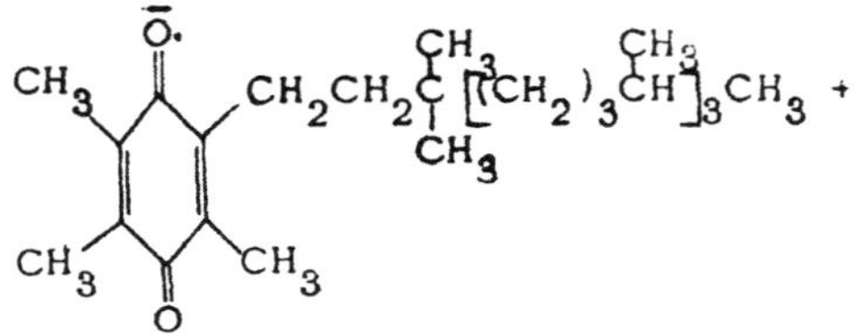	$CHON(CH_3)_2$	25	7,20	[241]

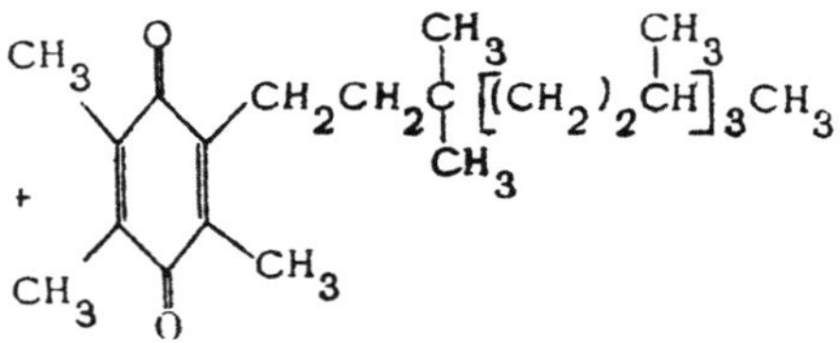

(exchange of ē)

Table 109 (continued)

Reaction	Conditions	t^{o}	log k	log A	E	Ref.
$C_{12}H_{10}^{-} + C_2H_5OH_2^{+} \longrightarrow C_{12}H_{11}^{\cdot} + C_2H_5OH$	C_2H_5OH	25	10,52			[238]
$C_{14}H_{10}^{\dot{-}} + C_{14}H_{10}$ (exchange of $\bar{e}$)	$CHON(CH_3)_2$	25	9,08	11,26	3,0	[234]
	- " -	25	8,68			[240]
$C_{14}H_{10}^{\dot{-}} + CH_3OH \longrightarrow C_{14}H_{11}\cdot + CH_3O^{-}$	CH_3OH	25	4,91			[238]
$C_{14}H_{10}^{\dot{-}} + C_2H_5OH \longrightarrow C_{14}H_{11}\cdot + C_2H_5O^{-}$	C_2H_5OH	25	4,36			[238]
$C_{14}H_{10}^{\dot{-}} + n\text{-}C_3H_7OH \longrightarrow C_{14}H_{11}\cdot + HC_3H_7O^{-}$	$n\text{-}C_3H_7OH$	25	4,38			
$C_{14}H_{10}^{\dot{-}} + (CH_3)_2CHOH \longrightarrow C_{14}H_{11}\cdot + (CH_3)_2CHO^{-}$	$(CH_3)_2CHOH$	25	3,56			
$C_{14}H_{10}^{\dot{-}} + C_2H_5OH_2^{+} \longrightarrow C_{14}H_{11}\cdot + C_2H_5OH$	C_2H_5OH	25	10,57			[238]
$C_{18}H_{14}^{\dot{-}} + CH_3OH \longrightarrow C_{18}H_{15}\cdot + CH_3O^{-}$	CH_3OH	25	2,60			
$C_{18}H_{14}^{\dot{-}} + C_2H_5OH \longrightarrow C_{18}H_{15}\cdot + C_2H_5O^{-}$	C_2H_5OH	25	2,30			

$C_{18}H_{14}^{-} + C_2H_5OH_2^{+} \longrightarrow C_{18}H_{15}\cdot + C_2H_5OH$	- " -	25	10,28			[238]
$(CN)_2C{=}C(CN)_2^{-}\cdot + (CN)_2C{=}C(CN)_2$ (exchange of ē)	$\overline{CH_2(CH_2)_2CH_2O}$	25	8,32			[242]
$C_6H_5(CH_3)CH$– 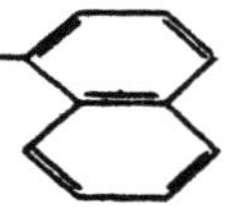 $^{-}\cdot$ +	$CHON(CH_3)_2$	25	7,90	8,34	0,6	[243]
+ $C_6H_5(CH_3)CH$– 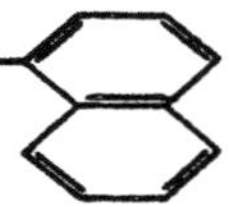 (exchange of ē)	$\overline{CH_2(CH_2)_2CH_2O}$	25	7,28	8,52	1,7	[243]
$C_6H_5NO_2^{-}\cdot + C_6H_5NO_2$ (exchange of ē)	$CHON(CH_3)_2$	25	7,48			[240]
	$CHON(CH_3)_2$ + 10% H_2O	25	5,505			[240]
p-$NO_2C_6H_4NO_2^{-}\cdot$ + p-$NO_2C_6H_4NO_2$ (exchange of ē)	$CHON(CH_3)_2$ + 10% H_2O	25	8,78			[240]

Table 109 (continued)

Reaction	Conditions	t°	log k	log A	E	Ref.
m-$NO_2C_6H_4NO_2^{\overline{\cdot}}$ + m-$NO_2C_6H_4NO_2$ (exchange of ē)	$CHON(CH_3)_2$ + + 10% H_2O	25	8,715			[240]
p-$ClC_6H_4NO_2^{\overline{\cdot}}$ + p-$ClC_6H_4NO_2$ (exchange of ē)	- " -	25	7,90			[240]
m-$ClC_6H_4NO_2$ + m-$ClC_6H_4NO_2$ (exchange of ē)	- " -	25	7,945			[240]
m-$Cl_2C_6H_3NO_2^{\overline{\cdot}}$ + m-$Cl_2C_6H_4NO_2$ (exchange of ē)	- " -	25	8,205			[240]
(p-$NO_2C_6H_4)_3C^-$ + (p-$NO_2C_6H_4)_3C\cdot$ (exchange of ē)	$CHON(CH_3)_2$	25	9,64	11,09	2,0	[244]
	$CH_2(CH_2)_2CH_2O$ (cyclic)	25	9,66	10,24-11,55	0,8-2,6	[244]
	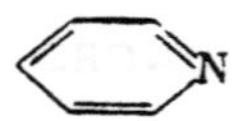	25	9,83	11,28	2,0	[244]
	CH_3CN	25	10,21	10,94	1,0	[244]

TABLE 110. Reactions between Ion-Radicals

Reaction	Solvent	t°	log k	log A	E	Method	Ref.
$O^{-} + O^{-} \longrightarrow O_2^{2-}$	H_2O pH 13	23	9,30			PR	[184a]
$O_2^{-}\cdot + O_2^{-}\cdot \longrightarrow O_2^{2-} + O_2$	H_2O; pH 5–7	23	7,53			PR	[245]
	H_2O; pH 5–8	23	7,48			PR	[246]
	H_2O; pH 5,5	23	7,46			PR	[247]
$HO\cdot + O_2^{-}\cdot \longrightarrow HO^{-} + O_2$	H_2O; pH 0,5–6,7	23	9,85			PR	[248]
$C_{14}H_{10}^{-}\cdot Na^{+} + C_6H_5CH{=}CH_2^{-} Na^{+} \longrightarrow$ dimer	$\overline{CH_2(CH_2)_2CH_2O}$	25	4,0			JM	[249]
$2(C_6H_5)_2\dot{C}{=}CH_2^{-}Na^{+} \longrightarrow$ dimer	- " -	25	6,4			JM	[250]
$2(C_6H_5C{\equiv}CC_6H_5)^{-}\cdot Li^{+} \longrightarrow$ dimer	- " -	-30 to +20	0,52 (0°)	12,52	15,0	JM	[251]
$2(C_6H_5C{\equiv}CC_6H_5)^{-}\cdot Na^{+} \longrightarrow$ dimer	- " -	-30 to +20	0,87 (0°)	15,67	18,5	JM	[251]
$2(C_6H_5C{\equiv}CC_6H_5)^{-}\cdot K^{+} \longrightarrow$ dimer	- " -	-30 to +20	1,115 (0°)	12,64	14,4	JM	[251]

Table 110 (continued)

Reaction	Solvent	t°	log k	log A	E	Method	Ref.
$2(C_6H_5C{=}CC_6H_5)^{\overline{\cdot}}Cs^+ \longrightarrow$	$\overline{CH_2(CH_2)_2CH_2O}$	-30 to +20	0,70 (0°)	12,08	14,2	JM	[251]
$(C_6H_5C{=}CC_6H_5)^{\overline{\cdot}}Na^+ +$ + Na^+ phenanthrene$^{\overline{\cdot}}$ $\longrightarrow$ $\longrightarrow Na^+(C_6H_5C{=}CC_6H_5)^{=}Na^+ +$ + phenanthrene	$\overline{CH_2(CH_2)_2CH_2O}$	-36 to +4	1,94 (0°)	17,95	20,0	JM	[252]
$(C_6H_5C{=}CC_6H_5)^{\overline{\cdot}}Na^+ +$ + Na^+ naphthalene$^{\overline{\cdot}}$ $\longrightarrow$ $\longrightarrow Na^+(C_6H_5C{=}CC_6H_5)^{2-}Na^+ +$ + naphthalene	$\overline{CH_2(CH_2)_2CH_2O}$	-32 to +5	1,97 (0°)	13,98	15,0	JM	[252]

TABLE 111. Reactions of Solvated Electrons with Molecules and Radicals in Aqueous Solutions at Room Temperature

Compound	pH	log k	Ref.
H_2	7	7,0	[256]
$H^{\cdot}$	10,5	10,40	[257, 258]
$\bar{e}_{aq}$	10,5	9,65	[258]
	13,3	9,74	[258]
H_2O	8,3	1,20 (sec^{-1})	[256, 259]
D_2O	10	< 2,3 (sec^{-1})	[260]
HO°	10,5	10,48	[257, 258]
O_2	7	10,27	[257, 261]
	CH_3OH	10,28	[152]
H_2O_2	7	10,09	[262]
CO	7	9,00	[263]
CO_2	7	9,89	[257]
NO	7	10,49	[257]

Table 111 (continued)

Compound	pH	log k	Ref.
N_2O	7	9,75	[256, 261, 264]
	12,4 M KF	9,38	[265]
N_2H_4	-	< 8,0	[256]
NH_2OH	-	< 7,3	[256]
I_2	7	10,71	[266]
CH_4	-	< 7,0	[263]
C_2H_4	-	< 8,4	[267]
C_6H_6	7	< 6,84	[268]
	11	7,14	[269]
$C_6H_5CH_3$	11	7,08	[269]
$C_6H_5CH{=}CH_2$	7	10,175	[263, 268]
	12,7	10,04	[268]

$CH_2 = CH{-}CH = CH_2$	7	9,90	[268]
$C_{10}H_8$	11	9,73	[269]
$C_6H_5 - C_6H_5$	C_2H_5OH	9,63	[270]
$C_{10}H_8$	C_2H_5OH	9,73	[270]
$C_{18}H_{12}$	C_2H_5OH	10,01	[270]
CH_3OH	-	< 4,0	[152]
C_2H_5OH	-	< 5,0	[263]
$CH_2 = CH{-}CH_2OH$	11,5	< 6,0	[271]
$C_6H_5CH_2OH$	11	8,12	[269]
C_6H_5OH	11	< 6,6	[269]
$p{-}HOC_6H_4OH$	13	< 7,0	[268]
$\alpha{-}C_{10}H_7OH$	11	8,98	[269]
$\beta{-}C_{10}H_7OH$	11	9,08	[269]
	11	8,255	[268]
CH_2O	7	7,0	[257]
CH_3CHO	6–11	9,545	[257,263, 268]
	5–8	9,73	[272]
CH_3COCH_3	7	9,77	[257, 268]
	11	9,84	[261]

Table 111 (continued)

Compound	pH	log k	Ref.
CH_3COCH_3	11	9,75	
	14	9,72	[273]
	3 M OH^-	9,62	
	12,4 M KF	9,20	[265]
cyclo-$C_6H_{10}O$	5,5–6	9,90	[152]
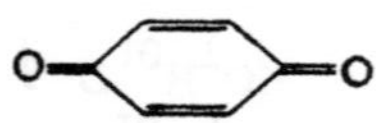	6,6	9,13	[268]
HCOOH	5	8,15	[257]
CH_3COOH	5,4	8,255	[257]
$(CH_3)_3CCOOH$	5,0	7,99	[152]
$HOOCCH_2COOH$	6,0	8,38	[152]
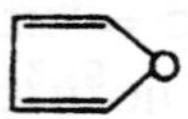	7,9	6,48	[274]
$C_2H_5OC_2H_5$	-	7.0	[263]
CH_3COOC_2H	-	<7,0	[263]

	6,5	7,77	
$CH_3CH_2COOCH_3$	6,8	7,955	
$(CH_3)_3CCOOCH_3$	5,9	7,36	[152]
$HOCH_2COOCH_3$	10,7	8,88	
FCH_2COCH_3	11	8,98	
FC_6H_5	11	7,785	
o-FC_6H_4OH	11	8,53	[269]
m-FC_6H_4OH	11	8,30	
p-FC_6H_4OH	11	8,08	
F_3CCOCF_3	11	7,82	[152]
$C_6H_5CF_3$	11	9,255	[269]
$ClCH_2CH_2CH_3$	6,3	8,64	[274]
$Cl(CH_2)_3CH_3$	7,3	8,65	[274]
	11	8,505	[275]
$CH_3CHClC_2H_5$	6,6	8,71	[274]
	11	8,71	[275]
$ClCH_2CH(CH_3)_2$	5,8	8,71	[274]

Table 111 (continued)

Compound	pH	log k	Ref.
$C_6H_5CH_2Cl$	11	9,725	[269]
ClC_6H_5	11	8,70	[269]
p-$ClC_6H_4CH_3$	11	8,65	[269]
$CHCl_3$	7	10,30	[257]
	7	10,48	[268]
CCl_4	7	10,48	[268]
$ClCH_2CH_2OH$	11	8,61	[275]
o-ClC_6H_4OH	11	8,30	[269]
m-ClC_6H_4OH	11	8,70	[269]
p-ClC_6H_4OH	11	8,81	[269]
$CClH_2COOH$	1	9,81	[276]
o-ClC_6H_4Cl	11	9,67	[269]
m-ClC_6H_4Cl	11	9,72	[269]
p-ClC_6H_4Cl	11	9,70	[269]
$C_6H_5CCl_3$	11	9,92	[269]

CH_3CH_2Br	7,1	10,08	
$CH_3CH_2CH_2Br$	6,1	9,93	[274]
$CH_3(CH_2)_2CH_2Br$	6,6	10,0	
C_6H_5Br	11	9,635	[269]
$Br\ CH_2CH_2OH$	11	9,20	[275]
o-Br C_6H_4OH	11	9,28	[269]
m-Br C_6H_4OH	11	9,43	[269]
p-Br C_6H_4OH	5,5	10,08	[277]
	11	9,46	[269]
CH_3I	6,8	10,23	
ICH_2CH_3	6,5	10,18	[274]
$ICH_2CH_2CH_3$	6,2	10,11	
$ICH_2(CH_2)_2CH_3$	7,6	10,08	
C_6H_5I	11	10,08	[269]
$C_6H_5CH_2I$	11	10,11	[269]
CH_3CONH_2	10,9	7,23	[152]
$CO(NH_2)_2$	-	5,48	[152]

Table 111 (continued)

Compound	pH	log k	Ref.
$NH{=}C(NH_2)_2$	6,1	8,40	
	11,1	8,28	[278]
	11,9	8,20	
CH_3CHNOH	7,7	8,48	[152]
$CH_2{=}CHCONH_2$	7	10,255	[257, 268]
	7	10,30	[279]
	7	10,52	[152]
$C_6H_5CONH_2$	11	10,23	[269]
$C_6H_5SO_2NH_2$	11	10,20	[269]
CH_3CN	7,2	77,48	
CH_3CH_2CN	10,9	8,18	[152]
$CH_2{=}CHCH_2CN$	7,0	8,96	
C_6H_5CN	11	10,20	
α-$C_{10}H_7CN$	11	10,32	[269]
β-$C_{10}H_7CN$	11	10,32	
o-CNC_6H_4OH	11	9,91	

CH_3NH_2	7,6	8,25	[278]
$C_6H_5NH_2$	11,9	7,30	[263,268]
o-$CNC_6H_4NH_2$	10	10,04	[152]
$NCCH_2COOCH_3$	10,9	8,50	[152]
CH_2—CO / CH_2—CO (ring, NH)	8,0	9,86	[274]
$(CN)_2C{=}C(CN)_2$	7	10,175	[263, 268]
CH_3NO_2	11	9,70	[152]
$C(NO_2)_4$	6	10,66	[152]
	7	10,79	[280]
$C_6H_5NO_2$	7	10,48	[263, 268, 269]
$(CH_3)_2C(NO_2)CH_2OH$	10	10,32	[152]
$CH_3C(NO_2)(CH_2OH)_2$	10	10,11	[152]
p-$CH_3C_6H_4NO_2$	11	10,28	[269]
o-$NO_2C_6H_4OH$	11	10,30	[269]
m-$NO_2C_6H_4OH$	11	10,40	[269]
p-$NO_2C_6H_4OH$	11	10,40	[269]
C_5H_5CN	7	9,00	[263, 268]
	7	9,57	[152]

Table 111 (continued)

Compound	pH	log k	Ref.
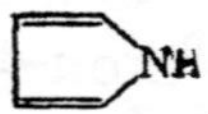	10,3	5,78	[274]
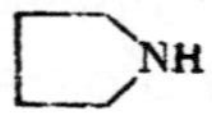	12,1	6,62	
	7,8	7,11	
$[(HOOCCH)_2N]_2C{=}C[N(CH_2COOH)_2]_2$	8–11	6,18	[152]
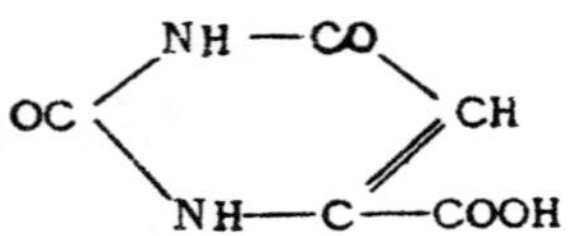	6,5	10,175	[152]
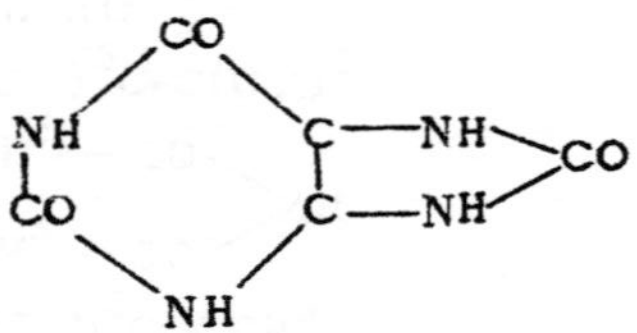	5	9,78	[263]
CH_3SH	4–7	10,255	[281]

	6,7	7,81	[274]
C_6H_5SH	11	7,67	[269]
$SC(NH_2)_2$	6,4	9,48	[263, 268, 282]
CS_2	7,7	10,49	[257,268]
Arabinose		< 7,0	[263]
Glucose		⩽ 6,0	[283, 284]
Ribose	-	7,0	[263]
Saccharose	6,5	7,28	[285]
	7,6	7,15	[286]
Alanine	6,4	6,77	[285]
	6,7	7,25	[287]
	3,0	8,90	[287]
-Alanine [sic]	6,8	⩽ 6,48	[278]
β-Alanine	6,8	< 6,6	[278]
Arginine	6,1	8,18	[278]
	7,8	8,08	[278]

Table 111 (continued)

Compound	pH	log k	Ref.
Arginine	8,7	8,08	[278]
	8,9	8,04	
	9,9	7,81	
	11,5	7,80	
Asparagine	4,7	8,30	[152]
	7,3	8,175	[278]
	8,6	8,04	[152]
	11,7	7,38	[278]
Valine	-	6,52	[285]
	6,4–9,5	6,48	[278, 152]
Glycine	3	8,65	[287]
	6,7	7,145	[287]
	6,4	6,915	[278, 285]
Histidine	5,9	9,59	[278]
	6,7	9,15	[278]

	7,8	8,30	[152]
	8,6	7,65	[278]
	11,1	7,08	[278]
Leucine	6,5	< 7,0	[278]
Lysine	7,8	~ 7,3	[278]
	8,8	< 7,7	[278]
Methionine	6,0	7,545	[278]
Norleucine	–	6,52	[285]
Proline	6,7	7,30	[279]
	10,1	6,0	[279]
Sarcosine	6,3	7,28	[285]
	7,6	7,15	[286]
Serine	7,1	7,18	[285]
	6,1	< 7,48	[278]
Threonine	-	7,30	[285]
	6,2	< 7,0	[278]

Table 111 (continued)

Compound	pH	log k	Ref.
Threonine	9,0	8,48	[278]
Tryptophan	7,3	8,415	[285]
	6,8	8,61	[278]
	8,9	8,49	
	11,5	8,175	
Tyrosine	5,8	8,20	[278]
	7,8	8,60	[285]
Phenylalanine	6,3	8,04	[278]
	6,7	8,175	[278, 285]
	8,6	7,945	[278]
	11,2	7,23	[278]
Cysteine	6,3	9,94	[278]
	11,6	7,875	[278]

Cystine	6,1	10,115	[278]
	10,7	9,40	[278]
	12,0	9,53	[263, 268]
Glycylglycine	6,38	8,40	[288]
	11,75	7,70	
Glycyl-*dl*-alanine	6,22	8,46	
dl-Alanylglycine	6,22	8,32	
Glycycl-*l*-leucine	6,46	8,45	[288]
	8,74	7,845	
	8,94	7,81	
dl-Leucylglycine	6,09	8,04	[288]
Glycyl-*dl*-valine	5,97	8,415	
dl-Alanyl-*dl*-leucine	6,46	8,115	
l-Leucyl-*l*-alanine	6,1	8,22	
dl-Alanyl-*dl*-alanine	6,27	8,115	
l-Leucyl-*l*-leucine	5,97	7,955	
Glycyl-*l*-asparagine	5,33	8,73	
	11,41	7,90	

Table 111 (continued)

Compound	pH	log k	Ref.
Glycyl-*l*-proline	6,66	9,04	[288]
Glycyl-*l*-tyrosine	6,13	8,61	[288]
Glycyl-*l*-tryptophan	6,37	8,65	[288]
l-Phenylalanyl-*l*-phenylalanine	5,66	8,65	[288]
Histidylhistidine	7,3	9,115	[288]
Glycylglycylglycine	6	8,95	[288]
	11,1	7,95	[288]
l-Leucylglycylglycine	6,93	8,45	[288]
	9,5	7,70	[288]
Acetylglycine	5,95	7,30	[288]
Acetyl-*dl*-alanine	8,6–9,0	7,00	[288]
Histidylhistidine	5,51	9,90	[288]
	11,0	7,71	[288]
Adenine	7,7	9,79	[152]

Adenosine	5	10,49	[194]
	12,0	10,0	[268]
Adenosine-5-phosphate	7,0	9,58	[194]
Adenylic acid	5,8	9,72	[152]
Imidazole	-	9,63	[278]
	6,3	9,53	[152]
	8,7	8,15	[152]
	9,1	7,57	[274]
	11,5	7,34	[152]
Indole	-	~8,85	[256]
NAD (nicotine adenine dinucleotide)		10,415	[152]
NAD - H_2		9,725	[152]
Purine	7,2	10,23	[263, 268]
Thiazole	6,6	9,40	[274]
Thymine	5,5	10,255	[194]
	6,0	10,23	[268]
	11	9,60	[261]
Uracil	6,4	9,89	[263, 268]
	12,2	9,36	[263, 268]

Table 111 (continued)

Compound	pH	log k	Ref.
Hemin	-	9,78	[152]
Hemoglobin	-	10,415	[152]
Cytidine	12,0	10,08	[263, 268]
Cytosine	8,0	~ 9,85	[263]
$C_6H_5COO^-$	14	9,46	[273]
	3 M OH^-	9,38	[273]
	-	9,23	[265]
p-Br $C_6H_4COO^-$	11	9,89	[269]
o-$ClC_6H_4COO^-$	11	9,08	[269]
m-$ClC_6H_4COO^-$	11	9,74	[269]
p-$ClC_6H_4COO^-$	11	9,78	[269]
p-$CNC_6H_4COO^-$	11	10,0	[269]
o-$FC_6H_4COO^-$	11	9,49	[269]
m-$FC_6H_4COO^-$	11	9,825	[269]
p-$FC_6H_4COO^-$	11	9,58	[269]

o-$HOC_6H_4COO^-$	11	9,505	[269]
m-$HOC_6H_4COO^-$	11	9,04	
p-$HOC_6H_4COO^-$	11	8,60	
o-$IC_6H_4COO^-$	11	9,66	
m-$IC_6H_4COO^-$	11	10,115	
p-$IC_6H_4COO^-$	11	9,96	
$C_6H_5CH_2COO^-$	5,4	7,71	[274]
	12,3	7,15	[274]
p-$NO_2C_6H_4CH_2COO^-$	11	10,30	[152]
o-$^-OOCC_6H_4COO^-$	5,8	9,79	[274]
	13	9,255	[268, 284, 289]
m-$^-OOCC_6H_4COO^-$	13	9,48	[289]
p-$^-OOCC_6H_4COO^-$	13	9,86	[289]
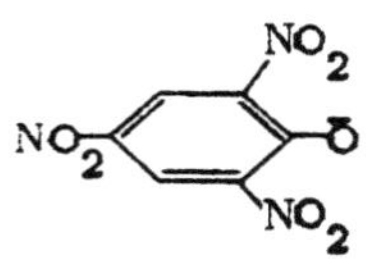	13	10,545	[263, 268]

Table 111 (continued)

Compound	pH	log k	Ref.
$o\text{-}CH_3C_6H_4COO^-$	11	8,43	
$m\text{-}CH_3C_6H_4COO^-$	11	9,415	[269]
$p\text{-}CH_3C_6H_4COO^-$	11	9,48	
$p\text{-}NH_2C_6H_4COO^-$	11	9,32	
$HCOO^-$	10	6,0	[257]
CH_3COO^-	10	6,0	[257]
FCH_2COO^-	11	6,30	
$ClCH_2COO^-$	11	9,08	
$BrCH_2COO^-$	11	9,79	
ICH_2COO^-	11	10,08	
$CH_3CHClCOO^-$	11	9,15	[275]
$ClCH_2CH_2COO^-$	11	8,60	
$CH_3CHBrCOO^-$	11	9,725	
$BrCH_2CH_2COO^-$	11	9,43	
$ICH_2CH_2COO^-$	11	9,82	
$C_2O_4^{2-}$	-	7,0	[263, 268]

Cl_3CCOO^-	11	9,93	[275]
F_3CCOO^-	11	6,415	[275]
$NH_2CH_2CH_2CH_2COO^-$	6,6	$\leqslant$9,95	[279]
	11,3	~6,5	[152]
$N(CH_2COO^-)_3$	10	6,60	[152]
$NCCH_2COO^-$	11	7,60	[273]
$CH_2{=}C(CH_3)COO^-$	10,1	9,90	[263, 268]
cis $^-OOCCH{=}CHCOO^-$	6,5	10,08	[263, 268]
	8,4	9,23	[268]
	12,7	9,34	[263, 268]
trans $^-OOCCH{=}CHCOO^-$	12–13	9,875	[263, 268]
$^-OOCCH_2CH_2COO^-$	11,0	7,49	[263]
$C_6H_5COO^-$	5,4	9,73	[274]
	7,4	9,49	[274]
	11	9,49	[269]
	12,3	9,56	[274]

Table 111 (continued)

Compound	pH	log k	Ref.
$C_6H_5SO_3^-$	11	9,60	[269]
p-$NH_2C_6H_4SO_3^-$	11	8,66	[269]
$(NO_2)_3C^-$	7	10,48	[290]
$C_6H_5CH{=}CHCOO^-$	7,2	9,83	[274]
$C_6H_5CH_2CH_2COO^-$	5,4	7,69	[274]
	12,1	7,04	[274]
CH_3COCOO^-	12,7	9,83	[268]
$CH_3CH(OH)COO^-$	11	< 6,3	[152]
α-$C_{10}H_7COO^-$	11	9,785	[269]
β-$C_{10}H_7COO^-$	11	9,98	[269]
$HSCH_2COO^-$	10	7,91	[152]

TABLE 112. Rate Constants for Reactions of $\bar{e}_{aq}$ with Ions in Aqueous Solutions at Room Temperature (20-25°)

Ion	pH	log k	Ref.
H^+	4–5	10,37	[291, 257]
	4,1–4,7	10,35	[292]
BF_4^-	5,8	< 5,36	[152]
HCO_3^-	–	< 6,0	[266]
CO_3^{2-}	–	< 6,0	[266]
CN^-	11,0	< 8,0	[273]
CNO^-	11	6,115	[152]
CNS^-	7	< 6,0	[266]
NH_4^+	5,3	6,115	[152]
NO_2^-	7,0	9,66	[266]
	–	9,545	[205, 256]
	7,0	9,91	[257]
NO_3^-	7,0	10,04	[257, 266]
	–	9,915	[205, 256]
	12,4 M KF	9,28	[265]
N_3^-	11	< 6,75	[152]
O^-	13	10,34	[259]
HO_2^-	13	9,545	[262]
F^-	7,2	< 4,3	[263]
HF	5	7,78	[293]
HF_2^-	5	7,48	[294]
Na^+	–	< 6,0	[265]
Al^{3+}	6,8	9,3	[273]
	11,2	8,6	[273]

Table 112 (continued)

Ion	pH	log k	Ref.
$Al(OH)_4^-$	14	6,74	[273]
$Al(EDTA)^-$	12	7,48	
SiF_6^{2-}	5,9	⩽6,18	
$H_2PO_2^-$	6,8	<5,0	[152]
$H_2PO_3^-$	6,7	6,86	
$H_2PO_4^-$	-	7,23	[295]
	-	6,74	[294]
	7–9	7,175	[296]
	6,7	6,89	[152]
SO_3^{2-}	10	< 6,12	[152]
	7	< 6,0	[256, 266]
$S_2O_3^{2-}$	11,9	< 8,0	[266]
	7	8,785	[152]
$S_2O_8^{2-}$	7	10,03	[266]
	7	9,88	[205, 256]
Cl^-	10	< 5,0	[297]
ClO^-	10	9,86	[152]
ClO_3^-	10	8,545	[205, 256]
	11	7,965	[152]
	10	6,60	[266]
	7	8,34	[152]
ClO_4^-	–	< 5,0	[152]
	7	7,08	[152]
K^+		< 4,48	[265]
$Se(EDTA)^-$	11,5	7,545	
$Ti(EDTA)^-$	11,2	<8,6	[152]
TiO_3^{2-}	11,5	< 6,7	

Table 112 (continued)

Ion	pH	log k	Ref.
TiF_6^{2-}	(0,1 MKF)6,5	9,875	[152]
VO_3^-	11	9,69	[152]
Cr^{2+}	6,9	10,62	[273]
	11,2	10,28	[273]
	3 M OH^-	8,34	[152]
Cr^{3+}	-	10,60	[204]
	10,9	10,66	
	14	8,30	[273]
	3 M OH^-	8,34	
	7,0	10,78	
$Cr(NH_3)_5Cl^{2+}$	6,7	10,755	[152]
CrF_6^{3-}	10	10,15	[273]
CrF_6^{4-}	8,5	9,61	[152]
$Cr(CN)_6^{3-}$	10,6	10,175	[273]
$Cr(CN)_6^{4-}$	11,1	10,15	[152]
$Cr(NCS)_6^{3-}$	10	8,38	[152]
$Cr(En)_3^{3+}$	6,8	10,725	[274]
	-	10,875	[205]
$Cl(En)_2Cl_2^+$	5,5	10,85	[274]
	5,6	10,62	[274]
$Cr(EDTA)^-$	4,9	10,415	
$Cr(C_2O_4)_3^{3-}$	4,8-6,1	10,255	[274]
cis-$Cr(C_2O_4)_2(H_2O)_2^-$	6,4	10,115	
trans-$Cr(C_2O_4)_2(H_2O)_2^-$	6,2	10,175	
$Cr_2O_7^{2-}$	7	10,52	[266]

Table 112 (continued)

Ion	pH	log k	Ref.
CrO_4^{2-}	13	10,73	[266]
	-	10,255	[205]
Mn^{2+}	-	7,89	[256]
	-	7,58	[205]
$Mn(EDTA)^{-2}$	12	< 6,34	[152]
$Mn(CN)_6^{4-}$	10	10,40	[298]
MnO_4^-	7	10,34	[266]
	13	10,57	[266]
	7	10,48	[205]
	7	10,64	[152]
Fe^{2+}	-	8,545	[256]
	-	8,20	[205]
	5,0	8,08	[152]
$Fe(EDTA)^{2-}$	12	9,00	[152]
FeF_6^{3-}	6,6	10,04	[152]
$Fe(EDTA)^-$	12	10,36	[152]
$Fe(CN)_6^{3-}$	7; 10,3	9,48	[292, 257]
	7	9,64	[152]
$Fe(CN)_6^{4-}$		< 5,0	[298]
$Fe(CN)_5NH_3^{3-}$	9,8	< 7,0	[152]
$Fe(CN)_5NO^{2-}$	11	10,38	[152]
Co^{2+}		10,08	[256,205]
$Co(EDTA)^{2-}$	12	8,71	[152]

Table 112 (continued)

Ion	pH	log k	Ref.
$Co(NH_3)_6^{3+}$	11,1	10,955	[205, 256, 273]
	6,5	10,91	[299]
	4	10,93	[152]
$Co(NH_3)_5H_2O^{3+}$	-	10,79	[204, 256]
	7,0	10,90	[152]
	10	10,78	[152]
$Co(NH_3)_4(H_2O)_2^{3+}$	-	10,64	[205]
$Co(NH_3)_5F^{2+}$	6,3	10,88	[152]
	4	10,875	[152]
$Co(NH_3)_5Cl^{2+}$	-	10,73	[205, 256]
	-	10,785	[152]
	4	10,89	[152]
$Co(NH_3)_5Br^{2+}$	7,5	10,79	[152]
	4	10,905	[152]
$Co(NH_3)_5CN^{2+}$	6,4	10,80	[152]
	4	10,87	[152]
$Co(NH_3)_5N_3^{2+}$	8,2	10,80	[152]
$Co(NH_3)_4(H_2O)CN^{5+}$	6,1	10,75	[152]
$Co(NH_3)_5$ p-OOCC$_6H_4$COO)$^-$	-	10,78	[260]
$Co(CN)_6^{3-}$	10	9,61	[152]
	10	9,555	[273]
	-	9,43	[205, 256]

Table 112 (continued)

Ion	pH	log k	Ref.
$Co(CN)_5Cl^{3-}$	–	10,255	[205]
$Co(CN)_5OH^{3-}$	–	10,04	[205]
$Co(CN)_5N_3^{3-}$	–	10,115	[205]
$Co(CN)_5NO_2^{3-}$	–	9,90	[205]
$Co(NO_2)_6^{3-}$	–	10,76	[205, 256]
	10	10,505	[152]
$Co(C_2O_4)_3^{3-}$	–	10,115	[205]
$Co(EDTA)^-$	–	10,78	[152]
	–	10,46	[205]
$Co(En)_3^{3+}$	–	10,915	[205]
	6,5	10,86	[274]
$Co(En)_2Cl_2^+$	–	10,505	[205, 256]
	5,5	10,85	[274]
$Co(En)_2CO_3^+$	7,5	10,69	[152]
cis-$Co(En)_2(CNS)_2^+$	6,0	10,84	[274]
trans-$Co(En)_2(CNS)_2^+$	6,5	10,73	[274]
Ni^{2+}	–	10,34	[205, 256]
	7	10,46	[152]
$Ni(EDTA)^{2-}$	12	8,00	[152]
NiF^+	8,5	9,86	[152]
$Ni(CN)_4^{2-}$	11	9,74	[152]
$Ni(En)_3^{2+}$	11	9,70	[152]

Table 112 (continued)

Ion	pH	log k	Ref.
Cu^{2+}	7	10,48	[256, 300]
	7	10,52	[291, 273]
	7	10,60	[152]
$Cu(EDTA)^{2-}$	12	10,00	[152]
$Cu(OH)_4^{2-}$	14	9,76	[273]
	3 M OH^-	9,65	[273]
	5 M OH^-	8,53	[273]
$Cu(NH_3)_4^{2+}$	11,1	10,235	[273]
$Cu(CN)_2^-$	10	8,48	[273]
Zn^{2+}	7	9,175	[205]
	7	9,115	[152]
	7,1	9,00	[273]
	9,7	8,75	[273]
	12	8,30	[273]
$Zn(EDTA)^{2-}$	12	$\leq$ 6,25	[152]
$Zn(OH)_4^{2-}$	14	7,20	[273]
	3 M OH^-	6,875	[273]
$Zn(NH_3)_4^{2+}$	11,1	8,81	[273]
$Zn(CN)_4^{2-}$	10,8	8,255	[273]
$Ca(EDTA)^-$	11	7,89	[152]
AsO_2^-	10,6	8,77	[152]
$HAsO_4^{2-}$	11	8,32	[273]
AsF_6^-	7	9,955	[273]
SeO_3^{2-}	10,8	7,08	[152]

Table 112 (continued)

Ion	pH	log k	Ref.
SeO_4^{2-}	11	9,04	[152]
Br_2^-	7	10,115	[199]
BrO^-	13	10,36	[184]
BrO_3^-	7	9,89	[152]
	7	9,32	
	11	9,58	[273]
	14	9,76	
	3 M OH^-	9,725	
Y^{3+}	-	8,30	[205, 256]
$Y(EDTA)^-$	12	7,04	
$Pu(NH_3)_6^{3+}$	7	10,87	
$Pu(CN)_6^{4-}$	10,6	<10	[152]
$Ph(NH_3)_6^{3+}$	7	10,90	
$PdCl_4^{2-}$	7,1	10,08	[273]
$Pd(CN)_4^{2-}$	10	9,30	[273]
	10,6	9,45	[152]
Ag^+	7	10,505	[166, 256, 258]
$Ag(NH_3)_2^+$	11,1	10,90	[273]
$Ag(CN)_2^-$	10	9,175	[273]
$Ag(EDTA)^{3-}$	12	9,205	[152]
Cd^{2+}	7	10,715	[205, 256]
$Cd(EDTA)^{2-}$	12	8,59	[273]
$Cd(H_2O)_4^{2+}$	6,5	10,68	[273]

Table 112 (continued)

Ion	pH	log k	Ref.
$Cd(NH_3)_4^{2+}$	11,1	10,49	
$CdCl(H_2O)_3^{+}$	6,8	10,04	[273]
$CdI(H_2O)_3^{+}$	7,2	10,206	
$Cd(CN)_4^{2-}$	10,0	8,145	
$In(EDTA)^{-}$	12	8,61	[152]
Sn^{2+}	11	9,53	[273]
$Sn(EDTA)^{2-}$	12	9,15	[152]
SnO_2^{2-}	14	9,79	[273]
SnO_2^{2-}	3 M OH^{-}	9,83	[273]
SnO_3^{2-}	11	8,80	
SnF_3^{-}	10	9,97	[152]
SnF_6^{2-}	6,5	9,61	
Sb^{5+}	11	10,08	[273]
SnO_3^{-}	11	10,115	[152]
TeO_3^{2-}	7	8,78	[260]
	12,6	8,75	[301]
	14	9,48	[301]
	10,9	9,04	[152]
TeO_4^{2-}	11	10,20	[152]
I_3^{-}	7	10,30	[302]
IO_3^{-}	7	9,885	
	11	9,92	
	14	9,98	[273]
	3 M OH^{-}	9,91	

Table 112 (continued)

Ion	pH	log k	Ref.
IO_4^-	7	10,04	[273]
	11	10,28	[273]
	14	10,33	[273]
	3 M OH^-	10,21	[273]
Ce^{3+}	-	< 9,0	[256]
$Ce(EDTA)^-$	11,5	⩽ 7,50	[152]
Pr^{3+}	6,1	8,46	[266]
		7,0	[205, 256]
$Pr(EDTA)^-$	11,5	6,555	[152]
Sm^{3+}	5,9	10,40	[266]
$Sm(EDTA)^-$	11,5	7,415	[152]
Eu^{3+}	5,5	10,79	[266]
$Eu(EDTA)^-$	11,5	9,75	[152]
Gd^{3+}	6,0	8,74	[266]
$Gd(EDTA)^-$	12	6,78	[152]
Tb^{3+}	6,1	8,57	[266]
		7,23	[205, 256]
$Tb(EDTA)^-$	12	6,725	[152]
Dy^{3+}	5,9	8,66	[266]
$Dy(EDTA)^-$	12	6,97	[152]
Ho^{3+}	5,9	9,38	[266]
		7,82	[205, 303]
$Ho(EDTA)^-$	12	6,99	[152]
Er^{3+}	-	7,845	[205, 256]

Table 112 (continued)

Ion	pH	log k	Ref.
Er (EDTA)	12	7,04	[152]
Yb^{3+}	6,0	10,63	[266]
		10,57	[205, 256]
$Yb(EDTA)^-$	12	9,30	[152]
Lu^{3+}	6,2	8,40	[266]
$Lu(EDTA)^-$	12	7,18	[152]
La^{3+}	7,0	8,53	[266]
	-	8,84	[205, 256]
$La(EDTA)^-$	12	$\leqslant$ 6,08	[152]
$Os(NH_3)_6^{3+}$	7	10,86	[152]
$Os(CN)_6^{4-}$	10,5	< 7,0	[152]
$IrCl_6^{2-}$	10	10,30	[152]
$IrCl_6^{3-}$	10	9,97	[152]
$Ir(NH_3)_6^{3+}$	7	10,115	[152]
$PtCl_4^{2-}$	-	9,955	[205, 256]
	6,8	10,08	[273]
$PtCl_6^{2-}$	10	10,30	[152]
$Pt(CN)_4^{2-}$	10	9,505	[273]
	10,6	9,59	[152]
$Au(CN)_2^-$	10,6	9,62	[152]
$Hg(CN)_4^{2-}$	10	8,28	[152]
$Hg(EDTA)^{2-}$	12	9,71	[152]

Table 112 (continued)

Ion	pH	log k	Ref.
Tl^{+}	-	10,04	[256]
	7	10,48	[205]
	7	10,67	[152]
Pb^{2+}	7,0	10,59	[273, 290]
	11,2	10,115	[273]
Pb(EDTA)	12	9,58	[152]
PbO_2^{2-}	14	10,00	[273]
	3 M OH^{-}	9,965	[273]
$PbCl_4^{2-}$	7,1	10,08	[273]
UO_2^{2+}		10,87	[205]

CHAPTER XIII

IONIC OXIDATION-REDUCTION REACTIONS

§1. Methods for Measuring Reaction Rates for Electron Transfer from Ion to Ion

The exchange of an electron between two ions of a given metal that differ only in charge (e.g., between Fe^{2+} and Fe^{3+}) is studied mainly by the isotope method (IM). A radioactive isotope of one of the ions (e.g., Fe^{2+}) is introduced into the system being studied, samples are taken at predetermined time intervals, one of the ions is precipitated, and the radioactive isotope content of the precipitate is determined. The kinetics of conversion of the metal isotope from one state to the other is used to determine the rate constant for electron transfer from one ion to the other. Control experiments are run to check the isotope exchange taking place during precipitation and separation of the ions.

EPR and NMR methods have been used to a much lesser extent in measuring the electron exchange rate (see Chapter IV).

If an oxidation-reduction reaction between two ions is slow, the kinetics are studied by the conventional methods described in Chapter I, such as consumption of starting material (CSM) and end-product formation (KEPF). In addition to these methods, ionic reactions may be studied to determine the rate constants for electron transfer from the kinetics of racemization (Kin. Racem.) and from studies of ionic catalytic reactions (Catal.).

Fast ionic oxidation-reduction reactions are studied by the same methods used for free radical reactions (Chapter IV), such as the method of competing reactions (MCR), jet method (JM), pulse radiolysis (PR), flash method (FM), and EPR and NMR methods. A temperature-jump (TJ) method is also used to study fast ionic reactions [304]. In this method, the temperature is changed rapidly in a system where an equilibrium ionic reaction is taking place, and the approach of the system to a new equilibrium position is followed by high-speed recording. The system temperature is changed within a time increment $\sim 10^{-6}$ sec. The reaction is followed from the change in electrical conductivity as measured by a recording oscillograph, or by high-speed spectrophotometry.

Polarographic methods (Pol.) have been applied to ionic reactions.

§2. Electron Exchange Reactions

See Table 113

§3. Oxidation-Reduction Reactions between Ions

See Table 114

TABLE 113. Rate Constants for Electron Exchange (IM Method)

Reaction	Medium	t°	log k	log A	E	Ref.
		Chromium				
$Cr^{3+} + Cr^{2+}$	$\mu = 1$	24,5	$\leqslant -4,7$			[305]
$CrOH^{2+} + Cr^{2+}$	$\mu = 1$	0–24,5 –34,5	$\sim -0,16$ (24,5°)	16	22	[305]
$CrCl^{2+} + Cr^{2+}$	$\mu = 1$; $[H^+] = 0,95$ M	0	1,04			[306]
$CrF^{2+} + Cr^{2+}$	$\mu = 1$; $[H^+] = 0,93$ M	0–43	–2,585 (0°)	7,8	14,2	[306]
$CrBr^{2+} + Cr^{2+}$	$\mu = 1$; 1 M $HClO_4$	0	> 1,78			[306]
$CrN_3^{2+} + Cr^{2+}$	$\mu = 0,5$; 0,5 M $HClO_4$	0–36,4	0,785 (25°)	8,28	10,2	[307]
cis-$Cr(N_3)_2^{+} + Cr^{2+}$	$\mu = 0,5$	0	~ 1,778			[308]
$CrNCS^{2+} + Cr^{2+}$	$\mu = 1$; 1 M $HClO_4$	27	–3,744			[306]
$Cr(H_2PO_2)^{2+} + Cr^{2+}$, $\frac{k}{[H^+]}$	$\mu = 1$	25	–3,215	11,7	20,3	[309]
$Cr[SC(NH_2)_2]_6^{3+} + Cr[SC(NH_2)_2]_6^{2+}$	CH_3OH	56	–4,1			[310]

Table 113 (continued)

Reaction	Medium	t°	log k	log A	E	Ref.
$Cr(OOCCH_3)^{2+} + Cr^{2+}$	CH_3OH	56	−4,1			[310]
$Cr(OOCCH_3)^{2+} + Cr^{2+}$ $\frac{k}{[H^+]}$	μ = 1,0	15–35	−3,236(25°)	11,1	19,5	[311]
		Vanadium				
$V^{2+} + V^{3+}$	μ = 2	0–40	−2 (25°)			[312]
$V^{2+} + VOH^{2+}$	μ = 2	0–40	~0,255 (25°)			[312]
$VOH^{2+} + VO^{2+}$, $\frac{k}{[H^+]}$	μ = 2,5; $[H^+]$ = 0,5–2,0M	25–40	~0 (25°)			[313]
		Manganese				
$Mn^{3+} + Mn^{2+}$	3 M $HClO_4$	25	~−3,52*			[314]
$MnO_4^- + MnO_4^{2-}$	0,16 N NaOH	0,1–21,7	2,845 (0,1°)	11,2	10,4	[315]
$MnO_4^- + MnO_4^{2-}$	0,57 N KOH	12,1–30	4 (30°)[2*]	10,1	8,3	[316]
		Iron				
$Fe^{3+} + Fe^{2+}$	μ = 0,5	5–25	0,602 (25°)	8,7	11,1	[317]
	μ = 0,55; D_2O	7,1	−0,155			[318]
	μ = 0,55	0–21,6	−0,06(0°)	7,2	9,9	[319]
	μ = 0,5; D_2O	5	−0,4±0,1			[317]

$Fe^{3+} + Fe^{2+}$	0,548 M $HClO_4$	20,5	0,915			[320]
	0,548 M $HClO_4$ 0,58 M MeOH	20,5	0,2			[320]
	0,548 M $HClO_4$; 0,49 M C_2H_5OH	20,5	0,082			[320]
	0,548 M $HClO_4$; 0,49 M n-C_3H_7OH	20,5	0,256			[320]
	Dimethyl sulfoxide; μ = 0,20; 0,5 M $HClO_4$	20–45	1,398 (25°)	8,9	10,2	[321]
	Eutectic μ = 0,55; $HClO_4$–0,55M	-78–22	-2,68 (-78°)	6,8	8,4	[322]
$FeOH^{2+} + Fe^{2+}$	μ = 0,55	0–21,6	3 (0°)	8,95	7,4	[319]
	μ = 0,5	5–25	3,476 (25°)	10,13	9,03	[317]
$FeOD^{2+} + Fe^{2+}$	μ = 0,55	7	2,884			[318]
	μ = 0,5; D_2O	5–25,2	3,483 (25,2°)	12,4	12,1	[317]
$FeF^{2+} + Fe^{2+}$	μ = 0,5	0–25	0,986 (0°)	8,3	9,1	[323]
$FeF_2^{+} + Fe^{2+}$	μ = 0,5	0–25	0,398 (0°)	8	9,5	[323]
$FeF_3 + Fe^{2+}$	μ = 0,5	0–25	-0,3 (0°)			[323]
$FeCl^{2+} + Fe^{2+}$	μ = 0,55	20	1,462	8	8,9	[319]
	μ = 0,5	0–20	0,732 (0°)	9,9	11,5	[324]

Table 113 (continued)

Reaction	Medium	t°	log k	log A	E	Ref.
$FeCl^{2+} + Fe^{2+}$	μ = 0,5; D_2O	10–20	0,959 (20°)	10,8	13,2	[324]
	μ = 3	25	1,760			[325]
$FeCl_2^+ + Fe^{2+}$	μ = 3	25	2,2			[325]
$FeBr^{2+} + Fe^{2+}$	μ = 0,55	0–20	0,69 (0°)	7,5	8,5	[320]
$FeBr_2^+ + Fe^{2+}$	μ = 0,55	0–20	1,278 (0°)	12,5	14	[320]
$FeSCN^{2+} + Fe^{2+}$	μ = 0,55	0–20	0,618 (0°)	8,4	9,7	[320]
	μ = 0,5	0–25	1,086 (0°)	7,4	7,9	[326]
$Fe(SCN)^+$	μ = 0,55	0–20	1,615 (0°)	10,7	11,3	[320]
	μ = 0,5	0–25	0,3 (0°)	7,2	8,6	[326]
$FeN_3^{2+} + Fe^{2+}$	μ = 0,55	0–15	3,676 (10°)	14,5	13,9	[327]
	μ = 0,55; D_2O	0–15	3,58 (10,1°)	14,9	14,6	[327]
$FeC_2O_4^+ + Fe^{2+}$	μ = 0,55	10–20	3,33 (20°)	9,2	9,2	[328]
	μ = 0,5	0–30	3,518 (20°)	19,2	21	[329]
$Fe(C_2O_4)_2^- + Fe^{2+}$	μ = 0,55	20	3,652			[328]
$FeSO_4^+ + Fe^{2+}$	μ = 0,25	25	2,84			[330]
	μ = 0,5	0–25,1	2,54 (25,1°)	9,2	9,0	[331]

	μ = 1,0	25	2,47	12,8	13,8	[332]
$Fe(SO_4)_2^- + Fe^{2+}$	μ = 1,0	0–28	2,737 (28°)	12,6	13,6	[329]
	μ = 0,25	25	4,288			[330]
$FeOH^{2+} + FeSO_4$	μ = 1,10	25	4,243	15,7	15,6	[332]
	μ = 0,25	25	6,3			[330]
$FeHPO_4^+ + Fe^{2+}$	μ = 0,5	0–20	3,631 (20°)	14,8	15	[329]
$Fe(CN)_6^{3-} + Fe(CN)_6^{4-}$	μ = 0,2 to 0,4M	25–65	4,944 (25°)^{2*}	8,0	4,2	[333]
	0,01N KOH	0,1	2,55	6,3	4,6	[334]
	0,01 M $(CH_3)_4N^+$	0,1–20,6	3,1 (0,1°)	3,34	0,3	[335]
	0,01 M $(C_2H_5)_4N^+$	0,1–20,6	2,398 (0,1°)	2,88	0,6	
	0,01 M $(n\text{-}C_3H_7)_4N^+$	0,1–20,6	1,613 (0,1°)	5,79	5,2	
	0,01 M $(n\text{-}C_4H_9)_4N^+$	0,1–20,6	1,362 (0,1°)	5,37	5,0	
	0,01 M $(n\text{-}C_5H_{11})_4N^+$	0,1–20,7	1,204 (0,1°)	7,46	7,8	
	0,01 M $(C_6H_5)_4As^+$	0,1–31,1	1,447 (0,1°)	6,50	6,3	

Table 113 (continued)

Reaction	Medium	t°	log k	log A	E	Ref.
$Fe(CN)_6^{3-} + Fe(CN)_6^{4-}$	0,01 M $Co(C_5H_5)_2^+$	0,1–20,6	3,022 (0,1°)	2,22	–1,0	[335]
	0,1 M K^+	0,1–20,8	2,354 (0,1°)	7,16	6,0	[335]
	[cation] ⟶ 0	0,1–20,7	0,778 (0,1°)	8,09	9,1	[335]
$Fe(PDTA)^- + Fe(PDTA)^{2-}$	H_2O	25	~ 4,6			[336]
$Fe(C_5H_5)_2^+ + Fe(C_5H_5)_2$	CH_3OH	–75	5,954			[337]
	CD_3COCD_3	26	>5[2*]			[338]
$Fe(Phen)_3^{3+} + Fe(Phen)_3^{2+}$	3N H_2SO_4	25	>5[2*]			[338, 339]
$Fe[4,7\text{-}(CH_3)_2Phen]_3^{3+} + Fe[4,7\text{-}(CH_3)_2Phen]_3^{2+}$	H_2O	25	>5[2*]			[340]
$Fe[5,6\text{-}(CH_3)_2Phen]_3^{3+} + Fe[5,6\text{-}(CH_3)_2Phen]_3^{2+}$	H_2O	25	>5[2*]			[340]

$Fe[3,4,7,8\text{-}(CH_3)_4Phen]_3^{3+} + H_2O + Fe[3,4,7,8\text{-}(CH_3)_4Phen]_3^{2+}$		25	5^{2*}			[340]
$Fe[3,5,6,8\text{-}(CH_3)_4Phen]_3^{3+} + D_2O + Fe[3,5,6,8\text{-}(CH_3)_4Phen]_3^{2+}$		25	5^{2*}			[340]
Cobalt						
$Co^{3+} + Co^{2+}$	$\mu = 1$; 1 N $HClO_4$	3,2–24,93	–0,076 (32°)			[341]
	$\mu = 3$; $NaClO_4$	0	0			[342]
	$\mu = 3$; $NaClO_4$; D_2O	0	–0,328			[342]
	$\mu = 1$; $NaClO_4$	0,15–18,35	0,146 (10,6°)	9,7	12,4	[343]
	$\mu = 0,5$; $LiClO_4$	0,15–18,35	–0,07 (10,6°)	8,4	10,9	[343]
	$\mu = 0,5$; $NaClO_4$	0,15–18,35	–0,05 (10,6°)	8,5	11,0	[343]

Table 113 (continued)

Reaction	Medium	t°	log k	log A	E	Ref.
$Co^{3+} + Co^{2+}$, $\frac{k}{[H^+]}$	$\mu = 1$; $NaClO_4$	3,2	–0,854			[341]
	$\mu = 3$; $NaClO_4$	0	–0,347			[342]
	$\mu = 3$; $NaClO_4$; D_2O	0	–0,638			[342]
	$\mu = 1$; $NaClO_4$	0,15–18,35	–0,495 (10,6°)	11,8	15,9	[343]
	$\mu = 1$; $LiClO_4$	0,15–18,35	–553 (10,6°)	13,8	18,6	[343]
	$\mu = 0,5$; $NaClO_4$	0,15–18,35	–0,6 (10,6°)	13,2	19,1	[343]
$CoOH^{2+} + Co^{2+}$	$\mu = 1$; $NaClO_4$	0,15–18,35	1,87 (10,6°)	6,4	5,9	[343]
	$\mu = 0,5$; $LiClO_4$	0,15–18,35	1,3 (10,6°)	8,0	8,6	[343]
	$\mu = 0,5$; $NaClO_4$	0,15–18,35	1,17 (10,6°)	8,2	9,1	[343]

$CoSO_4^{+} + Co^{2+}$	$\mu = 0,5$; $LiClO_4$; $[H^{+}] = 0,2$ M	0,15–18,35	1,12 (10,4°)	10,4	14,8	[343]
$Co(NH_3)_6^{3+} + Co(NH_3)_6^{2+}$	Liquid NH_3	24–45	–4,22 (25°)	12,8	23	[344]
$Co(NH_3)_5NH_2^{2+} + Co(NH_3)_6^{2+}$	$\mu = 1,0^{4*}$	25	–5,47	5,0	13,5	[345]
$Co(NH_3)_6(OH)^{2+} + Co(NH_3)_6^{2+}$	$\mu = 1,0^{4*}$	25	–4,21	6,2	13,5	[345]
$Co(NH_3)_6^{3+}Cl^{-} + Co(NH_3)_n^{2+}$	$\mu = 1,0^{5*}$	64,5	–3,134			[346]
$Co(NH_3)_6^{3+}OH^{-} + Co(NH_3)_n^{2+}$	$\mu = 1,0^{5*}$	45–80,1	–2,234 (64,5°)	6,0	12,6	[346]
$Co(NH_3)_5H_2O^{3+} + Co(NH_3)_5H_2O^{2+}$	H_2O	60	–3,125			[347]
$Co(En)_3^{3+} + Co(En)_3^{2+}$	$\mu = 0,98$; KCl; K_2SO_4	25–98	–4,114 (25°)^{3*}	6,3	14,1	[348]
	$\mu = 0,98$	25–45,1	–4,284 (25°)	6,1	14,2	[349]
$Co(C_2O_4)_3^{3-} + Co(C_2O_4)_3^{4-}$	H_2O	25	–6,046			[350]
$Co(C_2O_4)_2^{-} + Co(C_2O_4)_2^{2-}$	H_2O	25	–4,854			[350]
$Co(EDTA)^{-} + Co(EDTA)^{2-}$	pH 2	60–85	–4,7 (60°)	9,7	22	[351]

Table 113 (continued)

Reaction	Medium	t^o	log k	log A	E	Ref.
$Co(Dipy)_3^{3+} + Co(Dipy)_3^{2+}$	$\mu = 0$	0	0,322			[352]
	$\mu \sim 0$; pH 7	45	$\sim$1,272			[353]
$Co(Phen)_3^{3+} + Co(Phen)_3^{2+}$	$\mu \approx 0$	0–12,8	0,04 (0°)	14,1	17,5	[352]
	pH $\approx$ 7	20	$\sim$0,65			[353]
$Co(Tripy)_2^{3+} + Co(Tripy)_2^{2+}$	$\mu \approx 0$	0	$\sim$1,09			[352]
$CoO_4W_{12}O_{36}(III)^{5-}$ + $+ CoO_4W_{12}O_{36}(II)^{6-}$	$\mu = 0$	0–25	-2,346 (0°)	11,6	18,0	[354]
	Various Metals					
$Mo(CN)_8^{3-} + Mo(CN)_8^{4-}$	pH 2–11	0	> 3,0			[355]
$Cu^{2+} + Cu^+$	12 M HCl	20	$7,7^{2*}$			[356]
$Ag^{2+} + Ag^+$	5,87 M $HClO_4$	0,2	< 1,0			[357]
$SnCl_2 + SnCl_4$	C_2H_5OH	25,1–42,2	-4,284 (25°)	13,1	23,7	[358]
Ce(IV) + Ce(III)	μ = 6,18; HNO_3+ + $NaNO_3$	−9,7 to +25	-1,108 (k_1, 25°)	4,5	7,7	[359]

$k_1 + \frac{k_2}{[H^+]^2}$			$-0{,}988$ (k_2, 25°)	18,6	24	[359]
$Ce^{4+} + Ce^{3+}$	6,18 *M* $HClO_4$	0–15	$-3{,}369$ (k_1, 15°)	11,4	19,4	[359]
$W = k_1[Ce^{3+}] + k_2[Ce^{3+}] \cdot [Ce^{4+}]$	H_2O		$-1{,}533$ (k_2, 15°)	11,3	16,8	[359]
$Eu^{3+} + Eu^{2+}$	$\mu = 2{,}0$; ClO_4^-; $[H^+] = 1$ *M*; Cl^-	32–50	$-3{,}488$ (25°)	11,4	20,8	[360]
$W(CN)_8^{2-} + W(CN)_8^{3-}$	0,0001	0	$> 4{,}6$			[361]
	0,04 *M*	20	$\leqslant 8{,}6^{6*}$			[362]
$IrCl_6^{2-} + IrCl_6^{3-}$	1 N HCl	1	$> 2{,}462$			[363]
$Hg^{2+} + Hg^+$	$\mu = 0{,}125$	0–30,5	$-2{,}244$ (0°)	8,8	14,0	[364]
		Thallium				
$Tl^{3+} + Tl^+$	$\mu = 6$	25–41,8	$-4{,}585$ (25°)	6,2	14,7	[365]
	$\mu = 6$; 0,4 *M* HCl + 5,6 *M* $HClO_4$	31,8–41,6	$-1{,}34$ (31,8°)	20	29,6	[365]

Table 113 (continued)

Reaction	Medium	t°	log k	log A	E	Ref.
$Tl^{3+} + Tl^{+}$	$\mu = 3{,}68$	9,9–50,1	–4,37 (k_1, 25°)	8,5	17,6	[366]
$k_1 + k_2/[H^+] +$			–4,71 (k_2, 25°)	2,8	10,3	[366]
$+ k_3[NO_3^-]$			–3,352 (k_3, 25°)	8,35	16	[366]
$Tl^{3+} + Tl^{+}$	$\mu = 0{,}5$; $[H^+] = 0{,}5$ M	30	–0,126			[367]
	$\mu = 3{,}68$; $[H^+] = 2{,}5$ M	37,3	–0,187			[367]
	$\mu = 2{,}0$; $[H^+] = 1{,}0$ M	12,7–50	–0,42 (20°)	9,4	13,2	[367]
	$\mu = 3{,}0$; $[H^+] = 1{,}0$ M	25	–0,602			[367]
	$\mu = 3{,}68$; $[H^+] = 2{,}9$ M	24,9	–0,688			[367]

$TlCl^{2+} + Tl^{+}$	μ = 3,0; [H^{+}] =1 M	25	–2,0			[367]
$TlCl_2^{+} + Tl^{+}$	- " -	25	–3,0			
$TlCl_3 + Tl^{+}$	- " -	25	-1,7			
$TlCl_4^{-} + Tl^{+}$	- " -	25	0			
$TlNO_3^{2+} + Tl^{+}$	μ = 3,68; [H^{+}] =2,5 M	37,3	0,176			
$Tl^{3+} + TlNO_3$	- " -	37,3	0,176			
$TlNO_3^{2+} + TlNO_3$	μ = 3,68; [H^{+}] = 2,5 M	37,3	0,322			
$TlN_3^{2+} + Tl^{+}$	μ = 2,0; [H^{+}] =1 M	12,7–50	–0,468 (20°)	8,4	11,8	[367]
$Tl(N_3)_2^{+} + Tl^{+}$	- " -	12,7–50	–0,523 (20°)	9,2	13,0	
$Tl(N_3)_3 + Tl^{+}$	- " -	12,7–50	0,04 (20°)	9,6	12,8	
$TlN_3^{2+} + TlN_3$	- " -	12,7–50	1,544 (20°)	7,4	8,4	
$Tl(N_3)_2^{+} + TlN_3$	- " -	12,7–50	–0,523 (20°)	9,2	13,0	
$Tl(N_3)_3 + TlN_3$	- " -	12,7–50	0,3 (20°)	9,1	11,7	

Table 113 (continued)

Reaction	Medium	t°	log k	log A	E	Ref.
$Tl(CN)^{2+} + Tl^{+}$	μ = 0,5; [H⁺] = 0,5 M	30	-1,0			
$Tl(CN)_2^{+} + Tl^{+}$	- " -	30	-2,0			
$Tl(CN)_3 + Tl^{+}$	- " -	30	-1,0			
$Tl(CN)_4^{-} + Tl^{+}$	- " -	30	0			
$Tl(NTA)(III) + Tl^{+}$	μ = 2,0; [H⁺] = 1,0 M	20	-2,3			[367]
$Tl(NTA)_2(III) + Tl^{+}$	μ = 2,0; [H⁺] = 1,0 M	20	-3,7			
$Tl(NTA)(III) + Tl(NTA)(I)$	- " -	20	-1,0			
$Tl(NTA)_2(III) + Tl(NTA)(I)$	- " -	20	-1,7			
$Tl(NTA)(III) + Tl(NTA)_2(I)$	- " -	20	-0,126			
$Tl(NTA)_2(III) + Tl(NTA)_2(I)$	- " -	20	-1,155			
		Neptunium				
Np(VI) + Np(V)	1 M $HClO_4$	0–10	1,462 (0°)	8,1	8,3	[368]
$NpO_2^{2+} + NpO_2^{+}$	3 M $HClO_4$	0–9,84	1,948 (0°)	10,5	10,7	[369]

$NpO_2Cl^+ + NpO_2^+$	3 *M* $HClO_4$	0–9,84	2,332 (0°)	14,7	15,4	[369]
$NpO_2Cl_2 + NpO_2^+$	3 *M* $HClO_4$	0–9,84	1,934 (0°)	14,2	15,3	[369]
$NpO_2^{2+} + NpO_2^+$	0,106 *M* $HClO_4$; water + saccharose	0	1,310			[370]
	0,106 *M* $HClO_4$; water + ethylenediamine	0	1,346			[370]
		Plutonium				
$Pu^{4+} + Pu^{3+}$	$\mu = 2$	0–25	2,262	8,5	7,7	[371]
$PuOH^{3+} + Pu^{3+}$	$\mu = 2$	0–25	4,124	6,3	2,8	[371]

* Caluclated on basis of rate constants for reactions $Fe^{3+} + Fe^{2+}$ and $Mn^{3+} + Fe^{2+}$ by means of Marcus theory.

2* NMR method.

3* Method using racemization kinetics.

4* Reaction rate is inversely proportional to $[H^+]$. In the calculation, it was postulated in the first case that the reacting particle is $Co(NH_3)_5(NH_2)^{2+}$, and in the second case the outer sphere complex $Co(NH_3)_6(OH)^{2+}$.

5* n = 3 + 6. Rate constant is independent of n.

6* EPR method.

TABLE 114. Oxidation-Reduction Reactions between Ions

Reaction	Medium	t^{o}	log k	log A	E	Method	Ref.
	Titanium						
$Ti^{3+} + ClO_4^-$	μ=0,5	40–50	-1,05 (k_1, 40°)	17,7	26,6	CSM	[372]
$k_1 + k_2[H^+]$			-1,468 (k_2, 40°)	3,2	6,6		
$TiCl^{2+} + ClO_4^-$	μ=0,5	40–50	-1,456 (k_1, 40°)	16,5	25,6	CSM	[372]
$k_1 + k_2[H^+]$			-1,866 (k_2, 40°)	5,6	10,6		
$Ti^{3+} + ClO_4^-$	μ=1; $[H^+]$=0,2--0,6 M	15–40	-3,724 (k_1, 25°)	12,6	22,1	CSM	[373]
$k_1 + k_2[H^+]$			-3,903 (k_2, 25°)				
$Ti^{3+} + NH_3OH^+$	0,2 M $H_2C_2O_4$	25	1,624	7,4	7,9	Pol.	[374]

$Ti^{3+} + ClO_3^-$	0,1 M $H_2C_2O_4$; 0,2 M H_2SO_4	25	4,416			Pol.	[375]
	0,2 M H_2SO_4	25	4,724			Pol.	[375]
Vanadium							
$V^{2+} + VO^{2+}$	$\mu=2$	0,4–34,5	0,176 (25°)	9,656	12,9	CSM	[376]
$VO^{2+} + V^{2+}$	0,4 M H_2SO_4; 0,15 M $NaHSO_4$	25	1,086			CSM	[376]
	$\mu=0,8$; pH 0,2	25	1,0			Pol.	[377]
$V^{3+} + VO_2^+$	$\mu=2$	0,2-34,2	0,634 (k_1, 25°)	15,0	19,5	CSM	[378]
$\frac{k_1}{[H^+]^2} + \frac{k_2}{[H^+]} + k_3 + k_4[H^+]$			1,392 (k_3, 25°)	8,9	8,9		
			1,834 (k_4, 25°)	11,6	13,2		

Table 114 (continued)

Reaction	Medium	t^{o}	log k	log A	E	Method	Ref.
$V^{3+} + Cr^{2+}$	$\mu = 2$; 1 M $HClO_4$	25	−0,07			CSM	[379]
$VO^{2+} + Cr^{2+}$	$\mu = 1$	5	>3,9			CSM	[380]
$VO_2^{+} + Fe^{2+}$	$\mu = 1$	0,2–55,6	3,531 (k_3, 25°)	5,08	2,11	CSM	[381]
$\frac{k_1}{[H^+]} + k_2 + k_3[H^+]$	$\mu = 1$	25	$k_1 = 0$; 1,778 (k_2)			CSM	[381]
$VO^{2+} + Cu^{+}$, $k_1[H^+]$	$\mu = 1$	1,6–34,2	2,184 (25°)	4,19	2,74	CSM	[382]
$V^{3+} + Eu^{2+}$	$\mu = 2$	25	−2,046	6,75	12,0	CSM	[383]
$VOH^{2+} + Eu^{2+}$	$\mu = 2$	25	0,3	5,3	6,3	CSM	[383]
$V(IV) + Eu^{2+}$	$\mu = 2$; $[H^+] = 0,2$	25	~3			CSM	[383]
$VO_2^{+} + (Ta_6Br_{12})^{2+}$; $W = k_1[(Ta_6Br_{12})^{2+}][VO_2^{+}] + k_2[(Ta_6Br_{12})^{3+}][VO_2^{+}] + k_3[(Ta_6Br_{12})^{3+}][H^+][VO_2^{-}]^2$	$\mu = 0,5$	25	2,322 (k_1); 0,056 (k_2); 0,068 (k_3)			CSM	[384]

$VO_2^+ + (Ta_6Br_{12})^{3+}$ $W = k_2[(Ta_6Br_{12})^{3+}][VO_2^+] + k_3[(Ta_6Br_{12})^{3+}][H^+][VO_2^+]^2$	$\mu = 0,5$	25	0,056 (k_2) 0,068 (k_3)			CSM	[384]
$VO^{2+} + UO_2^+$	$[H^+] = 1$; $\mu = 2$	25	4,7			Catal.	[371]
$V(V) + U(IV)$, $k[H^+]$	$\mu = 1$	25	4,3	13,8	13		[385]
		Chromium					
$CrF^{2+} + Cr^{2+}$, $\frac{k_1}{[H^+]} + k_2[H^+]$	$\mu = 2$	40	−4,36 (k_2) −5,1 (k_1)			CSM	[386]
$CrCl^{2+} + Cr^{2+}$, $\frac{k}{[H^+]}$	$\mu = 2$	25	−3,495	11,8	20,8	CSM	[386]
	$\mu = 2$	25–40	−3,585 (25°)	12,7	22,2	CSM	[387]
$CrCl(OH)^+ + Cr^{2+}$	$\mu = 2$	20	1,3	~10,9	~12,8	CSM	[387]
cis-$CrCl_2^+ + Cr^{2+}$	$\mu = 1$	16–34,2	2,46 (25°)	7,1	6,3	JM	[388]
trans-$CrCl_2^+ + Cr^{2+}$	$\mu = 1$	16–34,2	2,614 (25°)	6,7	5,5	JM	[388]
$CrBr^{2+} + Cr^{2+}$, $\frac{k}{[H^+]}$	$\mu = 1$	25	−2,76	11,1	18,8	JM	[386]

Table 114 (continued)

Reaction	Medium	t°	log k	log A	E	Method	Ref.
$CrI^{2+} + Cr^{2+}$, $\frac{k}{[H^+]}$	$\mu=1$	25	−1,67	11,2	17,6	JM	[386]
	$\mu=1$	15–30	−1,575 (25°)	10,6	16,6	JM	[388]
$CrI^{2+} + Cr^{2+}$, $k[F^-]$	$\mu=2,0$	25	4,724			JM	[389]
$CrI^{2+} + Cr^{2+}$, $k[Cl^-]$	$\mu=2,0$	25	2,431			JM	[389]
$CrI^{2+} + Cr^{2+}$, $k[Br^-]$	$\mu=2,0$	25	1,799			JM	[389]
cis-$Cr(N_3)_2^+ + Cr^{2+}$	0,2 M $HClO_4$	0–25	0,278 (0°)	7,2	8,6	JM	[390]
$CrNH_3^{3+} + Cr^{2+}$	$\mu=2$	25–70	−4,6 (k_1, 28°)	6,5	15,1	JM	[391]
$k_1 + \frac{k_2}{[H^+]}$			−4,23 (k_2, 25°)	12,1	22,2		
$CrNH_3^{3+} + Cr^{2+}$, $\frac{k}{[H^+]}$	$\mu=2$	25	−4,14	11,4	21,1	JM	[386]
$CrCN^{2+} + Cr^{2+}$, $\frac{k}{[H^+]}$	$\mu=1$	55	−2,55			JM	[392]

$CrSCN^{2+} + Cr^{2+}$	$\mu=1$; 1 M $HClO_4$	25	1,8			JM	[393]
$Cr(OOCCH_3)^{2+} + Cr^{2+}$	$\mu=1{,}0$	25–55	−4,807 (25°)	15,2	27,3	CSM	[311]
$Cr(OH)(OOCCH_3)^{+} + Cr^{2+}$	$\mu=1{,}0$	25–55	−0,07 (25°)	12,1	16,6	CSM	[311]
$Cr(NH_3)_5F^{2+} + Cr^{2+}$	$\mu=1{,}0$	27–41	−3,57 (25°)	6,7	14	KEPF	[394]
$Cr(NH_3)_5Cl^{2+} + Cr^{2+}$	$\mu=1{,}0$	10,5–38,9	−1,3 (25°)	7,3	11,7	KEPF	[394]
$Cr(NH_3)_5Br^{2+} + Cr^{2+}$	$\mu=1$	12,4–39,1	−0,49 (25°)	6,21	9,1	KEPF	[394]
$Cr(NH_3)_5I^{2+} + Cr^{2+}$	$\mu=1$	25	0,74			KEPF	[394]
cis-$Cr(En)_2(H_2O)Cl^{2+} + Cr^{2+}$	$\mu=1$	25	−1,936	6	10,8	CSM	[395]
cis-$Cr(En)_2Cl_2^{+} + Cr^{2+}$	$\mu=1$	25	−1,71	5,3	9,6	CSM	[395]
trans-$Cr(En)_2(H_2O)Cl^{2+} + Cr^{2+}$	$\mu=1$	25	−0,362	6,0	8,6	CSM	[395]
trans-$Cr(En)_2Cl_2^{+} + Cr^{2+}$	$\mu=1$	25	−0,294	6,1	8,7	CSM	[395]
$CrSCN^{2+} + V^{2+}$	$\mu=1$; 1 M $HClO_4$	25	0,85			JM	[396]
$Cr(OOCCH_3)^{2+} + V^{2+}$ $k_1 + \frac{k_2}{[H^+]}$	$\mu=1{,}0$	15–35	−0,788 (k_1, 25°) −2,174 (k_2, 25°)	~9,1	~13,5	CSM	[311]

Table 114 (continued)

Reaction	Medium	t°	log k	log A	E	Method	Ref.
$Cr^{3+} + Ag^{2+}$	3 M $HClO_4$	15–30	0,68 (25°)	9,9	12,6	CSM	[397]
$Cr^{3+} + Eu^{2+}$	μ=2	25	~–4,8			CSM	
$CrCl^{2+} + Eu^{2+}$	μ=2	25	~–2,66			CSM	[383]
$Cr^{2+} + ClO_3$	μ=2	5–25	0,59 (20°)	9,25	11,8	CSM	
Manganese							
$Mn^{3+} + Fe^{2+}$	μ=3,04	15	3,8	12,7	12,1	JM	[399]
	1 M $HClO_4$	25	4,222			JM	[314]
$Mn^{3+} + Fe\text{-}(Phen)_3^{2+}$	1 M $HClO_4$	25	3,267			JM	[314]
$Mn^{3+} + Fe\text{-}(5\text{-}CH_3\text{-}Phen)_3^{2+}$	1 M $HClO_4$	25	3,505			JM	
$Mn^{3+} + Fe\text{-}(5\text{-}Cl\text{-}Phen)_3^{2+}$	1 M $HClO_4$	25	2,596			JM	
$Mn^{3+} + Fe\text{-}(5\text{-}NO_2\text{-}Phen)_3^{2+}$	1 M $HClO_4$	25	1,954			JM	[314]
$Mn(H_2P_2O_7)_3^{3-} + Fe\ Phen)_3^{2+}$	μ=0,5; pH 1,0	25	1,857			JM	
$Mn(H_2P_2O_7)_3^{3-} + Fe\text{-}(5\text{-}CH_3\text{-}Phen)_3^{2}$	– " –	25	2,146			JM	

$Mn(H_2P_2O_7)_3^{3-}$ + $Fe-(5-Ph-Phen)_3^{2+}$	- " -	25	1,832			JM	[314]
$Mn(H_2P_2O_7)_3^{3-}$ + $Fe-[4,7-(CH_3)_2-Phen]_3^{2+}$	- " -	25	3,146			JM	[314]
$Mn(H_2P_2O_7)_3^{3-}$ + $Fe-[5,6-(CH_3)_2-Phen]_3^{2+}$	- " -	25	2,59			JM	[314]
$Mn(H_2P_2O_7)_3^{3-}$ + $Fe-[3,5,6,8-(CH_3)_4-Phen]_3^{2+}$	- " -	25	2,633			JM	[314]
MnO_4^- + $Fe(CN)_6^{4-}$	0,16 N NaOH	0,1	~4			IM	[315]
	0,1 N NaOH	25	4,127			JM	[400]
MnO_4^- + $KFe(CN)_6^{3-}$	μ=0,0933; pH 5,77	10–30	5,072 (25°)	21,3	22,1	CSM	[401]
MnO_4^- + $H_2Fe(CN)_6^{2-}$	μ=0,0933; pH~2	10–25	5,318 (25°)	8,9	4,9	CSM	[401]
MnO_4^- + $HFe(CN)_6^{3-}$	μ=0,093; pH 2,0	15	5,24			CSM	[401]
Mn^{3+} + Hg(O)	μ=4,5; [H^+]=3 M	50	~3,7			CSM	[402]
Mn(IV) + Hg_2(I)	μ=4,5; [H^+]=3 M	50	>1,7			CSM	[402]

Table 114 (continued)

Reaction	Medium	t°	log k	log A	E	Method	Ref.
	Iron						
$FeCl^{2+} + Fe^{2+}$	$\mu = 3,0$	25	0,792 (k_1)			JM	[403]
$k_1 + \frac{k_2}{[H^+]}$			1,17 (k_2)				
$FeNCS^{2+} + Fe^{2+}$	$\mu = 3,0$	25	1,021			JM	[404]
$Fe(Phen)_3^{3+} + Fe^{2+}$	$\mu = 0,5$; ClO_4^-	15– 25–35	4,568	5,15	0,8	JM	[405]
	0,5 N H_2SO_4	25	5,477			JM	[406]
$Fe\text{-}(5\text{-}CH_3\text{-}Phen)_3^{3+} + Fe^{2+}$	$\mu = 0,5$; ClO_4^-	25	4,3			JM	[406]
	0,5 N H_2SO_4	25	5,176			JM	[406]
$Fe\text{-}(5\text{-}NO_2\text{-}Phen)_3^{3+} + Fe^{2+}$	0,5 N $HClO_4$	25	6,04			JM	[406]
$Fe\text{-}(5\text{-}Cl\text{-}Phen)_3^{3+} + Fe^{2+}$	– " –	25	5,322			JM	[406]
	0,5 N H_2SO_4	25	6,176			JM	[406]

$Fe\text{-}(5\text{-}Ph\text{-}Phen)_3^{3+} + Fe^{2+}$	- " -	25	5,505	JM	
$Fe\text{-}[5,6\text{-}(CH_3)_2\text{-}Phen]_3^{3+} + Fe^{2+}$	0,5 N $HClO_4$	25	3,892	JM	
	0,5 N H_2SO_4	25	4,838	JM	
$Fe\text{-}[3,4,7,8\text{-}(CH_3)_4\text{-}Phen]_3^{3+} + Fe^{2+}$	- " -	25	3,278	JM	
$Fe(Dipy)_3^{3+} + Fe^{2+}$	0,5 N $HClO_4$	25	4,431	JM	
$Fe\text{-}[4,4'\text{-}(CH_3)_2\text{-}Dipy]_3^{3+} + Fe^{2+}$	0,5 N $HClO_4$	25	2,778	JM	[406]
$Fe(Dipy)_3^{3+} + Fe^{2+}$	0,5 N H_2SO_4	25	5,342	JM	
$Fe\text{-}[4,4'\text{-}(CH_3)_2\text{-}Dipy]_3^{3+} + Fe^{2+}$	- " -	25	3,77	JM	
$Fe(tripy)_2^{3+} + Fe^{2+}$	0,5 N $HClO_4$	25	4,83	JM	
	0,5 N H_2SO_4	25	5,869	JM	
$Fe(Phen)_3^{3+} + Fe(CN)_6^{4-}$	0,5 N H_2SO_4	25	> 8	JM	[400]
$Fe(Phen)_3^{3+} + Fe(Dipy)_3^{2+}$		18	> 5		
$Fe(Phen)_3^{3+} + Fe[4,4'\text{-}(CH_3)_2\text{-}Dipy]_3^{2+}$	0,5 N $HClO_4$	25	> 8	JM	[400]

Table 114 (continued)

Reaction	Medium	t^o	log k	log A	E	Method	Ref.
$Fe[4,4'-(CH_3)_2-Dipy]_3^{3+}$ + $Fe[3,4,7,8-(CH_3)_4-Phen]_3^{2+}$	0,5 N H_2SO_4	25	>8			JM	[400]
$Fe(CN)_6^{3-}$ + Hb(H_2O)	μ=0,1; pH 6	25	4,845			JM	[407]
$Fe(CN)_6^{3-}$ + Cyt C(Fe^{2+})	μ=0,1; pH 6	25	7,204			JM	[408]
	μ+0,17; pH 7	12	7,090			TJ	[409]
	μ=0,17	12	7,080			TJ	[409]
Cyt C(Fe^{3+}) + $Fe(CN)_6^{4-}$	μ=0,17	12	4,183			TJ	[409]
	μ=0,17	12	4,050			TJ	[409]
Fe^{3+} + V^{2+}	μ=1	25	4,255			JM	[396]
$FeCl^{2+}$ + V^{2+}	μ=1	25	5,663			JM	
$FeNCS^{2+}$ + V^{2+}	μ=1	25	5,82			JM	
FeN_3^{2+} + V^{2+}	μ=1	25	5,708			JM	

$Fe^{3+} + V^{3+}$	$\mu = 3$	15–30	−2,585 (k_0, 25°)	10,6	17,9	Catal.	[410]
$k_0 + \frac{k_1}{[H^+]} + \frac{k_2}{[H^+]^2}$			−2,482 (k_1, 25°)	15	23,8		
			−2,585 (k_2, 25°)	15,9	25,1		
$Fe(EDTA)^- + V(2,2'-Dipy)_3^{2+}$	$\mu = 0,1$	25	1,364			CSM	[411]
$Fe^{3+} + Cr^{2+}$	$\mu = 1$	25	3 362			JM	[412]
$FeOH^{2+} + Cr^{2+}$	$\mu = 1$	25	6,518			JM	[412]
$FeCl^{2+} + Cr^{2+}$	$\mu = 1$	25	7,3			JM	[412]
$FeNCS^{2+} + Cr^{2+}$	1 M $HClO_4$	25	≥7,3			JM	[412]
$FeN_3^{2+} + Cr^{2+}$	– " –	25	≥7,3			JM	[412]
$Fe^{3+} + Cr^{2+}$, k [SCN⁻]	1 M $HClO_4$; SCN^-	25	~5,3			JM	[412]
dl – $Fe(2,2'-Dipy)_3^{3+}$ + dl $[Co(-)PDTA]^{2-}$	$\mu = 0,5$; pH 4,5	25	4,908	9,5	6,3	JM	[413]
$Fe(Phen)_3^{3+} + Os(Dipy)_3^{2+}$	0,5 N $HClO_4$	25	>8			JM	[400]
$Fe(Dipy)_3^{3+} + Os(Dipy)_3^{2+}$	3 N H_2SO_4	D	>6,4			CSM	[339]

Table 114 (continued)

Reaction	Medium	t^o	log k	log A	E	Method	Ref.
$Fe^{3+} + Sn^{2+}$	0,136 M Cl^-	25	−1,4			CSM	[413a]
	0,068 M Br^-	25	−0,7			CSM	[413a]
	0,088 M I^-	25	0,78			CSM	[413a]
$Fe-[4,7-(CH_3)_2-Phen]_3^{3+} + IrCl_6^{3-}$	μ=0,1	10–30	9,0 (10°)	13,7	6	TJ	[414]
$Fe-[5,6-(CH_3)_2-Phen]_3^{3+} + IrCl_6^{3-}$	μ=0,10	10	9,447			TJ	[415]
$Fe-[4,4'-(CH_3)_2-Phen]_3^{3+} + IrCl_6^{3-}$	μ=0,10	10	8,3			TJ	[415]
$Fe-[5,6-(CH_3)_2-Phen]_3^{3+} + IrBr_6^{3-}$	μ=0,30	10	9,278			TJ	[415]
$Fe-[4,4'-(CH_3)_2-Phen]_3^{3+} + IrBr_6^{3-}$	μ=0,10	10	9,278			TJ	[415]
	μ=0,51	10	9,322			TJ	[415]
$Fe(CN)_6^{3-}$ + Sb(III)	μ=0,10 0,14–0,14M NaOH	25	−0,050			IM	[416]
$Fe^{3+} + Np^{3+}$	μ=2	0,5–36,7	2,78 (k_1, 25°)	13,88	15,1	KEPF	[417]
$\frac{k_1}{[H^+]} + k_0$			1,56 (k_0, 25°)	5,0	4,7		

Fe(III) + U(IV)	μ=1	3,1–24,8	0,474 (k'_0, (25°)	17,0	22,5	CSM	[418]
$k=2(k'_0[H^+]+k''_0)/(K_1+[H^+])(K_2+[H^+])$ *			1,025 (k''_0, 25°)	18,8	24,2		
Fe^{3+} + Np(IV), $\frac{k}{[H^+]^3}$	μ=1; $HClO_4$ + $NaClO_4$	24,3–35,4	−1,244 (25°)	24,7	35,2	CSM	[419]
	μ=1; HNO_3 + $NaNO_3$	25	−1,456			CSM	[419]
Fe^{3+} + Cu^+	1 M $HClO_4$	25	5,08			JM	[420]
Fe^{3+} + UO_2^+	μ=2; $[H^+]$=1	25	5,4			Catal.	[421]
Fe^{3+} + HO_2^-	μ=0,435	20–45	5,3 (20°)	26,2	28	SUNR	[422]
$Fe(Dipy)_3^{3+}$ + HO_2^-	H_2O	20	7,15	12,4	7	CSM	[423]
$Fe(Tripy)_3^{3+}$ + HO_2^-	– " –	20	9,78	18	11	CSM	[423]
$Fe(CN)_5CO^{2-}$ + HO_2^-	– " –	20	8,3	15	9	CSM	[423]
$Fe(SCN)^{2+}$ + SCN^-	μ=1,28; $[H^+]$ = 0,967 M	14,8–29,9	−4,745 (25°)	14,5	26,2	KEPF	[424]
$Fe(SCN)_2^+$ + SCN^-	– " –	14,8–29,9	−3,853 (25°)	16	27	KEPF	[424]

* K_1 and K_2 are constants for hydrolysis of Fe(III) and U(IV), respectively.

Table 114 (continued)

Reaction	Medium	t^o	log k	log A	E	Method	Ref.
$Fe^{2+} + S_2O_8^{2-}$	H_2O	25	1,845	11,1	12,6		[425]
	μ=0,07; 0,01 M $HClO_4$	25	1,478			CSM	[426]
	μ=0	7–40	2,02 (25°)	8	8,1	KEPF	[427]
	μ=0	0–10	1,285 (0°)	11, 0	12,2	CSM	[428]
$Fe(CN)_6^{4-} + S_2O_8^{2-}$	μ=0,06	25	−2,036			CSM	[429]
$KFe(CN)_6^{3-} + KS_2O_8^{-}$	μ=0	25	0,376	7,4	9,6	KEPF	[430]
$Fe(Phen)_3^{2+} + S_2O_8^{2-}$	pH 7	20–40	−0,968 (25°)	8,9	13,5	CSM	
$Fe\text{-}(5\text{-}NO_2\text{-}Phen)_3^{2+} + S_2O_8^{2-}$	pH 7	20–40	−0,970 (25°)	19,2	27,5	CSM	
$Fe\text{-}(5\text{-}CH_3\text{-}Phen)_3^{2+} + S_2O_8^{2-}$	pH 7	20–40	−1,108 (25°)	7,6	11,9	CSM	[431]
$Fe\text{-}(5\text{-}Cl\text{-}Phen)_3^{2+} + S_2O_8^{2-}$	pH 7	20–40	−1,377 (25°)	13,9	20,7	CSM	
$Fe\text{-}(5\text{-}CH_3\text{-}6\text{-}NO_2\text{-}Phen)_3^{2+} + S_2O_8^{2-}$	pH 7	20–40	−0,788 (25°)	19,1	27	CSM	

$Fe-[3,6-(CH_3)_2-Phen]_3^{2+} + S_2O_8^{2-}$	pH 7	20–40	–1,758 (25°)	8	13,3	CSM	[431]
$Fe-[4,7-(CH_3)_2-Phen]_3^{2+} + S_2O_8^{2-}$	pH 7	20–40	0,272 (25°)	5,3	6,9	CSM	[431]
$Fe-[3,5,6,8-(CH_3)_4-Phen]_3^{2+} + S_2O_8^{2-}$	pH 7	20–40	–0,7 (25°)	6	9,1	CSM	[431]
$Fe(Dipy)_3^{2+} + S_2O_8^{2-}$	pH 7	20–40	–0,388 (25°)	7,2	10,3	CSM	[431]
$Fe(Dipy)_2^{2+} + S_2O_8^{2-}$	pH 3,90	0	~1,43			CSM	[428]
$Fe[4,4'-(CH_3)_2-Dipy]_3^{2+} + S_2O_8^{2-}$	pH 7	20–40	0,778 (25°)	5,6	6,6	CSM	[431]
$Fe[5,5'-(CH_3)_2-Dipy]_3^{2+} + S_2O_8^{2-}$	pH 7	20–40	0,182 (25°)	5,9	7,8	CSM	[431]
$Fe(Tripy)_2^{2+} + S_2O_8^{2-}$	pH 7	20–40	–0,293 (25°)	7,6	10,7	CSM	[431]
$Fe^{2+} + Br_3^-$	$\mu=1$	20,4–29,8	1,338 (20,4°)	7,63	8,4	CSM	[431]
	Ti (III)						
$Co(NH_3)_5F^{2+} + Ti^{3+}$	$\mu=1$; ClO_4^-	15–35	2,3 (25°)			CSM	[373]
$Co(NH_3)_5Cl^{2+} + Ti^{3+}$	– " –	15–35	1,064 (25)			CSM	[373]
$Co(NH_3)_5Br^{2+} + Ti^{3+}$		15–35	0,176 (25°)			CSM	[373]

Table 114 (continued)

Reaction	Medium	t °	log k	log Λ	E	Method	Ref.
$Co(NH_3)_5I^{2+} + Ti^{3+}$	$\mu = 1$; ClO_4^-	15–35	0,591 (25°)			CSM	[373]
$Co(NH_3)_5H_2O^{3+} + Ti^{3+}$	- " -	25	-2,046			CSM	[373]
$Co(NH_3)_5SO_4^{+} + Ti^{3+}$	- " -	25	-1,036			CSM	[373]
$Co(NH_3)_5Fe^{2+} + Ti^{3+}$	$[H^+]$=0,2; 0,5; μ=0,52; SO_4^{2-}	25	2,076			CSM	[373]
$Co(NH_3)_5Cl^{2+} + Ti^{3+}$ $k_1 + k_2[H^+]$	- " -	25	0,591 (k_1) -0,638 (k_2)			CSM	[373]
$Co(NH_3)_5Br^{2+} + Ti^{3+}$ $k_1 + k_2[H^+]$	- " -	25	0,08 (k_1) -1,046 (k_2)			CSM	[373]
$Co(NH_3)_5I^{2+} + Ti^{3+}$ $k_1 + k_2[H^+]$	$[H^+]$ =0,2+ 0,5; μ=0,52; SO_4^{2-}	25	0,532 (k_1) -0,602 (k_2)				[373]
$Co(NH_3)_5H_2O^{3+} + Ti^{3+}$	- " -	25	-1,658 (k_1) -1,222 (k_2)			CSM	[373]

$Co(NH_3)_5SO_4^+ + Ti^{3+}$		25	−1,214 (k_1) −1,796 (k_2)			CSM	[373]
	Cobalt-vanadium						
$Co^{3+} + VO^{2+}$	4 M $HClO_4$	25	0,136			CSM	[432]
$Co^{3+} + VO^{2+}$, $k_1 + \frac{k_2}{[H^+]}$	μ=3,0	−2,1 to + +20	−0,33 (k_1, 20°)	15,9	21,8	CSM	[433]
			0,17 (k_2, 20°)	12,0	15,9		
Co(III) + V^{2+}	1 M $HClO_4$	0	>2,48			CSM	[434]
$Co(NH_3)_5NH_3^{3+} + V^{2+}$	μ=0,4	25–37	−2,432 (25°)	4,7	9,7	CSM	[435]
$Co(NH_3)_5NH_3^{3+} + V^{2+}$	μ=1	25	−2,357			CSM	[436]
$Co(NH_3)_5NH_3^{3+} + V^{2+}$	μ=1; D_2O	37	−2,4			CSM	[435]
$Co(NH_3)_5NH_3^{3+} + V^{2+}$, $k[F^-]$	μ=0,4; F^-	25	+1,962			CSM	[436]
$Co(NH_3)_5NH_3^{3+} + V^{2+}$, $k[Cl^-]$	μ=0,4 Cl^-	25–37	−1,678 (25°)	8,1	14,7	CSM	[435]
	μ=0,4	25	−1,456			CSM	[436]
	μ=1; D_2O; Cl^-	37	−1,488			CSM	[435]
$Co(NH_3)_5NH_3^{3+} + V^{2+}$, $k[SO_4^{2-}]$	μ=0,4	25	0,83			CSM	[436]

Table 114 (continued)

Reaction	Medium	t°	log k	logA	E	Method	Ref.
$Co(NH_3)_5H_2O^{3+} + V^{2+}$	$\mu=1,0$	12–37,5	–0,276 (25°)	6,2	8,8	CSM	[436]
	$\mu=0,4$	25	–0,538			CSM	[436]
$Co(NH_3)_5H_2O^{3+} + V^{2+}$, $k[F^-]$	$\mu=1,0$; F^-	25	3,653			CSM	[436]
	$\mu=0,4$; F^-	25	3,653			CSM	[436]
$Co(NH_3)_5H_2O^{3+} + V^{2+}$, $k[Cl^-]$	$\mu=1,0$; Cl^-	12–37,5	0,4 (25°)	11,2	13,9	CSM	[436]
	$\mu=0,4$; Cl^-	25	0,4			CSM	[436]
$Co(NH_3)_5H_2O^{3+} + V^{2+}$, $k[I^-]$	$\mu=1,0$; I^-	25	–0,495			CSM	[436]
$Co(NH_3)_5H_2O^{3+} + V^{2+}$, $k[SO_4^{2-}]$	$\mu=1,0$; SO_4^{2-}	25	2,222			CSM	[436]
$Co(NH_3)_5F^{2+} + V^{2+}$	$\mu=0,52$; SO_4^{2-}; $[H^+]=0,2\div0,5$ M	25	1,03			CSM	[373]
$Co(NH_3)_5Cl^{2+} + V^{2+}$	– " –	25	1,161			CSM	[373]
$Co(NH_3)_5Br^{2+} + V^{2+}$	– " –	25	1,888			CSM	[373]

$Co(NH_3)_5H_2O^{3+} + V^{2+}$	- " -	25	1,262			CSM	[373]
$Co(NH_3)_5SO_4^{+} + V^{2+}$	- " -	25	1,222			CSM	[373]
$Co(NH_3)_5H_2O^{3+} + V^{3+}$, k $[SO_4^{2-}]$	μ=0,4; SO_4^{2-}	25	2,724			CSM	[436]
$Co(NH_3)_5F^{2+} + V^{2+}$	μ=1,0	25	0,415			JM	[437]
	μ=0,52	22,5	0,462			CSM	[438]
$Co(NH_3)_5Cl^{2+} + V^{2+}$	μ=0,52	22,5	0,748			CSM	[438]
	μ=1,0; 74% D_2O	25	0,568			CSM	[435]
	μ=1,0	25	0,756			CSM	[435]
	μ=1,0	25	~0,7			JM	[437]
$Co(ND_3)_5Cl^{2+} + V^{2+}$	μ=1,0; 85% D_2O	25	0,415			CSM	[435]
$Co(NH_3)_5Br^{2+} + V^{2+}$	μ=0,52	22,5	1,164			CSM	[438]
	μ=1,0	25	1,4	8,5	9,7	JM	[437]
$Co(NH_3)_5I^{2+} + V^{2+}$	μ=1,0	25	2,08	~5,9	~5,2	JM	
$Co(NH_3)_5N_3^{2+} + V^{2+}$	μ=1,0	25	1,114	10,2	12,3	JM	[437]
$Co(NH_3)_5NCS^{2+} + V^{2+}$	μ=1,0	25	−0,523			JM	

Table 114 (continued)

Reaction	Medium	t°	log k	log A	E	Method	Ref.
$Co(NH_3)_5SO_4^+ + V^{2+}$	μ=1,0	10–19,5	0,892 (25°)	5,8	6,7	CSM	[437]
$Co(NH_3)_5PO_4 + V^{2+}$	μ=1,0	25	7,146			JM	[437]
$Co(NH_3)_5HPO_4^+ + V^{2+}$	μ=1,0	25	2,204			JM	[437]
$Co(NH_3)_5H_2PO_4^{2+} + V^{2+}$	μ=1,0	25	0,362			JM	[437]
$Co(NH_3)_5H_3PO_4^{3+} + V^{2+}$	μ=5,0	25	0,654			JM	[437]
$Co(En)_3^{3+} + V^{2+}$	μ=1,0	25	~ −3,7			JM	[437]
$Co(En)_2SO_4 + V^{2+}$	μ=1,0; $[H^+]$ = 0,2 to 0,6 M	19–30	0,82 (25°)	8,7	10,8	CSM	[439]
$Co(En)_2SO_4^+ + V^{2+}$, $k[H^+]$	– " –	25	0,968			CSM	[439]
$Co(En)_2(NH_3)(SO_4)^+ + V^{2+}$	– " –	10–24	0,792	7,9	9,7	CSM	[439]
$(Tetren)Co(SO_4)^+ + V^{2+}$	μ=1,0	10–19,5	0,814 (25°)	7,2	8,7	CSM	[439]
$Co(NH_3)_5(OOCCH_3)^{2+} + V^{2+}$	μ=1,0	25	−0,366			CSM	[440]
	μ=1,0	25	−0,368	4,3	6,4	CSM	[441]
$Co(NH_3)_5(OOCCH_2Cl)^{2+} + V^{2+}$	μ=1,0	25	0,096	7,4	10,0	CSM	[441]

$Co(NH_3)_5(OOCCHCl_2)^{2+} + V^{2+}$	$\mu = 1,0$	25	0,012	7,5	10,2	CSM	[441]
$Co(NH_3)_5(OOCCH_2CN)^{2+} + V^{2+}$	$\mu = 1,0$	25	0,053	7,4	10,0	CSM	[441]
cis-$Co(NH_3)_4(OOCCH_3)_2^{+} + V^{2+}$ $k_1 + k_2[H^+]$	$\mu = 1,0$; $[H+]$ = =0,2to0,6 M	13–25	1,04 (k_1, 25°) 1,342 (k_2)	3,8	3,7	CSM	[442]
trans-$Co(NH_3)_4(OOCCH_3)_2^{+} + V^{2+}$ $k_1 + k_2[H^+]$	$\mu = 1,0$; $[H^+]$ = =0,2 to0,6 M	25	1,04 (k_1) 0,902 (k_2)			CSM	[442]
$Co(NH_3)_5(OOCCOOH)^{2+} + V^{2+}$ $k_1 + k_2[H^+]$	$\mu = 1,0$; $[H^+]$ = =0,2 to0,6 M	25	1,243 (k_1) 1,255 (k_2)			CSM	[442]
$Co(NH_3)_5(OOCCOO)^{+} + V^{2+}$	$\mu = 1,0$; $[H^+]$ = =0,013 to 1,0M	25	1,654			KRR	[443]
$Co(NH_3)_5(OOCCOOH)^{2+} + V^{2+}$	– " –	15–37	1,097 (25°)	10,5	12,8	KRR	[443]
$Co(NH_3)_4(OOCCOO)^{+} + V^{2+}$ $k_1 + k_2[H^+]$	$\mu = 1,0$; $[H^+]$ = =0,2to0,6M	13–23 25	1,634 (k_1, 25°) 1,342 (k_2)	10,5	12,0	KRR	[442]

Table 114 (continued)

Reaction	Medium	t^{o}	log k	logA	E	Method	Ref.
$Co(NH_3)_4[OCC(CH_2)_2COO]^{+} + V^{2+}$ $k_1 + k_2[H^{+}]$	$\mu = 1{,}0$; $[H^{+}] = 0{,}2$ to $0{,}6$ M	25–43 25	1,04 (k_1, 25°) 0,902 (k_2)	8,1	9,6	KRR	[442]
$Co(NH_3)_5[OOC(CH_2)_2COOH]^{2+} + V^{2+}$	$\mu = 1{,}0$	25	−0,05 (k_2)			KRR	[440]
$Co(NH_3)_5(cis\text{-}OOCCH{=}CHCOOH)^{2+} + V^{2+}$ $k_1 + k_2[H^{+}]$	$\mu = 1{,}0$; $[H^{+}] =$ 0,2 to 0,6 M	25	1,113 (k_1) 1,0 (k_2)			KRR	[442]
$Co(NH_3)_4(cis\text{-}OOCCH{=}CHCOO)^{2+} + V^{2+}$ $k_1 + k_2[H^{+}]$	$\mu = 1{,}0$; $[H^{+}] =$ 0,2 to 0,6 M	25	1,08 (k_1) 0,978 (k_2)	3,3	3,0	KRR	[442]
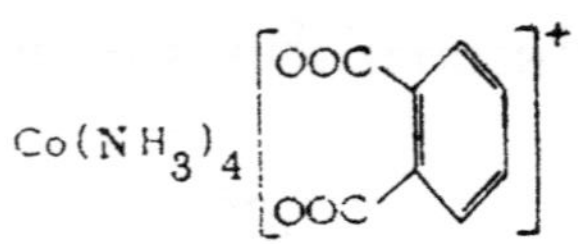 $+ V^{2+}$ $k_1 + k_2[H^{+}]$	$\mu = 1{,}0$; $[H^{+}] =$ 0,2 to 0,6 M	10–25 25	0,653 (k_1, 25°) 1,04 (k_2)	6,0	7,2	CSM	[442]

$Co(NH_3)_5$ 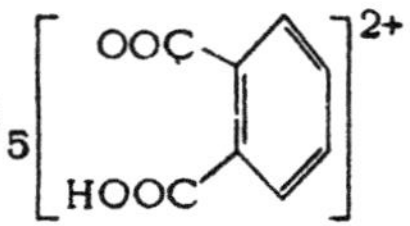 + V^{2+}	$\mu = 1,0$	25	0,004	7,9	10,8	CSM	[441]
$Co(NH_3)_5$ 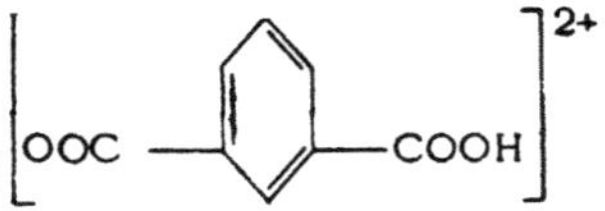 + V^{2+}	$\mu = 1,0$	25	$\bar{1}$,78	6,8	9,6	CSM	[441]
$Co(NH_3)_5$ 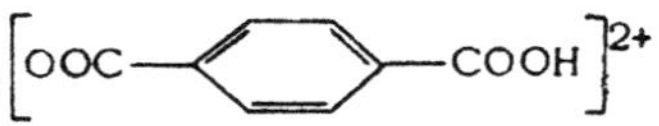 + V^{2+} $k_1 + k_2[H^+]$	$\mu = 1,0$; $[H^+]$ = 0,2 to 0,6 M	21	−0,367 (k_1) −0,356 (k_2)			CSM	[444]
$Co(NH_3)_5(OOCC_6H_5)^{2+} + V^{2+}$	$\mu = 1,0$	25	−0,284	5,1	7,3	CSM	[441]
$Co(NH_3)_5(OOC\text{-}p\text{-}C_6H_4CHO)^{2+} + V^{2+}$ $k_1 + k_2[H^+]$	$\mu = 1,0$; $[H^+]$ = 0,2 to 0,6 M	25	−0,244 (k_1) −0,77 (k_2)			CSM	[444]
$Co(NH_3)_5(OOC\text{-}m\text{-}C_6H_4Cl)^{2+} + V^{2+}$	$\mu = 1,0$	25	−0,244	8,9	11,1	CSM	[441]
$Co(NH_3)_5(OOC\text{-}p\text{-}C_6H_4Cl)^{2+} + V^{2+}$	$\mu = 1,0$	25	−0,322	6,0	8,6	CSM	[441]
$Co(NH_3)_5(OOC\text{-}m\text{-}C_6H_4I)^{2+} + V^{2+}$	$\mu = 1,0$	25	−0,046	12,0	16,4	CSM	[441]
$Co(NH_3)_5(OOC\text{-}p\text{-}C_6H_4I)^{2+} + V^{2+}$	$\mu = 1,0$	25	−0,432	6,6	9,6	CSM	[441]

Table 114 (continued)

Reaction	Medium	t°	log k	log A	E	Method	Ref.
$Co(NH_3)_5(OOC\text{-}m\text{-}C_6H_4OH)^{2+} + V^{2+}$	$\mu = 1,0$	25	-0,194	7,1	9,9	CSM	
$Co(NH_3)_5(OOC\text{-}p\text{-}C_6H_4OH)^{2+} + V^{2+}$	$\mu = 1,0$	25	-0,276	7,0	9,9	CSM	[441]
$Co(NH_3)_5(OOC\text{-}p\text{-}C_6H_4CN)^{2+} + V^{2+}$	$\mu = 1,0$	25	-0,056	7,9	10,8	CSM	
$Co(NH_3)_5(OOC\text{-}CHBr\,COOCH_3)^{2+} + V^{2+}$	$\mu = 1,0$	25	-0,215			CSM	[145]
$Co(NH_3)_5[OOC(CH_2)_2COOCH_3]^{2+} + V^{2+}$	$\mu = 1,0$	25	0			CSM	[445]
$Co(NH_3)_5(OOCCHO)^{2+} + V^{2+}$	$\mu = 1,0$	25	0,916	7,6	11,6	CSM	
$Co(NH_3)_5(OOCCOCH_3)^{2+} + V^{2+}$	$\mu = 1,0$	25	1,008	10,0	12,2	CSM	[443]
$Co(NH_3)_5[OOCCOC(CH_3)_3]^{2+} + V^{2+}$	$\mu = 1,0$	25	0,324	8,9	12,5	CSM	
$Co(NH_3)_5(OOCONH_2)^{2+} + V^{2+}$	$\mu = 1,0$	25	1,316	7,7	12,2	CSM	
$Co(NH_3)_5$ 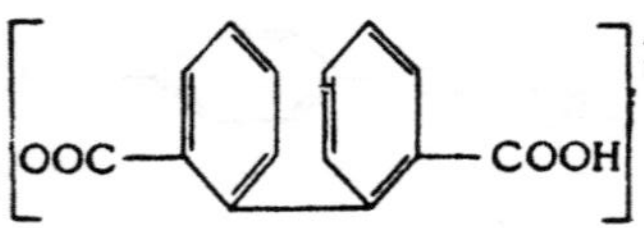 $^{2+}$ + V^{2+}	$\mu = 1,0$	21	-0,92 (k_1)			CSM	[444]
$k_1 + k_2[H^+]$	$\mu = 1,0$; $[H^+]$ = 0,2 to 0,6 M		-0,854 (k_2)				

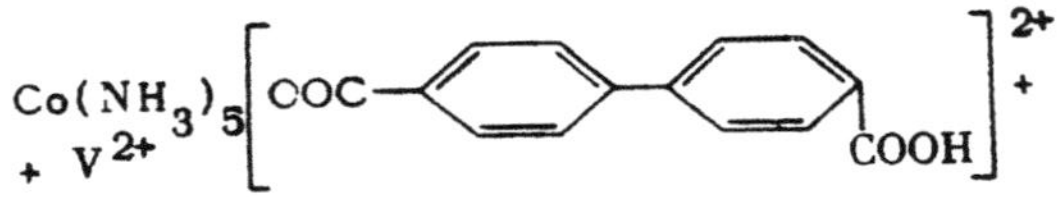 $k_1 + k_2[H^+]$	$\mu = 1,0$ $[H^+] =$ 0,2 to 0,6 M	21	−0,854 (k_1) −0,824 (k_2)	CSM	[444]
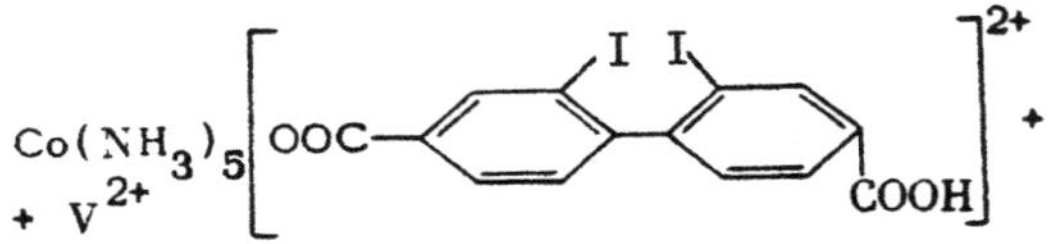	$\mu = 1,0$	25	−0,388	CSM	[444]
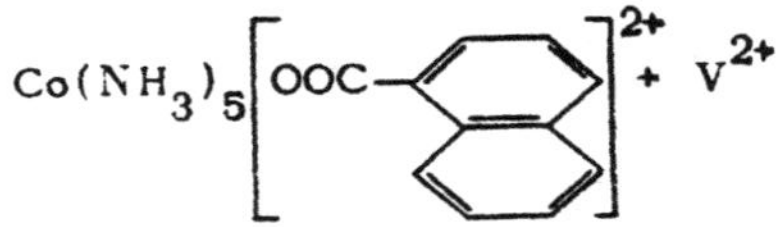	$\mu = 1,0$	21	−0,854	CSM	[444]
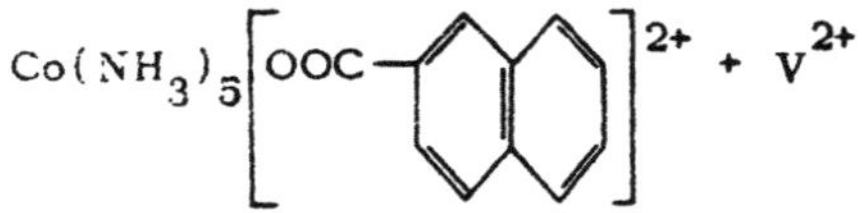	$\mu = 1,0$	21	−0,7	CSM	[444]
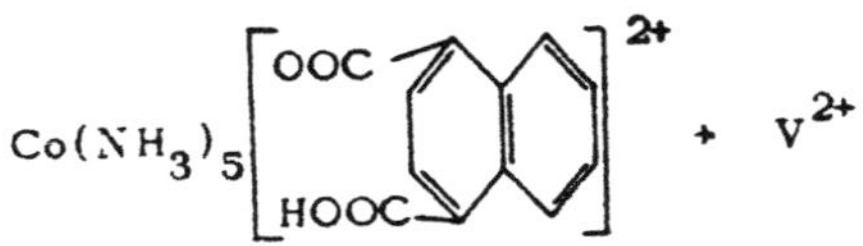 $k_1 + k_2[H^+]$	$\mu = 1,0$; $[H^+] =$ 0,2 to 0,6 M	21	−0,824 (k_1) −0,96 (k_2)	CSM	[444]

Table 114 (continued)

Reaction	Medium	t°	log k	log A	E	Method	Ref.
$Co(NH_3)_5$ 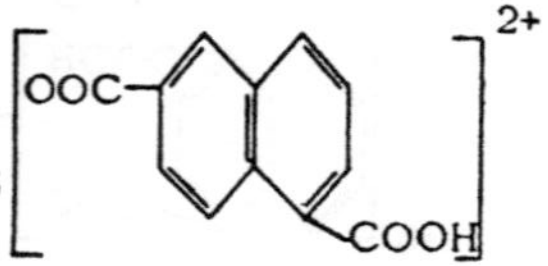 + V^{2+}	μ=1,0	21	-0,77			CSM	[444]
$Co(NH_3)_5$ 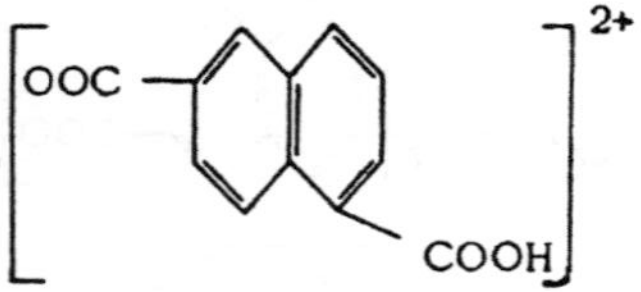 + V^{2+}	μ=1,0	21	-0,7			CSM	[444]
$(NH_3)_4CoNH_2Co(NH_3)_4^{5+} + V^{2+}$	μ=0,4	25	-0,154			CSM	
$(NH_3)_4CoNH_2Co(NH_3)_4^{5+} + V^{2+}$, $k[Cl^-]$	μ=0,4	25	0,320			CSM	
$(NH_3)_4CoNH_2Co(NH_3)_4^{5+} + V^{2+}$, $k[F^-]_2$	μ=0,4	25	3,834			CSM	
$(NH_3)_4CoNH_2Co(NH_3)_4^{5+} + V^{2+}$, $k[F^-]^2$	μ=0,4	25	5,85			CSM	[446]
$(NH_3)_4CoNH_2Co(NH_3)_4^{5+} + V^{2+}$ $k[SO_4^{2-}]$	μ=0,4	25	3,0			CSM	
$(NH_3)_4CoNH_2Co(NH_3)_4^{5+} + V^{2+}$ $k[SO_4^{2-}]^2$	μ=0,4	25	5,096			CSM	

$[(NH_3)_4Co(OH)_2Co(NH_3)_4]^{4+} + V^{2+}$	$\mu = 1,0$	25	−1,0			CSM	[447]
$Co(NH_3)_5Cl^{2+} + V(Dipy)_3^{2+}$	$\mu = 0,25$	25	−2,328			CSM	[411]
$Co(NH_3)_5(OOCCH_3)^{2+} + V(Dipy)^{2+}$	$\mu = 0,25$	25	−3,222			CSM	[411]
$Co(NH_3)_5(OOCCH{=}CHCH_3)^{2+} + V(Dipy)^{2+}$	$\mu = 0,25$	25	−3,0			CSM	[411]
$Co(EDTA)^- + V(Dipy)_3^{2+}$	$\mu = 1,0$	25	−1,265			CSM	[411]
$Co(Dipy)_3^{3+} + V(Dipy)_3^{2+}$	$\mu = 0,1$	25	0,946			CSM	[411]
$Co(Tripy)_2^{3+} + V(Tripy)_2^{2+}$	$\mu = 0,1$	25	2,42			CSM	[411]
Cobalt-chromium							
$Co^{3+} + Cr^{2+}$	1 M $HClO_4$	0	> 2,48			CSM	[434]
$Co(NH_3)_5NH_3^{3+} + Cr^{2+}$	$\mu = 0,4$	25–37	−4,05 (25°)	7,2	15,3	CSM	[435]
	$\mu = 0,4$	37	−3,745			CSM	[435]
	$\mu = 2,60$	25	−2,143			CSM	[448]
	$\mu = 0,4$; Cl^-	25	−2,57			CSM	[449]
$Co(NH_3)_5NH_3^{3+} + Cr^{2+}$, $k[Cl^-]$	$\mu = 0,4$; ClO_4^-; Cl^-	25–37	−0,82 (25°)	8,6	13,0	CSM	[435]

Table 114 (continued)

Reaction	Medium	t°	log k	log A	E	Method	Ref.
$Co(NH_3)_5NH_3^{3+} + Cr^{2+}$, $k[Cl^-]$	μ=0,4; ClO_4^-; Cl^-; D_2O	37	0,342			CSM	[435]
	μ=2,60; ClO_4^-; Cl^-	25	–0,222			CSM	[448]
$Co(NH_3)_5NH_3^{3+} + Cr^{2+}$, $k[Br^-]$	μ=2,60; ClO_4^-; Br^-	25	–0,578			CSM	[448]
$Co(NH_3)_5H_2O^{3+} + Cr^{2+}$	μ=1,1	25,0–10,3	–0,495 (25°)	5,8	8,7	CSM	[450]
	μ=1,2	10–20	–0,3 (k_1, 20°)	2,3	3,5	CSM	[451]
$k_1 + \frac{k_2}{[H^+]}$			0,195 (K_2, 20°)	11,1	14,8		
$Co(NH_3)_5D_2O^{3+} + Cr^{2+}$	μ=1,2; D_2O	20	–0,886			CSM	[451]
$Co(NH_3)_5OH^{2+} + Cr^{2+}$	μ=1	25	~5,52			CSM	[450]
	μ=1,2	10–20	6,176	8,4	5,2	CSM	[451]

$Co(NH_3)_5OD^{2+} + Cr^{2+}$	μ=1,2; H_2O	20	5,82	CSM	[451]
$Co(NH_3)_5F^{2+} + Cr^{2+}$	μ=1,0	25	$>$6,3	JM	[437]
	μ=0,1	25	5,4	JM	[452]
	μ=1,0	25	5,954	JM	[452]
$Co(NH_3)_5Cl^{2+} + Cr^{2+}$	μ=1,0	25	$>$6,3	JM	[437]
	μ=1,0	25	6,415	JM	[452]
	μ=0,1	25	5,778	JM	[452]
$Co(NH_3)_5Br^{2+} + Cr^{2+}$	μ=1,0	25	$>$6,3	JM	[437]
	μ=0,1	25	6,146	JM	[452]
$Co(NH_3)_5I^{2+} + Cr^{2+}$	μ=1,0	25	$>$6,3	JM	[437]
	μ=0,1	25	6,477	JM	[452]
$Co(NH_3)_5N_3^{2+} + Cr^{2+}$	μ=1,0	25	$\sim$5,48	JM	[437]
	μ=0,1	23	$\sim$1,23	CSM	[453]
cis-$Co(NH_3)_4(N_3)_2^{+} + Cr^{2+}$ *	-	0	$>$3	CSM	[390]
cis-$Co(NH_3)_4(N_3)_2^{+} + Cr^{2+}$ 2*	-	0	$>$3	CSM	[390]

* Bridge-activated complex with two N_3^- ions.
2* Bridge-activated complex with one N_3^- ion.

Table 114 (continued)

Reaction	Medium	t°	log k	log A	E	Method	Ref.
$Co(NH_3)_5NCS^{2+} + Cr^{2+}$	$\mu = 1,0$	15,2–35,1	1,278 (25°)	4,2	7,5	JM	[437]
$Co(NH_3)_5CN^{2+} + Cr^{2+}$	1 M $HClO_4$	15	1,785			CSM	[454]
$Co(NH_3)_5NO_2^{2+} + Cr^{2+}$	$\mu = 1,0$	25	2,44			CSM	[455]
$Co(NH_3)_5NO_3^{2+} + Cr^{2+}$	$\mu = 1,0$	25	~1,95			JM	[437]
$Co(NH_3)_5SO_4^{+} + Cr^{2+}$	$\mu = 1,0$	15–25–35	1,254	6,3	6,8	JM	[437]
	$\mu = 1,0$	9–30	1,362 (25°)	7,9	8,9	CSM	[439]
$Co(NH_3)_5PO_4 + Cr^{2+}$	$\mu = 1,0$	25	9,681			JM	[437]
$Co(NH_3)_5HPO_4^{+} + Cr^{2+}$	$\mu = 1,0$	25	3,918			JM	[437]
$Co(NH_3)_5H_2PO_4^{2+} + Cr^{2+}$	$\mu = 1,0$	25	–0,523			JM	[437]
$Co(NH_3)_5H_3PO_4^{3+} + Cr^{2+}$	$\mu = 1,0$	25	–0,523			JM	[437]
cis–$Co(NH_3)_4(H_2O)_2^{3+}$ $k_1 + \frac{k_2}{[H^+]}$	$\mu = 1,0$	25	0,844 (k_1) 0,924 (k_2)			CSM	[456]

$(NH_3)_4Co(\mu\text{-}OH)_2Co(NH_3)_4^{4+} + Cr^{2+}$	$\mu = 1$	25–45	–2,762 (25°)	4,1	9,3	CSM	[447]
$[(NH_3)_4Co(OH)_2]_3Co^{6+} + Cr^{2+}$	$\mu = 1$	25–45,2	–2,495 (25°)	4,6	9,7	CSM	[447]
$Co(NH_3)_5(OOCH)^{2+} + Cr^{2+}$	1 M $HClO_4$	25	~0,85			CSM	[457]
$Co(NH_3)_5(OOCCH_3)^{2+} + Cr^{2+}$	$\mu = 1{,}65$	25	–0,468			CSM	[448]
	$\mu = 0{,}1$	~ 23	–0,77			CSM	[453]
	$\mu = 1{,}0$	25,1–37,7	–0,746 (25°)	2,3	4,1	CSM	[458]
cis-$Co(NH_3)_4(H_2O)(OOCCH_3)^{2+} + Cr^{2+}$	$\mu = 1{,}0$	25	1,672 (k_1)			CSM	[456]
$k_1 + \frac{k_2}{[H^+]}$			0,447 (k_2)				
trans-$Co(NH_3)_4(OOCCH_3)_2^{+} + Cr^{2+}$	$\mu = 1{,}0$	25	1,176 (k_1)	7,4	8,5	CSM	[456]
$k_1 + k_2[H^+]$			1,7 (k_2)	7,1	7,4		

Table 114 (continued)

Reaction	Medium	t^o	log k	log A	E	Method	Ref.
cis-$Co(NH_3)_4(OOCCH_3)_2^+ + Cr^{2+}$	μ = 1,0	10	1,255 < <log k < <1,602			CSM	[442]
$Co(NH_3)_5(OOCCH_3)^{2+} + Cr^{2+}$, k[$Cl^-$]	μ = 1,65	25	−0,7			CSM	[448]
$Co(NH_3)_5(OOCCH_3)^{2+} + Cr^{2+}$, k[$Br^-$]	μ = 1,65	25	−1,194			CSM	[448]
$Co(NH_3)_5[OOCC(CH_3)_3]^{2+} + Cr^{2+}$	μ = 1,5	25	−2,057			CSM	[452]
$Co(NH_3)_5[OOCC(C_2H_5)_3]^{2+} + Cr^{2+}$	μ = 1,5	25	−2,62			CSM	[452]
$Co(NH_3)_5(OOCCH_2OCH_3)^{2+} + Cr^{2+}$	μ = 1,0	19,6–40	−0,246 (30°)	8,2	11,9	CSM	
$Co(NH_3)_5(OOCCH_2OC_2H_5)^{2+} + Cr^{2+}$	μ = 1,0	25	−0,388			CSM	[457]
$Co(NH_3)_5[OOCCH_2O$- 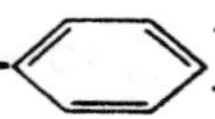 $]^{2+} + Cr^{2+}$	μ = 1,0	25	−0,83			CSM	
$Co(NH_3)_5(OOCCH_2Cl)^{2+} + Cr^{2+}$	μ = 1,0	25	−1,0	5,3	8,5	CSM	[441]
$Co(NH_3)_5(OOCCH_2CN)^{2+} + Cr^{2+}$	μ = 1,0	25	−0,96	2,4	4,6	CSM	[441]

$Co(NH_3)_5(OOCCHCl_2)^{2+} + Cr^{2+}$	$\mu = 1{,}0$	25	-1,13	1,2	3,1	CSM	[441]
$Co(NH_3)_5(OOCNH_2)^{2+} + Cr^{2+}$	$\mu = 1$	25	0,96 (k_1)			CSM	[459]
$k_1 + k_2[H^+]$			2,03 (k_2)				
$Co(NH_3)_5(OOCCH_2NH_2)^{2+} + Cr^{2+}$	$\mu = 1$	24	-0,276			CSM	[459]
	$\mu = 1$	15–25	-0,821 (25°)	8,6	12,8	CSM	[459]
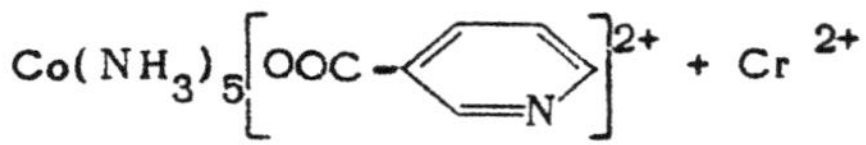	$\mu = 1{,}0$	17–30	-1,184 (17°)	2,8	5,3	CSM	[459]
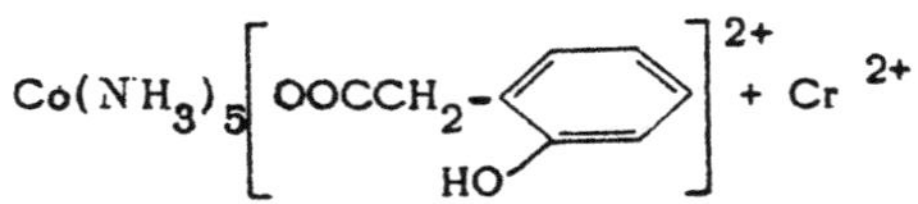	$\mu = 1{,}5$	25	-0,538			CSM	[460]
$Co(NH_3)_5(OOCCH_2OCH_2COOH)^{2+} + Cr^{2+}$	$\mu = 1{,}5$	25	-0,77			CSM	[460]
$Co(NH_3)_5(OOCCH_2OCOCH_3)^{2+} + Cr^{2+}$	$\mu = 1{,}5$	25	-0,77			CSM	[460]
$Co(NH_3)_5(OOCC_3H_7)^{2+} + Cr^{2+}$	0,1 M $HClO_4$	~23	-1,076			CSM	[453]
$Co(NH_3)_5(OOCCH_2CH_3)^{2+} + Cr^{2+}$	$\mu = 1{,}5$	25	-0,854			CSM	[461]
$Co(NH_3)_5(OOCC\equiv CH)^{2+} + Cr^{2+}$	$\mu = 1{,}5$	25	0,012			CSM	[460]

Table 114 (continued)

Reaction	Medium	t°	log k	log A	E	Method	Ref.
$Co(NH_3)_5(OOCCHO)^{2+} + Cr^{2+}$	$\mu = 1,0$	25	>3,844			JM	[443]
$Co(NH_3)_5(OOCCOCH_3)^{2+} + Cr^{2+}$	$\mu = 1,0$	8–25	4,04	8,7	6,4	JM	[443]
$Co(NH_3)_5[OOCCOC(CH_3)_3]^{2+} + Cr^{2+}$	$\mu = 1,0$	15–34	3,968 (25°)	6,5	8,5	JM	[443]
$Co(NH_3)_5(OOCCONH_2)^{2+} + Cr^{2+}$	$\mu = 1,0$	8–35	2,278 (25°)	9,2	9,4	CSM	[443]
$Co(NH_3)_5(OOCCH_2OH)^{2+} + Cr^{2+}$	$\mu = 1,0$	25	0,486	7,5	9,6	CSM	[457]
$Co(NH_3)_5(OOCCH_2SCH_2COOH)^{2+} + Cr^{2+}$	$\mu = 1,5$		0,954 (k_1)				
$k_1 + \frac{k_2}{[H^+]}$	$\mu = 1,5$; $[H^+] =$ =0,012 to 1,2	25	1,146 (k_2)			CSM	[461]
$Co(NH_3)_5(OOCCH_2SCH_2SCH_2COOH)^{2+} + Cr^{2+}$	$\mu = 1,5$	25	0,204			CSM	[461]
$Co(NH_3)_5(OOCCH_2SC_6H_5)^{2+} + Cr^{2+}$	$\mu = 1,5$	25	–0,042			CSM	[461]
$Co(NH_3)_5(OOCCH_2SCH_2C_6H_5)^{2+} + Cr^{2+}$	$\mu = 1,5$	25	0,716			CSM	[460]

$Co(NH_3)_5[OOCCH(OH)CH_3]^{2+} + Cr^{2+}$	$\mu = 1,0$	25	0,822			CSM	[457]
	$\mu = 1,0$; $[H^+] = 0,65$ *M*	24,8	−0,01			CSM	[441]
	$\mu = 1,0$; $[H^+] = 0,20$ *M*	24,8	0,342			CSM	[441]
$Co(NH_3)_5[OOCCH(OH)CH_2CH_3]^{2+} + Cr^{2+}$	$\mu = 1,0$; $[H^+] = 0,20$ *M*	14,7–36,2	1,06 (25°)	8,2	9,7	CSM	[457]
$Co(NH_3)_5[OOC{-}CH(OH){-}CH(OH){-}CH(OH){-}CH_2OH]^{2+} + Cr^{2+}$	$\mu = 1,5$	25	0,431			CSM	[460]
$Co(NH_3)_5(OOCCH{=}CHCH_3)^{2+} + Cr^{2+}$	0,1 *M* $HClO_4$	23	−0,658			CSM	[453]
$Co(NH_3)_5(OOCCH_2COOH)^{2+} + Cr^{2+}$	$\mu = 1,0$	25	−0,468 (k_1)			CSM	[462]
$k_1 + \frac{k_2}{[H^+]} + k_3[H^+]$			0,032 (k_2)				
			−0,027 (k_3)				
$Co(NH_3)_5(OOCCH_2COOH)^{2+} + Cr^{2+}$	$\mu = 1,0$	25	−0,538 (k_1)			CSM	[457]
$k_1 + \frac{k_2}{[H^+]} + k_3[H^+]$			0,093 (k_2)				
			0 (k_o)				

Table 114 (continued)

Reaction	Medium	t°	log k	log A	E	Method	Ref.
$Co(NH_3)_5[OOC(CH_3)_2COOH]^{2+} + Cr^{2+}$	$\mu = 1,0$	25	-1,155 (k_1)			CSM	[457]
$k_1 + \frac{k_2}{[H^+]}$			-1,3 (k_2)				
$Co(NH_3)_5[OOCCH(C_2H_5)COOH]^{2+} + Cr^{2+}$	$\mu = 1,0$	25–35,6	–0,745 (k_1,25°)	3,6	5,9	CSM	[457]
$k_1 + \frac{k_2}{[H^+]}$		14,7–35,6	–0,7 (k_2,25°)	8,3	12,3		
$Co(NH_3)_5[OOCCH(NH_2)(CH_2)_2SCH_3]^{2+} + Cr^{2+}$	$\mu = 1,5$	25	-1,96			CSM	[461]
$Co(NH_3)_5(OOCCOOH)^{2+} + Cr^{2+}$	$\mu = 1$; $[H^+] = 0,2$ to 0,6 M	25	2,654 (k_1)			CSM	[442]
$k_1 + k_2[H^+]$			3,0 (k_2)				
$Co(NH_3)_5(OOCCOO^-)^{+} + Cr^{2+}$	$\mu = 1,0$	8–25	4,662 (25°)	6,8	2,9	JM	[443]

$Co(NH_3)_5(OOCCOOH)^{2+} + Cr^{2+}$	$\mu = 1,0$	25	2,0			JM	[443]
$Co(NH_3)_5(OOCCHCOOH)^{2+} + Cr^{2+}$	$\mu = 1,0$	25	−0,004 (k_1)			CSM	[457]
$k_1 + \frac{k_2}{[H^+]}$			−0,338 (k_2)				
$Co(NH_3)_5(OOCCH_2COOCH_3)^{2+} + Cr^{2+}$	$\mu = 2$	7,3–35,5	−0,3 (25°)	7,8	10,7	CSM	[445]
$Co(NH_3)_5(OOCCH_2COOC_2H_5)^{2+} + Cr^{2+}$	$\mu = 2$	7,0–38,3	−0,174 (25°)	7,3	10,2	CSM	[445]
$Co(NH_3)_5[OOCCH(OH)COOH]^{2+} + Cr^{2+}$	$\mu = 1,0$	25	0,173 (k_1)			CSM	[457]
$k_1 + \frac{k_2}{[H^+]}$			−1,222 (k_2)				
$Co(NH_3)_5(OOCCH_2CH_2COOH)^{2+} + Cr^{2+}$	$\mu = 1$; $[H^+] = 0,2$ to $0,6$ M	25–40	1,475 (k_1, 25°)	7,7	8,5	CSM	[442]
$k_1 + k_2[H^+]$			1,874 (k_2, 25°)				
$Co(NH_3)_5(OOCCH_2CH_2COOH)^{2+} + Cr^{2+}$	$\mu = 1,0$	25	−0,57			CSM	[441]
	0,1 M $HClO_4$	23	−0,886			CSM	[453]

Table 114 (continued)

Reaction	Medium	t°	log k	log A	E	Method	Ref.
$Co(NH_3)_5(OOCCH_2CH_2COOCH_3)^{2+} + Cr^{2+}$	$\mu = 2$; $[H^+] = 1$	25	−0,783			CSM	[458]
	$\mu = 1$	25	−0,77			CSM	[458]
$Co(NH_3)_5(O_2CCH_2CHOHCOOH)^{2+} + Cr^{2+}$	$\mu = 1,0$	19	−0,77			CSM	[441]
$Co(NH_3)_5(O_2CCHOHCH_2COOH)^{2+} + Cr^{2+}$	$\mu = 1,0$; $[H^+] = 0,1$ to 0,65 M	19	−0,046 (k_1)			CSM	[441]
$k_1 + \frac{k_2}{[H^+]}$			−0,745 (k_2)				
$Co(NH_3)_5[OOCCH(OH)CH_2COOH]^{2+} + Cr^{2+}$	$\mu = 1,0$	25	0,43 (k_1) −0,523 (k_2)			CSM	[457]
$Co(NH_3)_5(OOCCHOHCH_2COOH)^{2+} + Cr^{2+}$	$\mu = 1,0$	19	−0,098 (k_1)			CSM	[440]
$k_1 + \frac{k_2}{[H^+]}$			−0,64 (k_2)				

$Co(NH_3)_3(OOCCH_2CHOHCOOH)^{2+} + Cr^{2+}$	$\mu \approx 1,0$	25	$-0,444\ (k_1)$	CSM	[457]
$k_1 + \frac{k_2}{[H^+]}$			$-1,523\ (k_2)$		
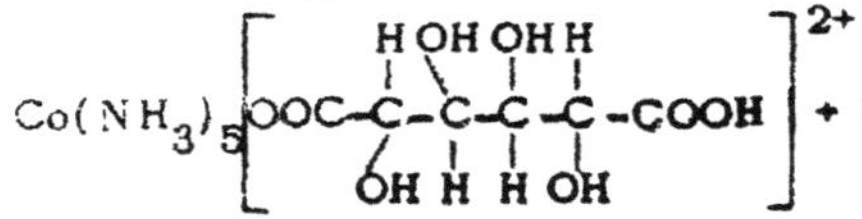 $+ Cr^{2+}$	$\mu = 1,5$	25	~ 0	CSM	[461]
$Co(NH_3)_5[OOC(CH_2)_2S(CH_2)_2COOH]^{2+} + Cr^{2+}$	$\mu = 1,5$	25	1,04	CSM	[461]
$Co(NH_3)_5[OOC(CH_2)_8COOH]^{2+} + Cr^{2+}$	$\mu = 3,0$	25	$-0,553$	CSM	[463]
$Co(NH_3)_5[OOC(CH_2)_2COCH_3]^{2+} + Cr^{2+}$	$\mu = 3,0$	25	$-0,456\ (k_1)$	CSM	[463]
$k_1 + k_2[H^+]$	$[H^+] = 0,16$ to $2,90\ M$		$0\ (k_2)$		
$Co(NH_3)_5(cis\text{-}OOCCH{=}CHCOOH)^{2+} + Cr^{2+}$	$\mu = 1$	25	$1,3\ (k_1)$	CSM	[442]
$k_1 + \frac{k_2}{[H^+]}$	$[H^+] = 0,2$ to $0,6\ M$		$2,0\ (k_2)$		
$Co(NH_3)_5(cis\text{-}OOCCH{=}CHCOOH)^{2+} + Cr^{2+}$	$0,1\ M\ HClO_4$	~ 23	$> 1,4$	CSM	[453]
	$\mu = 1,5$	25	2,3	CSM	[461]
$Co(NH_3)_5(cis\text{-}OOCCH{=}CClCOOH)^{2+} + Cr^{2+}$	$\mu = 1$	24,9	$1,23\ (k_1)$	CSM	[441]
$k_1 + k_2[H^+]$	$[H^+] = 0,1$ to $0,6\ M$		$1,3\ (k_2)$		

Table 114 (continued)

Reaction	Medium	t°	log k	log A	E	Method	Ref.
$Co(NH_3)_5-(cis-OOCCCl=CHCOOH)^{2+} + Cr^{2+}$	μ = 1	24,9	−0,92 (k_1)			CSM	[441]
$k_1 + \frac{k_2}{[H^+]}$	$[H^+]$ = 0,1 to 0,6 M		−1,155 (k_2)				
$Co(NH_3)_5-(cis-OOCCH=CHCOOCH_3)^{2+} + Cr^{2+}$	μ = 1	23	> 2			CSM	[458]
$Co(NH_3)_4(OOCCH=CHCOO)^{+} + Cr^{2+}$	μ = 1	25,13–24	2,093	5,0	3,9	CSM	[439]
$k_1 + k_2 [H^+]$	μ = 1; $[H^+]$ = 0,2 to 0,6 M	25	1,7			CSM	[442]
$Co(NH_3)_5-(trans-OOCCH=CCOOH)^{2+} + Cr^{2+}$ $k_1 + k_2 [H^+]$	μ = 1	25	0,04 (k_1) 0,556 (k_2)			CSM	[464]

$Co(NH_3)_5trans\text{-}OOCCH{=}CHCOOH)^{2+} + Cr^{2+}$	$\mu = 1$; D_2O	1,9	−0,398 (k_1)			CSM	[458]
$k_1 + k_2[H^+]$			0,39 (k_2)				
$Co(NH_3)_5(trans\text{-}OOCCH{=}CHCOOH)^{2+} + Cr^{2+}$	$\mu = 1$	1,9–25,1	0,12 (k_1, 25°)	5,5	7,3	CSM	[458]
$k_1 + \frac{k_2 k_D}{[H^+]} + k_3[H^+]$		5,4	0,04 (k_2)				
		1,9–25,1	0,542 (k_3, 25°)	3,1	3,5	CSM	[458]
$Co(NH_3)_5(trans\text{-}OOCCH{=}CHCOOH)^{2+} + Cr^{2+}$	$\mu = 3,4$; ClO_4^-; Cl^-; pH 1,73	25	0 (k_1)			CSM	[448]
$k_1 + k_2[Cl^-]$ *			−0,16 (k_2)				
$Co(NH_3)_5(trans\text{-}OOCCH{=}CHCOOH)^{2+} + Cr^{2+}$	$\mu = 3.4$; ClO_4^-; Br^-; pH 1,73	25	0 (k_1)			CSM	[448]
$k_1 + k_2[Br^-]$ 2*			−0,523 (k_2)				

* k_1 depends on pH. At pH 1.73, contribution of pH-dependent part is minimal.

2* In these two cases k_1 depends on pH. At pH 1.73, contribution of pH-dependent part is minimal.

Table 114 (continued)

Reaction	Medium	t^o	log k	log A	E	Method	Ref.
$Co(NH_3)_5(trans\text{-}OOCCH{=}CHCOOCH_3)^{2+} + Cr^{2+}$	$\mu = 1$	22	0,053 (k_1)			CSM	[464]
$k_1 + k_2[H^+]$			0,802 (k_2)				
$Co(NH_3)_5(trans\text{-}OOCCH{=}CHCOOCH_3)^{2+} + Cr^{2+}$	$\mu = 1$	2,5	−0,367 (k_1)			CSM	[458]
$k_1 + k_2[H^+]$			0,107 (k_2)				
$Co(NH_3)_5(trans\text{-}OOCCH{=}CHCOOC_6H_5)^{2+} + Cr^{2+}$		22	0,14 (k_1)			CSM	[464]
$k_1 + k_2[H^+]$			−0,222 (k_2)				
$Co(NH_3)_5(trans\text{-}OOCCH{=}CHCOOC_6H_5)^{2+} + Cr^{2+}$	$\mu = 1$	9–22	0,162 (25°)	3,8	4,9	CSM	[465]
$Co(NH_3)_5[OOCCH{=}C(C1)COOH]^{2+} + Cr^{2+}$	$\mu = 1$	20,2	0,096 (k_1)			CSM	[441]
$k_1 + k_2[H^+]$			0,08 (k_2)				
$Co(NH_3)_5(trans\text{-}OOCCC1{=}CHCOOH)^{2+} + Cr^{2+}$	$\mu = 1,0$	20,2–35,0	−1,046 (k_1, 25°)	4,1	7,0	CSM	[441]

$k_1 + k_2[H^+] + \frac{k_3}{[H^+]}$		25	$-0{,}3\ (k_2)$ $-1{,}098\ (k_3)$				
$Co(NH_3)_5(trans\text{-}OOCCH{=}CBr\,COOH)^{2+} + Cr^{2+}$	$\mu = 1{,}0$	25–35	$-0{,}156$ (k_1; 25°)	9,0	12,5	CSM	[441]
$k_1 + \frac{k_2}{[H^+]}$		25	$-1{,}098\ (k_2)$				
$Co(NH_3)_5(trans\text{-}OOCCBr{=}CHCOOH)^{2+} + Cr^{2+}$	$\mu = 1{,}0$	25–35	$-1{,}0\ (k_1)$	7,5	11,6	CSM	[441]
$k_1 + \frac{k_2}{[H^+]}$		25	$-1{,}4\ (k_2)$				
$Co(NH_3)_5(trans\text{-}OOCCH{=}CI\,COOH)^{2+} + Cr^{2+}$	$\mu = 1{,}0$	16,5–33,5	$-0{,}098$ (k_1, 25°)	4,8	6,6	CSM	[441]
$k_1 + \frac{k_2}{[H^+]}$		25	$-1{,}7\ (k_2)$				
$Co(NH_3)_5(trans\text{-}OOCCI{=}CHCOOH)^{2+} + Cr^{2+}$	$\mu = 1{,}0$	25	$-1{,}0\ (k_1)$			CSM	[441]
$k_1 + \frac{k_2}{[H^+]}$			$-1{,}347\ (k_2)$				
$Co(NH_3)_5(trans\text{-}OOCCCl{=}CClCOOH)^{2+} + Cr^{2+}$	$\mu = 1{,}0$	25	$-0{,}722\ (k_1)$			CSM	[441]
$k_1 + k_2[H^+] + (k_3/[H^+])$			$0{,}046\ (k_2)$ $-0{,}886\ (k_3)$				

Table 114 (continued)

Reaction	Medium	t^{o}	log k	log A	E	Method	Ref.
$Co(NH_3)_5(trans\text{-}OOCCBr{=}CBr\,COOH)^{2+} + Cr^{2+}$	$\mu = 1{,}0$	15–24	−0,854 (25°)	6,3	9,7	CSM	[441]
$Co(NH_3)_5(trans\text{-}OOCCI{=}CI\,COOH)^{2+} + Cr^{2+}$	$\mu = 1{,}0$	15–24	−1,0 (25°)	4,2	7,1	CSM	[441]
$Co(NH_3)_5[OOCCH{=}CHGH{=}CH(CH_3)]^{2+} + Cr^{2+}$	$\mu = 1{,}5$	25	−0,284			CSM	[461]
$Co(NH_3)_5(OOCCH{=}CHCH{=}CCOOH)^{2+} + Cr^{2+}$ $k_1 + k_2[H^+]$	$\mu = 1{,}5$ $[H^+] = 0{,}06\text{–}0{,}4$ M	25	0,342 (k_1) 1,176 (k_2)			CSM	[460]
$Co(NH_3)_5(trans\text{-}OOCCH_2CH{=}CHCH_2COOH)^{2+} + Cr^{2+}$	$\mu = 1{,}3$; $[H^+] = 1{,}2$ M	22	−0,7			CSM	[461]
$Co(NH_3)_5[OOC\text{-}CH<(CH_2)_2]^{2+} + Cr^{2+}$ (cyclopropane ring)	$\mu = 1{,}5$	25	−0,722			CSM	[460]
	$\mu = 1{,}0$	25–35	−0,904 (25°)	4,9	7,9	CSM	[459]

$Co(NH_3)_5[OOC\overline{CH(CH_2)_2CH_2}]^{2+} + Cr^{2+}$	$\mu = 1{,}0$	25–35	−1,3 (25°)	6,7	10,9	CSM	[459]
$Co(NH_3)_5[OOC\overline{CH(CH_2)_3CH_2}]^{2+} + Cr^{2+}$	$\mu = 1{,}0$	25–35	−1,143 (25°)	6,1	9,8	CSM	[459]
$Co(NH_3)_5[OOC\overline{CH(CH_2)_4CH_2}]^{2+} + Cr^{2+}$	$\mu = 1{,}0$	25–35	−1,6 (25°)	8,0	13,0	CSM	[459]
$Co(NH_3)_5[OOC\overline{CH(CH_2)_2CH_2}]^{2+} + Cr^{2+}$ $k_1 + \frac{k_2}{[H^+]}$	$\mu = 1{,}5$ $[H^+]$ = 0,15 to 0,9 **M**	25	−0,7 (k_1) −1,032 (k_2)			CSM	[460]
$Co(NH_3)_5$(trans-OOC-CH—CH-C_6H_5, with CH and CH bridged by CH_2)$^{2+}$ + Cr^{2+}	$\mu = 1{,}5$	25	−0,585			CSM	[461]
$Co(NH_3)_5$(cis-OOC-CH—CH-COOH, with CH and CH bridged by CH_2)$^{2+}$ + Cr^{2+}	$\mu = 1$	8–30,5	−0,786 (25°)	5,6	8,7	CSM	[465]
$Co(NH_3)_5$(trans-OOC-CH—CH-COOH, with CH and CH bridged by CH_2)$^{2+}$ + Cr^{2+}	$\mu = 1$	8–25	−0,824 (25°)	4,4	7,1	CSM	[465]

Table 114 (continued)

Reaction	Medium	t°	log k	log A	E	Method	Ref.
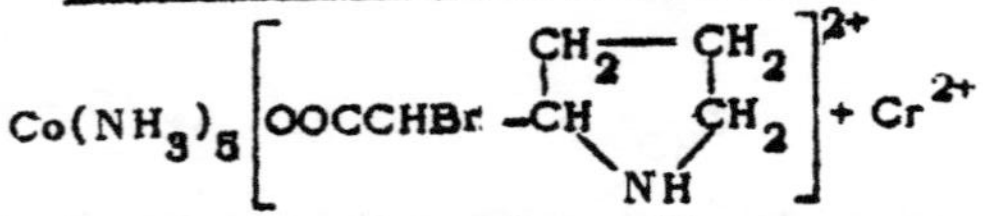	$\mu = 3{,}0$; $[H^+] = 0{,}16$ to $2{,}90\,M$	25	−1,57			CSM	[463]
$Co(NH_3)_5(OOCC_6H_5)^{2+} + Cr^{2+}$	$\mu = 1{,}0$	25	−0,854	3,2	5,5	CSM	[441]
$Co(NH_3)_5(OOC\text{-}o\text{-}FC_6H_4)^{2+} + Cr^{2+}$	$\mu = 3$	25	−0,824			CSM	[463]
$Co(NH_3)_5(OOC\text{-}o\text{-}ClC_6H_4)^{2+} + Cr^{2+}$	$\mu = 3$	25	−0,96			CSM	[463]
	$\mu = 1{,}0$	25	−1,13	3,7	6,6	CSM	[441]
$Co(NH_3)_5(OOC\text{-}p\text{-}C_6H_4Cl)^{2+} + Cr^{2+}$	$\mu = 1{,}0$	25	−0,678	7,1	10,6	CSM	[441]
$Co(NH_3)_5(OOC\text{-}o\text{-}BrC_6H_4)^{2+} + Cr^{2+}$	$\mu = 3{,}0$	25	−1,018			CSM	[463]
$Co(NH_3)_5(OOC\text{-}o\text{-}IC_6H_4)^{2+} + Cr^{2+}$	$\mu = 3{,}0$	25	−1,024			CSM	[463]
	$\mu = 1{,}0$	25	−1,086	1,4	3,4	CSM	[441]
$Co(NH_3)_5(OOC\text{-}o\text{-}HOC_6H_4)^{2+} + Cr^{2+}$	$\mu = 3{,}0$	25	−0,824			CSM	[463]
	$\mu = 1$; $[H^+] = 0{,}10\,M$	17	−0,824			CSM	[441]

$Co(NH_3)_5(OOC\text{-}o\text{-}HOC_6H_4)^{2+} + Cr^{2+}$	$\mu = 1$; $[H^+] = 0{,}15\ M$	17	−0,92			CSM	[441]
	$\mu = 1$; $[H^+] = 0{,}20\ M$	17	−1,046			CSM	[441]
	$\mu = 1$; $[H^+] = 0{,}60\ M$	17	−1,098			CSM	[441]
$Co(NH_3)_5(OOC\text{-}o\text{-}C_6H_4O)^{-} + Cr^{2+}$	$\mu = 3{,}0$	25	8,3			CSM	[463]
$Co(NH_3)_5(OOC\text{-}m\text{-}C_6H_4OH)^{2+} + Cr^{2+}$	$\mu = 3{,}0$	25	−0,7			CSM	[463]
$Co(NH_3)_5(OOC\text{-}p\text{-}C_6H_4OH)^{2+} + Cr^{2+}$	$\mu = 3{,}0$	25	−0,638			CSM	[463]
	$\mu = 1{,}0$	25	−0,886	6,6	10,2	CSM	[441]
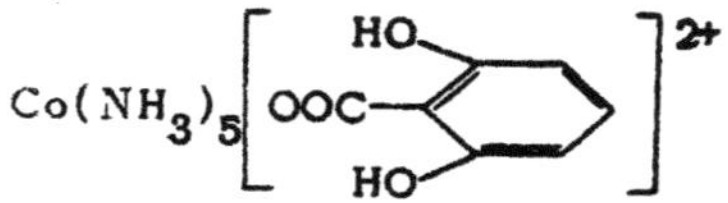 + Cr^{2+}	$\mu = 3{,}0$	25	1,024			CSM	[463]
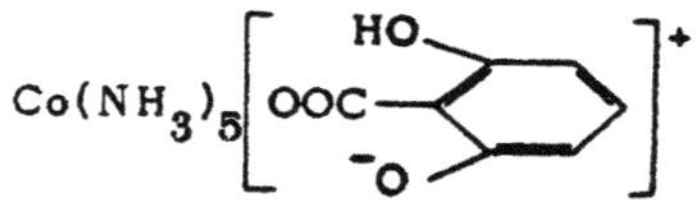 + Cr^{2+}	$\mu = 3{,}0$	25	5,7			CSM	[463]

Table 114 (continued)

Reaction	Medium	t °	log k	log A	E	Method	Ref.
$Co(NH_3)_5$ 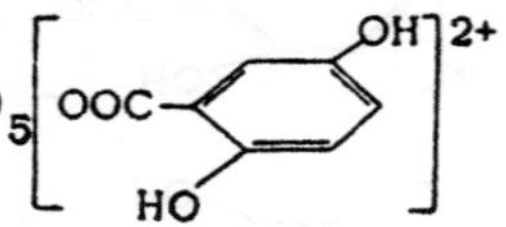 + Cr^{2+}	μ = 3,0	25	-0,538			CSM	[463]
$Co(NH_3)_5$ 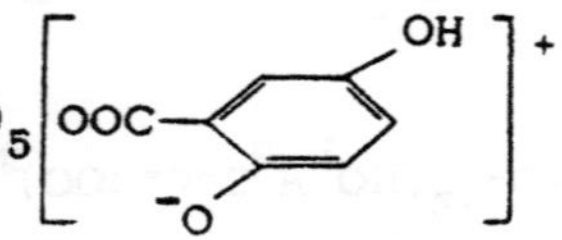 + Cr^{2+}	μ = 3,0	25	~6			CSM	[463]
$Co(NH_3)_5(OOC\text{-}p\text{-}C_6H_4OH)^{2+} + Cr^{2+}$	μ = 3,0	25	-0,886			CSM	[463]
$Co(NH_3)_5(OOC\text{-}m\text{-}C_6H_4OH)^{2+} + Cr^{2+}$	μ = 3,0	25	6,477			CSM	[463]
$Co(NH_3)_5(OOC\text{-}o\text{-}C_6H_4CHO)^{2+} + Cr^{2+}$	μ = 1,5	25	1,662			CSM	[460]
$Co(NH_3)_5(OOC\text{-}p\text{-}C_6H_4CHO)^{2+} + Cr^{2+}$ $k_1 + k_2[H^+]$	μ = 3,0 $[H^+]$ = = 0,16 to 0,90 M	25	1,462 (k_1) 2,732 (k_2)			CSM	[463]

$Co(NH_3)_5(OOC\text{-}p\text{-}C_6H_4CHO)^{2+} + Cr^{2+}$ $k_1 + k_2[H^+]$	$\mu = 1,5$ $[H^+] = 0,024$ to $0,36M$		1,8 (k_1) 2,6 (k_2)			CSM	[460]
$Co(NH_3)_5$ 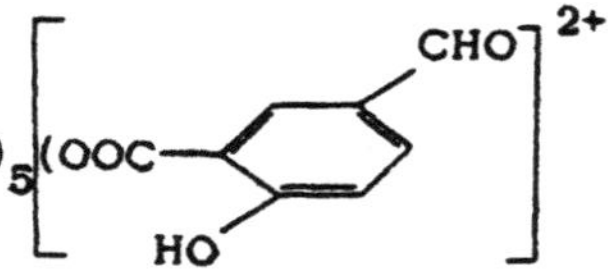 $+ Cr^{2+}$	$\mu = 3,0$	25	−0,26			CSM	[463]
$Co(NH_3)_5$ 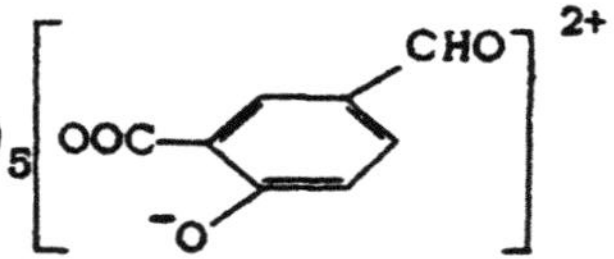 $+ Cr^{2+}$	$\mu = 3,0$	25	~9			CSM	[463]
$Co(NH_3)_5(OOC\text{-}o\text{-}C_6H_4OI)^{2+} + Cr^{2+}$	$\mu = 1,5$	25	−1,026			CSM	[460]
$Co(NH_3)_5(OOC\text{-}p\text{-}C_6H_4CN)^{2+} + Cr^{2+}$	$\mu = 1,0$	25	−0,745	5,2	8,1	CSM	[441]
$Co(NH_3)_5(OOC\text{-}p\text{-}C_6H_4SO_3H)^{2+} + Cr^{2+}$	$\mu = 1,0$	8–27	−0,796 (25°)	5,7	8,9	CSM	[465]
$Co(NH_3)_5(OOC\text{-}o\text{-}C_6H_4OCH_3)^{2+} + Cr^{2+}$	$\mu = 3,0$	25	−0,553			CSM	[463]
$Co(NH_3)_5(OOC\text{-}m\text{-}C_6H_4OCH_3)^{2+} + Cr^{2+}$	$\mu = 3,0$	25	−0,658			CSM	[463]
$Co(NH_3)_5(OOC\text{-}p\text{-}C_6H_4OCH_3)^{2+} + Cr^{2+}$	$\mu = 3,0$	25	−0,722			CSM	[463]
$Co(NH_3)_5(OOC\text{-}p\text{-}C_6H_4NH_3)^{3+} + Cr^{2+}$	$\mu = 3,0$	25	−0,854			CSM	[463]

Table 114 (continued)

Reaction	Medium	t°	log k	log A	E	Method	Ref.
$Co(NH_3)_5(OOC\text{-m-}C_6H_4NH_3)^{3+} + Cr^{2+}$	$\mu = 3{,}0$	25	−0,92			CSM	[480]
$Co(NH_3)_5(OOC\text{-o-}C_6H_4NH_2)^{2+} + Cr^{2+}$	$\mu = 1{,}5$	25	−1,28			CSM	[480]
$Co(NH_3)_5(OOC\text{-p-}C_6H_4N(CH_3)_2)^{2+} + Cr^{2+}$	$\mu = 1{,}5$	25	−0,77			CSM	[480]
$Co(NH_3)_5(OOC\text{-o-}C_6H_4NHC_6H_5)^{2+} + Cr^{2+}$	$\mu = 1{,}5$	25	−0,77			CSM	[480]
$Co(NH_3)_5$ 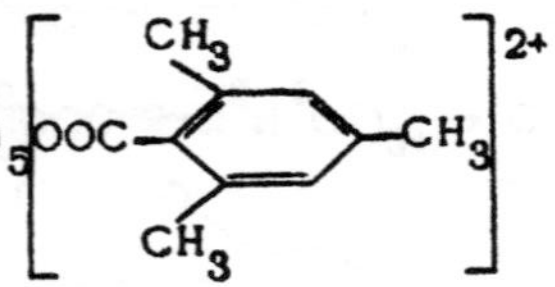 $+ Cr^{2+}$	$\mu = 1{,}5$	25	−1,347			CSM	[481]
$Co(NH_3)_5(OOC\text{-o-}C_6H_4COOH)^{2+} + Cr^{2+}$	$\mu = 1{,}0$	25	−1,125	3,1	5,7	CSM	[441]
	$\mu = 1$; $[H^+] = 0{,}2$ to $0{,}6$ M	25,6–35,6 25,6–	1,274	2,2	4,7	CSM	[466]
$Co(NH_3)_5(OOC\text{-m-}C_6H_4COOH)^{2+} + Cr^{2+}$	$\mu = 1{,}0$	25	−1,032	1,3	3,2	CSM	[441]
	$\mu = 1{,}0$	14,2–38,2	−0,04 (14,2°)	1,9	2,7	CSM	[458]
$Co(NH_3)_5(OOC\text{-o-}C_6H_4COOH)^{2+} + Cr^{2+}$	$\mu = 1{,}0$	14,2–25,1	−1,26 (k_1, 14,2°)	2,0	4,4	CSM	[458]
$k_1 + k_2\frac{k_D}{[H^+]}$			−0,15 (k_2, 14,2°)	9,7	13,4		

$Co(NH_3)_5(OOC\text{–}o\text{–}C_6H_4COOCH_3)^{2+} + Cr^{2+}$	$\mu = 1,0$	25,6–35,6	–1,276 (k_1, 25,6°)	5,07	8,65	CSM	[466]
$k_1 + k_2[H^+]$	$[H^+] =$ =0,2 to 0,6 M	25,6	–1,824 (k_2)				
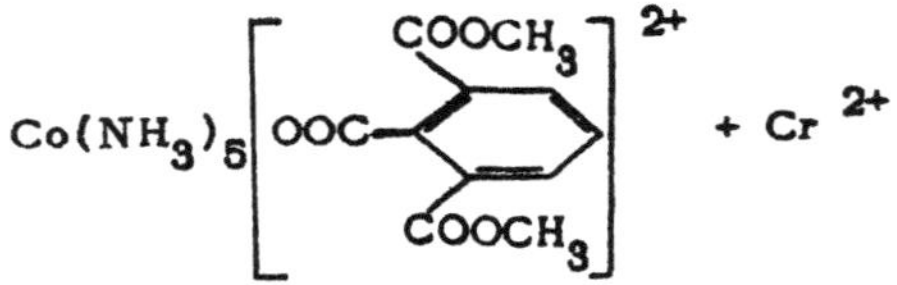	$\mu = 1,0$; $[H^+] =$ =0,2 to 0,6M	25,6–35,6	–1,4 (k_1)	5,7	9,7	CSM	[466]
$k_1 + k_2[H^+]$		25,6	–1,6 (k_2)				
$Co(NH_3)_5(OOC\text{–}p\text{-}C_6H_4COOH)^{2+} + Cr^{2+}$ $k_1 + k_2[H^+]$	$\mu = 1$	16,6	1,556 (k_1) 0,954 (k_2)			CSM	[458]
$Co(NH_3)_5(OOC\text{–}p\text{–}C_6H_4COOH)^{2+} + Cr^{2+}$	$\mu = 1,5$	25	–0,7			CSM	[460]
$Co(NH_3)_4(OOC\text{–}o\text{–}C_6H_4COO)^{+} + Cr^{2+}$	$\mu = 1,0$	10–25	1,114 (k_1, 25°)	5,7	6,3	CSM	[442]
$k_1 + k_2[H^+]$	$[H^+] =$ =0,2 to 0,6M		1,602 (k_2, 25°)				
$Co(NH_3)_5(OOC\text{–}p\text{-}C_6H_4COOCH_3)^{2+} + Cr^{2+}$	–	9–25	–0,77 (25°)	3,4	5,7	CSM	[465]

Table 114 (continued)

Reaction	Medium	t °	log k	log A	E	Method	Ref.
$Co(NH_3)_5(OOC{-}p{-}C_6H_4COOC_6H_5)^{2+} + Cr^{2+}$	-	9–27	–1,3 (25°)	–0,75	0,75	CSM	[465]
$Co(NH_3)_5(OOC{-}p{-}C_6H_4COOCH_2{-}p{-}C_6H_4OH)^{2+} + Cr^{2+}$	μ= 1,0	25	–1,4 (k_1)			CSM	[464]
$k_1 + k_2[H^+]$			–0,722 (k_2)				
$Co(NH_3)_5$ 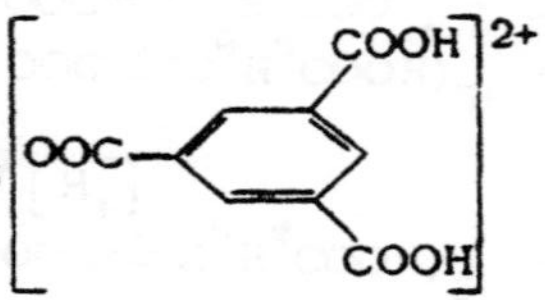 $+ Cr^{2+}$	μ = 3	25	–0,638			CSM	[463]
$Co(NH_3)_5$ 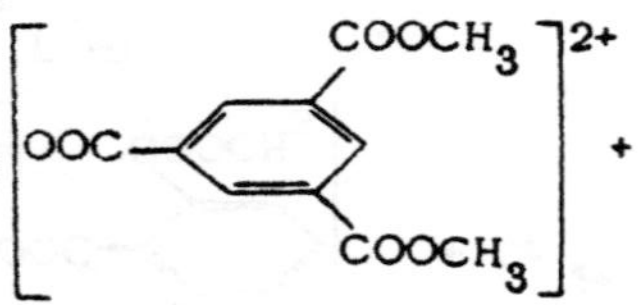 $+ Cr^{2+}$	μ = 1,0	25,6	–1,4 (k_1)			CSM	[466]
$k_1 + k_2[H^+]$	$[H^+]$ = =0,2 to 0,6 M		–1,6 (k_2)				

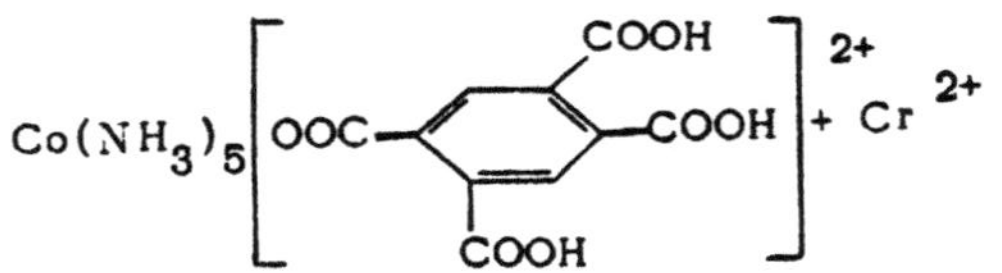	$\mu = 3$	25	−0,42	CSM	[463]
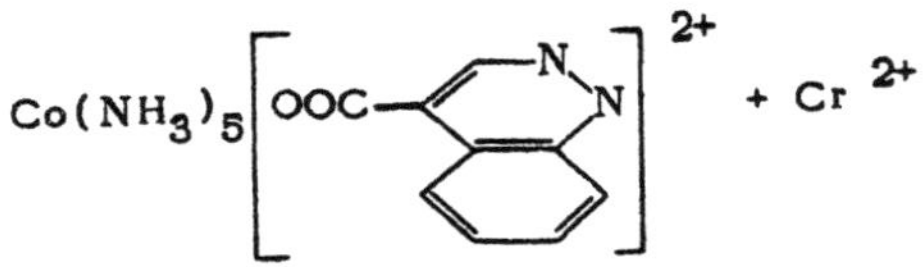	$\mu = 1,5$	25	$>3,04$	CSM	[460]
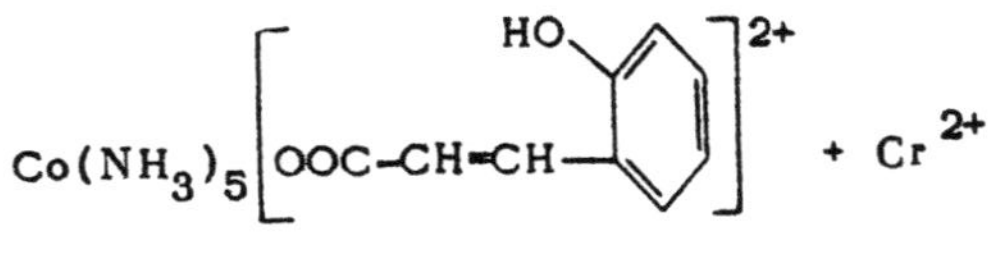	$\mu = 1,5$; $[H^+] = 0,05$ to $1,2M$	25	−0,41 (k_1)	CSM	[460]
$k_1 + k_2/[H^+]$			−1,854 (k_2)		
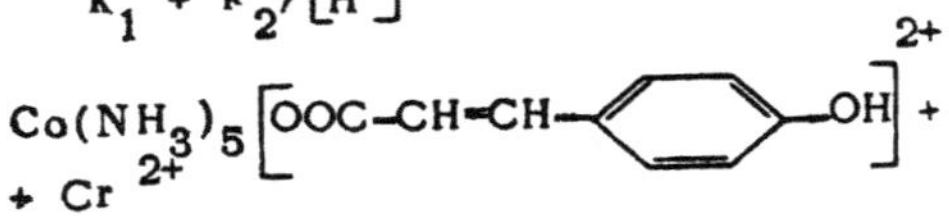	$\mu = 1,5$	25	−0,04	CSM	[460]
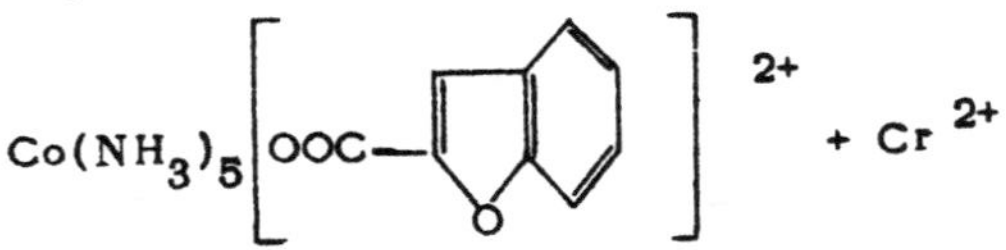	$\mu = 1,5$	25	−0,377	CSM	[461]

Table 114 (continued)

Reaction	Medium	t°	log k	log A	E	Method	Ref.
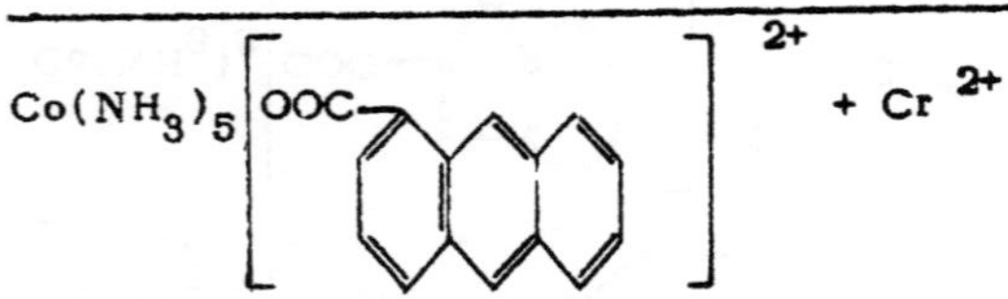	μ = 3,0	25	0,7			CSM	[463]
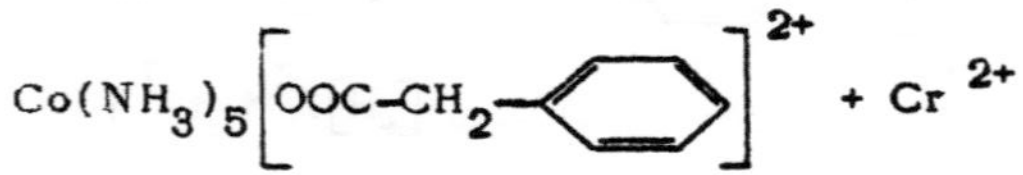	μ = 1,0	14,7–36,2	0,284 (25°)	7,9	10,3	CSM	[457]
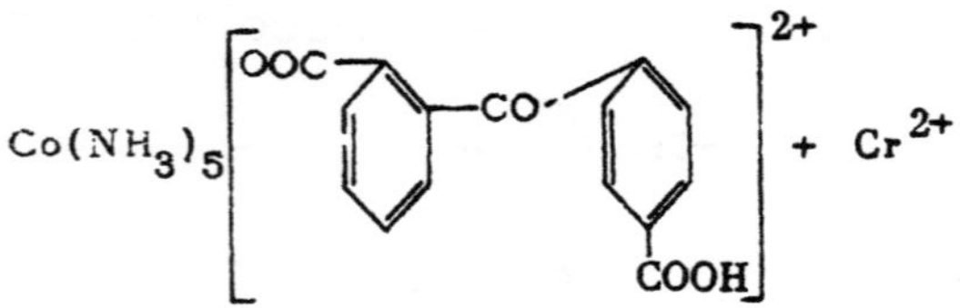	μ = 1,5	25	1,568			CSM	[460]
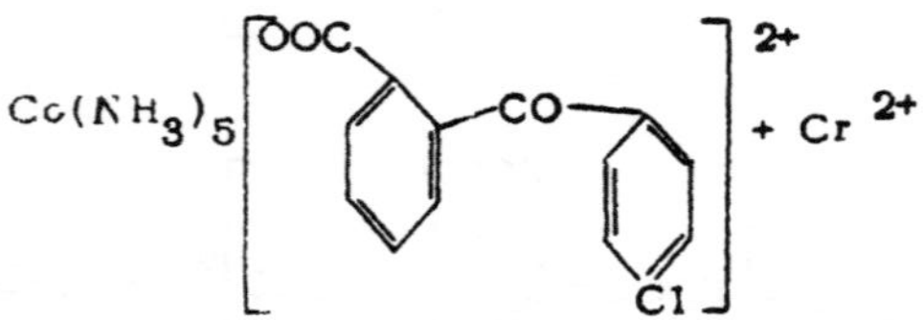	μ = 1,5	25	0,446			CSM	[460]

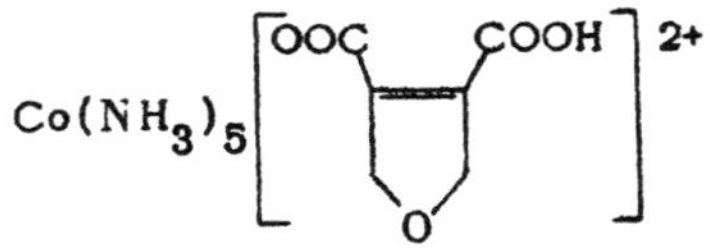 + Cr^{2+}	μ = 3; $[H^+]$ = 0,16 to 2,90M	25	−0,796	CSM	[463]
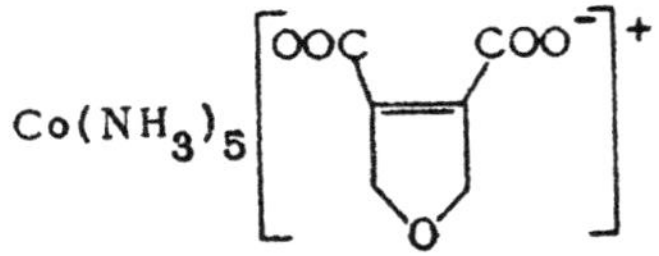 + Cr^{2+}	- " -	25	~1	CSM	[463]
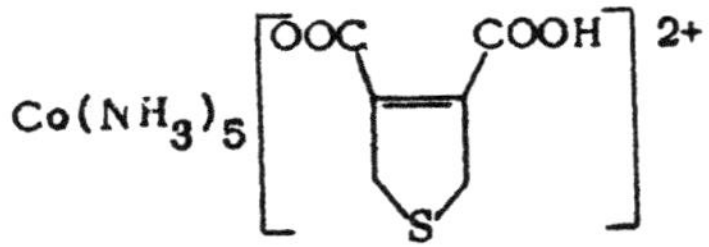 + Cr^{2+}	- " -	25	−1,126	CSM	[463]
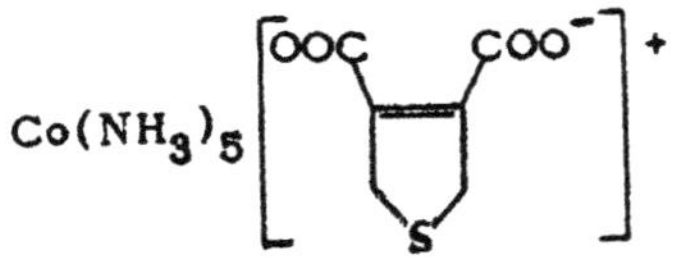 + Cr^{2+}	- " -	25	0,954	CSM	[463]
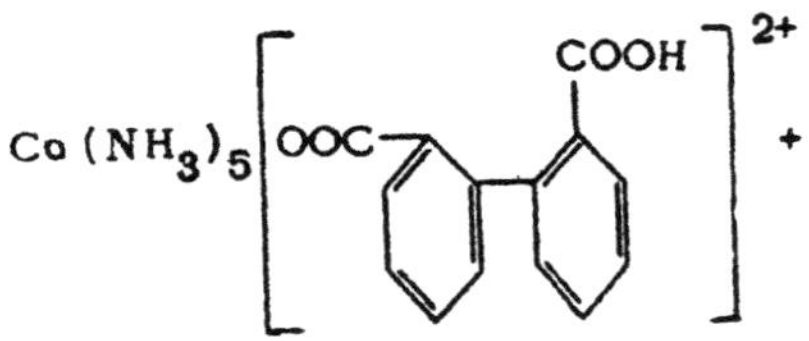 + Cr^{2+}	- " -	25	−1,108	CSM	[463]

Table 114 (continued)

Reaction	Medium	t°	log k	log A	E	Method	Ref.
$Co(NH_3)_5$ 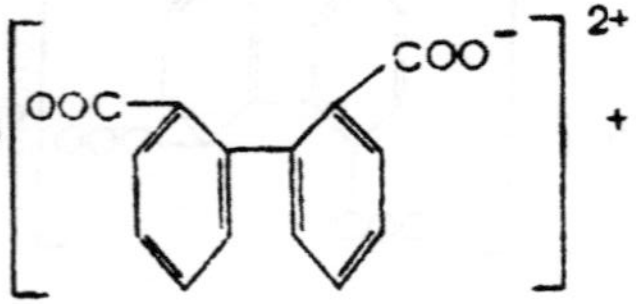 $+ Cr^{2+}$	$\mu = 3$; $[H^+] = 0,16$ to 0,90 M	25	0,322			CSM	[463]
$Co(NH_3)_5$ 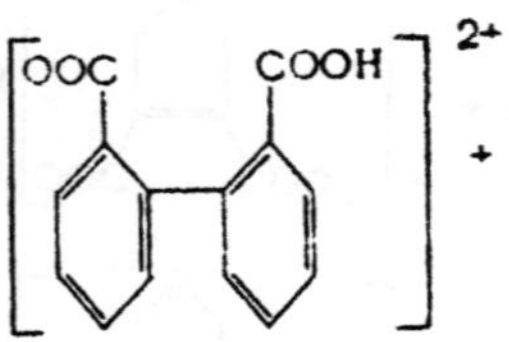 $+ Cr^{2+}$	$\mu = 1$	21	$-1,0$ (k_1)			CSM	[444]
$k_1 + k_2[H^+]$			$-0,432$ (k_2)				
$Co(NH_3)_5$ 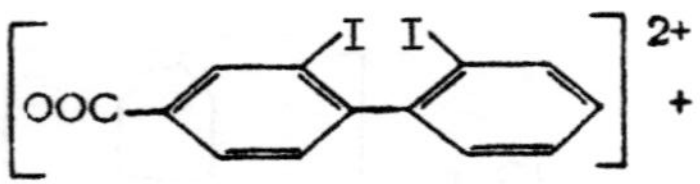 $+ Cr^{2+}$	$\mu = 1$	25	$-0,92$			CSM	[444]
$Co(NH_3)_5$ 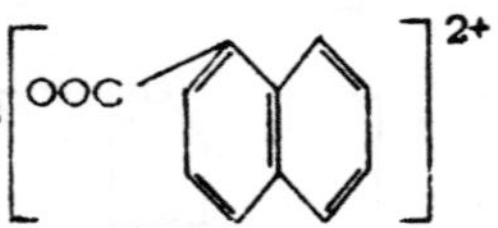 $+ Cr^{2+}$	$\mu = 1$	21	$-1,155$			CSM	[444]

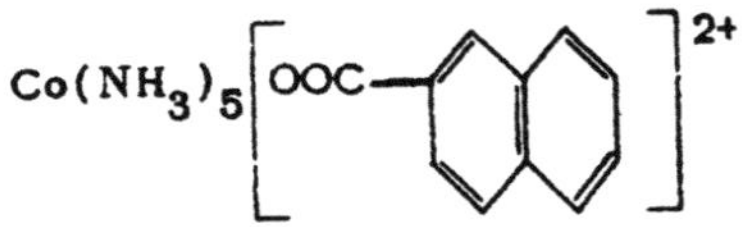 $+ Cr^{2+}$	$\mu = 1$	21	–0,96	CSM	[444]
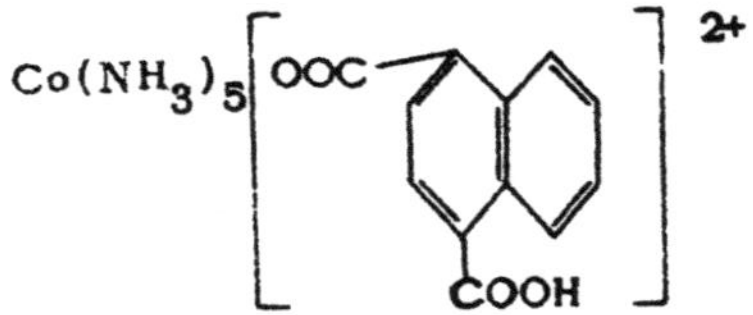 $+ Cr^{2+}$	$\mu = 1$	21	–1,098 (k_1)	CSM	[444]
$k_1 + k_2[H^+]$			0,53 (k_2)		
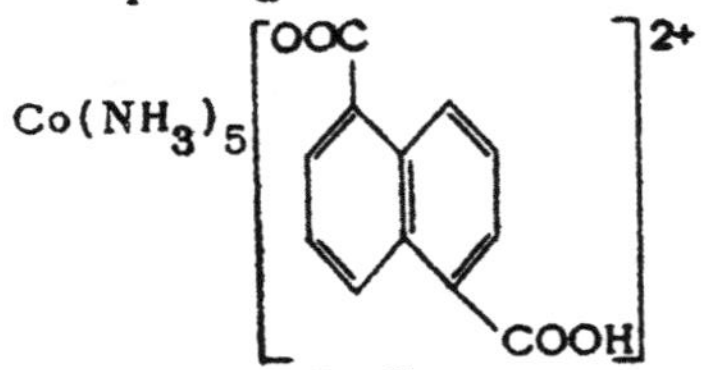 $+ Cr^{2+}$	$\mu = 1$	21	–1,098 (k_1)	CSM	[444]
$k_1 + k_2[H^+]$			0,447 (k_2)		
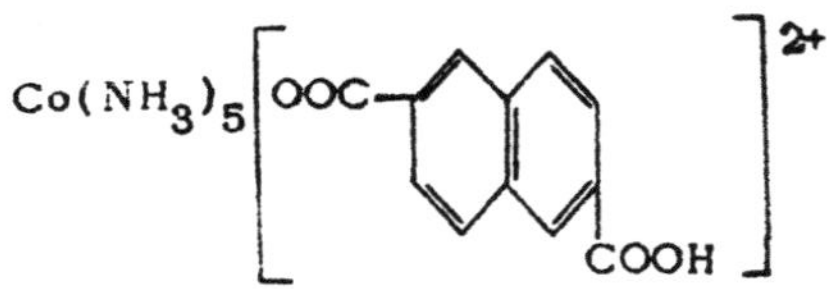 $+ Cr^{2+}$	$\mu = 1$	21	–1,098 (k_1)	CSM	[444]
$k_1 + k_2[H^+]$			0,43 (k_2)		

Table 114 (continued)

Reaction	Medium	t°	log k	log A	E	Method	Ref.
$Co(NH_3)_5$ 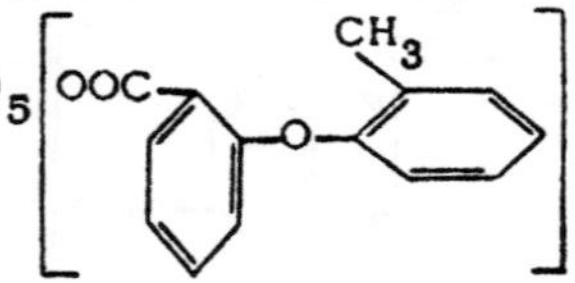 $^{2+}$ + Cr^{2+}	μ = 3,0	25	–0,96			CSM	[463]
$Co(NH_3)_5$ 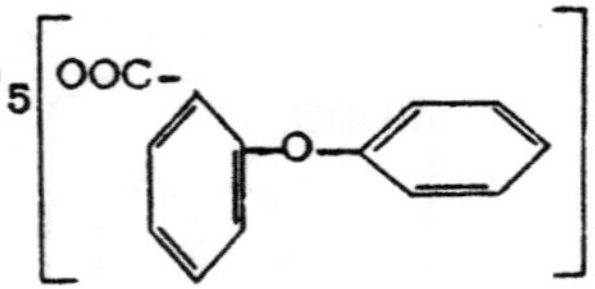 $^{2+}$ + Cr^{2+}	μ = 3,0	25	–0,77			CSM	[163]
$Co(NH_3)_5$ 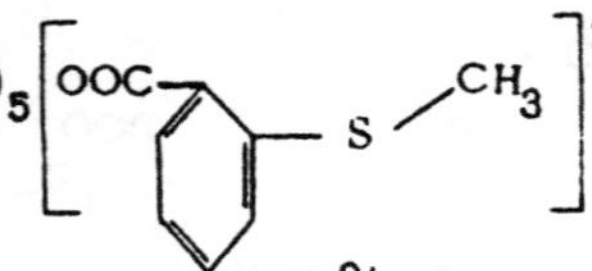 $^{2+}$ + Cr^{2+}	μ = 3	25	–0,42			CSM	[463]
$Co(NH_3)_5$ 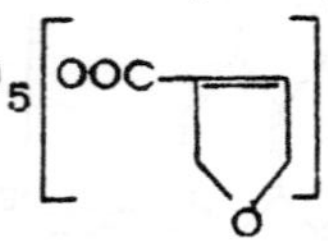 $^{2+}$ + Cr^{2+}	μ = 3	25	–0,523			CSM	[463]
$Co(NH_3)_5$ 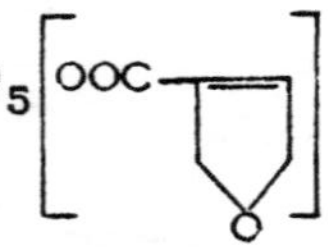 $^{2+}$ + Cr^{2+}	μ = 3	25	–0,638			CSM	[463]

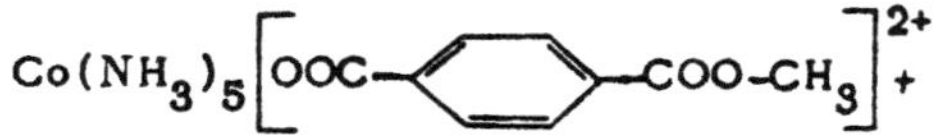 $+ Cr^{2+}$ $\mu = 3$	25	−0,7		CSM	[463]
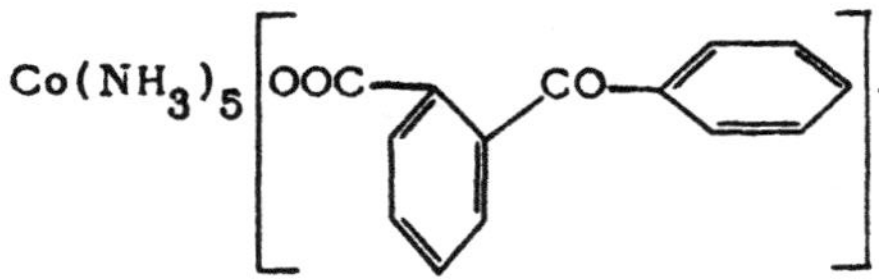 $+ Cr^{2+}$ $\mu = 3$	25	−0,098		CSM	[463]
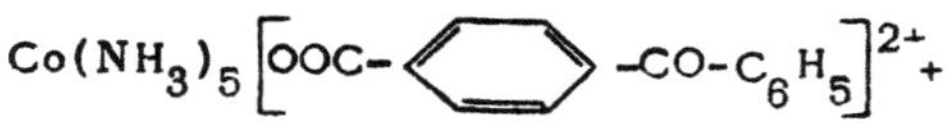 $+ Cr^{2+}$ $\mu = 3$	25	−0,495 (k_1)		CSM	[463]
$k_1 + k_2 [H^+]$		0,176 (k_2)			
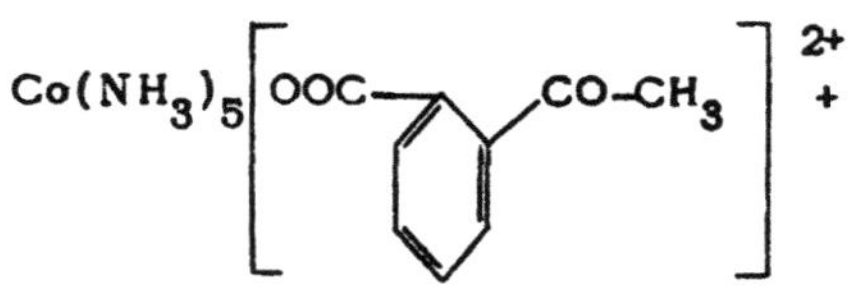 $+ Cr^{2+}$ $\mu = 3$	25	−0,098		CSM	[463]

Table 114 (continued)

Reaction	Medium	t^{o}	log k	log A	E	Method	Ref.
$Co(NH_3)_5$ 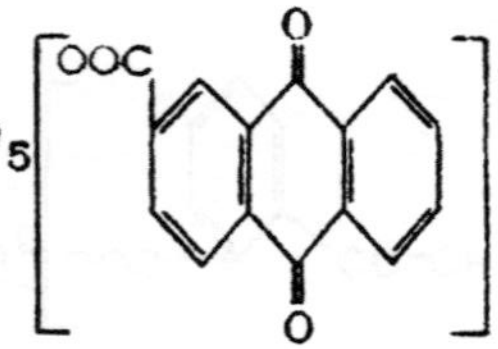 $^{2+}$ + Cr^{2+}	μ = 3	25	0,7			CSM	[463]
$Co(NH_3)_5(N(CH_3)_2{-}CHO)^{3+} + Cr^{2+}$	μ = 1,5	25	-2,046			CSM	[460]
$Co(NH_3)_5(NH_2COOC_2H_5)^{3+} + Cr^{2+}$	μ = 1,0	9,9	2,204			CSM	
$Co(NH_3)_5[NH_2(CH_2)_3COOH]^{3+} + Cr^{2+}$	μ = 1,0	25–35	-0,914 (25°)	6,8	10,5	CSM	
$Co(NH_3)_5[NH(C_2H_5)(CH_2)_3COOH]^{3+} + Cr^{2+}$	μ = 1,0	24,1–35,5	-1,77 (24,1°)	8,6	14,1	CSM	[459]
$Co(NH_3)_5(NH_2CH_2COOCH_3)^{3+} + Cr^{2+}$	μ = 1,0	24	0,602			CSM	
$Co(NH_3)_5$ 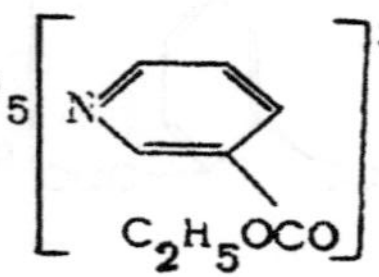 $^{3+}$ + Cr^{2+}	μ = 1,0	25–35,5	-1,87 (25°)	11,4	18,1	CSM	[459]

$Co(NH_3)_5$ 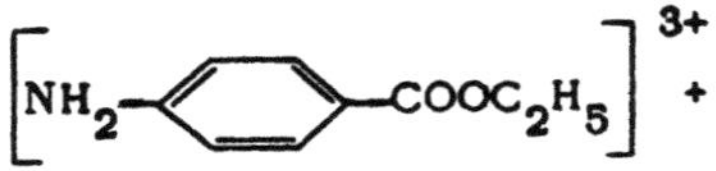 + Cr^{2+}	$\mu = 1,0$	10,5	2,255	CSM	[459]
$Co(NH_3)_5$ 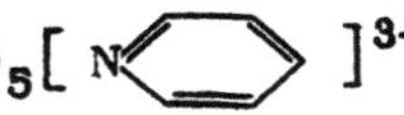 + Cr^{2+}	$\mu = 1,5$	25	−2,222	CSM	[460]
$Co(NH_3)_5$  + Cr^{2+}	$\mu = 1,0$	25	−2,398	CSM	[467]
$Co(NH_3)_5$ 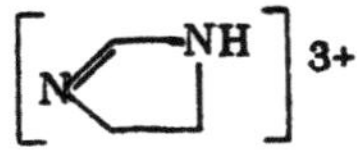 + Cr^{2+}	$\mu = 1,5$	25	−1,4	CSM	[461]
$Co(NH_3)_5(NC)^{2+} + Cr^{2+}$	$\mu = 1$	25	2	CSM	[468]
$Co(NH_3)_5(NCCH_3)^{3+} + Cr^{2+}$	$\mu = 1$	25	−1,77	CSM	
$Co(NH_3)_5(NC{-}CH_2{-}CN)^{3+} + Cr^{2+}$	$\mu = 1$	25	>1	CSM	
$Co(NH_3)_5(NC{-}CH{=}CH_2)^{3+} + Cr^{2+}$	$\mu = 1$	25	−1,538	CSM	
$Co(NH_3)_5$ 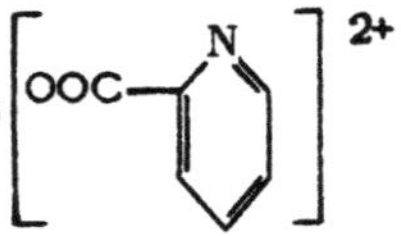 + Cr^{2+}	$\mu = 3,0$	25	5,3	CSM	[469]

Table 114 (continued)

Reaction	Medium	t^{o}	log k	log A	E	Method	Ref.
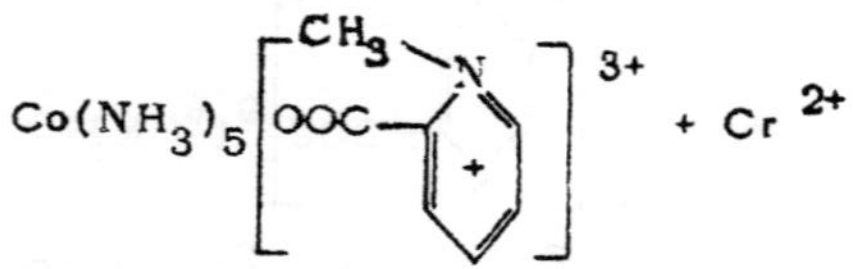	$\mu = 3{,}0$	25	-1,06			CSM	[469]
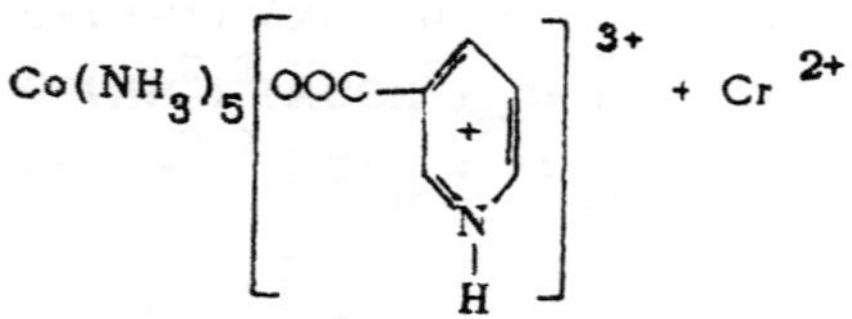	$\mu = 3{,}0$	25	-0,96			CSM	[469]
	$\mu = 3{,}0$	25	0,114			CSM	[469]
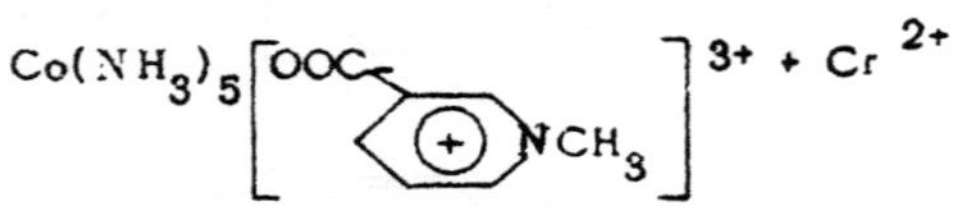	$\mu = 3{,}0$	25	-0,886			CSM	[469]
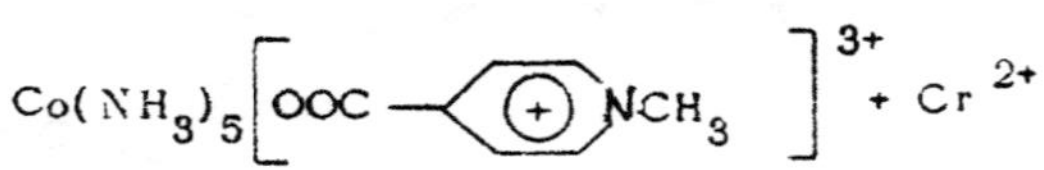	$\mu = 3{,}0$	25	0,146			CSM	[469]

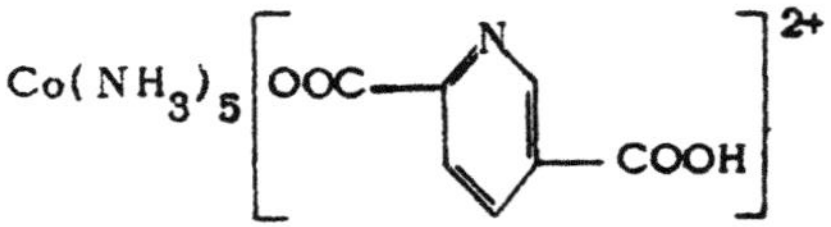	+ Cr^{2+}	$\mu = 3{,}0$	25	>2,3	CSM	[469]
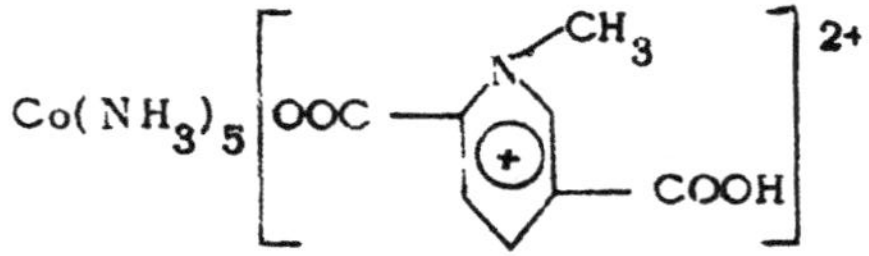	+ Cr^{2+}	$\mu = 3{,}0$	25	0,802	CSM	[469]
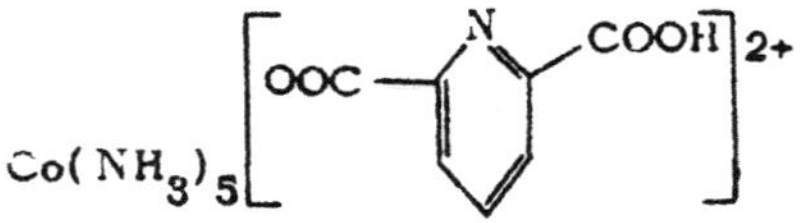	Cr^{2+}	$\mu = 3{,}0$	25	>2,176	CSM	[469]
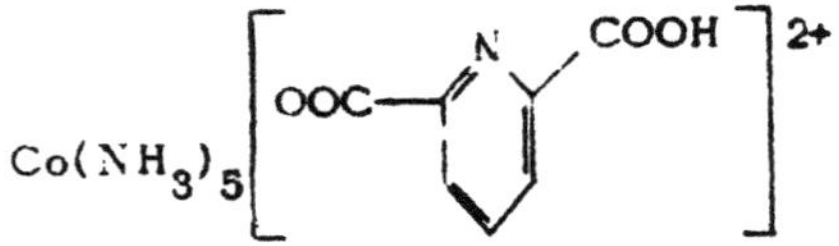	+ Cr^{2+}	$\mu = 1{,}5$	25	<1,477	CSM	[460]
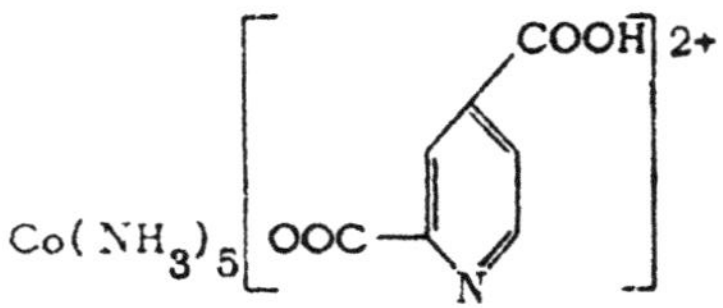	+ Cr^{2+}	$\mu = 1{,}5$	25	>3,78	CSM	[460]

Table 114 (continued)

Reaction	Medium	t^{o}	log k	log A	E	Method	Ref.
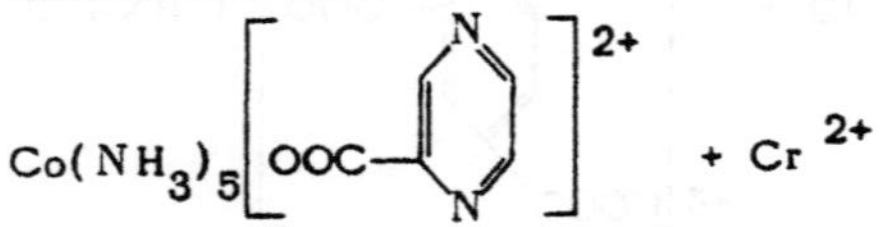	μ = 1,5	25	>3,78			CSM	[460]
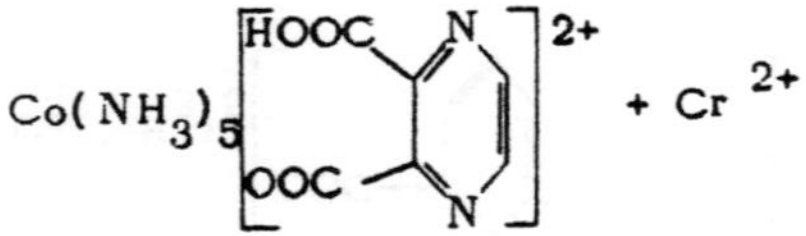	μ = 1,5	25	>2,844			CSM	[460]
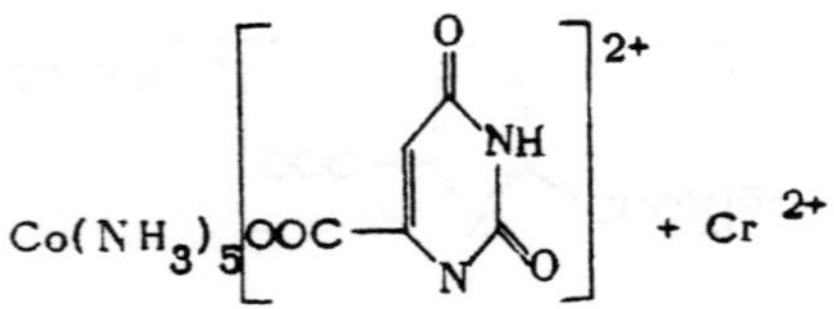	μ = 1,5	25	–0,7			CSM	[460]
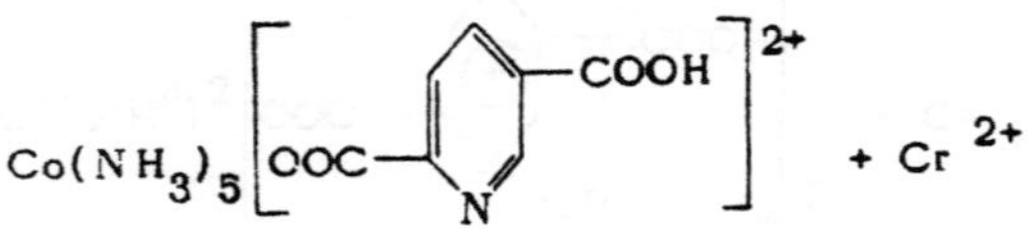	μ = 1,5		–0,7			CSM	[460]
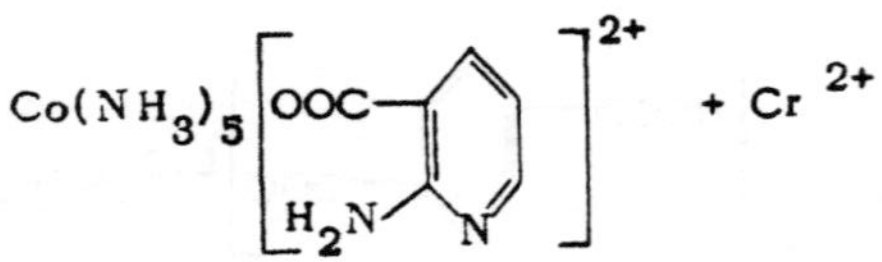	μ = 1,5		–1,32			CSM	[460]

$Co(NH_3)_5$ 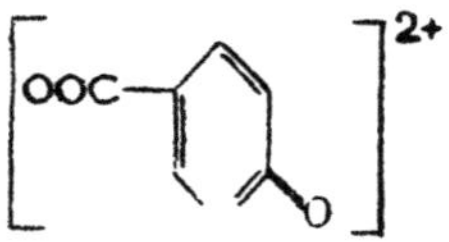 $+ Cr^{2+}$	$\mu = 1,5$		−0,886			CSM	[460]
$Co(NH_3)_5$ 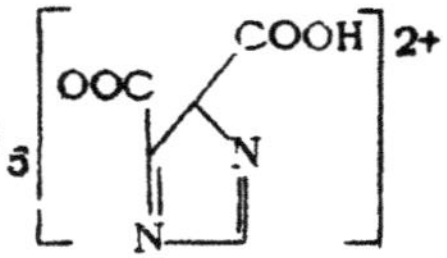 $+ Cr^{2+}$	$\mu = 1,5$	25	1,544			CSM	[460]
$Co(NH_3)_5$ 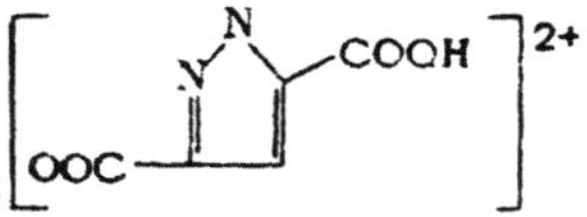 $+ Cr^{2+}$	$\mu = 1,5$	25	0,944			CSM	[460]
$Co(En)_2(SO_4)^+ + Cr^{2+}$	$\mu = 1,0$	25	1,544 (k_1)			CSM	[439]
$k_1 + k_2[H^+]$	$[H^+]$ = 0,2 to 0,6 M		1,748 (k_2)				
$Co(Tetren)(SO_4)^+ + Cr^{2+}$ *	$\mu = 1,0$	8–28	0,874 (25°)	8,4	10,3	CSM	[439]
$Co(Tetren)(SO_4)^+ + Cr^{2+}$ *	$\mu = 1,0$	21–28	0,204 (25°)	9,3	12,3	CSM	[439]
$Co(Tetren)(Cl)^{2+} + Cr^{2+}$ *	$\mu = 1,0$	10–21	0,732 (21°)	6,2	7,3	CSM	[471]

*In both cases, the slow reaction corresponds to the cis and the fast reaction to the trans configuration of the complexes, which resemble the structures of cis- and trans- $Co(En)_2(Cl)(X)$, respectively.

Table 114 (continued)

Reaction	Medium	t°	log k	log A	E	Method	Ref.
Co(Tetren)(Cl)$^{2+}$ + Cr^{2+} *	μ = 1,0	10–21	–0,07 (21°)	8,9	9,3	CSM	[471]
Co(Tetren)(H$_2$O)$^{2+}$ + Cr^{2+}	μ = 1,0	8–28	0,85 (k_1)			CSM	[439]
$k_1 + \frac{k_2}{[H^+]}$	$[H^+]$ = 0,2 to 0,6M		–0,032 (k_2)				
Co(En)$_2$(NH$_3$)(OOCCH$_3$)$^{2+}$ + Cr^{2+}	μ = 1,0	12,3–25	–0,722 (25°)	10,2	14,9	CSM	[439]
trans-Co(NH$_3$)$_4$(OOCCH$_3$)$_2^+$ + Cr^{2+}	μ = 1,0	25	1,176 (k_1)	7,4	8,5	CSM	[456]
$k_1 + k_2[H^+]$	$[H^+]$ = 0,2 to 0,6 M		1,7 (k_2)				
trans-Co(En)$_2$(OOCCH$_3$)$_2^+$ + Cr^{2+}	μ = 1,0	9–34,2	0,6 (k_1,25°)	7,6	9,5	CSM	[439]
$k_1 + k_2[H^+]$	$[H^+]$ = 0,2 to 0,6 M		1,13 (k_2)				
Co(En)$_2$(OOCCH=CHCOO)$^{2+}$ + Cr^{2+}	μ = 1,0	25	2,1 (k_1)			CSM	[439]

$k_1 + k_2[H^+]$	$[H^+] = 0,2$ to $0,6$ M		1,914 (k_2)				
$Co(En)_3^{3+} + Cr^{2+}$	$\mu = 0,4$	25	~ –4,7			CSM	[437]
$Co(En)_3^{3+} + Cr^{2+}$		~23	~ –3,08			CSM	[453]
cis–$Co(En)_2(H_2O)(OH)^{2+} + Cr^{2+}$	$\mu = 1,0$	7–30	5,898 (25,5°)	10,6	6,4	CSM	
trans–$Co(En)_2(H_2O)(OH)^{2+} + Cr^{2+}$	$\mu = 1,0$	7–30	6,415 (25,5°)	8,3	2,6	CSM	[470]
cis–$Co(En)_2(NH_3)(OH)^{2+} + Cr^{2+}$	$\mu = 1,0$	7–30	5,3 (25,5°)	9,9	6,2	CSM	
trans–$Co(En)_2(NH_3)(OH)^{2+} + Cr^{2+}$	$\mu = 1,0$	7–30	5,342 (25,5°)	7,1	2,4	CSM	
cis–$Co(En)_2(NH_3)(Cl)^{2+} + Cr^{2+}$	$\mu = 1,0$	10–29	–0,168 (19°)	7,1	9 7	CSM	[471]
trans–$Co(En)_2(NH_3)(Cl)^{2+} + Cr^{2+}$	$\mu = 1,0$	25	0,954			CSM	[471]
cis–$Co(En)_2(NH_3)(NCS)^{2+} + Cr^{2+}$	1 M $HClO_4$	25	0,490			JM	
trans–$Co(En)_2(NH_3)(NCS)^{2+} + Cr^{2+}$	– " –	25	0,580			JM	
cis–$Co(En)_2(NCS)(Cl)^{2+} + Cr^{2+}$	– " –	25	6,255			JM	[393]
trans–$Co(En)_2(NCS)(Cl)^{2+} + Cr^{2+}$	– " –	25	6,4			JM	
cis–$Co(En)_2(H_2O)(NCS)^{2+} + Cr^{2+}$	– " –	25	1,653			JM	
trans–$Co(En)_2(H_2O)(NCS)^{2+} + Cr^{2+}$	1 M $HClO_4$	25	3,146			JM	

Table 114 (continued)

Reaction	Medium	t °	log k	log A	E	Method	Ref.
$Co(En)_2(NCS)(OH)^{2+} + Cr^{2+}$	1M $HClO_4$	25	~7			JM	
cis-$Co(En)_2(NCS)_2^{+} + Cr^{2+}$	- " -	25	1,477			JM	[393]
trans-$Co(En)_2(NCS)_2^{+} + Cr^{2+}$	- " -	25	1,477			JM	
$Co(En)_2(NH_3)(SO_4)^{+} + Cr^{2+}$	$\mu = 1$	8–29	0,544 (25°)	8,6	11,0	CSM	[439]
$Co(En)_2(1,10\text{-}O\text{-}Phen)^{3+} + Cr^{2+}$	H_2SO_4	~23	~0,23			CSM	[453]
$Co(Tetren)(OOC\text{-}CH_3)^{2+} + Cr^{2+}$	$\mu = 1$	12,3–25	-0,745 (25°)	7,3	10,9	CSM	[439]
$Co(EDTA)^{-} + Cr^{2+}$	$\mu = 1$	20	5,204			MCR	
$Co(HEDTA)(H_2O) + Cr^{2+}$	$\mu = 1$	20	5,114			MCR	
$Co(EDTAOH)(H_2O) + Cr^{2+}$	$\mu = 0,3$	20	5,114			MCR	[472]
$Co(EDTAOH)(OH)^{-} + Cr^{2+}$	$\mu = 0,3$	20	> 8			MCR	
$Co(HEDTA)(Cl)^{-} + Cr^{2+}$	$\mu = 0,3$	20	6,4			MCR	
$Co(C_2O_4)_3^{3-} + Cr^{2+}$	$\mu = 1$	20	6,6			MCR	
	Cobalt + $Cr(Dipy)_3^{2+}$						
$Co(NH_3)_5NH_3^{3+} + Cr(Dipy)_3^{2+}$	$\mu = 0,4$	25	2,838			JM	[437]

$Co(NH_3)_5(NH_3)^{3+} + Cr(Dipy)_3^{2+}$	$\mu = 0,2$	25	2,398			CSM	[449]
	$\mu = 0,01$; NaCl	25	1,602	11,2	13,6	CSM	[382]
	$\mu = 0,05$; NaCl	25	2,053	10,8	11,9	CSM	[382]
	$\mu = 0,20$; NaCl	25	2,398	10,1	10,5	CSM	[382]
	$\mu = 0,20$; NaBr	23	2,778	11,2	11,4	CSM	[382]
	$\mu = 0,20$; NaCl; 94,8% D_2O	25	2,405			CSM	[382]
$Co(ND_3)_6^{3+} + Cr(Dipy)_3^{2+}$	$\mu = 0,20$; NaCl; 93,8% D_2O	25	2,262			CSM	[382]
$Co(NH_3)_5(H_2O)^{3+} + Cr(Dipy)_3^{2+}$	$\mu = 0,1$	25	4,7			JM	[437]
	$\mu = 0,05$; NaCl; pH 4	4	3,322			CSM	[382]
$Co(NH_3)_5(H_2O)^{3+} + Cr(Dipy)_3^{2+}$ *	$\mu = 0,05$; NaCl; D_2O; pH $\approx$ 4	4	2,902			CSM	[382]

* Measured in 93% D_2O. Constant is result of extrapolating to 100% D_2O.

Table 114 (continued)

Reaction	Medium	t °	logk	log A	E	Method	Ref.
$Co(NH_3)_5(H_2O)^{3+} + Cr(Dipy)_3^{2+}$	μ = 0,05; NaCl; pH 7,25	4	3,08			CSM	[382]
	μ = 0,01; NaCl; pH≈4	4	2,812			CSM	[382]
$Co(NH_3)_5OH^{2+} + Cr(Dipy)_3^{2+}$	μ = 0,2; NaCl	25	3,0			CSM	[382]
$Co(NH_3)_5F^{2+} + Cr(Dipy)_3^{2+}$	μ = 0,01	25	3,255			JM	[437]
$Co(NH_3)_5Cl^{2+} + Cr(Dipy)_3^{2+}$	μ = 0,1	25	5,952			JM	[437]
	μ = 0,01; NaCl	4	4,02			CSM	[382]
$Co(NH_3)_5Br^{2+} + Cr(Dipy)_3^{2+}$	μ = 0,1	25	6,7			JM	[437]
	μ = 0,0001; pH≈4	4	> 4,23			CSM	[382]

$Co(NH_3)_5N_3^{2+} + Cr(Dipy)_3^{2+}$	$\mu = 0{,}1$	25	4,612	JM	
$Co(NH_3)_5NCS^{2+} + Cr(Dipy)_3^{2+}$	$\mu = 0{,}1$	25	4,0	JM	
$Co(NH_3)_5SO_4^{+} + Cr(Dipy)_3^{2+}$	$\mu = 0{,}1$	25	4,652	JM	[437]
$Co(NH_3)_5S_2O_4^{+} + Cr(Dipy)_3^{2+}$	$\mu = 0{,}1$	25	4,902	JM	
$Co(NH_3)_5(OOCCH_3)^{2+} + Cr(Dipy)_3^{2+}$	$\mu = 0{,}1$	25	3,08	JM	
$Co(NH_3)_5(cis\text{-}OOCCH{=}CHCOOH)^{2+} + Cr(Dipy)_3^{2+}$	$\mu = 0{,}1$	25	3,0	JM	[437]
$Co(En)_3^{3+} + Cr(Dipy)_3^{2+}$	$\mu = 0{,}1$	25	2,255	JM	[437]
	$\mu = 0{,}1$; NaCl; pH ≈ 4	25	2,568	CSM	[382]
	$\mu = 0{,}2$; NaCl; pH ≈ 4	25	2,8	CSM	[382]
	Ccbalt + Fe(II)				
$Co^{3+} + Fe^{2+}$	3 M $HClO_4$	25	2,477	JM	[473]
	4 M $HClO_4$	25	2,518	JM	[432]
$CoCl^{2+} + Fe^{2+}$	$\mu = 3$	25	$\geqslant 3{,}7$	JM	[473]

Table 114 (continued)

Reaction	Medium	t^o	log k	log A	E	Method	Ref.
$Co(NH_3)_5F^{2+} + Fe^{2+}$	$\mu = 1,7$	15,2–36,2	–3,036 (25,5°)	8,9	16,2	CSM	[438]
	$\mu = 1,0$	25–45	–2,18 (25°)	8,3	14,3	CSM	[474]
$Co(NH_3)_5F^{2+} + Fe^{2+}$, k [$F^-$]	$\mu = 1,7$; F^-	25,8	1,114			CSM	[438]
$Co(NH_3)_5F^{2+} + Fe^{2+}$, k [$Cl^-$]	$\mu = 1,7$; Cl^-	25,8	–1,678			CSM	[438]
$Co(NH_3)_5F^{2+} + Fe^{2+}$, k [$Br^-$]	$\mu = 1,7$; Br^-	25,8	–2,456			CSM	[438]
$Co(NH_3)_5Cl^{2+} + Fe^{2+}$	$\mu = 1,7$	15,2–36,2	–2,796 (25,5°)	8,3	15,1	CSM	[438]
	$\mu = 1,0$	25–45	–2,87 (25°)	6,7	13,1	CSM	[474]
$Co(NH_3)_5Br^{2+} + Fe^{2+}$	$\mu = 1,7$	15,2–36,2	–2,12 (25,5°)	8,2	14,0	CSM	[438]
	$\mu = 1,0$	25–45	–3,134 (25°)	7,1	13,9	CSM	[474]
$Co(NH_3)_5Br^{2+} + Fe^{2+}$, k [$F^-$]	$\mu = 2,0$	25,1	0,602			CSM	[438]

$Co(NH_3)_5Br^{2+} + Fe^{2+}$, k [$Cl^-$]	μ = 2,0; Cl^-	25,1	≤ −3			CSM	[438]
$Co(NH_3)_5(N_3)^{2+} + Fe^{2+}$	μ = 0,26	25	−2,062			CSM	[475]
cis-$Co(NH_3)_4(H_2O)N_3^{2+} + Fe^{2+}$	μ = 0,5	25	−0,45			CSM	[476]
trans-$Co(NH_3)_4(H_2O)N_3^{2+} + Fe^{2+}$	μ = 0,5	25	1,38			CSM	[476]
cis-$Co(NH_3)_4(N_3)_2^{+} + Fe^{2+}$	μ = 0,26	25	−0,733			CSM	[475]
trans-$Co(NH_3)_4(N_3)_2^{+} + Fe^{2+}$	μ = 0,26	25	−1,134 (k_1)			CSM	[475]
$k_1 + k_2[H^+]$	$[H^+]$ = 0,072 to 0,219 M		0,136 (k_2)				
$Co(NH_3)_5NCS^{2+} + Fe^{2+}$	μ = 1,0	25	< −2,523			CSM	[474]
$Co(NH_3)_5(OOCCH_3)^{2+} + Fe^{2+}$	μ = 1,0	25	< −1,3			CSM	[474]
$Co(NH_3)_5(OOCCOOH)^{2+} + F^{2+}$, $\frac{k}{[H^+]}$	μ = 1,0	25	0,58	9,8	12,6	CSM	[474]
$Co(C_2O_4)_3^{3-} + Fe^{2+}$	μ = 0	5,4–20,3	3,176 (25°)	12,4	12,6	CSM	[477]
		25	1,278				[472]

Table 114 (continued)

Reaction	Medium	t°	log k	log A	E	Method	Ref.
cis-$Co(En)_2(NH_3)(Cl)^{2+} + Fe^{2+}$	μ = 1,0	25	–4,745			CSM	
trans-$Co(En)_2(NH_3)(Cl)^{2+} + Fe^{2+}$	μ = 1,0	25	–4,18			CSM	
cis-$Co(En)_2(NCS)(Cl)^{+} + Fe^{2+}$	μ = 1,0	25	–3,77			CSM	
trans-$Co(En)_2(NCS)(Cl)^{+} + Fe^{2+}$	μ = 1,0	25	–3,886			CSM	
cis-$Co(En)_2(H_2O)(Cl)^{2+} + Fe^{2+}$	μ = 1,0	25	–3,347			CSM	
trans-$Co(En)_2(H_2O)(Cl)^{2+} + Fe^{2+}$	μ = 1,0	25	–0,82			CSM	[478]
cis-$Co(En)_2Cl_2^{+} + Fe^{2+}$	μ = 1,0	25	–2,796			CSM	
trans-$Co(En)_2Cl_2^{+} + Fe^{2+}$	μ = 1,0	25	–1,495			CSM	
trans-$Co(En)_2(Br)(Cl)^{+} + Fe^{2+}$	μ = 1,0	25	–1,444			CSM	
trans-$Co(En)_2Br_2^{+} + Fe^{2+}$	μ = 1,0	25	–1,658			CSM	
trans-$Co(En)_2(N_3)(Cl)^{+} + Fe^{2+}$	μ = 1,0	25	–1,207			CSM	
$Co(EDTA)^{-} + Fe^{2+}$	μ = 1	20–30	–3,222 (25°)	5,4	11,8	CSM	[479]
$Co(HEDTA)(H_2O) + Fe^{2+}$	μ = 1	5,6–25	–2,07 (25°)	6,05	11,1	CSM	[479]
$Co(EDTA)(H_2O)^{-} + Fe^{2+}$	μ = 1	5,6–25	–0,048 (25°)	16,15	22,1	CSM	[479]
$Co(EDTAOH)(H_2O) + Fe^{2+}$	μ = 1	20–35	–2,335 (25°)	8,1	14,2	CSM	[479]

$Co(EDTAOH)(OH)^{-}$ – + Fe^{2+}	μ = 1	20–35	4,384 (25°)	14,5	13,7	CSM	[479]
$Co(HEDTA)(Cl)^{-}$ + Fe^{2+}	μ = 3,0; $[HClO_4]$ = =2,2 to 2,8 M	25	0,278			JM	[480]
	μ = 0,90; $[HClO_4]$ = =0,2 M	20	0,124			JM	[481]
$Co(HEDTA)(Cl)^{-}$ + Fe^{2+}	μ = 1,0; $[H^{+}]$ = =0,05 to 0,98 M	15–30	0,276 (30,2°)	8,4	11,3	JM	[482]
log k = log k + A $[H^{+}]$			A=0,156			JM	[482]
d-$Co(EDTA)^{-}$ + *l*-Fe(–)$PDTA^{2-}$	μ = 0,5; pH ≈ 4,5	25	1,097	7,3	6,5	JM	[413]
l-$Co(EDTA)^{-}$ + *l*-Fe(–)$PDTA^{2-}$	– " –	25	1,114	8,1	7,1	JM	[413]
dl-cis-$Co(En)_2(NH_3)(Cl)^{2+}$ + + *l*-Fe(–)$PDTA^{2-}$	– " –	25	2,3	11,5	10,8	JM	[413]
d-$Co(EDTA)(Cl)^{2-}$ + + *l*-Fe(–)$PDTA^{2-}$	– " –	25	2,59	6,9	7,7	JM	[413]

Table 114 (continued)

Reaction	Medium	t^{o}	log k	log A	E	Method	Ref.
l-$Co(EDTA)(Cl)^{2-}$ + l $[Fe(-)PDTA^{2-}]$	μ = 0,5;	25	2,59	7,1	7,8	JM	[413]
dl-$[Co(C_2O_4)_2]^{3-}$ + l-$Fe(-)PDTA^{2-}$	pH ≈ 4,5	25	2,342	3,6	5,0	JM	
$Co(NH_3)_5Cl^{2+}$ + dl-$Fe(PDTA)^{2-}$	- " -	25	3,342			JM	
$Co(NH_3)_5Br^{2+}$ dl-$Fe(EDTA)^{2+}$	- " -	25	3,255			JM	
$Co(NH_3)_5N_3^{2+}$ + dl-$Fe(PDTA)^{2-}$	- " -	25	3,845			JM	
$Co(NH_3)_5NCS^{2+}$ dl-$Fe(PDTA)^{2-}$	- " -	30	-0,569			JM	
$(EDTA)Co(III)-NC-Fe(CN)_5(II)^{5-}$ + $Fe(CN)_6^{3-} \underset{k_2}{\overset{k_1}{\rightleftharpoons}} Fe(CN)_6^{4-}$ +	μ = 0,6	25	3,3 (k_1)			JM, TJ	[483]
+ $(EDTA)Co(III)-CN-Fe(CN)_5(II)^{4-}$	μ = 0,6	25	4,377 (k_2)			JM, TJ	[483]
	Cobalt						
$Co(NH_3)_6^{3+}$ + $Co(En)_3^{2+}$	μ = 0,98	25	-1,692			IM	[349]
$Co(NH_3)_5NH_3^{3+}$ + $Co(CN)_5^{3-}$; k $[CN^-]$	μ = 0,2; CN^-	25	4,924			JM	[484]

$Co(NH_3)_5OH^{2+} + Co(CN)_5^{3-}$	$\mu = 0,2$	25	4,968	JM	
$Co(NH_3)_5F^{2+} + Co(CN)_5^{3-}$	$\mu = 0,2$	25	3,255	JM	
$Co(NH_3)_5F^{2+} + Co(CN)_5^{3-}$, k $[CN^-]$	$\mu = 0,2$; CN^-	25	4,23	JM	
$Co(NH_3)_5Cl^{2+} + Co(CN)_5^{3-}$	$\mu = 0,2$	25	~7,7	JM	[484]
$Co(NH_3)_5N_3^{2+} + Co(CN)_5^{3-}$	$\mu = 0,2$	25	6,204	JM	
$Co(NH_3)_5NCS^{2+} + Co(CN)_5^{3-}$	$\mu = 0,2$	25	6,04	JM	
$Co(NH_3)_5SO_4^{+} + Co(CN)_5^{3-}$ k $[CN^-]$	$\mu = 0,2$; CN^-	25	4,602	JM	[484]
$Co(NH_3)_5PO_4 + Co(CN)_5^{3-}$ k $[CN^-]$	$\mu = 0,2$; CN^-	25	2,7	JM	[484]
$Co(NH_3)_5CO_3^{+} + Co(CN)_5^{3-}$ k $[CN^-]$	$\mu = 0,2$; CN^-	25	~3	JM	[484]
$Co(NH_3)_5(OOCCH_3)^{2+} + Co(CN)_5^{3-}$ k $[CN^-]$	$\mu = 0,2$	25	4,04	JM	[484]
$Co(NH_3)_5$(trans-OOCCH=CHCOOH)$^{2+}$ + + Cr^{2+}, k $[CN^-]$	$\mu = 0,2$	25	4,08	JM	[484]

Table 114 (continued)

Reaction	Medium	t^{o}	log k	logA	E	Method	Ref.
$Co(NH_3)_5(OOCCOOH)^{2+} + Co(CN)_5^{3-}$ k[CN^-]	μ = 0,2	25	4,0			JM	[484]
$Co(NH_3)_5(cis\text{-}OOCCH{=}CHCOOH)^{2+}$ + + $Co(CN)_5^{3-}$, k [CN^-]	μ = 0,2	25	3,775			JM	[484]
$Co(NH_3)_5[OOC(CH_2)_2COOH]^{2+}$ + $Co(CN)_5^{3-}$, k [CN^-]	μ = 0,2	25	3,778			JM	[484]
$Co(EDTA)^- + Co(Dipy)_3^{2+}$	μ = 0,1	25	~-1,7			CSM	[411]
$Co(Dipy)_3^{3+} + Co(Tripy)_2^{2+}$	μ = 0,05	25	1,806	8,43	9,0	JM	
$Co(Phen)_3^{3+} + Co(Tripy)_2^{2+}$	μ = 0,05	25	2,447	7,95	7,5	JM	[485]
$Co(Phen)_3^{3+} + Co(Tripy)_2^{2+}$	μ = 0,05; 95% C_2H_5OH	25	1,682	10,7	12,3	JM	
$Co[3,5,8,8\text{-}(CH_3)_4\text{-}Phen]_3^{3+}$ + + $Co(Tripy)_2^{2+}$	μ = 0,02; 95% C_2H_5OH	25	1,832	10,05	11,2	JM	[485]
$Co(Dipy)_2^{3+} + Co(Tripy)_2^{2+}$	μ = 0,05	25	4,114	7,6	4,8	JM	
$Co(Phen)_2^{3+} + Co(Tripy)_2^{2+}$	μ = 0,05	25	4,476	8,1	4,9	JM	
$Co(Dipy)^{3+} + Co(Tripy)_2^{2+}$	μ = 0,05	25	2,832	11,2	11,4	JM	[485]
$Co^{3+} + Co(Tripy)_2^{2+}$	μ = 0,05	25	4,867	2,9	4,0	JM	
$CoOH^{2+} + Co(Tripy)_2^{2+}$	μ = 0,05	25	4,23	7,2	5,1	JM	

trans-Co[]$(Cl_2)^+$ + + $Co(tripy)_2^{2+}$	μ = 0,00016	25	5,476	9,7	5,8	JM	[485]
trans-Co[]$(Cl_2)^+$ + + $Co(Dipy)_3^{2+}$	μ = 0,00016	25	4,04	8,3	5,8	JM	[485]
trans-Co[]$(Cl_2)^+$ + + $Co(Phen)_3^{2+}$	μ = 0,00016	25	3,959	10,6	9,0	JM	[485]
$Co(Dipy)^{3+}$ + $Co(Phen)^{2+}$	μ = 1,55	25	−1,222	5,6	9,3	JM	[485]
$Co(Phen)^{3+}$ + $Co(Tripy)_2{}^{2+}$	μ = 0,05	25	3,146	11,6	11,5	JM	
$Co(Tripy)_2{}^{3+}$ + $Co(Dipy)_3^{2+}$	μ = 0,05	25	1,432	8,2	9,3	JM	
	Cobalt + europium (II)						
$Co(NH_3)_5NH_3^{3+}$ + Eu^{2+}	μ = 0,1	25	−1,7			JM	[437]
$Co(NH_3)_5H_2O^{3+}$ + Eu^{2+}	μ = 1,0	25	−0,824			CSM	[486]
$Co(NH_3)_5F^{2+}$ + Eu^{2+}	μ = 1,0	25	4,415			JM	[437]
$Co(NH_3)_5Cl^{2+}$ + Eu^{2+}	μ = 1,0	15–35	2,59 (25°)	6,7	5,6	JM	[437]

Table 114 (continued)

Reaction	Medium	t°	log k	log A	E	Method	Ref.
$Co(NH_3)_5Br^{2+} + Eu^{2+}$	$\mu = 1,0$	15–35	2,398 (25°)	6,3	5,3	JM	[437]
$Co(NH_3)_5N_3^{2+} + Eu^{2+}$	$\mu = 1,0$	15–35	2,278 (25°)	6,8	6,1	JM	[437]
$Co(NH_3)_5NO_2^{2+} + Eu^{2+}$	$\mu = 1,0$	5–30	1,755 (25°)	7,6	8,0	CSM	[488]
$Co(NH_3)_5ONO^{2+} + Eu^{2+}$	$\mu = 1,0$	25	1,874			CSM	[488]
$Co(NH_3)_5ONO_2^{2+} + Eu^{2+}$	$\mu = 1,0$	25	2,235			CSM	[488]
$Co(NH_3)_5ONO_2^{2+} + Eu^{2+}$	$\mu = 1,0$	25	~2,0			JM	[437]
$Co(NH_3)_5NCS^{2+} + Eu^{2+}$	$\mu = 1,0$	25	~−0,155			JM	[437]
$Co(NH_3)_5SO_4^{+} + Eu^{2+}$	$\mu = 1,0$	15–35	2,146 (25°)	7,5	7,3	JM	[437]
$Co(NH_3)_5HPO_4^{+} + Eu^{2+}$	$\mu = 1,0$	25	2,7			JM	[437]
$Co(NH_3)_5H_2PO_4^{2+} + Eu^{2+}$	$\mu = 1,0$	25	0,778			JM	[437]
$Co(NH_3)_5H_3PO_4^{3+} + Eu^{2+}$	$\mu = 5,0$	25	~0,477			JM	[437]

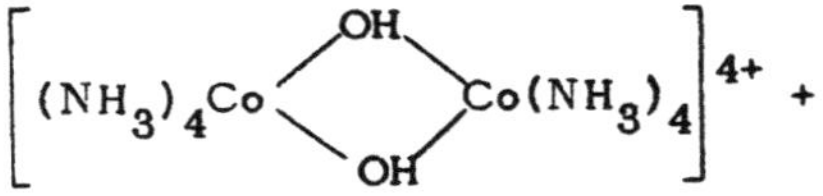 + Eu^{2+}	μ = 1	25–43	−1,456 (25°)	4,0	7,4	CSM	[447]
$Co(En)_3^{3+} + Eu^{2+}$	μ = 1,0	25	∼−2,3			JM	[437]
$Co(NH_3)_5(OOCCH_3)^{2+} + Eu^{2+}$	μ = 1,0	25	−0,745			CSM	[440]
$Co(NH_3)_5(OOCCH_3)^{2+} + Eu^{2+}$	μ = 1,0	25	−0,745	2,9	5,0	CSM	[441]
$Co(NH_3)_5(OOC{-}CH_2Cl)^{2+} + Eu^{2+}$	μ = 1,0	25	0,5	7,7	9,9	CSM	[441]
$Co(NH_3)_5(OOC{-}CHCl_2)^{2+} + Eu^{2+}$	μ = 1,0	25	0,318	5,8	7,5	CSM	[441]
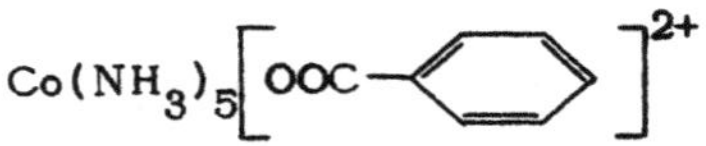 + Eu^{2+}	μ = 1,0	25	−0,62	4,4	6,8	CSM	[441]
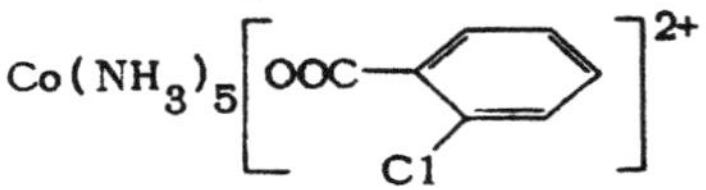 + Eu^{2+}	μ = 1,0	25	−0,41	2,4	3,8	CSM	[441]
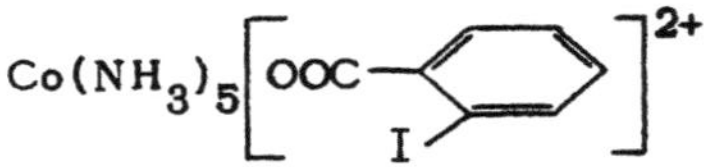 + Eu^{2+}	μ = 1,0	25	−0,553	2,2	3,9	CSM	[441]
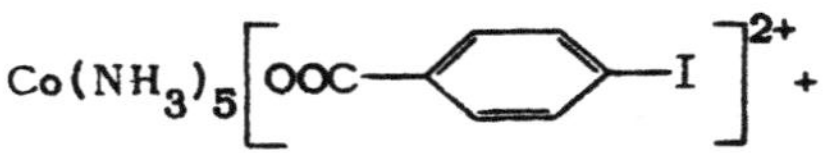 + Eu^{2+}	μ = 1,0	25	−0,569	1,4	2,7	CSM	[441]

Table 114 (continued)

Reaction	Medium	t°	log k	log A	E	Method	Ref.
$Co(NH_3)_5$ 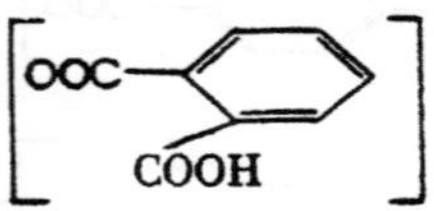 $^{2+}$ + + Eu^{2+}	μ = 1,0	25	0,334	8,8	8,8	CSM	[441]
$Co(NH_3)_5$ 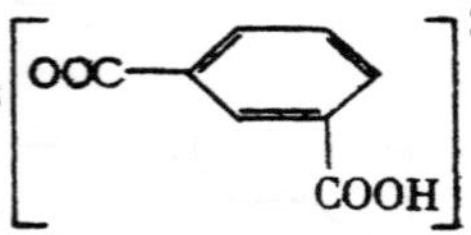 $^{2+}$ + + Eu^{2+}	μ = 1,0	25	-0,194	6,1	8,5	CSM	[441]
$Co(NH_3)_5(OOCCH_2CH_2COOH)^{2+}$ + Eu^{2+}	μ = 1,0	25	0,415			CSM	[441]
$Co(NH_3)_5(OOCCH_2CH_2COOCH_3)^{2+}$ + + Eu^{2+}	μ = 1,0	25	0,398			CSM	[441]
$Co(NH_3)_5(OOCCH_2COOCH_3)^{2+}$ + Eu^{2+}	μ = 1,2	25	1,146			CSM	[445]
	Cobalt + $Ru(NH_3)_6^{2+}$						
$Co(NH_3)_5NH_3^{3+}$ + $Ru(NH_3)_6^{2+}$	μ = 0,2; $[H^+]$ = =0,015 M	25	-1,96			CSM	[488a]

	$\mu = 0{,}2$; pH 9,2	25	$\leqslant -1{,}523$	CSM	[488a]
$Co(NH_3)_5H_2O^{3+} + Ru(NH_3)_6^{2+}$	$\mu = 0{,}22$; $[H^+] = 10^{-3}$ *M*	25	0,477	CSM	[488a]
$Co(NH_3)_5OH^{2+} + Ru(NH_3)_6^{2+}$	$\mu = 0{,}22$; pH 9,0	25	-1,398	CSM	[488a]
$Co(NH_3)_5Cl^{2+} + Ru(NH_3)_6^{2+}$	$\mu = 0{,}18$; $[H^+] = 2\cdot10^{-4}$ *M*	25	2,415	CSM	[488a]
$Co(NH_3)_5Br^{2+} + Ru(NH_3)_6^{2+}$	$\mu = 0{,}18$; $[H^+] = 2\cdot10^{-4}$ *M*	25	3,204	CSM	[488a]
$Co(NH_3)_5I^{2+} + Ru(NH_3)_6^{2+}$	– " –	25	3,826	CSM	[488a]
$Co(NH_3)_5N_3^{2+} + Ru(NH_3)_6^{2+}$	$\mu = 0{,}22$; $[H^+] = 10^{-3}$ *M*	25	-0,080	CSM	[488a]
$Co(NH_3)_5ONO_2^{2+} + Ru(NH_3)_6^{2+}$	$\mu = 0{,}2$; $[H^+] = 3\cdot10^{-3}$ *M*	25	1,513	CSM	[488a]

Table 114 (continued)

Reaction	Medium	t°	log k	log A	E	Method	Ref.
$Co(NH_3)_4(CO_3)^+ + Ru(NH_3)_6^{2+}$	$\mu = 0{,}19$; $[H^+] = 10^{-3}$ M	25	0,954			CSM	[488a]
cis-$Co(NH_3)_4(H_2O)_2^{3+} + Ru(NH_3)_6^{2+}$	$\mu = 9{,}1 \times 10^{-3}$; $[H^+] = 5{,}1 \cdot 10^{-3}$ M	25	1,748			CSM	[488a]
	$\mu = 0{,}18$; $[H^+] = 10^{-3}$ M	25	2,431			CSM	[488a]
cis-$Co(En)_2Cl_2^+ + Ru(NH_3)_6^{2+}$	$\mu = 0{,}15$; $[H^+] = 10^{-4}$ M	25	2,944			CSM	[488a]
trans-$Co(En)_2Cl_2^+ + Ru(NH_3)_6^{2+}$	$\mu = 0{,}15$; $[H^+] \approx 10^{-4}$ M	25	3,954			CSM	[488a]

trans-$Co(NH_3)_4(OOCCH_3)_2^{+}$ + + $Ru(NH_3)_6^{2+}$ $k_1 + k_2 [H^+]$	μ = 0,23–0,3; $[H^+]$ = =0,02÷0,14 M	25	–0,31 (k_1) 1,114 (k_2)			CSM	[488a]
$Co(NH_3)_5(OOCCH_2CH_2CH_3)^{2+}$ + + $Ru(NH_3)_6^{2+}$	μ = 0,16; $[H^+]$ = =10^{-2} M	25	–0,456			CSM	[488a]
$Co(NH_3)_5(OOCCOOH)^{2+}$ + $Ru(NH_3)_6^{2+}$	μ = 0,31; $[H^+]$ = =$4 \cdot 10^{-3}$ M	25	–0,222			CSM	[488a]
$Co(NH_3)_5(OOCCOOH)^{2+}$ + $Ru(NH_3)_6^{2+}$	μ = 1,0	14,6–35,2	–0,3 (25°)	12,0	16,7	CSM	
$Co(NH_3)_5(OOCCOO^-)^{+}$ + $Ru(NH_3)_6^{2+}$	μ = 1,0	14,6–35,2	–1,0 (25°)	4,15	7,0	CSM	[443]
$Co(NH_3)_5(OOCCHO)^{2+}$ + $Ru(NH_3)_6^{2+}$	μ = 1,0	15,2–35,3	–1,05 (25°)	9,0	13,7	CSM	
$Co(NH_3)_5(OOCCOCH_3)^{2+}$ + $Ru(NH_3)_6^{2+}$	μ = 1,0	15,1–35	–0,7 (25°)	11,4	16,5	CSM	
$Co(NH_3)_5[OOCCOC(CH_3)_3]^{2+}$ + + $Ru(NH_3)_6^{2+}$	μ = 1,0	15,3–35	–0,59 (25°)	9,5	13,7	CSM	[443]

Table 114 (continued)

Reaction	Medium	t°	log k	log A	E	Method	Ref.
$Co(NH_3)_5(cis\text{-}OOCCH{=}CHCOOH)^{2+}$ + + $Ru(NH_3)_6^{2+}$	μ = 0,20 [H^+] = = 10^{-3} to 10^{-2} M	25	2,3			CSM	[488a]
$Co(NH_3)_5(trans\text{-}OOCCH{=}CHCOOH)^{2+}$ + + $Ru(NH_3)_6^{2+}$	μ = 0,2; [H^+] = = 10^{-3} to 10^{-2} M	25	−0,347			CSM	[488a]
	μ = 0,19; [H^+] = =0,18 M	25	−0,222			CSM	[488a]
$Co(NH_3)_5$[OOC— 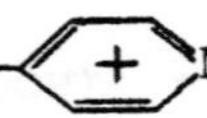 N—$CH_3]^{3+}$ + + $Ru(NH_3)_6^{2+}$	μ = 0,25; [H^+] = = 10^{-3} to 10^{-2} M	25	−0,7			CSM	[488a]
	Cobalt						
Co^{3+} + Mn^{2+}	3 M $HClO_4$	25	2,0			JM	[489]
Co^{3+} + Ag^+	3 M $HClO_4$	25	2,040	15,7	18,6	CSM	[397]

$Co^{3+} + [Hg(I)]_2$, $k_1 + \frac{k_2}{[H^+]}$	$\mu = 3,0$	5–19,9	–1,18 (k_1, 20°)	15,3	22,1	CSM	[433]
			–0,58 (k_2, 20°)	21,2	29,2	CSM	
$Co^{3+} + Te^+$	$\mu = 2,7$	0–25	–2,602	16,8±1,5	26,4±2	CSM	[490]
$CoOH^{2+} + Te^+$	$\mu = 2,7$ $[H^+] = 0,25$ to 2M	25	~–2			CSM	[490]
$CoSO_4^+ + Te^+$	$\mu = 2,7$; $[H^+] = 2,5$ M	25	–2,432			CSM	[490]
$CoOH^{2+} + CeClO_4^{2+}$	$\mu = 1,0$	14–32	1,977 (25°)	16,4	19,6	CSM	[491]
$CoOH^2 + CeNO_3^{2+}$	$\mu = 1,0$	20–40	1,968 (25°)	12,7	14,6	CSM	[492]
$CoOH^{2+} + CeF^{2+}$	$\mu = 1,0$	25	3,93			CSM	[492]
$Co(EDTA)^- + IrCl_6^{4-}$	$\mu = 1$; pH 3,9	22	3,6			JM	[493]
$Co^{3+} + Np(V)$	$\mu = 2,1$; $[H^+] = 1,28$ M	25	2,50 (k)				[494]
$\kappa = k_o e^{-\beta [H^+]}$	$\mu = 2,1$	5–25	2,55 (k_o, 25°)	12,0	12,9	CSM	[494]

Table 114 (continued)

Reaction	Medium	t°	log k	log A'	E	Method	Ref.
$Co^{3+} + Np(V) \cdot Cr(III)$	$\mu = 2,1$; $[H^+] = 1,02$ M	25	0,360 (k)			CSM	[494]
$k = k_o[H^+]$	$\mu = 2,1$	3–25	0,358 (k_o, 25°)	13,8	18,2	CSM	[494]
$Co(NH_3)_4(H_2O)Cl^{2+} + (Ta_6Cl_{12})^{2+}$	$\mu = 0,05$	25–34,7	–0,020 (25°)	10,3	14,1	CSM	[106в]
$Co(NH_3)_5I^{2+} + I^-$, $k_1 + k_2[I^-]$	$\mu = 1,0$	45	–4,176 (k_1) 1,068 (k_2)			KEPF	[495]
$Co(NH_3)_5H_2O^{3+} + I^-$	$\mu = 1,0$	45	–4,9			KEPF	[495]
$Co^{3+} + HO_2^-$	$\mu = 2,0$	0–25,1	12,3 (20°)	26,0	30,0	CSM	[496]

	Copper						
$Cu^{2+} + V^{3+}$, $k_1 + \frac{k_2}{[H^+]}$	$\mu = 3,0$ $[H^+] = 0,7$ to $0,6$ M	10–25	−1,027 (k_1, 25°)	14,4	21	Catal.	[410]
			−0,569 (k_2, 25°)	13,3	19		
$Cu^{2+} + V^{2+}$	$\mu = 11$	25	1,447			CSM	[420]
$Cu^{2+} + VO^{2+}$, $k[H^+]$	$\mu = 3$	25	2,59			CSM	[420]
$Cu^{2+} + Cr^{2+}$	$\mu = 1$	15,6–24,6	−0,77 (k_1, 24,6°)	8,9	13,1	CSM	[497]
$k_1 + \frac{k_2}{[H^+]}$			−0,232 (k_2, 24,6°)	12,8	17,7		
$Cu^{2+} + Eu^{2+}$		25	~0,477			CSM	[420]
$Cu^{2+} + (Ta_6Cl_{12})^{2+}$	1 M $HClO_4$	25	−1,569			CSM	[106в]
$Cu(II) + S_2O_8^{2-}$	$\mu = 1$; chloride medium		3,146			Pol.	[498]
	Arsenic						
$AsO_3^{3-} + TeO_4^{2-}$	$\mu = 0,05$ M	89,6–120	−2,44 (89,6°)	6,3	14,4	CSM	[499]
$TeO_4^{3-} + AsO_3^{4-}$	– " –	89,6	−5,92			CSM	[499]

Table 114 (continued)

Reaction	Medium	t°	log k	log A	E	Method	Ref.
$Mo(CN)_8^{3-} + Os(2,2'\text{-}Dipy)_3^{2+}$	$\mu = 0,5$	10	9,3			TJ	[500]
$Mo(V)_2 + NH_3OH^+$	3 M HCl	25–35	−4,4 (25°)	12,5	23,0	CSM	[501]
$[Mo(V)]_2 + I_3^-$	$\mu = 0,51$; pH=2,0	2,5–40	−3,952 (30°)	13,6	24,3	CSM	[502]
$Mo(VI) + ClO_3^-$	1 M H_2SO_4	25	8,114			Pol.	[503]
$Mo(VI) + NO_3^-$	– " –	25	2,968			Pol.	[503]
$Mo(VI) + ClO_4^-$	– " –	25	1,477			Pol.	[503]
	Ruthenium						
$Ru(NH_3)_6^{3+} + V^{2+}$	$\mu = 0,176$	25	1,903			CSM	[488a]
	$\mu = 5,3 \cdot 10^{3-}$	25	1,623			CSM	[488a]
$Ru(NH_3)_6^{3+} + V(Dipy)_3^{2+}$	$\mu = 0,1$	25	−0,932			CSM	[411]
$Ru(NH_3)_6^{3+} + V(Phen)_3^{2+}$	μ = 0,16 to 0,17	25	−0,836			CSM	[411]

$Ru(NH_3)_6^{3+} + V(Tripy)_2^{2+}$	$\mu = 0,1$ $C_2H_5OH/H_2O = 0,1$	25	0,285			CSM	[411]
$Ru(NH_3)_6^{3+} + Cr^{2+}$	$\mu = 0,2$	25	2,3			CSM	[488a]
$Ru(Dipy)_3^{3+} + Fe^{2+}$	0,5N $HClO_4$	25				JM	[400]
$Ru(Dipy)_3^{3+} + Fe(Dipy)_3^{2+}$		18	>5			CSM	[502a]
$Ru(Dipy)_3^{3+} + Fe(CN)_6^{4-}$		18	>5			CSM	[502a]
$Ru(Dipy)_3^{3+} + Fe(Phen)_3^{2+}$	0,1N H_2SO_4	0	>6,48			CSM	[339]
$Ru(Dipy)_3^{3+} + Fe(5\text{-}C_6H_5\text{-}Phen)_3^{2+}$	5N H_2SO_4	25	>8			JM	[400]
$Ru(Dipy)_3^{3+} + Fe[3,4,7,8\text{-}(CH_3)_4\text{-}Phen]_3^{2+}$	0,5N H_2SO_4	25	>8			JM	[400]
$Ru(NH_3)_6^{3+} + Cu^{+}$	$\mu = 0,2$	25	4,342			CSM	[488a]
$RhCl_6^{2-} + Ru(Phen)_3^{2+}$	$\mu = 0,1$	10	>9,4			TJ	[415]
$Ru(Dipy)_3^{3+} + Os(Dipy)_3^{2+}$	0,5N $HClO_4$	25	>8			JM	[400]
$Ru(Phen)_3^{3+} + RhCl_6^{3-}$	$\mu = 0,1$		>8,43			TJ	[415]
$Ru_{2+} + H_2O$, $W = k\,[Ru(II)]$	2,5 M KCl; pH 1,5	30–44,5	−4,334 (30°)	14,8	26,4	CSM	[503a]

Table 114 (continued)

Reaction	Medium	t°	log k	log A	E	Method	Ref.
Ru(II)+D_2O, W=k [Ru(II)]	2,5 M KCl; pH 1,5	30	–4,42			CSM	[503a]
Silver							
Ag(III)+Ag^+	5,87 M $HClO_4$	–14,8–11,4	3,008 (0,2°)	13,0	12,5	IM	[357]
Ag(II)+V(IV)	4 M $HClO_4$	25	3,7			JM	[432]
Ag(II)+Cr^{3+}	3 M $HClO_4$	15–30	1,681 (25)°	10,9	12,6	CSM	[397]
Ag(II)+Mn(II)	4 M $HClO_4$	25	4,477			JM	[432]
Ag(II)+Fe^{2+}	4 M $HClO_4$	25	5,602			JM	[432]
$Ag(Dipy)_2^{2+}$+Fe^{2+}	0,5 N H_2SO_4	25	6,146			JM	[400]
$Ag(Phen)_2^{2+}$+Fe^{2+}	0,5 N H_2SO_4	25	6,23			JM	[400]
Ag(II)+Co^{2+}	3 M $HClO_4$	15–30	2,176 (25°)	7	6,6	CSM	[397]
Ag(II)+Co^{2+}	4 M $HClO_4$	25	3,243			JM	[432]
Ag^+(II)+$S_2O_3^{2-}$	μ=0,02	25–40	–2,013 (25°)		17,9	CSM	[504]

	$C_2H_5OH:H_2O = 1:1$						
$Ag^{+} + S_2O_8^{2-}$	$\mu = 1$	35–50	−2,149 (35°)	6,6	12,3	CSM	[505]
$AgC_2O_4^{-} + S_2O_8^{2-}$	$\mu = 0,27$	25	−1,547			CSM	[506]
$Ag(II) + S_2O_6^{-}$, k [H^{+}]	$\mu = 3,5$		−1,26			CSM	[507]
$Sn(II) + S_2O_8^{2-}$	0,468 M HCl	15–30	−0,168 (25°)	8,6	11,9	CSM	[508]
$SbCl_5 + SbCl_3$	CCl_4	50,1–81	−6,92 (k_1, 50,1°)	6	19,0	IM	[509]
$W = k_1[SbCl_3] + k_2[SbCl_3] \cdot [SbCl_5]^2$			−3,8 (k_2, 50,1°)	6,4	15,0		
Sb(V) + Sb(III)	11,7 M HCl	25	−2,222			IM	[509]
	Cerium						
$Ce(IV) + Mn^{2+}$	2 N $HClO_4$ + 0,1 N H_2SO_4	50	−2,0			CSM	[510]
$Ce(IV) + Fe^{2+}$	0,5 N H_2SO_4	25	6,146			JM	[511]
$Ce(IV) + Fe^{2+}$	0,5 N H_2SO_4	25	6,114	13,0	9,5	JM	[512]
$Ce^{4+} + Fe^{2+}$	$\mu = 2$ $HClO_4$ + $NaClO_4$	0,3	3,77			JM	[513]

Table 114 (continued)

Reaction	Medium	t°	log k	log A	E	Method	Ref.
$CeOH^{3+} + Fe^{2+}$	μ=1 $HClO_4$+ +$NaClO_4$	0,3–15,1	2,602 (0,3°)	10,1	9,4	JM	[513]
$Ce(OH)_2^{2+} + Fe^{2+}$	μ=1 $HClO_4$+ +$NaClO_4$	0,3	3,505			JM	[513]
$CeSO_4^{2+} + Fe^{2+}$	μ=0,23 ($HClO_4$)	0	3,7			JM	[513]
Ce(IV) + $Fe(CN)_6^{4-}$	1N H_2SO_4	25	6,279			JM	[511]
Ce(IV) + $Fe(Dipy)_3^{2+}$	1 M $HClO_4$	18	> 5			CSM	[502a]
Ce(IV) + $Fe(Phen)_3^{2+}$	0,5 N H_2SO_4	25	5,152			JM	[512]
Ce(IV) + + Fe $(5\text{-}NO_2\text{-}Phen)_3^{2+}$	1 M $HClO_4$	18	~3,6			CSM	[502a]
	0,5 N H_2SO_4	25	3,59			JM	[512]
Ce(IV) + + Fe $(5\text{-}Cl\text{-}Phen)_3^{2+}$	0,5 N H_2SO_4	25	4,398			JM	[512]
Ce(IV) + + Fe $(5\text{-}CH_3\text{-}Phen)_3^{2+}$	0,5 N H_2SO_4	25	5,342			JM	[512]

Ce(IV) + + Fe $(5\text{-Ph-Phen})_3^{2+}$	0,5 N H_2SO_4	25	5,08			JM	[512]
Ce(IV) + + Fe $[5,6\text{-}(CH_3)_2\text{-Phen}]_3^{2+}$	0,5 N H_2SO_4	25	5,633			JM	[512]
Ce(IV) + + Fe $[3,4,7,8\text{-}(CH_3)_4\text{-Phen}]_3^{2+}$	0,5 N H_2SO_4	25	6,204			JM	[512]
$Ce^{4+} + Ag^{+}$	μ = 4,5	9,8–30	−0,988 (25°)	13,7	20	Catal.	[410]
	2 N $HClO_4$ + +0,1 N H_2SO_4	50	−2,222			Catal.	[510]
$Ce^{4+} + Ru(Dipy)_3^{2+}$	1 M $HClO_4$	25	3,602			CSM	[502]
$Ce^{4+} + Ru(Phen)_3^{2+}$	0,5 N H_2SO_4	5–40	3,763 (25°)	7,4	4,9	JM	[514]
$Ce^{4+} + [Hg(I)]_2^{2+}$	2 N $HClO_4$	50–70	−0,854 (50°)	8,9	14,4	CSM	[515]
$Ce^{4+} + Ta_6Cl_{12}^{2+}$	1N $HClO_4$	25	7,176			JM	[106в]
Ce^{4+} + U(IV)	μ = 2; $HClO_4$	2,4–15,6	3,608 (15,6°)	14,7	14,6	CSM	[515a]
$Ce(SO_4)_2 + H_3PO_2 \rightleftharpoons Ce(SO_4)_2H_3PO_2 \longrightarrow CeSO_4^{+} + H_2PO_2^{-} + HSO_4^{-}$	1 M H_2SO_4	30–35	−1,352(k)	7,2	11,8	CSM	[516]

Table 114 (continued)

Reaction	Medium	t°	log k	logA	E	Method	Ref.
$Ce^{4+} + H_3PO_2$	μ = =1,3 to 4,4	25	$-2{,}914(k_1)$			CSM	[517]
$W = [H^+]\sum_1^3 k_n \times [Ce(H_2PO_2)_n^{+(4-n)}]$	$[H^+]$ =1,185 to 4,195	0,05–25	$-4{,}103$ (k_2, 25°)	12,8	23,0	CSM	[517]
		0,05–25	$-3{,}738$ (k_3, 25°)	11,7	21,0	CSM	[517]
$Ce^{4+} + NH_3OH^+$	3,93 M H_2SO_4	17,6–25	0,82	11,7	14,8	CSM	[518]
Europium							
$Eu^{3+} + Cr^{2+}$	μ = 2	25	~ $-4{,}854$			MCR	[383]
Osmium							
$Os(Dipy)_3^{3+} + Fe^{2+}$	0,5 N $HClO_4$	25	3,155			JM	[400]
	0,5 N H_2SO_4	25	4,13			JM	[400]
$Os(Dipy)_3^{3+} + Fe(CN)_6^{4-}$	0,5 N $HClO_4$	25	> 8			JM	[400]

$OsCl_6^{2-} + Fe(CN)_6^{4-}$	H_2O	25	−0,748			JM	[400]
$Os(Dipy)_3^{3+} + Mo(CN)_8^{4-}$	$\mu = 0,5$	25	9,602			TJ	[500]
$Os(Dipy)_3^{2+} + S_2O_8^{2-}$	$\mu = 0$	10,2–25,5	1,716 (25,5°)	8,45	9,4	CSM	[519]
	Iridium						
$IrCl_6^{2-} + Fe^{2+}$	0,5 N $HClO_4$	15–35	6,505 (25°)	7,9	1,9	JM	[400]
$IrCl_6^{2-} + Fe(CN)_6^{4-}$	H_2O	25	6,08			JM	[400]
$IrCl_6^{2-} + Fe(CN)_6^{4-}$	0,5 N $HClO_4$	25	2,612			JM	[400]
$IrCl_6^{2-}$ + + $Fe[4,4'-(CH_3)_2-Dipy]_3^{2+}$	$\mu = 0,10$	10	8,982			TJ	[415]
$IrCl_6^{2-}$ + + $Fe[4,7-(CH_3)_2-Phen]_3^{2+}$	$\mu = 0,10$	10–30	9,04*		0,5	TJ	[414]
$IrCl_6^{2-}$ + + $Fe[5,6-(CH_3)_2-Dipy]_3^{2+}$	$\mu = 0,10$	10	8,342			TJ	[415]

* Authors of (414) give this value of constant for entire interval 10-30°C.

Table 114 (continued)

Reaction	Medium	t°	log k	log A	E	Method	Ref.
$IrBr_6^{2-}$ + + $Fe[4,4'-(CH_3)_2-Dipy]_3^{2+}$	μ = 0,10	10	8,832			TJ	[415]
$IrBr_6^{2-}$ + + $Fe[4,7-(CH_3)_2-Phen]_3^{2+}$	μ 0,10	10	8,204			TJ	[415]
$IrBr_6^{2-}$ + $Fe[5,6-(CH_3)_2-Phen]_3^{2+}$	μ = 0,30	10	8,204			TJ	[415]
$IrCl_6^{2-}$ + $Os(Dipy)_3^{2+}$	0,5 N $HClO_4$	25	>8			JM	[400]
Hg^{2+} + $Ta_6Cl_{12}^{2+}$	1 N $HClO_4$		0,539			CSM	[106в]
	Thallium						
$TlOH^{2+}$ + Fe^{2+}	μ = 3; 0,5–2 N $HClO_4$	25–45	–0,854 (25°)	12,4	18,0	CSM	[520]
Tl(III) + Fe^{2+}, $\frac{k}{[H^+]^2}$	μ = 3; 0,5–2 N $HClO_4$	25	–0,92			CSM	[520]

$TlOH^{2+} + Fe(Phen)_3^{2+}$	2,7 M $HClO_4$	25–34	–0,354 (25°)	12,7	18,4	CSM	[502a]
$Tl^{3+} + Os(Dipy)_3^{2+}$	μ = 2,99; $[H^+]$ = =0,52–2,99 M	17,5–35	1,556 (24,5°)	6,6	6,9	CSM	[521]
$TlOH^{2+} + Os(Dipy)_3^{2+}$	μ = 2,99; $[H^+]$ = =0,52–2,99 M	17,5–35	1,167 (24,5°)	9,6	11,5	CSM	[521]
$Tl(III) + U(IV)$, $\frac{k}{[H^+]}$	μ = 2,9	15–50	–1,821 (25°)	19,2	28,6	CSM	[522]
$Tl(III) + U(IV)$, $\frac{k}{[H^+]^2}$	μ = 2,9	15–50	–1,845 (25°)	17,7	26,5	CSM	[522]
$Tl(III) + U(IV)$	μ = 2,9	16–25	–1,676 (k_1, 25°)	16,8	25,2	CSM	[523]
$\frac{k_1}{[H^+]} + \frac{k_2}{[H^+]^2}$			–1,672 (k_2, 25°)	14,7	22,3		
$Tl(III) + Ru(Dipy)_3^{2+}$	4,75 M $HClO_4$	20–40	–0,722 (25°)	15,9	22,8	CSM	[524]
$Tl(III) + Ru[4,4'-(CH_3)_2-Dipy]_3^{2+}$	- " -	20–40	0,526 (25°)	14,5	19,3	CSM	[524]
$Tl(III) + Ru(Tripy)_2^{2+}$	- " -	20–40	–0,081 (25°)	7	9,6	CSM	[524]
$Tl(III) + Ru(Phen)_3^{2+}$	- " -	20–40	–0,244 (25°)	12	16,7	CSM	[524]

Table 114 (continued)

Reaction	Medium	t°	log k	log A	E	Method	Ref.
Tl(III) + $Ru(5\text{-}Cl\text{-}Phen)_3^{2+}$	4,75 M $HClO_4$	25–45	–0,984 (25°)	12,5	18,6	CSM	
Tl(III) + $Ru(5\text{-}CH_3\text{-}Phen)_3^{2+}$	- " -	20–40	0,305 (25°)	13	17,5	CSM	
Tl(III) + $Ru[4,7\text{-}(CH_3)_2\text{-}Phen]_3^{2+}$	- " -	20–40	0,174 (25°)	10	13,5	CSM	
Tl(III) + $Ru[5,5'\text{-}(CH_3)_2\text{-}Dipy]_3^{2+}$	- " -	10–30	1,62 (25°)	16	19,7	CSM	[524]
Tl(III) + $Ru[5,6\text{-}(CH_3)_2\text{-}Phen]_3^{2+}$	- " -	10–30	0,555 (25°)	8,1	10,4	CSM	
Tl(III) + + $Ru[3,4,7,8\text{-}(CH_3)_4\text{-}Phen]_3^{2+}$	- " -	20–35	0,59 (25°)	6,7	8,6	CSM	
		20–35	0,222 (25°)	21,7	29,5	CSM	
Pb(IV) + Mn^{2+}	CH_3COOH	25	–0,366	9,9	14	CSM	[525]
Pb(IV) + Co^{2+}	- " -	25	–0,252	10,0	14	CSM	[525]
Pb(IV) + Ce^{4+}	- " -	25	–0,62	14,8	21	CSM	

	Uranium						
$UO_2^{2+} + U^{4+}$	$\mu = 2$	25,1–47,1	−6,672 (25,1°)	21,3	38,1	IM	[526]
$UO_2^{+} + UO_2^{+}$, $k[H^+]$	$\mu = 0,4$	25	2,113			CSM	[527]
	$\mu = 2,1$	8–36	2,64 (25°)	10,6	10,8	CSM	[528]
	$[H^+] = 0,0059$ to 0,0153 M	25	2,415			CSM	[529]
	$\mu = 1$	5–35	2,474 (25°)	6,6	4,2	CSM	[530]
	$\mu = 2$; $[U(IV)] \to 0$	0,3–25,1	2,38 (25,1°)	9,0	9,0	KEPF	[531]
U(V) + U(V), $k[H^+]$	$\mu = 1$, sulfate	5–35	4,403 (25°)	10,65	8,5	CSM	[530]
	$\mu = 1$, citrate	5–35	4,322 (25°)	6,4	2,8	CSM	[530]
	$\mu = 1$, tartrate	5–35	4,694 (25°)	5,65	1,3	CSM	[530]
$UO_2^{2+} + V^{2+}$	$\mu = 2,02$; $LiClO_4$	0,6–36,8	1,87 (25,1°)	7,5	7,7	KEPF	[532]
	$\mu \to 0$; $LiClO_4$	0,6–36,8	0,465 (25,1°)	6,37	6,7	KEPF	[532]
$UO_2^{2+} + V^{3+}$, $\frac{k}{[H^+]^{1,8}}$	$\mu = 2$	0,2–34,8	−0,553 (25,2°)	15,6	22,1	Catal.	[533, 385]

Table 114 (continued)

Reaction	Medium	t °	log k	log A	E	Method	Ref.
U(VI) + Sn(II)	6 M HCl	35–55	–4,824 (35°)	8,4	18	CSM	[534]
U(IV) + $HClO_3$	0,1–0,75 M $HClO_4$	25	–2,523			CSM	[535]
U(IV) + HNO_2 W = k [U(IV)]	$[HNO_2]$ = 7,4 x 10^{-3} mole/liter μ = 2; 2 M HNO_3	25	–3,914			CSM	[536]
	$[HNO_2]$ = 9,1 x 10^{-3} mole/liter μ = 0,54; 0,5 M HNO_3	25	–4,08			CSM	[536]
U(IV) + ClO_2	$[H^+]$ = 0,25 M	25	0,812			CSM	[537]
U(IV) + ClO^-	$[H^+]$ = 0,25 M	25	0,613			CSM	[537]
U(IV) + ClO_2^-	$[H^+]$ = 0,25 M	0	0,634			CSM	[537]
	$[H^+]$ = 0,25 M	25	1,342			CSM	[537]
U(IV) + BrO_3^-	μ = 4	15–35	–0,384 (k_1, 25°)			CSM	[538]

$k_1 + k_2 [H^+]^2$			−0,857 (k_2,25°)				
$U(IV) + S_2O_8^{2-}$	0,5 M H_2SO_4		−2,097			CSM	[539]
$U(III) + NO_3^-$	0,1 M KCl	25	6,204	6,4	6,5	Pol.	[540]
	Neptunium						
$NpO_2^+ + Np^{3+}$, $k [H^+]$	$\mu = 2$	5,53–25	1,633 (25°)	6,4	6,5	KNP	[541]
	D_2O		1,536 (25°)	5,65	5,6		
$NpO_2^{2+} + Np^{4+}$	$\mu = 2$; $NaClO_4 + HClO_4$	17,4–34,6	−1,18 (k_1,25°)	17	24,9	CSM	[542]
$\frac{k_1}{[H^+]^2} + \frac{k_2}{[H^+]^3}$		17,4–34,6	−2,22 (k_2,25°)	20,8	31,5		
$NpO_2^{2+} + Np^{4+}$, $\frac{k}{[H^+]^2}$	$\mu = 2$	15,4–34,8	−1,35 (25°)	17,15	25,6	CSM	[543]
	$\mu = 2{,}2$	24,9	−1,24			CSM	[543]
	$\mu = 2{,}2$; 95% D_2O	24,9	−1,92			CSM	[543]
$NpO_2^{2+} + Np^{4+}$ $k_1 + k_2 [NO_3^-]$ with $[H^+]$ = const	$[H^+] = 2$ M	25	−0,864 (k_1)				
$k_2 = \frac{k_2'}{[H^+]^2}$	$\mu = 2$	18,2–32,4	−0,078 (k_2', 25°)	16,6	22,8	CSM	[544]

Table 114 (continued)

Reaction	Medium	t°	log k	log A	E	Method	Ref.
$NpO_2^{2+} + NpSO_4^{2+}$	$\mu = 2{,}2$	25	–0,593 (k_1)			CSM	[545]
$k_1 [H^+]^2 + \frac{k_2}{[H^+]^3}$			–0,918 (k_2)				
$NpO_2SO_4 + NpSO_4^{2+}$	$\mu = 2{,}2$	25	–0,365 (k_1)			CSM	[545]
$\frac{k_1}{[H^+]^2} + \frac{k_2}{[H^+]^3}$			–0,69 (k_2)				
$NpO_2^{2+} + V^{3+}$	$\mu = 1$	17–34	0,8 (k_1, 25°)	14,5	18,6	CSM	[546]
$k_1 + \frac{k_2}{[H^+]}$			1,31 (k_2, 25°)	17,9	22,6		
$NpO_2^{+} + V^{3+}$	0,37 M $HClO_4$	15,05–35	–1,42 (25°)	9,0	14,2	CSM	[547]
$k_1 + \frac{k_2}{[H^+]}$	$\mu = 3$; 0,37 M $HClO_4$		–0,57 (25°)	10,3	14,8		

$NpO_2^{+} + Cr^{2+}$	μ = 0,207	5–25	3,059 (k_o, 25°)	4,86	2,45	KEPF	[548]
$k_o[H^+]\,e^{-\beta[H^+]}$	0,02–0,2 M $HClO_4$	25	–1,142 (β)			KEPF	[548]
$Np^{4+} + Cr^{2+}$	μ = 1,0	5–25	–0,368 (k_1, 25°)	12,1	18,4	CSM	[549]
$\frac{k_1}{[H^+]} + \frac{k_2}{[H^+]^2}$	$[H^+]$ = 0,166–1,0 M	25	–0,244 (k_2, 25°)	12,6	17,5		
$NpO_2^{+} + Fe^{2+}$	1 M $HClO_4$	25	–1,08			CSM	[419]
$NpO_2^{2+} + UOH^{3+}$	μ = 1,0	5–25	2,964 (25°)	8,7	7,8	CSM	[435]
NpO_2^{+} + U(IV)	$[H^+]$=0,1 M; μ = 1; ClO_4^-	20–34,3	–3,076 (25°)	20,5	32,2	KEPF	[550]
NpO_2^{+} + U(IV)	$[H^+]$=7,05 x x 10^{-2}; μ = 0,15; Cl^-	13–40	–5,83 (25°)	14,5	27,8	CSM	[550]
Np(IV) + U(IV)	$[H^+]$=0,1; μ = 1; ClO_4^-	20–34,3	–2,273 (25°)	8,9	15,2	CSM	[550]
	$[H^+]$=7,05 x x 10^{-2}; μ = 0,14; Cl^-	13–40	–3,5 (25°)	21	33,3	CSM	[550]

Table 114 (continued)

Reaction	Medium	t°	log k	log A	E	Method	Ref.
$NpO_2^+ + I^-$ $W = k_1[H^+]^2[NpO_2^+][I^-] + k_2[H^+]^3[NpO_2^+][I^-]$	3,29 M HCl	25	−3,518 (k_1) −2,046 (k_2)			CSM	[551]
Plutonium							
$PuO_2^{2+} + Pu^{3+}$	μ = 1,0	5–35	0,417 (25°)	3,95	5,4	CSM	[552]
$PuO_2^+ + Pu^{3+}$ $k[H^+]$	μ = 0,47; 0,1–0,3 M HNO_3	25	−1			CSM	[553]
	0,5 M HCl	1	−1,237			CSM	[554]
$PuO_2^+ + Pu^{4+}$	μ = 1	5–35	1,572 (25°)	12,0	14,2	CSM	[552]
$Pu^{4+} + Pu^{4+}$, $\frac{k}{[H^+]^3}$	μ = 1; $HClO_4$	25	−4,514			CSM	[555]
	μ = 1; HCl	6,3–45,1	−3,804 (25°)	24,8	39	CSM	[556]
	$\mu \approx 0,4$; HNO_3	25	−4,584			CSM	[553]

$PuO_2^{+} + PuO_2^{+}$, $k[H^+]$	$\mu = 1,0$	10,3–35	–2,444 (25°)	12,0	19,6	CSM	[557]
$PuO_2^{2+} + Ti^{3+}$, $\frac{k}{[H^+]}$	$\mu = 2$	2,5–25	1,408 (2,5°)	10,1	10,9	CSM	[558]
$Pu^{4+} + Ti^{3+}$, $\frac{k}{[H^+]}$	$\mu = 2,02$	2,4	1,102			CSM	[559]
	$\mu = 1,02$	2,4–25	0,78 (2,4°)	14,5	17,3	CSM	[559]
$PuO_2^{2+} + V^{3+}$, $\frac{k_1}{[H^+]} + \frac{k_2}{[H^+]^2}$	$\mu = 1,0$	25	0,257 (k_1) –0,7 (k_2)			CSM	[560]
	$\mu = 2,0$	25	0,326 (k_1) –0,642 (k_2)			CSM	[560]
$Pu^{4+} + V^{3+}$, $\frac{k_1}{[H^+]} + \frac{k_2}{[H^+]^2}$	$\mu = 2,0$	2,4–20	1,074 (k_1, 20°)	14,3	17,7	CSM	[561]
			1,302 (k_2, 20°)	17,1	22,1		
$PuO_2^{2+} + Fe^{2+}$	0,42 M $HClO_4$	0,2	2,434			CSM	[562]
	$\mu = 2$; $[H^+] = 1$ M	0,1	2,7			CSM	[562]
$PuO_2^{+} + Fe^{2+}$	1 M $HClO_4$	0,2	1,204			CSM	[562]

Table 114 (continued)

Reaction	Medium	t°	log k	log A	E	Method	REf.
$Pu^{4+} + Fe^{2+}$, $\frac{k}{[H^+]}$	$\mu = 2{,}03$	2,5–20,2	0,494 (2,5°)	16,1	19,7	CSM	
$Pu^{4+} + Fe^{2+}$, $k[Cl^-]$	$\mu = 2{,}03$	2,5–20,2	2,172 (20,2°)			CSM	[563]
$Pu^{4+} + Fe^{2+}$, $\frac{k[Cl^-]}{[H^+]}$	$\mu = 2{,}03$	2,5–20,5	1,944 (20,2°)	13,3	15	CSM	
$FuO_2^{2+} + Sn^{2+}$	$HClO_4$ + + HCl; $\mu = 2$	2,4–29,6	3,186 (k_1, 20,2°)	18,3	16,3	CSM	[564]
$k_1[Cl^-]^3 + k_2[Cl^-]^4$			3,711 (k_2, 20,2°)	14,2	14,1		
$Pu^{4+} + U^{4+}$, $\frac{k}{[H^+]^2}$	$\mu = 2{,}03$; $[H^+]$=0,4 to 2,0 M	2,8–29,8	0,365 (2,8°)	20,2	24,9	CSM	[565]
$Pu^{3+} + HNO_2$, $k[H^+]$	3,6 M HCl	24	1,255	4,4	6,0	CSM	[85]
$Pu^{3+} + HNO_2$, $k[H^+][NO_3^-]$	2,06 M HNO_3	24	0,176	10,5	14,0	CSM	[85]

Various oxidation-reduction reactions

$XeO_3 + Pu^{3+}$	2 M $HClO_4$	15,6–44,3	−1,742 (25°)	10	15,9	CSM	[566]
$Ta_6Cl_{12}^{2+} + Ta_6Br_{12}^{2+}$	1 M $HClO_4$	25	≥8			JM	[106 в]
$AmO_2 + AmO_2$ $k_1[H^+]^2 + k_2[H^+]^3$	μ = 2	60–85	−3,154 (k_1, 75,7°)	7,6	17,2	CSM	[567]
			−3,338 (k_2, 75,7°)	3,7	11,2		
$AmO_2^+ + AmO_2^+$	μ = 6; 2 M $NaClO_4$ + + 4 M $HClO_4$	25–75	−2,954 (25°)	11,2	19,3	CSM	[568]
Am(V) + Am(V)	4 M H_2SO_4	25	−1,246			CSM	[568]
	6 M HNO_3	25	−2,544			CSM	[568]

LITERATURE CITED (PART III)

1. R. N. Mehrotra, J. Chem. Soc., B, 1968, p. 1123.

2. J. R. Jones and W. A. Waters, J. Chem. Soc., 1962, p. 2068.

3. T. J. Kemp and W. A. Waters, Proc. Roy. Soc. (London), A274:480 (1963).

4. R. Shanker and S. N. Joshi, Indian J. Chem., 1:289 (1963).

5. J. S. Littler, A. I. Mallet, and W. A. Waters, J. Chem. Soc., 1960, p. 2761.

6. J. R. Jones, W. A. Waters, and J. S. Little, J. Chem. Soc., 1961, p. 630.

7. J. R. Jones and W. A. Waters, J. Chem. Soc., 1963, p. 352.

8. J. S. Littler and W. A. Waters, J. Chem. Soc., 1959, p. 3014.

9. J. R. Jones and W. A. Waters, J. Chem. Soc., 1962, p. 1629.

10. T. J. Kemp and W. A. Waters, J. Chem. Soc., 1964, p. 1610.

11. G. V. Bakore and R. Shanker, Can. J. Chem., 44:1717 (1969).

12. T. J. Kemp and W. A. Waters, J. Chem. Soc., 1964, p. 1192.

13. U. S. Mahnot, R. Shanker, and S. N. Swami, Z. Physik. Chem. (Leipzig), 222:240 (1963).

14. R. N. Mehrotra, J. Chem. Soc., B, 1968, p. 642.

15. G. V. Bakore, R. Shanker, and S. S. Dua, Z. Physik. Chem. (Leipzig), 236:129 (1967).

16. W. F. Pickering and A. McAuley, J. Chem. Soc., A, 1968, p. 1173.

17. M. M. Taqui Khan and A. E. Martell, J. Am. Chem. Soc., 90:6011 (1968).

18. R. N. Mehrotra, J. Chem. Soc., B, 1968, p. 1563.

19. P. S. Radhakrishna Murti and S. C. Pati, J. Indian Chem. Soc., 45:640 (1968).

20. P. J. Andrulis, M. J. S. Dewar, R. Dietz, and R. L. Hunt, J. Am. Chem. Soc., 88:5473 (1966).

21. P. J. Andrulis and M. J. S. Dewar, J. Am. Chem. Soc., 88:5483 (1966).

22. T. J. Kemp and W. A. Waters, J. Chem. Soc., 1964, p. 339.

23. G. V. Bakore and M. S. Bararia, Z. Physik. Chem. (Leipzig), 229:245 (1965).

24. M. A. Beg and F. Ahmad, J. Indian Chem. Soc., 44:947, (1967).

25. H. Taube, J. Am. Chem. Soc., 70:1216 (1948).

26. T. J. Kemp and W. A. Waters, J. Chem. Soc., 1964, p. 1489.

27. A. Y. Drummond and W. A. Waters, J. Chem. Soc., 1955, p. 497.

28. P. Levesley and W. A. Waters, J. Chem. Soc., 1955, p. 217.

29. R. Shanker and G. V. Bakore, Current Sci. (India), 31:376 (1962).

30. V. M. Gol'dberg and L. K. Obukhova, Neftekhimiya, 7:88 (1967).

31. V. M. Gol'dberg and L. K. Obukhova, Izv. Akad. Nauk SSSR, Ser. Khim., 1967, p. 1419.

32. R. N. Mehrotra, Z. Physik. Chem. (Frankfurt), 43:140, (1964).

33. J. K. Thomas, G. Trudel, and S. Bywater, J. Phys. Chem., 64:51 (1960).

34. D. I. Metelitsa and E. T. Denisov, Kinetika i Kataliz, 9:733 (1968).

35. J. H. Baxendale and H. R. Hardy, Trans. Faraday Soc., 50:808 (1964).

35a. W. C. E. Higginson and P. Wright, J. Chem. Soc., 1955, p. 1551.

36. J. S. Littler and I. G. Sayce, J. Chem. Soc., 1964, p. 2545.

37. T. J. Wallace, J. Org. Chem., 31:3071 (1966).

38. M. M. Taqui Kham and A. E. Martel, J. Am. Chem. Soc., 90:3386 (1968).

39. C. E. H. Bawn and A. G. White, J. Chem. Soc., 1951, p. 331.

40. J. H. Baxendale and C. F. Wells, Trans. Faraday Soc., 53:800 (1957).

41. C. E. H. Bawn and J. A. Sharp, J. Chem. Soc., 1957, p. 1854.

42. T. A. Cooper and W. A. Waters, J. Chem. Soc., B, 1967, p. 687.

43. D. G. Hoare and W. A. Waters, J. Chem. Soc., 1964, p. 2560.

44. D. G. Hoare and W. A. Waters, J. Chem. Soc., 1962, p. 965.

45. D. G. Hoare and W. A. Waters, J. Chem. Soc., 1964, p. 2552.

46. K. Jijie and M. Santappa, Proc. Indian Acad. Sci., 65:155 (1967).

47. D. G. Hoare and W. A. Waters, J. Chem. Soc., 1962, p. 971.

48. T. A. Cooper and W. A. Waters, J. Chem. Soc., 1964, p. 1538.

49. G. Hargreaves and L. H. Sutcliffe, Trans. Faraday Soc., 51:786 (1955).

50. T. A. Cooper and W. A. Waters, J. Chem. Soc., B, 1967, p. 687.

51. C. E. H. Bawn and A. G. White, J. Chem. Soc., 1951, p. 339.

52. A. A. Clifford and W. A. Waters, J. Chem. Soc., 1965, p. 2796.

53. P. R. Sharan, P. Smith, and W. A. Waters, J. Chem. Soc., B, 1968, p. 1322.

53a. W. A. Waters, Discussions Faraday Soc., 46:158 (1968).

54. K. Jijie and M. Santappa, Indian J. Chem., 6:262 (1968).

55. J. Hill and A. McAuley, J. Chem. Soc., A, 1968, p. 1169.

56. C. F. Wells, Trans. Faraday Soc., 63:156 (1967).

57. V. M. Solyanikov and E. T. Denisov, Neftekhimiya, 6:97 (1966).

58. V. S. Martem'yanov and E. T. Denisov, Zh. Fiz. Khim., 40:2315 (1966).

59. V. S. Martem'yanov and E. T. Denisov, Zh. Fiz. Khim., 41:687 (1967).

60. R. B. Wiberg and W. G. Nigh, J. Am. Chem. Soc., 87:3849 (1965).

61. P. S. R. Murti and S. C. Pati, Chem. Ind. (London), 1967, p. 702.

62. S. S. Muhammad and K. V. Rao, Bull. Chem. Soc. Japan, 36:943 (1963).

63. S. S. Muhammad and K. V. Rao, Bull. Chem. Soc. Japan, 36:949 (1963).

64. M. Ardon, J. Chem. Soc., 1957, p. 1811.

65. B. Sethuram and S. S. Muhammad, Acta Chim. Acad. Sci. Hungary, 16:115 (1965).

66. M. Rangaswamy and Santappa, Current Sci. (India), 35:332 (1966); 34:282 (1965).

67. H. L. Hintz and D. C. Johnson, J. Org. Chem. 32:556 (1967).

68. J. S. Littler, J. Chem. Soc., 1959, p. 4135.

69. A. A. Katai, V. K. Kulshrestha, and R. H. Marchessault, J. Phys. Chem. 68:522 (1964).

70. F. R. Duke and A. A. Forist, J. Am. Chem. Soc., 71:2790 (1949).

71. G. G. Guilbault and W. H. McCurdy, J. Phys. Chem., 67:283 (1963).

72. G. Mino, S. Kaizerman, and E. Rasmussen, J. Am. Chem. Soc., 81:1494 (1959).

73. G. Hargreaves and L. H. Sutcliffe, Trans. Faraday Soc., 51:1105 (1955).

74. N. Rangaswamy and M. Santappa, Acta Chim. Sci. Hungary, 56:413 (1968).

75. J. Shorter, J. Chem. Soc., 1962, p. 1868.

76. R. N. Mehrotra, Z. Physik. Chem. (Leipzig), 230:221 (1965).

77. V. H. Dodson and A. H. Black, J. Am. Chem. Soc., 79:3657 (1957).

78. K. K. Sengupta, J. Proc. Inst. Chemists (India), 36:138 (1964).

79. K. K. Sengupta and S. Aditya, Z. Physik. Chem. (Frankfurt), 38:25 (1963).

80. R. L. Yadav and W. V. Bhagwat, J. Indian Chem. Soc., 41:389 (1964).

81. A. McAuley, J. Chem. Soc., 1965, p. 4054.

82. G. V. Bakore, R. Dayal, and P. Nath, Z. Physik. Chem. (Leipzig), 227:19 (1964).

83. B. Krishna and K. C. Tewari, J. Chem. Soc., 1961, p. 3097.

84. K. K. Sengupta, Indian J. Chem., 2:267 (1964).

85. K. B. Bhargava, R. Shanker, and S. N. Joshi, J. Sci. Ind. Res. (India), 21B:573.

86. A. McAuley and C. H. Brubaker, J. Chem. Soc., A, 1966, p. 960.

87. K. K. Sengupta, S. Aditya, and B. N. Ghosh, J. Indian Chem. Soc., 40:823 (1963).

88. S. K. Misra and R. C. Mehrotra, J. Indian Chem. Soc., 44:927 (1968).

89. R. N. Mehrotra, and S. Gosh, Z. Physik. Chem. (Leipzig), 230:231 (1965).

90. K. K. Sengupta, J. Indian Chem. Soc., 41:423, 427 (1964).

91. R. N. Mehrotra and S. Gosh, Z. Physik. Chem. (Leipzig), 224:57 (1963).

92. R. Cecil and J. S. Littler, J. Chem. Soc., B, 1968, p. 1420.

93. A. M. Posner, Trans. Faraday Soc., 49:382 (1953).

94. M. Cher and N. Davidson, J. Am. Chem. Soc., 77:793 (1955).

95. L. K. Lepin' and B. P. Matssevskii, Dokl. Akad. Nauk SSSR, 173:1336 (1967).

96. P. George, J. Chem. Soc., 1954, p. 4349.

97. S. Utsumi and K. Murosima, J. Chem. Soc. Japan, Pure Chem. Sect., 86:593 (1965); abstracted in Ref. Zh. Khim., 1966, p. 3B689).

98. R. E. Huffman and N. Davidson, J. Am. Chem. Soc., 78:4836 (1956).

99. A. B. Lamb and L. W. Elder, J. Am. Chem. Soc., 53:137 (1931).

100. G. S. Hammond and Chin-Hua S. Wu, International Symposium on Oxidation, Vol. 2, San Francisco, American Chemical Society, 1967, p. 413.

101. É. A. Kutner, Candidate's Dissertation, Riga (1968).

102. J. B. Ramsey, R. Sugimoto, and H. De Vorkin, J. Am. Chem. Soc., 63:3480 (1941).

103. J. Simplicio and R. G. Wilkins, J. Am. Chem. Soc., 84:6092 (1967).

104. C. Tanford, D. C. Kirk, and M. K. Chantooni, J. Am. Chem. Soc., 76:5325 (1954).

105. H. Nord, Acta Chem. Scand., 9:430 (1955).

105a. A. Zuberbuehler, Helv. Chim. Acta, 50:466 (1967).

105b. J. Pecht and M. Anbar, J. Chem. Soc., A, 1968, p. 1902.

105c. J. Halpern and J. G. Smith, Can. J. Chem., 34:1419 (1956).

106. G. R. Hill, J. Am. Chem. Soc., 70:1306 (1948).

107. B. Chance, J. Franklin Inst., 229:737 (1940).

107a. G. Davies, L. J. Kirschenbaum, and K. Kustin, Inorg. Chem., 7:146 (1968).

108. J. H. Baxendale, M. G. Evans, and G. S. Park, Trans. Faraday Soc., 42:155 (1946).

109. W. G. Barb, J. H. Baxendale, P. George, and K. H. Hargrave, Trans. Faraday Soc., 47:462 (1951).

110. F. S. Dainton and H. C. Sutton, Trans. Faraday Soc., 49:1011 (1953).

111. J. P. Keene, Radiation Res., 22:14 (1964).

112. Z. Pospisil, Collection Czech. Chem. Commun., 18:337 (1953).

113. T. Rigg, W. Taylor, and J. Weiss, J. Chem. Phys., 22:575 (1954).

114. T. J. Hardwick, Can. J. Chem., 35:428 (1957).

115. C. F. Wells and M. A. Salem, Trans. Faraday Soc., 63:620 (1967).

116. W. G. Barb, J. H. Basendale, P. George, and K. R. Hargrave, Trans. Faraday Soc., 51:935 (1955).

116a. V. M. Berdnikov, Yu. M. Kozlov, and A. P. Purmal, Khim. Vysokikh Energii, 3:321 (1969).

116b. I. M. Kolthoff and R. Woods, J. Electroanal. Chem. Interfac. Chem., 12:385 (1966).

117. R. Woods, I. M. Kolthoff, and E. J. Meehan, J. Am. Chem. Soc., 85:3334 (1963).

118. G. Czapcki, B. H. J. Bielski, and N. Sutin, J. Phys. Chem., 67:201 (1963).

119. F. B. Baker and T. W. Newton, J. Phys. Chem., 65:1897 (1961).

120. A. J. Zielen, J. C. Sullivan, D. Cohen, and J. C. Hindman, J. Am. Chem. Soc., 80:5632 (1958).

121. A. A. Zaitsev, V. N. Kosyakov, A. G. Rykov, Yu. P. Sobolev, and G. N. Yakovlev, Radiokhimiya, 2:348 (1960).

122. W. L. Reynolds and R. J. W. Lumry, J. Chem. Phys., 23:2460 (1955).

123. W. L. Reynolds and I. M. Kolthoff, J. Phys. Chem., 60:969 (1956).

124. J. W. L. Fordham and H. L. Williams, J. Am. Chem. Soc., 73:1634 (1950).

125. J. W. L. Fordham and H. L. Williams, J. Am. Chem. Soc., 72:4465 (1950).

126. R. Hiatt, K. C. Irwin, and C. W. Gould, J. Org. Chem., 33:1430 (1968).

127. W. L. Reynolds and I. M. Kolthoff, J. Phys. Chem., 60:996 (1956).

128. R. J. Orr and H. L. Williams, J. Phys. Chem., 57:925 (1953).

129. R. J. Orr and H. L. Williams, Can. J. Chem., 30:985 (1952).

130. S. Hasegawa and N. Nishimura, Bull. Chem. Soc. Japan, 33:775 (1960).

131. S. Hasegawa, N. Nishimura, and K. Miura, Bull. Chem. Soc. Japan, 33:1323 (1960).

132. S. Hasegawa, N. Nishimura, S. Mitsumoto, and K. Yokoyama, Bull. Chem. Soc. Japan, 36:522 (1963).

133. W. H. Richardson, J. Am. Chem. Soc., 87:247 (1965).

134. W. H. Richardson, J. Am. Chem. Soc., 87:1096 (1965).

135. K. Onuma, K. Wada, J. Yamashita, and H. Hashimoto, Bull. Chem. Soc. Japan, 40:2900 (1967).

136. H. Hock and H. Kropf, J. Prakt. Chem., 16:113 (1962).

137. H. Kropf, Ann. Chem., 637:111 (1960).

138. T. M. Bulgakova, Z. K. Maizus, and I. P. Skibida, Kinetika i Kataliz, 7:332 (1966).

139. A. E. Woodward and R. B. Mesrobian, J. Am. Chem. Soc., 75:6189 (1953).

140. Y. Kamiya, S. Beaton, A. Lafortune, and K. U. Ingold, Can. J. Chem., 41:2020 (1963).

141. Y. Kamiya, S. Beaton, A. Lafortune, and K. U. Ingold, Can. J. Chem., 41:2034 (1963).

142. E. Dyer, K. R. Carle, and D. E. Weiman, J. Org. Chem., 23:1464 (1958).

143. J. Tomiška, Collection Czech. Commun., 27:1549 (1962).

144. J. A. Sharp, J. Chem. Soc., 1957, p. 2026.

145. W. H. Richardson, J. Am. Chem. Soc., 88:975 (1966).

146. A. G. Korsun, V. Ya. Shlyapintokh, and N. M. Émanuél', Izv. Akad. Nauk SSSR, Otd. Khim. Nauk, 1961, p. 788.

147. B. G. Bal'kov and I. P. Skibiba, Izv. Akad. Nauk SSSR, Ser. Khim., 1969, p. 95.

148. C. Copping and N. Uri, Discussions Faraday Soc., 46:202 (1968).

149. D. G. Knorre, Z. K. Maizus, M. I. Markin, and N. M. Émanuél', Zh. Fiz. Khim., 33:398 (1959).

150. G. Czapski, J. Jortner, and G. Stein, J. Phys. Chem., 63:1769 (1959).

151. F. S. Dainton and S. Sills, Proc. Chem. Soc. (London), 1962, p. 223.

152. M. Anbar and P. Neta, International J. Appl. Radiation Isotopes, 18:493 (1967).

152a. W. G. Rothschild and A. O. Allen, Radiation Res., 8:101 (1958).

153. P. Riesz and E. J. Hart, J. Phys. Chem., 63:858 (1959).

154. H. A. Schwarz, J. Phys. Chem., 67:2827 (1963).

155. H. A. Schwarz, J. Am. Chem. Soc., 79:534 (1957).

156. J. W. Boyle, S. Weiner, and C. J. Hochanadel, J. Phys. Chem., 63:892 (1959).

157. J. P. Sweet and J. K. Thomas, J. Phys. Chem., 68:1363 (1964).

158. J. H. Baxendale, R. S. Dixon, and D. A. Stott, Trans. Faraday Soc., 64:2398 (1968).

159. G. Czapski, J. Rabani, and G. Stein, Trans. Faraday Soc., 58:2160 (1962).

160. A. Appleby, G. Scholes, and M. Simic, J. Am. Chem. Soc., 85:3891 (1963).

160a. E. Hayon and M. Moreau, J. Chim. Phys., 1965, p. 391.

160b. P. Airey and G. S. Dainton, Proc. Roy. Soc. (London), A291:478 (1966).

161. G. Navon and G. Stein, J. Phys. Chem., 69:1390 (1965).

162. M. Anbar and D. Meyerstein, Nature, 206:818 (1965).

163. J. H. Baxendale and D. H. Smithies, Z. Physik. Chem., (Frankfurt), 7:242 (1956).

163a. J. Halpern and J. Rabani, J. Am. Chem. Soc., 88:699 (1966).

164. S. Nehari and J. Rabani, J. Phys. Chem., 67:1609 (1963).

165. M. Danield, J. Phys. Chem., 66:1475 (1962).

166. J. P. Baxendale, E. M. Fielden, and J. P. Keene, in: Pulse Radiolysis (M. Ebert, ed.), New York, Academic Press (1965), p. 207.

167. J. W. Boyle, S. Weiner, and C. J. Hochanadel, J. Phys. Chem., 63:892 (1959).

168. B. Čerček and M. Ebert, Trans. Faraday Soc., 63:1687 (1967).

169. J. Jortner and J. Rabani, J. Phys. Chem., 66:2078, 2801 (1962).

170. M. S. Matheson, Radiation Res. Suppl., 4:1 (1964).

171. J. Rabani and G. Stein, J. Chem. Phys., 37:1865 (1962).

172. M. Anbar and P. Neta, Trans. Faraday Soc., 63:141 (1967).

173. J. K. Thomas and E. J. Hart, J. Phys. Chem., 68:2414 (1964).

174. J. Rabani, J. Am. Chem. Soc., 84:868 (1962); J. Phys. Chem., 66:361 (1962).

175. M. Anbar, D. Meyerstein, and P. Neta, J. Chem. Soc., B, 1966, p. 742.

175a. M. Anbar and D. Meyerstein, J. Phys. Chem., 68:3184 (1964).

176. D. M. Brown, F. S. Dainton, D. C. Walker, and J. P. Keane, in: Pulse Radiolysis (M. Ebert, ed.), New York, Academic Press (1965), p. 221.

176a. F. S. Dainton and T. J. Hardwick, Trans. Faraday Soc., 53:333 (1957).

177. D. Bunn, F. S. Dainton, G. A. Salmon, and T. J. Hardwick, Trans. Faraday Soc., 55:1760 (1959).

178. H. A. Schwarz, J. Phys. Chem., 66:255 (1962).

179. A. K. Pikaev, P. Ya. Glazunov, and V. I. Sinitsyn, Izv. Akad. Nauk SSSR, Ser. Khim., 1965, p. 401.

180. B. Chutny, Collection Czech. Chem. Commun., 31:358 (1966).

181. J. Rabani and G. Stein, Trans. Faraday Soc., 58:2150 (1962).

182. J. K. Thomas, Trans. Faraday Soc., 61:702 (1965).

183. R. W. Mattews and D. F. Sangster, J. Phys. Chem., 69:1938 (1965).

184. I. Kraljič and C. N. Trumbore, J. Am. Chem. Soc., 87:2547 (1965).

184a. J. Rabani and M. S. Matheson, J. Am. Chem. Soc., 86:3175 (1964).

185. J. H. Baxendale, E. M. Fielden, and J. P. Keene, in: Pulse Radiolysis (M. Ebert, ed.), New York Academic Press (1965), p. 217.

185a. B. Čerček, M. Ebert, and A. J. Swallow, J. Chem. Soc., A, 1966, p. 612.

186. T. J. Sworski, Radiation Res., 4:483 (1956); J. Am. Chem. Soc., 79:3655 (1957).

187. A. E. Cahill and H. Taube, J. Am. Chem. Soc., 74:2312 (1952).

188. F. S. Dainton and S. A. Sills, Proc. Chem. Soc., 1962, p. 223.

189. H. A. Schwarz and A. O. Allen, J. Am. Chem. Soc., 77:1324 (1955).

189a. G. E. Adams, J. W. Boag, and B. D. Michael, Trans. Faraday Soc., 61:1417 (1965).

190. G. E. Adams and J. W. Boag, Proc. Chem. Soc., 1964, p. 112.

190a. T. W. Woodward and H. C. Sutton, Trans. Faraday Soc., 62:70 (1966).

191. T. J. Sworski, Radiation Res., 6:645 (1957); J. Phys. Chem., 67:2858 (1962) sic .

192. C. J. Hochanadel, Radiation Res., 17:286 (1962).

192a. J. C. Muller and C. Ferradini, J. Chim. Phys., 1965, p. 654.

193. G. E. Adams, J. W. Boag, J. Currant, and B. D. Michael, in: Pulse Radiolysis (M. Ebert, ed.), New York, Academic Press (1965), p. 117.

194. G. Scholes, P. Shaw, R. L. Wilson, and M. Ebert, in: Pulse Radiolysis (M. Ebert, ed.), New York, Academic Press (1965), p. 151.

195. J. F. Ward and L. S. Myers, Radiation Res., 26:483 (1965).

195a. D. G. Marketos and N. Th. Rakintzis, Z. Physik. Chem. (Frankfurt), 44:285 (1965).

196. M. Burton, K. C. Kurien, J. Phys. Chem., 63:899 (1959).

197. M. Anbar and J. K. Thomas, J. Phys. Chem., 68:3829 (1964).

197a. C. Ferradini and A. M. Koulkes-Pujo, J. Chim. Phys., 1963, p. 1310.

198. A. Hummel and A. O. Allen, Radiation Res., 17:302 (1962).

199. H. C. Sutton, G. E. Adams, J. W. Boag, and B. D. Michael, in: Pulse Radiolysis (M. Ebert, ed.), New York, Academic Press (1965), p. 61.

200. M. S. Matheson, W. A. Mulac, J. L. Weeks, and J. Rabani, J. Phys. Chem., 70:2092 (1966).

201. A. O. Allen, Radiation Res. Suppl., 4:54 (1964).

202. M. Anbar, D. Meyerstein, and P. Neta, J. Chem. Soc., B, 1966, p. 742.

203. Z. D. Draganić, M. M. Kosanić, and M. T. Nenadović, J. Phys. Chem., 71:2390 (1967).

204. G. V. Buxton, F. S. Dainton, and F. Wilkinson, Chem. Commun., 1966, p. 320.

205. J. H. Baxendale, E. M. Fielden, and J. P. Keene, Proc. Roy. Soc. (London), A286:320 (1965).

206. J. L. Weeks and J. Rabani, J. Phys. Chem., 70:2100 (1966).

207. J. P. Keene, Y. Raef, and A. J. Swallow, in: Pulse Radiolysis (M. Ebert, ed.), New York, Academic Press (1965), p. 99.

208. M. Anbar, D. Meyerstein, and P. Neta, J. Phys. Chem., 70:2660 (1966).

209. J. P. Keene, Radiation Res., 22:14 (1964).

210. J. H. Baxendale, Radiation Res., 17:312 (1962).

211. J. H. Baxendale, J. P. Keene, and D. A. Stott, in: Pulse Radiolysis (M. Ebert, ed.), New York, Academic Press (1965), p. 107.

212. H. C. Sutton, G. E. Adams, J. W. Boag, and B. D. Michael, in: Pulse Radiolysis (M. Ebert, ed.), New York, Academic Press (1965), p. 61.

213. B. Čerček, M. Ebert, C. W. Gilbert, and A. J. Swalow, in: Pulse Radiolysis (M. Ebert, ed.), New York, Academic Press, (1965), p. 83.

214. E. M. Tochina, L. M. Postnikov, and V. Ya. Shlyapintokh, Izv. Akad. Nauk SSSR, Ser, Khim., 1968, p. 1489.

215. L. M. Postnikov, E. M. Tochina, and V. Ya. Shlyapintokh, Dokl. Akad. Nauk SSSR, 172:651 (1967).

216. V. M. Gol'dberg and L. K. Obukhova, Dokl. Akad. Nauk SSSR, 165:860 (1965).

217. V. P. Shcheredin and E. T. Denisov, Izv. Akad. Nauk SSSR, Ser. Khim., 1967, p. 1428.

218. A. L. Aleksandrov and E. T. Denisov, Izv. Akad. Nauk SSSR, Ser. Khim., 1969, p. 1652.

219. J. K. Kochi, A. Bemis, and C. L. Lenkins, J. Am. Chem. Soc., 90:4616 (1968).

220. E. Collinson, F. S. Dainton, B. Mile, S. Tazuke, and D. R. Smith, Nature, 198:26 (1963).

221. E. Collinson, F. S. Dainton, and G. S. McNaughton, Trans. Faraday Soc., 53:489 (1957).

222. E. Collinson, F. S. Dainton, D. R. Smith, G. J. Trudel, and S. Tazuke, Discussions Faraday Soc., 29:188 (1960).

223. F. S. Dainton and W. D. Sisley, Trans. Faraday Soc., 59:1377 (1963).

224. K. W. Chambers, E. Collinson, F. S. Dainton, W. A. Seddon, and F. Wilkinson, Trans. Faraday Soc., 63:1699 (1967).

225. C. H. Bamford, A. D. Jenkins, and R. Johnston, Proc. Roy. Soc. (London), A239:214 (1957).

226. N. Colebourne, E. Collinson, D. J. Currie, and F. S. Dainton, Trans. Faraday Soc., 59:1357 (1963).

227. W. I. Bengough and W. H. Fairservice, Trans. Faraday Soc., 61:1206 (1965).

228. R. M. Davydov, Zh. Fiz. Khim., 42:2639 (1968).

229. C. E. H. Bawn and D. Verdin, Trans. Faraday Soc., 56:519 (1960).

230. R. I. Zusman, Yu. I. Naberukhin, Yu. N. Molin, and V. V. Voevodskii, Kinetika i Kataliz, 7:414 (1966).

231. V. V. Voevodskii, R. I. Zusman, G. I. Shamovskaya, and Yu. N. Molin, Izv. Sibirsk. Otd. Akad. Nauk SSSR, 3:127 (1968).

232. L. M. Dorfman and M. S. Matheson, Progress in Reaction Kinetics, 3:237 (1965).

233. G. E. Adams, J. W. Boag, and D. B. Michael, Nature, 205:898 (1965).

234. G. L. Malinoski and W. H. Bruning, J. Am. Chem. Soc., 89:5063 (1967).

235. R. Ghang and C. S. Johnston, J. Am. Chem. Soc., 88:2338 (1966).

236. P. J. Zandstra and S. I. Weissman, J. Am. Chem. Soc., 84:4408 (1962).

237. R. L. Ward and S. I. Weissman, J. Am. Chem. Soc., 79:2086 (1957).

238. S. Arai and L. M. Dorfman, J. Chem. Phys., 41:2190 (1964).

239. N. Hirota and S. I. Weissman, J. Am. Chem. Soc., 86:2537 (1964).

240. T. Layloff, T. Miller, R. N. Adams, H. Fäh, A. Horsfield, and W. Proctor, Nature, 205:382 (1965).

241. J. M. Fritsch, S. V. Tatwawadi and R. N. Adams, J. Phys. Chem., 71:338 (1967).

242. W. D. Phillips, J. C. Rowell, and S. I. Weissman, J. Chem. Phys., 33:626 (1960).

243. W. Bruning and S. I. Weissman, J. Am. Chem. Soc., 88:373 (1966).

244. M. T. Jones and S. I. Weissman, J. Am. Chem. Soc., 84:4269 (1962).

245. G. Czapski and L. M. Dorfman, J. Phys. Chem., 68:1169 (1964).

246. G. Czapski and B. H. J. Bielski, J. Phys. Chem., 67:2180 (1963).

247. K. Z. Schmidt, Z. Naturforschung, 16:206 (1961).

248. K. Sehested, O. L. Rasmussen, and H. Fricke, J. Phys. Chem., 72:626 (1968).

249. S. C. Chadha, J. Jagur-Grodzinski, and M. Szwarc, Trans. Faraday Soc., 63:2994 (1967).

250. M. Matsuda, J. Jagur-Grodzinski, and M. Szwarc, Proc. Roy. Soc. (London), A288:212 (1965).

251. D. A. Dadley and A. G. Evans, J. Chem. Soc., B, 1967, p. 418.

252. D. A. Dadley and A. G. Evans, J. Chem. Soc., B, 1968, p. 107.

253. K. H. Schmidt and W. L. Buck, Science, 151:70 (1966).

254. J. Jortner, Radiation Res. Suppl., 4:139 (1964).

255. F. M. Fielden and E. J. Hart, Trans. Faraday Soc., 63:2975 (1967).

256. J. H. Baxendale, E. M. Fielden, C. Capellos, J. M. Francis, et al., Nature, 201:468 (1964).

257. S. Gordon, E. J. Hart, M. S. Matheson, J. Rabani, and J. K. Thomas, Discussions Faraday Soc., 36:193 (1963).

258. M. S. Matheson and J. Rabani, J. Phys. Chem., 69:1324 (1965).

259. E. J. Hart, S. Gordon, and E. M. Fielden, J. Phys. Chem., 70:150 (1966).

260. D. M. Brown, F. S. Dainton, J. P. Keene, and D. C. Walker, Proc. Chem. Soc., 1964, p. 266.

261. E. J. Hart and E. M. Fielden, Solvated Electron, Advances in Chemistry Series, 50:253 (1965).

262. W. D. Felix, B. H. Gall, and L. M. Dorfman, J. Phys. Chem., 71:384 (1967).

263. E. J. Hart, J. K. Thomas, and S. Gordon, Radiation Res. Suppl., 4:74 (1964).

264. J. P. Keene, Radiation Res., 22:1 (1964).

265. M. Anbar and E. J. Hart, J. Phys. Chem., 69:1244 (1965).

266. J. K. Thomas, S. Gordon, and E. J. Hart, J. Phys. Chem., 68:1524 (1964).

267. C. F. Cullis, J. M. Francis, and A. J. Swallow, Proc. Roy. Soc. (London), A287:15 (1965).

268. E. J. Hart, S. Gordon, and J. K. Thomas, J. Phys. Chem., 68:1271 (1964).

269. M. Anbar and E. J. Hart, J. Am. Chem. Soc., 86:5633 (1964).

270. S. Arai and L. M. Dorfman, J. Chem. Phys., 41:2190 (1964).

271. G. Scholes and M. Simic, J. Phys. Chem., 68:1731, 1738 (1964).

272. J. T. Allen and C. M. Beck, J. Am. Chem. Soc., 86:1483 (1964).

273. M. Anbar and E. J. Hart, J. Phys. Chem., 69:973 (1965).

274. A. Szutka, J. K. Thomas, S. Gordon, and E. J. Hart, J. Phys. Chem., 69:289 (1965).

275. M. Anbar and E. J. Hart, J. Phys. Chem., 69:271 (1965).

276. E. Hayon and A. O. Allen, J. Phys. Chem., 65:2181 (1961).

277. M. Anbar, Z. B. Alfassi, and H. Bregman-Reisler, J. Am. Chem. Soc., 89:1263 (1967).

278. R. Braams, Radiation Res., 27:319 (1966).

279. K. Chambers, E. Collinson, F. S. Dainton, and W. Seddon, Chem. Commun. 1966, p. 498.

280. K. D. Asmus and A. Henglein, Ber. Buns. Phys. Chem., 68:348 (1964).

281. D. A. Armstrong and V. G. Wilkening, Can. J. Chem., 42:2631 (1964).

282. A. Charlesby, P. J. Fyedelor, P. M. Kopp, J. P. Keene, and A. J. Swallow, in: Pulse Radiolysis, (M. Ebert, ed.), New York, Academic Press (1965), p. 193.

283. J. V. Davies, W. Griffiths, and G. O. Philips, in: Pulse Radiolysis (M. Ebert, ed.), New York Academic Press (1965), p. 181.

284. G. O. Philips, W. Griffiths and J. V. Davies, J. Chem. Soc., B, 1966, p. 194.

285. J. V. Davies, M. Ebert, and A. J. Swallow, in: Pulse Radiolysis, (M. Ebert, ed.), New York, Academic Press (1965), p. 165.

286. R. Braams, in: Pulse Radiolysis, (M. Ebert, ed.), New York, Academic Press (1965), p. 171.

287. R. L. S. Willix, and W. M. Garrison, Radiation Res., 32:452 (1967).

288. R. Braams, Radiation Res., 31:8 (1967).

289. S. Gordon, E. J. Hart, and J. K. Thomas, J. Phys. Chem., 68:1262 (1964).

290. J. Rabani, W. A. Mulac, and M. S. Matheson, J. Phys. Chem., 69:53 (1965).

291. S. Gordon, E. J. Hart, M. S. Matheson, J. Rabani, and J. K. Thomas., J. Am. Chem. Soc., 85:1375 (1963).

292. L. M. Dorfman and I. A. Taub, J. Am. Chem. Soc., 85:2370 (1963).

293. M. Anbar and P. Neta, Trans. Faraday Soc., 63:141 (1967).

294. J. Rabani, Solvated Electron, Advances in Chemistry Series, 50:242 (1965).

295. A. Appleby, G. Scholes, and M. Simic, J. Am. Chem. Soc., 85:3891 (1963).

296. M. Halmann and I. Platzner, J. Phys. Chem., 70:2281 (1966).

297. M. Anbar and J. K. Thomas, J. Phys. Chem., 68:3829 (1964).

298. M. Anbar, Chem. Commun., 1966, p. 416.

299. G. Gzapski, J. Jornter, and G. Stein, J. Phys. Chem., 63:1769 (1959).

300. J. H. Baxendale, E. M. Fielden, J. P. Keene, in: Pulse Radiolysis (M. Ebert, ed.), New York, Academic Press (1965), p. 217.

301. F. S. Dainton and R. Rumfeldt, Proc. Roy. Soc. (London), A287:44 (1965).

302. F. S. Dainton and S. R. Logan, Proc. Roy. Soc. (London), A287:81 (1965).

303. J. P. Keene, E. J. Land, A. J. Swallow, in: Pulse Radiolysis (M. Ebert, ed)., New York, Academic Press (1965), p. 227.

304. E. F. Caldin, Fast Reactions in Solution [Russian translation], Moscow, "Mir" (1966). Original English language edition; New York, Wiley.

305. A. Anderson and N. A. Bonner, J. Am. Chem. Soc., 76:3826 (1954).

306. D. L. Bull and E. L. King, J. Am. Chem. Soc., 80:1091 (1958).

307. R. Snellgrove and E. L. King, Inorg. Chem., 3:288 (1964).

308. R. Snellgrove and E. L. King. J. Am. Chem. Soc., 84:4609 (1962).

309. K. A. Schroeder and J. H. Espenson, J. Am. Chem. Soc., 89:2548 (1967).

310. E. K. Morsessian and D. R. Stranks, Cited in: Modern Chemistry of Coordination Compounds, Moscow, IL, 1963, p. 158.

311. E. Deutsch and H. Taube, Inorg. Chem., 7:1532 (1968).

312. K. V. Krishnamurty and A. C. Wahl, J. Am. Chem. Soc., 80:5921 (1958).

313. S. C. Furman and C. S. Garner, J. Am. Chem. Soc., 74:2333 (1952).

314. H. Diebber and N. Sutin, J. Phys. Chem, 68:174 (1964).

315. J. C. Sheppard and A. C. Wahl, J. Am. Chem. Soc., 79:1020 (1957).

316. O. E. Myers and J. C. Sheppard, J. Am. Chem. Soc., 83:4739 (1961).

317. S. Fukushima and W. L. Reynolds, Talanta, 11:283 (1964).

318. J. Hudis and R. W. Dodson, J. Am. Chem. Soc., 78:911 (1956).

319. J. Silverman and R. W. Dodson, J. Phys. Chem., 56:846 (1952).

320. R. A. Horne, Thesis, Ann Arbor, Michigan (1957).

321. J. Menaschi, W. L. Reynolds, and G. Vananken, Inorg. Chem., 4:229 (1965).

322. R. A. Horne, J. Inorg. Nucl. Chem., 25:1139 (1963).

323. J. Hudis and A. C. Wahl, J. Am. Chem. Soc., 75:4153 (1953).

324. N. Sutin, J. K. Rowley, and R. W. Dodson, J. Phys. Chem., 65:1248 (1961).

325. R. J. Campion, T. J. Conocchioli, and N. Sutin, J. Am. Chem. Soc., 86:4591 (1964).

326. G. S. Laurence, Trans. Faraday Soc., 53:1326 (1957).

327. D. Bunn, F. S. Dainton, and S. Durchworth, Trans. Faraday Soc., 57:1131 (1961).

328. R. A. Horne, J. Phys. Chem., 64:1512 (1960).

329. J. C. Sheppard and L. C. Brown, J. Phys. Chem., 67:1025 (1963).

330. W. L. Reynolds and S. Fukushima, Inorg. Chem., 2:176 (1963).

331. R. L. S. Willix, Trans. Faraday Soc., 59:1315 (1963).

332. R. Bächman and K. H. Lieser, Z. Physik. Chem. (Frankfurt), 36:236 (1963).

333. A. Loewenstein and G. Ron, Inorg. Chem., 6:1604 (1967).

334. A. C. Wahl, Z. Elektrochem., 64:90 (1960).

335. R. J. Campion, C. F. Deck, P. King, and A. C. Wahl, Inorg. Chem., 6:672 (1967).

336. R. G. Wilkins and R. Yelin, Unpublished data, cited in: B. Grossman and R. G. Wilkins, J. Am. Chem. Soc., 89:4230 (1967).

337. D. R. Stranks, Discussions Faraday Soc., 29:73 (1960).

338. M. W. Dietrich and A. C. Wahl, J. Chem. Phys., 38:1591 (1963).

339. E. Eichler and A. C. Wahl, J. Am. Chem. Soc., 80:4145 (1958).

340. D. W. Larsen and A. C. Wahl, J. Chem. Phys., 41:908 (1964).

341. N. A. Bonner and J. P. Hunt, J. Am. Chem. Soc., 82:3826 (1960).

342. J. Shankar and B. C. DeSouza, J. Inorg. Nucl. Chem., 24:187 (1962).

343. H. S. Habib and J. P. Hunt, J. Am. Chem. Soc., 88:1668 (1966).

344. J. J. Grossman and C. S. Garner, J. Chem. Phys., 28:268 (1958).

345. N. S. Biradar, D. R. Stranks, M. S. Vaidya, and G. J. Weston, Trans. Faraday Soc., 55:1268 (1959).

346. N. S. Biradar, D. R. Stranks, and M. S. Vaidya, Trans. Faraday Soc., 58:2421 (1962).

347. D. R. Strans and D. J. Simpson, See: Modern Chemistry of Coordination Compounds, Moscow, IL, 1963, p. 158.

348. F. P. Dwyer and A. M. Sargeson, J. Phys. Chem., 65:1892 (1961).

349. W. B. Lewis, C. D. Coryell, and J. W. Irvine, J. Chem. Soc., 1949, p. 386.

350. F. S. Fainton, G. S. Laurence, W. Schneider, D. R. Stranks, and M. S. Vaidya, UNESCO Isotope Conference (Paris), 1957. Cited in: Modern Chemistry of Coordination Compounds, Moscow, IL, 1963.

351. A. W. Adamson and K. S. Vorres, J. Inorg. Nucl. Chem., 3:206 (1956).

352. B. R. Baker, F. Basolo, and R. M. Neumann, J. Phys. Chem., 63:371 (1959).

353. P. Ellis, M. J. G. Wilkins, and R. D. Williams, J. Chem. Soc., 1957, p. 4456.

354. P. G. Rasmissen and C. H. Brubaker, Inorg. Chem., 3:977 (1964).

355. R. L. Wolfgang, J. Am. Chem. Soc., 74:6144 (1952).

356. H. M. McConnell and H. E. Weaver, J. Chem. Phys., 25:307 (1956).

357. B. M. Gordon and A. C. Wahl, J. Am. Chem. Soc., 80:273 (1958).

358. E. G. Meyer and M. Kahn, J. Am. Chem. Soc., 73:4950 (1951).

359. J. W. Gryder and R. W. Dodson, J. Am. Chem. Soc., 73:2890 (1951).

360. D. J. Meier and C. S. Garner, J. Phys. Chem., 56:853 (1952).

361. E. L. Goodenow and C. S. Garner, J. Am. Chem. Soc., 77:5272 (1955).

362. S. J. Weissman and C. S. Garner, J. Am. Chem. Soc., 78:1072 (1956).

363. E. N. Sloth and C. S. Garner, J. Am. Chem. Soc., 77:1440 (1955).

364. R. L. Wolfgang and R. W. Dodson, J. Phys. Chem., 56:872 (1952).

365. G. Harbottle and R. W. Dodson, J. Am. Chem. Soc., 73:2442 (1951).

366. R. J. Prestwood and A. C. Wahl, J. Am. Chem. Soc., 71:3137 (1949).

367. G. Vogt and H. J. Born, Ber. Buns. Phys. Chem., 69:741 (1965).

368. D. Cohen, J. C. Sullivan, and J. C. Hindman, J. Am. Chem. Soc., 76:352 (1954).

369. D. Cohen, J. C. Sullivan, and J. C. Hindman, J. Am. Chem. Soc., 77:4964 (1955).

370. D. Cohen, J. C. Sullivan, E. S. Amis, and J. C. Hindman, J. Am. Chem. Soc., 78:1543 (1956).

371. T. K. Keenan, J. Phys. Chem., 61:1117 (1957).

372. F. R. Duke and P. H. Quinney, J. Am. Chem. Soc., 76:3800 (1954).

373. V. W. Cope, R. G. Miller, and R. T. M. Fraser, J. Chem. Soc., A, 1967, p. 301.

374. K. A. Blazek and J. Koryta, Collection Czech. Chem. Commun., 18:326 (1953).

375. J. Koryta, Collection Czech. Chem. Commun., 20:1125 (1955).

376. T. W. Newton and F. B. Baker, J. Phys. Chem., 68:228 (1964).

377. J. W. Olver and J. W. Ross, J. Phys. Chem., 66:1699 (1962).

378. N. A. Daugherty and T. W. Newton, J. Phys. Chem., 68:612 (1964).

379. A. G. Sykes, Chem. Commun., 1965, p. 442.

380. J. H. Espenson, Inorg. Chem., 4:1533 (1965).

381. N. A. Daugherty and T. W. Newton, J. Phys. Chem., 67:1090 (1963).

382. A. M. Zwickel and H. Taube, Discussions Faraday Soc., 29:42 (1960).

383. A. Adin and A. G. Sykes, Nature, 209:804 (1966).

384. J. H. Espenson, Inorg. Chem., 7:631 (1968).

385. T. W. Newton and F. B. Baker, Aqueous Oxidation-Reduction Reactions of Uranium, Neptunium, Plutonium, and Americium, in Advances in Chemistry Series, 71 (1967).

386. A. Adin, J. Dayle, and A. G. Sykes, J. Chem. Soc., A, 1967, p. 1504.

387. A. Adin and A. G. Sykes, J. Chem. Soc., A, 1966, p. 1518.

388. J. H. Espenson and S. G. Slocum, Inorg. Chem., 6:906 (1967).

389. D. E. Pennington and A. Haim, Inorg. Chem., 6:2138 (1967).

390. A. Haim, J. Am. Chem. Soc., 88:2324 (1966).

391. J. H. Espenson and D. W. Carlyle, Inorg. Chem., 5:586 (1966).

392. J. P. Birk and J. H. Espenson, Inorg. Chem., 7:991 (1968).

393. A. Haim and N. Sutin, J. Am. Chem. Soc., 88:434 (1966).

394. A. E. Ogard and H. Taube, J. Am. Chem. Soc., 80:1084 (1958).

395. D. E. Pennington and A. Haim, Inorg. Chem., 5:1887 (1966).

396. B. R. Baker, M. Orhanovic, and N. Sutin, J. Am. Chem. Soc., 87:4210 (1965).

397. J. B. Kirwin, P. J. Proll, and L. H. Sutcliffe, Trans. Faraday Soc., 60:119 (1964).

398. R. C. Thompson and G. Gordon, Inorg. Chem., 5:562 (1966).

399. M. Nicol and D. R. Rosseinsky, Chem. Ind. (London), 1963, p. 1166.

400. B. M. Gordon, L. L. Williams, and N. Sutin, J. Am. Chem. Soc., 83:2061 (1961).

401. J. A. Rawoof and J. R. Sutter, J. Phys. Chem., 71:2767 (1967).

402. D. R. Rosseinsky, J. Chem. Soc., 1963, p. 1181.

403. E. G. Moorhead and N. Sutin, Inorg. Chem., 6:428 (1967).

404. T. J. Conocchioli and N. Sutin, J. Am. Chem. Soc., 89:282 (1967).

405. N. Sutin and B. M. Gordon, J. Am. Chem. Soc., 83:70 (1961).

406. M. H. Ford-Smith and N. Sutin, J. Am. Chem. Soc., 83:1830 (1961).

407. N. Sutin, Nature, 190:438 (1961).

408. N. Sutin and R. Christman, J. Am. Chem. Soc., 83:1773 (1961).

409. D. H. Havsteen, Acta Chem. Scand., 19:227 (1965).

410. W. C. E. Higginson, D. R. Rosseinsky, J. B. Stead, and A. G. Sykes, Discussions Faraday Soc., 29:49 (1960).

411. L. E. Bennett and H. Taube, Inorg. Chem., 7:254 (1968).

412. A. Haim and N. Sutin, J. Am. Chem. Soc., 87:4210 (1965).

413. B. Grossman and R. G. Wilkins, J. Am. Chem. Soc., 89:4230 (1967).

413a. F. R. Duke and R. C. Pinkerton, J. Am. Chem. Soc., 73:3045 (1951).

414. J. Halpern, R. J. Legare, and R. Lumry, J. Am. Chem., Soc., 85:680 (1963).

415. R. Hurwitz and K. Kustin, Inorg. Chem., 3:823 (1964).

416. L. Meites and R. H. Schlossel, J. Phys. Chem., 67:2397 (1963).

417. T. W. Newton and N. A. Daugherty, J. Phys. Chem., 71:3768 (1967).

418. R. H. Betts, Can. J. Chem., 33:1780 (1955).

419. J. R. Huizenga and L. B. Magnusson, J. Am. Chem. Soc., 73:3202 (1951).

420. J. H. Espenson, J. Shaw, and O. J. Parker, J. Am. Chem. Soc., 89:5730 (1967).

421. T. W. Newton and F. B. Baker, J. Phys. Chem., 70:1943 (1966).

422. W. G. Barb, J. H. Baxendale, P. George, and K. H. Hargrave, Trans. Faraday Soc., 47:591 (1951).

423. J. H. Baxendale, The Kinetics and Mechanisms of Inorganic Reactions in Solution. The Chemical Society (London), Special Publ. 1 (1954), p. 40.

424. R. H. Betts and F. S. Dainton, J. Am. Chem. Soc., 75:5721 (1953).

425. W. L. Reynolds and R. W. Lumry, J. Chem. Phys., 23:2360 (1955).

426. R. Woods, I. M. Kolthoff, and E. J. Meehan, 85:2385 (1963).

427. S. A. Balabanova and A. M. Markevich, Zh. Fiz. Khim., 40:775 (1966).

428. J. W. L. Fordham and H. L. Williams, J. Am. Chem. Soc., 73:4855 (1951).

429. M. Kershaw and I. E. Prue, Trans. Faraday Soc., 63:1198 (1967).

430. R. W. Chlebek and M. W. Lister, Can. J. Chem., 44:437 (1966).

431. J. Burgess and R. H. Prince, J. Chem. Soc., A, 1966, p. 1772.

432. D. J. Huchital, N. Sutin, and B. Warnquist, Inorg. Chem., 6:838 (1968).

433. D. R. Rosseinsky and W. C. E. Higginson, J. Chem. Soc., 1960, p. 31.

434. L. F. Bennett and J. C. Sheppard, J. Phys. Chem. 66:1275 (1962).

435. J. C. Sullivan, A. J. Zielen, and J. C. Hindman, J. Am. Chem. Soc., 82:5288 (1960).

436. P. H. Dodel and H. Taube, Z. Physik. Chem. (Frankfurt), 44:92 (1965).

437. J. P. Candlin, J. Halpern, and D. L. Trimm, J. Am. Chem. Soc., 86:1019 (1964).

438. H. Diebler and H. Taube, Inorg. Chem., 4:1029 (1965).

439. R. T. M. Fraser, Inorg. Chem., 2:955 (1963).

440. R. T. M. Fraser, in: Advances in the Chemistry of the Coordination Compounds (S. Kirschner, ed.), New York, McMillan, (1961), pp. 187-195.

441. R. T. M. Fraser, J. Am. Chem. Soc., 84:3436 (1962).

442. R. T. M. Fraser, J. Am. Chem. Soc., 85:1747 (1963).

443. H. Price and H. Taube, Inorg. Chem., 7:1 (1968).

444. R. T. M. Fraser, J. Am. Chem. Soc., 83:4920 (1961).

445. D. J. Huchital and H. Taube, J. Am. Chem. Soc., 87:5371 (1965).

446. J. Doyle and A. G. Sykes, J. Chem. Soc., A, 1967, p. 795.

447. A. B. Hoffman and H. Taube, Inorg. Chem., 7:903 (1968).

448. P. V. Manning and R. C. Jarnagin, J. Phys. Chem., 67:2884 (1963).

449. A. M. Zwickel and H. Taube, J. Am. Chem. Soc., 81:2915 (1959).

450. R. K. Murmann, H. Taube, and F. A. Posey, J. Am. Chem. Soc., 79:262 (1957).

451. A. Zwickel and H. Taube, J. Am. Chem. Soc., 81:1288 (1959).

452. J. P. Candlin and J. Halpern, Inorg. Chem., 4:766 (1965).

453. H. Taube, J. Am. Chem. Soc., 77:4481 (1955).

454. J. H. Espenson and J. P. Birk, J. Am. Chem. Soc., 87:3280 (1965).

455. R. T. M. Fraser, J. Chem. Soc., 1965, p. 3641.

456. K. D. Kopple and R. R. Miller, Proc. Chem. Soc., 1962, p. 306.

457. R. D. Butler and H. Taube, J. Am. Chem. Soc., 87:5597 (1965).

458. D. K. Sebera and H. Taube, J. Am. Chem. Soc., 83:1785 (1961).

459. R. T. M. Fraser, Inorg. Chem., 3:1561 (1964).

460. E. S. Gould, J. Am. Chem. Soc., 87:4730 (1965).

461. E. S. Gould, J. Am. Chem. Soc., 88:2983 (1966).

462. G. Svatos and H. Taube, J. Am. Chem. Soc., 83:4172 (1961).

463. E. S. Gould and H. Taube, J. Am. Chem. Soc., 86:1318 (1964).

464. R. T. M. Fraser, J. Am. Chem. Soc., 83:564 (1961).

465. R. T. Fraser and H. Taube, J. Am. Chem. Soc., 83:2242 (1961).

466. R. T. M. Fraser, Inorg. Chem., 3:255 (1964).

467. F. Nordmeyer and A. M. Sargeson, Unpublished data, cited in [460].

468. R. B. Lordon, A. M. Sargeson, and H. Taube, Inorg. Chem., 5:1091 (1966).

469. E. S. Gould and H. Taube, J. Am. Chem. Soc., 85:3706 (1963).

470. R. D. Cannon and J. E. Earley, J. Am. Chem. Soc., 88:1872 (1966).

471. R. T. M. Fraser, Proc. Chem. Soc., 1963, p. 262.

472. P. B. Wood and W. C. E. Higginson, J. Chem. Soc., A, 1966, p. 1645.

473. T. J. Conocchioli, G. H. Nancollas, and N. Sutin, J. Am. Chem. Soc., 86:1453 (1964).

474. J. H. Espenson, Inorg. Chem., 4:121 (1965).

475. A. Haim, J. Am. Chem. Soc., 85:1016 (1963).

476. A. Haim, J. Am. Chem. Soc., 86:2352 (1964).

477. J. Barrett and J. H. Baxendale, Trans. Faraday Soc., 52:210 (1956).

478. P. Benson and A. Haim, J. Am. Chem. Soc., 87:3826 (1965).

479. P. B. Wood and W. C. E. Higginson, J. Chem. Soc., 1965, p. 2116.

480. A. Haim and N. Sutin, J. Am. Chem. Soc., 88:5343 (1966).

481. J. P. Candlin and J. Halpern, Inorg. Chem., 4:1086 (1965).

482. A. Pidcock and W. C. E. Higginson, J. Chem. Soc., 1963, p. 2790.

483. D. H. Huchital and R. G. Wilkins, Inorg. Chem., 6:1022 (1967).

484. J. P. Candlin, J. Halpern, and S. Nakamura, J. Am. Chem. Soc., 85:2517 (1963).

485. R. Farina and R. G. Wilkins, Inorg. Chem., 7:514 (1968).

486. A. Zwickel, Ph. D. Dissertation, Univ. Chicago (1959).

487. R. T. M. Fraser, R. N. Lee, and K. Hayden, J. Chem. Soc., A, 1967, p. 741.

488. R. G. Miller, D. E. Peters, and R. T. M. Fraser, Exchange Reactions. Intern. Atomic Energy Agency, Vienna (1965), p. 203. Cited in [248].

488a. J. F. Endicott and H. Taube, J. Am. Chem. Soc., 86:1686 (1964).

489. H. Diebler and N. Sutin, J. Phys. Chem., 68:174 (1964).

490. K. G. Ashurst and W. C. E. Higginson, J. Chem. Soc., 1956, p. 343.

491. L. H. Sutcliffe and I. R. Weber, Trans. Faraday Soc., 52:1225 (1956).

492. L. H. Sutcliffe and I. R. Weber, Trans. Faraday Soc., 55:1892 (1959).

493. K. Dyke and W. C. E. Higginson, J. Chem. Soc., 1963, p. 2802.

494. J. C. Sullivan and R. C. Thompson, Inorg. Chem., 6:1795 (1967).

495. R. G. Yalman, J. Am. Chem. Soc., 75:1842 (1953).

496. J. H. Baxendale and C. F. Wells, Trans. Faraday Soc., 53:800 (1957).

497. K. Shaw and J. H. Espenson, Inorg. Chem., 7:1619 (1967).

498. I. M. Kolthoff and R. Woods, J. Am. Chem. Soc., 88:137 (1966).

499. P. T. Stroup and V. W. Meloche, J. Am. Chem. Soc., 53:3331 (1931).

500. R. Champion, N. Purdie, and N. Sutin, J. Am. Chem. Soc., 85:3528.(1963).

501. G. P. Haight and A. C. Swift, J. Phys. Chem., 65:1921 (1961).

502. E. P. Guymon and I. T. Spence, J. Phys. Chem., 71:1616 (1967).

502a. P. George and D. H. Irvine, J. Chem. Soc., 1954, p. 587.

503. I. M. Kolthoff and I. Hodara, J. Elektroanal. Chem. Interfac. Chem., 4:369 (1962).

503a. G. A. Rechnitz and H. A. Catherino, Inorg. Chem., 4:112 (1965).

504. C. E. Bawn and D. Margerison, Trans. Faraday Soc., 51:925 (1955).

505. H. G. S. Sengar and Y. K. Gupta, J. Indian Chem. Soc., 43:223 (1966).

506. A. J. Kalb and T. L. Allen, J. Am. Chem. Soc., 86:5107 (1964).

507. G. Veith, E. Guthals, and A. Viste, Inorg. Chem., 6:667 (1967).

508. J. C. Gupta and S. P. Srivastava, Z. Physik. Chem., (Leipzig), 227:158 (1964).

509. F. B. Barker and M. Kahn, J. Am. Chem. Soc., 78:1317 (1956).

510. H. M. Neumann and H. Brown, J. Am. Chem. Soc., 78:1843 (1956).

511. H. Gerisher and W. Heim, Z. Physik. Chem., (Frankfurt), 46:345 (1965).

512. G. Dulz and N. Sutin, Inorg. Chem., 2:917 (1963).

513. M. G. Adamson, F. S. Dainton, and P. Glentworth, Trans. Faraday Soc., 61:689 (1965).

514. J. D. Miller and R. H. Prince, J. Chem. Soc., 1965, p. 5749.

515. W. H.McCurdy and G. G. Guilbault, J. Phys. Chem., 64:1825 (1960).

516. S. K. Mishra and Y. K. Gupta, J. Inorg. Nucl. Chem., 29:1643 (1967).

516a. F. B. Baker, T. W. Newton, and M. Kahn, J. Phys. Chem., 64:109 (1960).

517. R. L. Carroll and L. B. Thomas, J. Am. Chem. Soc., 88:1376 (1966).

518. W. A. Waters and I. R. Wilson, J. Chem. Soc., A, 1966, p. 534.

519. D. H. Irvine, J. Chem. Soc., 1958, p. 2166.

520. C. E. Johnson, J. Am. Chem. Soc., 74:959 (1952).

521. D. H. Irvine, J. Chem. Soc., 1957, p. 1841.

522. J. O. Wear, J. Chem. Soc., 1965, p. 5596.

523. A. C. Harkness and J. Halpern, J. Am. Chem. Soc., 81:3526 (1959).

524. J. D. Miller and R. H. Prince, J. Chem. Soc., A, 1966, p. 1048.

525. L. H. Sutcliffe and J. Walkley, Nature, 178:999 (1956).

526. B. J. Masters and L. L. Schwartz, J. Am. Chem. Soc., 83:2620 (1961).

527. E. F. Orlemann and D. M. H. Kern, J. Am. Chem. Soc., 75:3058 (1953).

528. H. Imai, Bull. Chem. Soc., Japan, 30:873 (1957).

529. F. R. Duke and R. C. Punkerton, J. Am. Chem. Soc., 73:2361 (1951).

530. J. Sobkowski, J. Inorg. Nucl. Chem., 27:2351 (1965).

531. T. W. Newton and F. B. Baker, Inorg. Chem., 4:1166 (1965).

532. T. W. Newton and F. B. Baker, J. Phys. Chem., 69:176 (1965).

533. T. W. Newton and F. B. Baker, J. Phys. Chem., 70:1943 (1966).

534. J. H. Baxendale, M. G. Evans, and G. S. Park, Trans. Faraday Soc., 42:155 (1946).

535. L. A. Fedorova, Radiokhimiya, 6:42 (1964).

536. V. S. Koltunov, Kinetics of Oxidation-Reduction Reactions of Uranium, Neptunium, and Plutonium, Moscow, Atomizdat (1965), p. 225.

537. G. Gordon and F. Feldman, Inorg. Chem., 3:1728 (1964).

538. A. G. Rykov, V. Ya. Vasil'ev, and G. N. Yakovlev, Radiokhimiya, 8:33 (1966).

539. E. A. Kanevskii and L. A. Fedorov, Radiokhimiya, 2:559 (1960).

540. J. Koryta, Collection Czech. Chem. Commun., 20:667 (1955).

541. J. C. Hindman, J. C. Sullivan, and D. Cohen, J. Am. Chem. Soc., 80:1812 (1958).

542. A. G. Rykov and G. N. Yakovlev, Radiokhimiya, 8:20 (1966).

543. J. C. Hindman, J. C. Sullivan, and D. Cohen, J. Am. Chem. Soc., 76:3278 (1954).

544. A. G. Rykov and G. N. Yakovlev, Radiokhimiya 8:27, (1966).

545. J. C. Hindman, J. C. Sullivan, and D. Cohen, J. Am. Chem. Soc., 79:4029 (1957).

546. J. C. Sheppard, J. Phys. Chem., 68:1190 (1964).

547. E. K. Appelman and J. C. Sullivan, J. Phys. Chem., 66:442 (1962).

548. R. C. Thompson and J. C. Sullivan, J. Am. Chem. Soc., 89:1098 (1967).

549. R. C. Thompson and J. C. Sullivan, J. Am. Chem. Soc., 89:1096 (1967).

550. N. K. Shastri and E. S. Amis, J. Inorg. Nucl. Chem., 27:2413 (1965).

551. N. K. Shastri, J. O. Wear, and E. S. Amis, J. Inorg. Nucl. Chem., 24:535 (1962).

552. S. W. Rabidean and R. J. Kline, J. Phys. Chem., 62:617 (1958).

553. P. I. Artyukhin, Zh. Neorg. Khim., 4:1324 (1959).

554. R. E. Connick, J. Am. Chem. Soc., 71:1528 (1949).

555. S. W. Rabidean, J. Am. Chem. Soc., 75:798 (1953).

556. S. W. Rabidean and H. D. Cowan, J. Am. Chem. Soc., 77:6145 (1955).

557. S. W. Rabidean, J. Am. Chem. Soc., 79:6350 (1957).

558. S. W. Rabidean and R. J. Kline, J. Phys. Chem., 63:1502 (1959).

559. S. W. Rabidean and R. J. Kline, J. Phys. Chem., 64:193 (1960).

560. S. W. Rabidean, J. Phys. Chem., 62:414 (1958).

561. S. W. Rabidean and R. J. Kline, J. Inorg. Nucl. Chem., 14:91 (1960).

562. T. W. Newton and F. B. Baker, J. Phys. Chem., 67:1425 (1963).

563. T. W. Newton and H. D. Cowan, J. Phys. Chem., 64:244 (1960).

564. S. W. Rabidean and B. J. Masters, J. Phys. Chem., 65:1256 (1961).

565. T. W. Newton, J. Phys. Chem., 63:1493 (1959).

566. I. M. Cleveland, Inorg. Chem., 6:1302 (1962).

567. J. S. Coleman, Inorg. Chem., 2:53 (1963).

568. A. A. Zaitsev, V. N. Kosyakov, A. G. Rykov, Yu. P. Sobolev, and G. N. Yakovlev, Radiokhimiya 2:339 (1960).

MIX
Papier aus verantwortungsvollen Quellen
Paper from responsible sources
FSC® C105338

If you have any concerns about our products,
you can contact us on
ProductSafety@springernature.com

In case Publisher is established outside the EU,
the EU authorized representative is:
Springer Nature Customer Service Center GmbH
Europaplatz 3, 69115 Heidelberg, Germany

Printed by Libri Plureos GmbH
in Hamburg, Germany